Nutraceuticals, Glycemic Health and Type 2 Diabetes

IFT PRESS

The *IFT Press* series reflects the mission of the Institute of Food Technologists – advancing the science and technology of food through the exchange of knowledge. Developed in partnership with Wiley-Blackwell, *IFT Press* books serve as leading edge handbooks for industrial application and reference and as essential texts for academic programs. Crafted through rigorous peer review and meticulous research, *IFT Press* publications represent the latest, most significant resources available to food scientists and related agriculture professionals worldwide.

WILEY-BLACKWELL

A John Wiley & Sons, Ltd., Publication

Nutraceuticals, Glycemic Health and Type 2 Diabetes

EDITORS

Vijai K Pasupuleti • James W Anderson

A John Wiley & Sons, Ltd., Publication

Edition first published 2008
© 2008 Blackwell Publishing and the Institute of Food Technologists

Blackwell Publishing was acquired by John Wiley & Sons in February 2007. Blackwell's publishing program has been merged with Wiley's global Scientific, Technical, and Medical business to form Wiley-Blackwell.

Editorial Office
2121 State Avenue, Ames, Iowa 50014-8300, USA

For details of our global editorial offices, for customer services, and for information about how to apply for permission to reuse the copyright material in this book, please see our website at www.wiley.com/wiley-blackwell.

Library of Congress Cataloguing-in-Publication Data

Nutraceuticals, glycemic health, and type 2 diabetes / editors, Vijai K. Pasupuleti and James W. Anderson. — 1st ed.
    p. ; cm. — (IFT Press series)
Includes bibliographical references and index.
ISBN 978-0-8138-2933-3 (alk. paper)
1. Diabetes—Diet therapy.   2. Diabetes—Prevention.   3. Functional foods. I. Pasupuleti, Vijai K. II. Anderson, James W.   III. Series.
[DNLM: 1. Diabetes Mellitus, Type 2—diet therapy. 2. Diabetes Mellitus, Type 2—prevention & control. 3. Dietary Supplements. 4. Prediabetic State—diet therapy. 5. Prediabetic State—prevention & control. WK 810 N976 2008]
RC661.F86N88 2008
616.4′620654—dc22

2008007431

e U.S. Library of Congress.

Titles in the *IFT Press* series

- *Accelerating New Food Product Design and Development* (Jacqueline H. Beckley, Elizabeth J. Topp, M. Michele Foley, J.C. Huang and Witoon Prinyawiwatkul)
- *Advances in Dairy Ingredients* (Geoffrey W. Smithers and Mary Ann Augustin)
- *Biofilms in the Food Environment* (Hans P. Blaschek, Hua H. Wang, and Meredith E. Agle)
- *Calorimetry and Food Process Design* (Gönül Kaletunç)
- *Food Ingredients for the Global Market* (Yao-Wen Huang and Claire L. Kruger)
- *Food Irradiation Research and Technology* (Christopher H. Sommers and Xuetong Fan)
- *Food Laws, Regulations and Labeling* (Joseph D. Eifert)
- *Food Risk and Crisis Communication* (Anthony O. Flood and Christine M. Bruhn)
- *Foodborne Pathogens in the Food Processing Environment: Sources, Detection and Control* (Sadhana Ravishankar and Vijay K. Juneja)
- *Functional Proteins and Peptides* (Yoshinori Mine, Richard K. Owusu-Apenten and Bo Jiang)
- *High Pressure Processing of Foods* (Christopher J. Doona and Florence E. Feeherry)
- *Hydrocolloids in Food Processing* (Thomas R. Laaman)
- *Microbial Safety of Fresh Produce: Challenges, Perspectives and Strategies* (Xuetong Fan, Brendan A. Niemira, Christopher J. Doona, Florence E. Feeherry and Robert B. Gravani)
- *Microbiology and Technology of Fermented Foods* (Robert W. Hutkins)
- *Multivariate and Probabilistic Analyses of Sensory Science Problems* (Jean-François Meullenet, Rui Xiong, and Christopher J. Findlay
- *Nondestructive Testing of Food Quality* (Joseph Irudayaraj and Christoph Reh)
- *Nanoscience and Nanotechnology in Food Systems* (Hongda Chen)
- *Nonthermal Processing Technologies for Food* (Howard Q. Zhang, Gustavo V. Barbosa-Cànovas, V.M. Balasubramaniam, Editors; C. Patrick Dunne, Daniel F. Farkas, James T.C. Yuan, Associate Editors)
- *Nutraceuticals, Glycemic Health and Type 2 Diabetes* (Vijai K. Pasupuleti and James W. Anderson)
- *Packaging for Nonthermal Processing of Food* (J. H. Han)
- *Preharvest and Postharvest Food Safety: Contemporary Issues and Future Directions* (Ross C. Beier, Suresh D. Pillai, and Timothy D. Phillips, Editors; Richard L. Ziprin, Associate Editor)
- *Processing and Nutrition of Fats and Oils* (Ernesto M. Hernandez, Monjur Hossen, and Afaf Kamal-Eldin)
- *Regulation of Functional Foods and Nutraceuticals: A Global Perspective* (Clare M. Hasler)
- *Sensory and Consumer Research in Food Product Design and Development* (Howard R. Moskowitz, Jacqueline H. Beckley, and Anna V.A. Resurreccion)
- *Sustainability in the Food Industry* (Cheryl J. Baldwin)
- *Thermal Processing of Foods: Control and Automation* (K.P. Sandeep)
- *Water Activity in Foods: Fundamentals and Applications* (Gustavo V. Barbosa-Cànovas, Anthony J. Fontana Jr., Shelly J. Schmidt, and Theodore P. Labuza)
- *Whey Processing, Functionality and Health Benefits* (Charles I. Onwulata and Peter J. Huth)

Contents

Contributors

James W Anderson, MD
Professor of Medicine and Clinical Nutrition, University of Kentucky, Room MN 524, Lexington, KY 40536-0298

Richard A Anderson, PhD
Diet, Genomics and Immunology Laboratory, Beltsville Human Nutrition Research Center, U.S. Department of Agriculture, Agricultural Research Service, Beltsville, MD 20705

Livia Augustin, PhD
Unilever Health Institute, Unilever Research and Development, Vlaardingen, The Netherlands

Monica S Banach, BSc
Clinical Nutrition and Risk Factor Modification Center, St. Michael's Hospital, Toronto, Ontario, Canada; Nutritional Sciences, University of Toronto, Toronto, Ontario, Canada

Alan Barclay, BSc
Diabetes Australia-NSW, University of Sydney, NSW, Australia

Jennie Brand-Miller, PhD
Professor J. Brand-Miller, Human Nutrition Unit (G08), University of Sydney NSW, Australia

Karen Chapman-Novakofski, RD, PhD
Professor Nutrition, Department of Food Science and Human Nutrition, Division of Nutritional Sciences, Department of Internal Medicine, and Illinois Extension, University of Illinois, Urbana-Champaign, IL

Anamarie Dascalu, MD, MSc
Risk Factor Modification Centre, St. Michael's Hospital, Toronto, Canada; Departments of Nutritional Sciences and Medicine, Faculty of Medicine, University of Toronto, Toronto, Canada

Philip Domenico, PhD
Nutrition Scientist/Health Educator, 40 W. 77th St. #3D New York, NY 10024-5128

Amin Esfahani, BSc
Clinical Nutrition and Risk Factor Modification Center, St. Michael's Hospital, Toronto, Ontario, Canada; Nutritional Sciences, University of Toronto, Toronto, Ontario, Canada

Kaye Foster-Powell, MSc
Diabetes Centre, Sydney West Area Health Service University of Sydney, NSW, Australia

Elvira González de Mejía, PhD
Department of Food Science and Human Nutrition, University of Illinois, Urbana-Champaign, IL

Frank Greenway, MD
Professor and Medical Director, Pennington Biomedical Research Center, Louisiana State University System, 6400 Perkins Road, Baton Rouge, LA 70808

Toyoshi Inoguchi, MD
Department of Medicine and Bioregulatory Science, Graduate School of Medical Sciences, Kyushu University, Maidashi 3-1-1, Higashi-ku, Fukuoka 812-8582, Japan

Alexandra L Jenkins, RD, PhD
Risk Factor Modification Centre, St. Michael's Hospital, Toronto, Canada; Departments of Nutritional Sciences and Medicine, Faculty of Medicine, University of Toronto, Toronto, Canada

David JA Jenkins, MD
Clinical Nutrition and Risk Factor Modification Center, St. Michael's Hospital, Toronto, Ontario, Canada; Department of Medicine, Division of Endocrinology and Metabolism, St. Michael's Hospital, Toronto, Ontario, Canada; Departments of Nutritional Sciences, Medicine, Faculty of Medicine, University of Toronto, Toronto, Ontario, Canada

Andrea R Josse, MSc
Clinical Nutrition and Risk Factor Modification Center, St. Michael's Hospital, Toronto, Ontario, Canada; Nutritional Sciences, University of Toronto, Toronto, Ontario, Canada

Cyril WC Kendall, PhD
Clinical Nutrition and Risk Factor Modification Center, St. Michael's Hospital, Toronto, Ontario, Canada; Departments of Nutritional Sciences, Medicine, Faculty of Medicine, University of Toronto, Toronto, Ontario, Canada

Joris Kloek, PhD
Scientist Nutrition and Health DSM Food Specialties, P.O. Box 1 2600 MA Delft, The Netherlands

James R Komorowski, MS
Nutrition 21, Inc., Purchase, NY 10577

Azadeh Lankarani-Fard, MD
Assistant Professor, Department of Internal Medicine, Hospitalist Division; Department of Veterans Affairs Greater Los Angeles Healthcare System and David Geffen School of Medicine at UCLA, Los Angeles, CA

Zhaoping Li, MD, PhD
Associate Professor, Center for Human Nutrition, David Geffen School of Medicine at UCLA and Department of Internal Medicine, Hospitalist Division; Department of Veterans Affairs Greater Los Angeles Healthcare System, CA

Mark F McCarty, BA
NutriGuard Research, Inc., 1051 Hermes Ave., Encinitas, CA 92024

Vijai K Pasupuleti, PhD
President, SAI International, 1436 Fargo Blvd., Geneva, IL 60134

L Raymond Reynolds, MD, FACP, FACE
Associate Professor of Internal Medicine, Fellowship Program Director, Division of Endocrinology and Molecular Medicine, University of Kentucky College of Medicine and Lexington VA Medical Center 800, Rose St. MN 524 UKMC, Lexington, KY 40536-0298

Rosalia Reynoso-Camacho, PhD
DIPA, Facultad de Quimica, UAQ, Queretaro, Qro, Mexico

Anne-Marie Roussel, PhD
INSERM,U884, Grenoble, F-38000, France; LBFA, Universite Joseph Fourier, Grenoble, F-38041, France

John L Sievenpiper, MD, PhD
Risk Factor Modification Centre, St. Michael's Hospital, Toronto, Canada; Departments of Nutritional Sciences and Medicine, Faculty of Medicine, University of Toronto, Toronto, Canada

Krishnapura Srinivasan, PhD
Senior Scientist, Department of Biochemistry and Nutrition, Central Food Technological Research Institute, Mysore 570 020, India

P. Mark Starvro, PhD
Risk Factor Modification Centre, St. Michael's Hospital, Toronto, Canada; Departments of Nutritional Sciences and Medicine, Faculty of Medicine, University of Toronto, Toronto, Canada

Luc JC Van Loon, PhD
Departments of Movement Sciences and Human Biology, Nutrition and Toxicology Research Institute Maastricht (NUTRIM), Maastricht University, The Netherlands

Vladimir Vuksan, PhD
Clinical Nutrition and Risk Factor Modification Centre, St. Michael's Hospital, #6 138-61 Queen Street East, Toronto, Ontario, Canada.

Rhonda S Witwer, BS, MBA
Business Development Manager, Nutrition National Starch Food Innovation, 10 Finderne Ave, Bridgewater, NJ 08807

Julia MW Wong, RD
Clinical Nutrition and Risk Factor Modification Center, St. Michael's Hospital, Toronto, Ontario, Canada; Nutritional Sciences, University of Toronto, Toronto, Ontario, Canada

Manan Jhaveri, MBBS
Graduate Student, University of Kentucky, Lexington, KY 40532.

Preface

The discovery of insulin by Banting and Best in 1921 and its subsequent commercialization moved diabetes from a death sentence to a chronic disease often associated with complications such as neuropathy, nephropathy, retinopathy, and cardiovascular problems. Subsequently there have remarkable improvements in types of insulin—moving from porcine sources to human insulin produced by recombinant microbial cultures—with development of fast, intermediate, and slow-acting forms. Many new pharmacological agents have been developed and approved by the FDA, but some have been discontinued or required serious warning labels. Recently, the outstanding prevention research results from the Diabetes Prevention Program studies from the United States and other countries across the world have ushered in a new era for the prevention and management of prediabetes and type 2 diabetes simply by lifestyle changes, diet, and exercise.

Despite all the advances in knowledge and progress in therapy, diabetes poses a greater challenge than ever. Fueled by the upsurge in overweight/obesity, the diabetes rates continue to rise all over the world. The most recent CDC study (2005) reports that diabetes has risen by over 14% in the last two years (2003 to 2005). Conservative projections suggest that by 2008, 24 million Americans— 7.9% of the U.S. population will have diabetes. A combined total of approximately 94 million (70 prediabetic and 24 million diabetic) puts almost one in three as either prediabetic or diabetic (CDC 2005). In addition to these alarming prevalence projections, the cost of diabetes was estimated to be $132 billion in 2002 (CDC 2005) and will escalate to an estimated $167 billion by 2008.

Preventing and controlling prediabetes and type 2 diabetes are important for improving the quality of life and reducing the economic burden. To illustrate the economics of using nutraceutical supplements, one study projected that use of its particular supplement would save $52.9 billion for diabetic patients and public healthcare system when that supplement was used as an adjunct to nutritional therapy. This clearly demonstrates that there is a great opportunity for using nutraceuticals to better control and manage diabetes. At the same time it also illustrates that there is a gap in understanding between consumers, physicians, and researchers from academia and industry as how nutraceuticals can assist in preventing and managing prediabetes and type 2 diabetes. As consumers navigate the ever-changing health-related nutrition messages promising lasting weight control and good health, they have discovered that good nutrition is not as simple as avoiding carbohydrates. The amount and type of carbohydrates in the diet are of vital importance when trying to combat the global epidemics of overweight, obesity, and type 2 diabetes. This book *Nutraceuticals, Glycemic Health and Type 2 Diabetes* draws experts from academia and industry to highlight the epidemiology, glycemia, and nutraceuticals from scientific/clinical and functional foods points of view to benefit type 2 diabetic patients.

Type 2 diabetes is a growing problem for the developed and developing countries, and it is a burden on healthcare systems as well as individuals. This book primarily focuses on the nutraceuticals that assist in preventing and managing prediabetes and type 2 diabetes. It provides an overview of glycemic health and highlights the use of novel and upcoming nutraceutical ingredients such as bioactive peptides, soy, fiber, traditional herbs from India, China, and Mexico, American ginseng, resistant starches, cinnamon, chromium, novel antioxidants, and others.

This book is a compilation and assessment on emerging concepts and nutraceuticals in the prevention and management of diabetes by the experts who presented valid, accurate, latest, and useful data based on evidence.

The book starts off with an introduction followed by epidemiology of diabetes around the world. In the first section, the glycemic concept is explored from physicians and scientific points of view. In the second section, emerging nutraceuticals including herbs are discussed in detail. The book ends with "Future Trends and Directions" and appendix with a comprehensive list of ingredients for diabetes.

The book is intended to bring out the latest concepts and nutraceuticals based on scientific evidence for the prevention and management of diabetes. We hope that we fulfilled these objectives for all those concerned with the role of nutraceuticals and glycemic health in the prevention and management of type 2 diabetes.

We hope that this book will fill a gap in providing an up-to-date reference from scientific point of view on the nutraceuticals and glycemic concept in the prevention and management of type 2 diabetes.

Vijai K Pasupuleti and James W Anderson

Acknowledgments

Our special thanks are due to the contributors; without their hard work and time to share their expertise, this book would not have been possible.

We also thank Mark Barrett[†] for coordinating and making it possible; Susan Engelken for providing editorial assistance; and Ronald D'Souza for going through the manuscripts meticulously.

VKP likes to thank his father P.V. Subba Rao; wife, Anita; and sons, Anoop and Ajai, for providing support, energy, and enthusiasm in completing the book.

JWA appreciates the love and support of his wife, Gay; children, Kathy and Steve; and five granddaughters who continue to inspire.

Nutraceuticals, Glycemic Health and Type 2 Diabetes

Chapter 1

Nutraceuticals and Diabetes Prevention and Management

James W Anderson, MD, and Vijai K Pasupuleti, PhD

Introduction

Individual bioactive chemicals or foods claimed to have health promoting, disease preventing, or medicinal effect on human health are called nutraceuticals. Such foods are also called functional foods. The impact of these nutraceuticals in diabetes prevention and management will be briefly discussed in this chapter and throughout the book.

Every nation on this planet at some stage in their history has adapted to new lifestyles in search of comfort, convenience, and taste. These changes appear to be the direct result of overweight, obesity, and related diseases such as type 2 diabetes. Ironically, in order to prevent, manage, or reverse diabetes the first step is the lifestyle change to eat healthy foods and exercise regularly (Diabetes Prevention Program 2002).

Diabetes is increasing in prevalence worldwide at an alarming rate. This is closely linked to the emergence of obesity in developed and developing countries. Zimmet (2007) has recently stated that this is the largest epidemic the world has ever faced. The estimated number of persons with diabetes in the world is projected to grow from present 246 million to 380 million in 2025. Because of frightening consequences of diabetes on individuals and national economies, on December 21, 2006, the United Nations General Assembly unanimously passed a resolution declaring diabetes as an international public health issue; after HIV/AIDS, diabetes is only the second disease to attain this unenviable designation.

Unfortunately, populations in developing areas have a greater genetic propensity for diabetes than populations in northern Europe. Consequently, African, Southern European, South American, and Asian populations are at a significantly greater risk for developing type 2 diabetes at lower body masses and with less weight gain than are indigenous populations, for example, England (Chapman-Novakofski 2008). Some believe that the risk for diabetes is much higher in these countries as well as in Pima Indians (prevalence and incidence rates of diabetes in Pima Indians are the highest in the world (Bogardus and Tataranni 2002), Nauruans are only next to Pima Indians and rank as the second highest incidence rates of type 2 diabetes in the world with 45% having type 2 diabetes and >75% of the adult population are overweight or obese (King and Rewers 1993)) and Australian aborigines (Guest and O'Dea 1992) because of thrifty genes (Neel 1962). However, the fundamental basis of the thrifty gene hypothesis has recently been challenged by Speakman (Speakman 2007).

Approximately half of all the individuals in the United States have diabetes, prediabetes, or are at a substantial risk for developing diabetes because of the presence of the metabolic syndrome. For 2008 the estimated prevalence of diabetes and related conditions for all ages in the U.S. population are: diabetes 7.9% or 24.1 million; prediabetes 23% or 70.3 million; and the metabolic syndrome 20.3% or 62.0 million. These estimates were derived from firm estimates in the literature (CDC 2005; Ford et al. 2002; Nathan 2007 and Weiss et al. 2004). Of those with diagnosed diabetes, approximately 90% have type 2 diabetes and ∼80% of these are obese (Anderson et al. 2003).

Prediabetes precedes type 2 diabetes and the term prediabetes was introduced in 2002 by Department of Health and Human Services and American Diabetes Association to bring awareness among physicians and general population. One of the reasons for renaming prediabetes from its former clinical name of impaired glucose tolerance (IGT) was to highlight the seriousness of the condition and to motivate people to get appropriate help. As the name suggests, prediabetes is defined as condition that precedes type 2 diabetes. Both IGT or impaired fasting glucose (IFG) are included in the term prediabetes. People with prediabetes have higher than normal blood glucose levels, but they are not elevated enough to be diagnosed as diabetes. A fasting plasma glucose value between 100 and 125 mg/dL or more indicates IFG. In an oral glucose tolerance test, plasma glucose values between 140 and 199 mg/dL at 2 h postglucose load indicates impaired glucose tolerance.

Prevention of type 2 diabetes should begin before or during the IFG and/or IGT stage. Both prevention and management of diabetes can be achieved by lifestyle changes, nutraceuticals, and/or drugs. In this short review, we cover the nutraceuticals and briefly touch upon the lifestyle and pharmacological approaches and how they can be potentially used to prevent and manage diabetes.

The main causative factors of diabetes are a genetic disposition, overweight or obesity, lack of physical activity, consumption of high-fat, low-fiber diets, oxidative stress, and possibly deficiencies in certain minerals (Figure 1.1).

Lifestyle

Diabetes prevention program studies in the United States, Europe, and Asia clearly demonstrated that by lifestyle changes one could prevent, manage, and

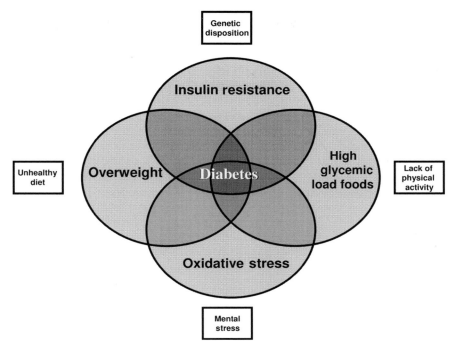

Figure 1.1. Contributing factors to prediabetes and type 2 diabetes.

reverse the prediabetes. U.S. Diabetes Prevention Program studies pointed out that lifestyle changes have achieved better results in prevention of diabetes compared to metformin (58% vs. 31%, respectively). The additional benefit of lifestyle change is that there are no side effects like those associated with the drugs. By adopting the lifestyle changes the human and economic costs of diabetes can be significantly reduced. It was estimated that in 2002 the total annual economic cost of diabetes was $132 billion or $1 for every $10 of healthcare dollars spent in America (French 2007) (ADA 2002). The goals for lifestyle changes are increased physical activity, improving dietary composition with fiber, and reducing the consumption of refined sugars, saturated fats, eliminating trans fats, avoiding smoking and excessive alcohol drinking, and losing body weight to a desirable level (Anderson et al. 2003) or at least 5–7% body weight. Hamman et al. (2006) followed up with DPP participants randomized to the intensive lifestyle intervention had a significantly reduced risk of diabetes compared to placebo participants. Weight loss is the dominant predictor for reduced diabetes incidence rates. Based on their studies it is predicted that patients who lose more weight than the DPP average of 5–7% and who meet physical activity and dietary fat goals could reduce their diabetes risk by greater than 90%. For every kilogram of weight loss, there was a 16% reduction in risk,

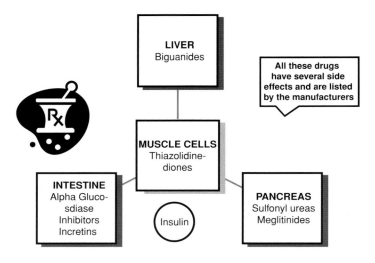

Currently there are no approved FDA drugs for the prevention of prediabetes and type 2 diabetes

Figure 1.2.   Action sites of different FDA approved drugs for type 2 diabetes.

adjusted for changes in diet and physical activity. Thus far it is evident and very impressive from all the preventive studies across the world that a simple lifestyle change makes a big difference. If this research can be implemented/translated to bring a change in individuals and in the communities we can prevent, manage, and reverse prediabetes, diabetes, and their associated risks in a considerable amount of population. This will not only improve the quality of life but also significantly reduce the financial burden on the individuals and Government. Greenway covers in detail lifestyle changes and its effect on diabetes in Chapter 3.

Pharmacologic Approaches

There are a number of FDA approved drugs for the management of diabetes, and each class of drugs is unique and acts on different sites Figure 1.2.

Active research is focused on new pharmacotherapies for prevention of type 2 diabetes. Some prior pharmacologic agents had serious side effects. An early agent in the biguanide class was withdrawn from the market because of renal toxicity and an association with lactic acidosis; the currently available metformin should not be used in persons with impaired renal failure or persons at risk for lactic acidosis. Troglitazone, a thiazolidinedione, was withdrawn from the market because of an association with liver failure; one of the currently available agents— rosiglitazone—is associated with an increased risk for congestive heart failure and, perhaps, for coronary heart disease.

Greenway (Chapter 3) reviews currently available pharmacologic agents that may have diabetes protective effects. Currently, metformin is the only agent that has documented protective effects related to type 2 diabetes (Diabetes Prevention Program Research Group 2002; Ramachandran et al. 2006). However, its protective effect is not as significant as with the lifestyle changes. To date, there are no FDA-approved drugs for the prevention of insulin resistance, prediabetes, and diabetes. Diet and exercise are the only currently accepted approaches for preventing the progression of prediabetes to type 2 diabetes.

Conventional Nutrition and Medical Nutrition Therapy

It may not be possible to come up with an optimal mix of macronutrients (carbohydrate, protein, and fat) as a universal diet for all the diabetics as the individual requirement varies from person to person (ADA 2006). Conventional or standard nutritional formulas are enriched with high carbohydrates, minerals, and vitamins and are low in fiber (Campbell and Schiller 1991). These may not be suitable to prediabetic and diabetic patients, as they tend to increase the blood glucose levels faster and compromise the glycemic control leading to several complications. Keeping this in mind diabetic specific formulas have been developed for use in conjunction with MNT as the comprehensive nutrition intervention (Coulston 1998). The diabetic specific ingredients may be combined with MNT formulas using a defined nutrient composition to achieve tighter glycemic control. Some of these specific nutrients are fiber, magnesium, soy protein and peptides, mono unsaturated fatty acids, antioxidants, chromium, and so on. Fiber, cinnamon, soy, antioxidants, resistant starch, and bioactive peptides are covered in detail in Chapters 7, 8, 9, 11, 16, and 17, respectively. The specific ingredients in MNT diabetic formulas assists in managing tighter glycemic control by delaying gastric emptying, intestinal absorption, producing smaller rise in glucose levels, increasing insulin sensitivity, and decreasing hepatic glucose output.

In the literature few clinical reviews have been published on the use of nutritional formulas for diabetes patients. Murakami et al. (2006) systematically reviewed published cohort studies on the effect of nutrient and food intake on the incidence of diabetes. They found significant inverse relationship with the intake of vegetable fat, poly unsaturated fat, dietary fiber, magnesium, and caffeine and direct relationship with intake of trans fatty acid and heme-iron, high glycemic index, and glycemic load foods to the incidence of diabetes.

Marinos et al. (2005) have done a systematic review and meta-analysis of diabetic specific formulas versus standard nutritional formulas. They looked at several parameters such as glycemia, lipidemia, nutritional status, medication requirements, quality of life, complications, and mortality. It was evident that the use of diabetes-specific formulas consistently resulted in significantly lower postprandial rise in blood glucose, peak blood glucose concentrations, and glucose area under the curve in diabetic patients.

The most important practice for diabetic individuals is to make a lifestyle change—eating a healthy diet, regular exercise, and using nutraceuticals to achieve an optimal blood glucose level. When it comes to ingredients in diabetic foods, there is no single magic bullet. However, there are ingredients that will slow down the absorption of glucose to avoid spikes, low glycemic index food ingredients that will yield relatively lower glucose levels, and some other ingredients increase insulin sensitivity and alleviate oxidative stress. The key to a successful diabetic diet is to combine all or some of these ingredients into foods and beverages with a great taste to appeal the consumers' palate (Pasupuleti 2007). Such a product coupled with regular exercise may replace or reduce the dosage of existing drugs without any side effects.

The American Diabetic Association (ADA 2006) issued nutrition recommendations and interventions for diabetes based on the best available scientific evidence. This is to bring awareness of beneficial nutrition intervention to the diabetic patients and healthcare providers for prevention and management of diabetes. The specific goals of MNT vary according to individual situations.

People with Prediabetes

The goal of MNT is to decrease the risk of diabetes and cardiovascular disease by lifestyle changes that include healthy food choices and physical activity leading to weight loss that is maintained. ADA (2006) recommended consumption of a minimum of 14 g fiber/1,000 kcal containing whole grains and other important nutrients. However, according to ADA there is no sufficient, consistent information to conclude that low glycemic load diets reduce risk for diabetes but low glycemic load diets that are rich in fiber and other important nutrients should be encouraged to prevent the progression of prediabetes to type 2 diabetes. Lifestyle changes, glycemia, and fiber have been covered in detail by Greenway, Wong, Reynolds, Foster-Powell, and Anderson in Chapters 3–7.

People with Type 2 Diabetes

The goals of MNT are to achieve and maintain blood glucose levels, lipid, lipoprotein profile, body weight, and blood pressure in the normal range or as close to normal as is safely possible; by changing the lifestyle and nutrition prevent or slow the rate of development of the chronic complications of diabetes; addressing the individual nutrition needs by taking into account personal and cultural preferences and willingness to change and to maintain the pleasure of eating by only limiting food choices when indicated by scientific evidence.

The most important aspect is to monitor blood glucose on a regular basis and, if needed, adjust the medication or diet by discussing with the healthcare professionals.

Minerals

Typically, micronutrient deficiencies are observed in uncontrolled diabetic patients. Therefore, it is essential to consume mineral requirements from natural food sources and if needed obtain from the supplements. There are reports suggesting that chromium, magnesium, zinc, vanadium, calcium, boron, and manganese assist in achieving tighter glycemic control. However, there is no clear evidence of benefit from mineral supplementation in people who do not have underlying deficiencies. Of all the minerals chromium has been extensively studied and it recently won a conditional health claim from FDA. Chromium and other minerals have been extensively reviewed by Domenico in Chapter 10.

Herbals

It is interesting to point out once again that certain ethnic groups such as Native Americans, Hispanics, and Asians are more vulnerable to diabetes and at the same time they have a long history of using traditional folk medicine in the form of herbs. More than 1,200 medicinal plants have been cited from Mexico, China, India, and other countries (Marles and Fransworth 1995). Some of the well-studied herbs like American ginseng, Chinese herbs, fenugreek, nopal, and traditional herbs from India and Mexico have been covered in detail by Sievenpiper, Lankarani-Fard, Srinivasan and Reynoso in Chapters 12, 13, 14, and 15, respectively. A list of all the herbs, and current concepts about their mechanisms and human studies are listed in Appendix.

Conclusions

From all the diabetes preventive studies one thing is evident that type 2 diabetes can be prevented or delayed by lifestyle modification. The screening tools for prediabetes and type 2 diabetes are readily available. However, the challenge is to implement the screening procedures and encourage high-risk individuals to modify their lifestyle. Further research is necessary to translate the excellent findings of DPP studies into practice in the communities to prevent or delay the onset of diabetes. At the same time intensive research efforts are ongoing across the globe to prevent and manage diabetes by nutraceuticals and pharmaceuticals. Anderson discusses future trends and directions in the prevention and management of diabetes in Chapter 18.

References

American Diabetes Association. 2002. Economic costs of diabetes in the USA in 2002. Diabetes Care 26:917–932.

American Diabetes Association. 2006. Nutrition recommendations and interventions for diabetes—2006. Diabetes Care 29(9):2140–2157.

Anderson JW, Kendall CWC, Jenkins DJA. 2003. Importance of weight management in type 2 diabetes: Review with meta-analysis of clinical studies. J Am Coll Nutr 22:331–339.

Campbell S, Schiller M. 1991. Considerations for enteral nutrition support of patients with diabetes. Top Clin Nutr 7:23–32.

CDC National Diabetes Fact Sheet. 2005. http:www.cdc.gov/diabetes/pubs/factshee05 htm. Accessed November 27, 2007.

Chapman-Novakofski K. 2008. In Epidemiology of Type 2 Diabetes, Nutraceuticals, Glycemic Health and Diabetes, edited by Pasupuleti V, Anderson JW. Ames, IA: Wiley-Blackwell Publishing.

Bogardus C, Tataranni PA. 2002. Reduced early insulin secretion in the etiology of type 2 diabetes mellitus in Pima Indians. Diabetes 51:S262–S264.

Coulston AM. 1998. Clinical experience with modified enteral formulas for patients with diabetes. Clin Nutr 17(Suppl 2):46–56.

Diabetes Prevention Program Research Group. 2002. Reduction in the incidence of type 2 diabetes with lifestyle intervention or metformin. N Engl J Med 346:393–403.

Ford ES, Giles WH, Dietz WH. 2002. Prevalence of the metabolic syndrome among US adults. J Am Med Assoc 287:356–359.

French S. 2007. Diabetes: An epidemic creates opportunities. Nat Prod Insid 12:46–47.

Guest C, O'Dea K. 1992. Diabetes in aborigines and other Australian populations. Aust J Public Health 16:349–349.

Hamman RF, et al. 2006. Effect of weight loss with lifestyle intervention on risk of diabetes. Diabetes Care 29(9):2102–2107.

King H, Rewers M, for World Health Organization Ad Hoc Diabetes Reporting Group. 1993. Diabetes in adults is now a Third World problem. Ethn Dis 3:S67–S74.

Murakami K, Okubo H, Sasaki S. 2006. Effect of dietary factors on incidence of type 2 diabetes: A systematic review of cohort studies. J Nutr Sci Vitaminol 51:292–310.

Marinos E, Antonio C, Heiner L, Alan JS, Meike E, Rebecca JS. 2005. Enteral nutritional support and use of diabetes-specific formulas for patients with diabetes. A systematic review and meta-analysis. Diabetes Care 29(9):2267–2279.

Marles R, Fransworth NR. 1995. Antidaibetic plants and their active constituents. Phytomedicine 2:137–189.

Nathan DM. 2007. Finding new treatments for diabetes—how many, how fast...how good? N Engl J Med 356:437–440.

Neel JV. 1962. Diabetes mellitus: A "thrifty" genotype rendered detrimental by "progress"?. Am J Hum Genet 14:353–362.

Pasupuleti VK. 2007. Food Technol 2:59–60.

Ramachandran A, Snehalatha C, Mary S, Mukesh B, Bhaskar AD, Vijay V. 2006. The Indian Diabetes Prevention Programme shows that lifestyle modification and metformin prevent type 2 diabetes in Asian Indian subjects with impaired glucose tolerance (IDPP-1). Diabetologia 49:289–297.

Reynolds LR. 2008. In Glycemia: Health Implications, Nutraceuticals, Glycemic Health and Diabetes, edited by Pasupuleti V, Anderson JW. Ames, IA: Wiley-Blackwell Publishing.

Speakman JR. 2007. A nonadaptive scenario explaining the genetic predisposition to obesity: The "predation release" hypothesis. Cell Metab 6(1):5–12.

Weiss R, Dzuira J, Burgert TS, et al. 2004. Obesity and the metabolic syndrome in children and adolescents. N Engl J Med 350:2362–2374.

Zimmet P. 2007. Diabetes—the biggest epidemic in human history. http//www.medscape.com/viewarticle/561261. pp. 39–40. Accessed November 28, 2007.

Chapter 2

Epidemiology of Type 2 Diabetes

Karen Chapman-Novakofski, PhD

Overview

Diabetes is a global concern. While some countries have higher prevalence rates than others, the increasing incidence and economic and social burden of the disease moves diabetes to the scientific forefront. Epidemiology provides a basis to examine similarities and differences in variables known to have significant roles in the development and progression of the disease. Causative factors include nonmodifiable variables such as genetics and ethnicity as well as modifiable environmental variables such as lifestyle and dietary habits. The complexity of these interactions must be investigated to successfully identify and implement strategies to reduce the burden of diabetes.

Prevalence of Diabetes

The prevalence of type 2 diabetes is increasing and quickly becoming a global concern. Worldwide diabetes prevalence was estimated to be 4.0% in 1995 with increases projected to 5.4% in 2025 using the World Health Organization's diabetes-related databases and the United Nations' demographic estimates. With this prediction, there will be 300 million adults with diabetes in 2025 (King et al. 1998). Other estimates predict that the number of persons with diabetes will double between 2000 and 2030 on the basis of population growth statistics alone (Wild et al. 2004).

Considering Europe, the United States, Canada, Australia, New Zealand, and Japan as "developed" countries, the prevalence data of developed countries were compared with those of all other "developing" countries. The prevalence of diabetes is higher in developed countries (6–7.5%) than in developing countries (3.3–4.9%) and is predicted to remain so. However, the increase in diabetes in developing countries is predicted to be steep (King et al. 1998). In developed countries, more women than men have diabetes, while in developing countries the incidence is

11

equal between genders. In general, increasing age is associated with increasing prevalence of diabetes. In developed countries, the diabetes prevalence is highest for those aged 65 and older. For developing countries, the age peak is younger, within the 45–65-year range (King et al. 1998; Wild et al. 2004).

Within the developed countries, the prevalence of 4.2% in England for diagnosed and undetected diabetes is lower than the prevalence reported in the United States or developed countries combined (Mainous et al. 2006). Similarly, the prevalence of diagnosed type 2 diabetes in France is only 3.3% (Mattout et al. 2006), while in Germany the prevalence is 6.45% among the adults who have health insurance (Stock et al. 2006).

Of course, the countries with the largest number of people are also the countries with the largest number of people who have diabetes, specifically China, India, and the United States. Projected estimates include these three countries as having the highest prevalence of diabetes for 2025 (King et al. 1998; Wild et al. 2004). Other interpretations of population estimates and prevalence of diabetes rank India as having the largest number of individuals with diabetes, having 31.7 million diagnosed cases in 2000 projected to 79.4 million in 2030, followed by Europe, Southeast Asia, and China (Wild et al. 2004).

Among the epidemiological studies, there is variation in many factors that could bias the estimated prevalence and projections for global health. Most studies base prevalence on diagnosed diabetes. In defining diabetes, however, there are two diagnostic indices. The World Health Organization (WHO) uses a fasting blood glucose value and a blood glucose level taken 2 h after a 75 g glucose challenge (Alberti and Zimmet 1998), whereas the American Diabetes Association (ADA) relies on fasting blood glucose values alone (Table 2.1). In addition, reference values used prior to 1999 are higher than those currently in use.

Studies comparing the WHO and ADA diagnostic indices have concluded that the ADA criteria underestimate both diabetes and impaired glucose tolerance (Botas et al. 2003; Gimeno et al. 1998; Harris et al. 1997; Mannucci et al. 1999; van Winkel et al. 2006). Moreover, studies using only diagnosed diabetes as assessed by either method cannot be compared with studies combining both impaired glucose tolerance and diagnosed diabetes for "total diabetes." For reproducible results and analyses, one diagnostic set of criteria is needed (Jorgensen et al. 2003).

In addition, there are various methods used for assessing the risk of having undiagnosed diabetes. Using either the WHO or ADA diagnostic criteria for individuals who have not been told they have diabetes and using this base incidence to project for populations is one method (Mainous et al. 2007). A number of risk algorithms exist for determining if a person is likely to develop diabetes. Among those individuals without diagnosed diabetes and without a blood glucose level elevated to the diagnostic criteria, a scoring system of related conditions may be used to classify individuals as "at risk." Generally those with increased waist circumference, elevated blood pressure, elevated blood triglycerides, and low high-density lipoproteins, or a combination of these conditions may be considered at risk for developing type 2 diabetes (Mainous et al. 2007). However, the specific cutoff

Table 2.1. Comparison of World Health Organization (WHO) and American Diabetes Association (ADA) criteria for normal and impaired blood glucose and diabetes.

	WHO (1985)	WHO (1999, 2006)	ADA (2005)
Normal			
Fasting	Not defined	<6.1 mmol/L	<5.6 mmol/L
Impaired fasting glucose	Not defined	≥6.1 and <7.0 mmol/L	5.6–6.9 mmol/L
Impaired glucose tolerance			
Fasting	<7.8 mmol/L	<7.8 mmol/L	≥7.8 mmol/L
2-h glucose[a]	≥7.8 mmol/L and <11.1 mmol/L	≥7.8 mmol/L and <11.1 mmol/L	<11.1 mmol/L
Diabetes			
Fasting	>7.8 mmol/L or	≥7.0 mmol/L or	≥7.0 mmol/L or
2-h glucose	≥11.1 mmol/L	≥11.1 mmol/L	≥11.1 mmol/L

[a]2-h glucose = blood glucose 2 h after ingestion of 75 g glucose load.

points determining "elevated" and "normal" may vary not only with ethnicity but also with cultural setting, especially for anthropometric measures (Ramachandran et al. 2005).

One additional caveat in interpreting diabetes prevalence on a global scale is the possibility that samples' statistic contributed from a particular country are not truly representative of the entire country's population. Indeed, extrapolation from one country's estimates to a neighboring country may under- or overestimate prevalence and population growth trends. Age categories are not always represented in each country's survey, and mathematical modeling by different methods to account for these missing data can give slightly different estimates (Wild et al. 2004). However, a broad perspective of the prevalence of type 2 diabetes globally must rely on available data regardless of limitations.

Asia-Pacific Region

Within the Asia-Pacific region, the countries in Southeast Asia considered as a group include Thailand, Malaysia, Indonesia, Singapore, the Philippines, Myanmar, Vietnam, Cambodia, and Laos. Epidemiological data from this area are scarce. Nevertheless, data from a decade ago indicate a diabetes prevalence range of 1.4 to nearly 12% (Cockram 2000).

Type 2 diabetes prevalence among Caucasians in Australia reflects on the rates of prevalence in other developed countries. However, Aboriginal people have a greatly increased prevalence of diabetes, in the range of 20–25% (Cockram 2000). Similarly, Polynesian New Zealanders have higher prevalence rates of type 2 diabetes than do Caucasian New Zealanders. The Pacific islands show great variation in prevalence rates, with some very high rates in Nauru and New Guinea (approximately 40%). Obesity is very common in the islands and certainly contributes to the diabetes epidemic (Simmons et al. 2001).

In contrast to the obesity-linked diabetes development is type 2 diabetes in Asians and Indians with much lower body mass index (BMI). This form of type 2 diabetes is accompanied by a higher percent of body fat, particularly abdominal body fat, than is normally seen in other cultures when matched for BMI. The higher degree of abdominal fat is subsequently associated with a higher degree of insulin resistance. The prevalence of diabetes in India ranges from 2.7% in rural India to 14% in urban India (Sanghera et al. 2006). A study conducted in southern India reported a known diabetes prevalence rate of 9%, equally distributed between genders (Menon et al. 2006).

A meta-analysis of prevalence rates in China found differences among individuals living in Mainland China, Hong Kong, and Taiwan. The variation in prevalence rates for those living in mainland China was large, with more recent studies reporting higher prevalence. The odds ratios comparing the prevalence of diabetes in Hong Kong and Taiwan to mainland China were 1.5 (95% confidence interval [CI] 1.4–1.7) and 2.0 (95% CI 1.8–2.2), respectively. There was a larger prevalence of undiagnosed diabetes in mainland China (68.6%) than in Taiwan and Hong Kong (52.6%) (Wong and Wang 2006).

Eastern Mediterranean and Middle East

The United Arab Emirates, Bahrain, Kuwait, and Oman have diabetes prevalence rates that rank in the top 10 countries with highest diabetes prevalence. However, mean prevalence rates for the entire region are more comparable to those of other areas at 7.0% (International Diabetes Federation 2003).

European Countries

Diabetes is a leading cause of death in the European Union, with estimated prevalence rates of 7.8% expected increase to ~9% by 2025 (International Diabetes Federation 2006). Prevalence rates were lowest for Iceland (2.0%) and Ireland (2.3%), but 9.0% or above for Poland, Malta, Lithuania, Czech Republic, Austria, Slovenia, Estonia, Hungary, Latvia, and Spain, with the highest rates in Germany (10.2%). Estimated prevalence rates for 2025 are 10% or above for these countries with the addition of Finland (International Diabetes Federation 2003).

The United States

The prevalence of diabetes was assessed by the National Health and Nutrition Examination Survey (NHANES) using data from 1999 to 2002. The overall prevalence rate was 6.5%, equal to that found in Mexican Americans. The prevalence of type 2 diabetes in non-Hispanic black adults was higher at 10.0% and lower for non-Hispanic whites at 5.6%. The overall prevalence was similar for men and women, but rose with age, reaching 15.8% for those aged 65 and older. Standardizing the prevalence rates for the U.S. census, age and gender estimates resulted in higher estimated prevalence rates for both non-Hispanic black (11%) and Mexican Americans (10.4%). When comparing the data from an earlier NHANES (1988–1994), the overall prevalence of diabetes rose even when standardized to account for an aging population (Cowie et al. 2006). Projected data suggest that the prevalence rate of diagnosed and undiagnosed diabetes could reach 14.5% by 2031 (Mainous et al. 2007).

Mexico

The prevalence of diabetes continues to grow in Mexico, from <3% in the 1960s to 6% in the 1980s and to 8.2% in 2000 (Garcia-Garcia et al. 2006). More than 11 million Mexicans are predicted to be diagnosed with diabetes by 2025 (Rull et al. 2005). Diabetes has become the main cause of death in Mexican adults (Ministry of Health 2003). The Mexican National Health Survey (2000) concluded that the prevalence of overweight and obesity greatly exceeded reports of previous surveys and were similar to U.S. surveys. Central Mexico (San Luis Potosi, Guanajuato, Queretaro) reports that diabetes and obesity are primary health problems. Even in normal-weight individuals, there is a growing incidence of abdominal obesity, which is expected to contribute to the rising prevalence of diabetes in Mexico (Sanchez-Castillo et al. 2005).

Latin America

Type 2 diabetes in Latin America and the Caribbean was estimated to be 3.2% in 2000 (Barcelo et al. 2003). In Central and South America, the prevalence rate was estimated to be 5.6% in 2003 with a projected rise to 7.2% in 2025 (International Diabetes Federation 2003).

Overall, by 2025 the world prevalence increase in type 2 diabetes is expected to be largest in East Asia and the Pacific, and the smallest increases are expected in sub-Saharan Africa (Narayan et al. 2006). Despite the limitations of prevalence studies previously discussed, it is clear that the burden of diabetes has grown and will continue to grow around the world. The cost to the individual can be measured in terms of quality of life as well as dollars, as the subsequent complications of diabetes, namely, nephropathy, retinopathy, neuropathy, and cardiovascular disease, compromise an individual's livelihood and longevity. The cost to society is certainly economic, but also compromises social structures and infrastructure development. Diabetes is a complex disease, with many contributing causative factors. While some factors are not readily modifiable, others promise a possible route to diminishing the prevalence of type 2 diabetes.

Causative Factors

Factors contributing to the increased growth of diabetes include population growth and aging, with concomitant rising trends in urbanization, inactivity, and obesity (Wild et al. 2004). Genetic predisposition is also an important causative factor. Rather than one single factor directing the development of type 2 diabetes, environmental and genetic factors interact such that over time, the molecular synchronization required to maintain a normal blood glucose level is disrupted and compensatory mechanisms fail.

Genetics

Although complicated because diabetes is a heterogeneous disease, genetics does play an important role in the development of type 2 diabetes. A positive family history of diabetes increases the likelihood of developing type 2 diabetes. The degree of influence that family history has is complicated by the number of factors that genetics may control as well as interaction with the environment. For instance, genetics may be involved with phenotypic trait of body size and strength, but environment may also influence body size and strength to a certain extent through exercise training. Observational studies have been used to estimate probabilities of diabetes developing among offspring. With one parent having diabetes, the probability of the offspring developing diabetes is about 38%. With both parents having type 2 diabetes, the probability increases to about 60% (Stumvoll et al. 2005). Family studies have assessed the heritability of a phenotypic trait to genetics for BMI (42–54%), fasting blood glucose (10–72%), and fasting insulin (8–37%) (Stumvoll et al. 2005). Familial clustering has been reported for several ethnicities (Li et al. 2002).

The candidate gene approach has been used to identify specific likely "candidate" genes that are associated with the disease. A candidate gene for a disease is one located in a chromosome region suspected of being involved in the disease. One possible candidate gene for type 2 diabetes is the gene for peroxisome proliferators-activated receptor gamma (PPARγ). PPARγ is specific to adipose tissue and has a role in adipocyte differentiation. Polymorphisms in these candidate genes could lead to a type 2 diabetes phenotype. Other genes that have been identified as having a role in type 2 diabetes development include sulfonylurea receptor 1, potassium inward rectifier 6-2, and insulin-like growth factors (Stumvoll et al.

2005). Genetic studies across populations have had varied results, due to either the methodology or the heterogeneity of type 2 diabetes (Sanghera et al. 2006).

Ethnicity

Diabetes prevalence is disproportionately higher in many minority and ethnic backgrounds, especially non-Hispanic black and Hispanic people in the United States. Statistical projections suggest that diabetes prevalence may reach 20% of Hispanic Americans by 2031 (Mainous et al. 2007).

Other ethnicities known to have high prevalence rates include South Asians, Micronesian Nauruans, Australian Aboriginals, and South African Blacks, as well as Pima Indians, non-Hispanic Blacks and Hispanic people in the United States (Davis et al. 2001; Mainous et al. 2007; Simmons et al. 2001; Zimmet 1992). Pacific Islanders and Maori peoples have prevalence rates above 20% (Simmons et al. 2001). In England, Black African/Caribbeans and Asians have higher prevalence rates of type 2 diabetes than do White Europeans (Forouhi et al. 2006). The prevalence rates for Asian Indians varies with locale, with rates as high as 22% for Asian Indians in Fiji (Sanghera et al. 2006). While Alaskan Eskimos have a relatively low prevalence of diabetes (3.8%), a higher prevalence of impaired glucose tolerance suggests a higher risk for type 2 diabetes in the future (Carter et al. 2005).

Homeostasis model assessment (HOMA) estimates insulin sensitivity and beta-cell function using blood glucose and insulin levels. In U.S. women, African Americans had higher beta-cell function than non-Hispanic whites, and Chinese and Japanese Americans had lower function. However, African American, Chinese American, and Japanese women had lower insulin sensitivity than non-Hispanic whites after correcting for waist circumference (Torrens et al. 2004). Within the United Kingdom Prospective Diabetes Study, Indian Asians had the highest insulin resistance but the greatest pancreatic beta-cell function while the African Carribeans had the highest insulin sensitivity and the lowest beta-cell function. However, glycemic control was similar for Indian Asians, African Carribeans, and white Caucasians over 9 years follow-up (Davis et al. 2001). The flux of ethnicities into or out of a country can obviously alter prevalence statistics for a country. Similarly, the cultural environment can increase or decrease the prevalence rate for an ethnicity.

Age

As with all chronic diseases, the prevalence of type 2 diabetes increases with age. In France, the prevalence of treated diabetes reaches its peak of 14% at 75–79 years. The peak prevalence for northern European countries is similarly estimated to be ~70 years (Fagot-Campagna et al. 2005). In southern India, the increase in type 2 diabetes prevalence is seen after 50 years, with a more substantial increase in females (Menon et al. 2006). Similar trends are seen in China (Wong and

Wang 2006). As previously discussed, the younger age peak is more apparent in developing countries than in developed countries. In addition, an increase in obesity and decrease in physical activity with age may amplify the impact of age on diabetes prevalence.

Urban/Rural Differences

Rural areas are assumed to have one-half to one-fourth as many cases of diabetes as urban counterparts (Wild et al. 2004). The expectation is that people in rural communities follow more traditional lifestyles, with higher physical activity levels and caloric intake more balanced with energy expenditures. Many studies have supported the concept that rural populations have a lower prevalence of diabetes than urban populations. For instance, in India the prevalence rates of diabetes have been reported to be two to four times higher in urban areas than in rural areas (Menon et al. 2006). Prevalence rates in Qingdao, China, were higher in urban areas than in rural areas, although the prevalence of undiagnosed diabetes was higher in rural than in urban setting (Dong et al. 2005). The influence of urban lifestyle may be overemphasized by a lack of screening and access to health care. Indeed, studies in the European Union countries reported no differences between urban and rural prevalence rates except in Turkey, Kazakhstan, Kyrgyzstan, Tajikistan, and Turkmenistan (International Diabetes Federation 2003).

Chronic Inflammation

Inflammation and dysregulation of cytokine production may have important roles in the pathogenesis of type 2 diabetes. Cytokines are secreted primarily from leukocytes and stimulate immune responses. The cytokine family of proteins includes interleukins, interferons, and tumor necrosis factors. Tumor necrosis factor α (TNFα) is expressed by adipose tissue, with higher levels of TNFα expression being associated with obesity and insulin resistance. TNFα has an important role in repressing genes involved in the uptake and storage of glucose and certain fatty acids, and may impair insulin signaling (Kershaw and Flier 2004; Rajala and Scherer 2003). Similar to TNFα, interleukin-6 is expressed by adipose tissue, and associated with obesity, impaired glucose tolerance, and insulin resistance (Kershaw and Flier 2004). Congruent with a chronic inflammation, acute phase reactants are increased and inflammation signaling pathways are activated in many cases of insulin resistance and type 2 diabetes. Why inflammatory responses are found in obese individuals is an important research question that is yet to be answered (Hotamisligil 2006).

Obesity

Many country profile predictions for diabetes prevalence assume that the level of obesity will remain constant (King et al. 1998; Wild et al. 2004). If this assumption is not true, and it most probably is not since the incidence of obesity is rising, then many diabetes prevalence estimates have been underestimated (Wild et al. 2004). For instance, in the United States, the prevalence of obesity is highly correlated with that of type 2 diabetes ($r = 0.64$, $P < 0.001$) (Muhammad 2004). This correlation has been explained by the increase in insulin resistance that accompanies obesity, especially abdominal obesity. However, not all obese people develop type 2 diabetes. While 54.1% of overweight U.S. adults had normal glucose metabolism, the remaining had diabetes, impaired glucose tolerance, and/or impaired fasting glucose that had not yet reached impairment to the point of diabetes diagnosis (Benjamin et al. 2003).

Nevertheless, waist circumference and BMI have both been used to predict type 2 diabetes in population studies, although waist circumference appears to be the better predictive variable (Diabetes Prevention Program Research Group 2006; Wei et al. 1997). Many studies combine both the BMI and waist circumference to achieve more discriminating relative risk results (Folsom et al. 2000; Meisinger et al. 2006; Wang et al. 2005). Central obesity has been shown to be a significant predictor of type 2 diabetes in several ethnicities, including African Americans, Mexican Americans, and Japanese (Ohnishi et al. 2006; Stevens et al. 2001; Wei et al. 1997). A lack of consensus on appropriate cut points in waist circumferences for Japanese and Asians adds complexity to comparative analyses (Oda 2006).

While obesity is an important factor in the development of type 2 diabetes, the interaction of genetics and environment is also important. For any group of people exposed to the same environment, the same food, or the same possibilities for physical activity, not everyone will remain lean nor will everyone become obese. Genetics may have a role in hormonal expression, body composition, height, appetite, and dexterity or inclination for physical activity.

Diet

Excess calories lead to weight gain and obesity, which is a risk factor for type 2 diabetes. However, certain dietary patterns have been investigated for their association with diabetes. Factor analysis identified two dietary patterns in Finnish adults. The prudent diet of higher fruit and vegetable intake was associated with a lower relative risk for type 2 diabetes than was the diet characterized by butter, potatoes, and whole milk after adjustment for nondietary confounding variables as well as energy intake and BMI (Montonen et al. 2005).

In a cross-sectional survey of Japanese self-defense officials, three dietary patterns emerged. The high-dairy, high-starch, high-fruit and -vegetable, low-alcohol pattern was associated with fewer glucose abnormalities, while a Japanese dietary pattern showed a positive trend with glucose intolerance and an animal meat pattern revealed no association (Mizoue et al. 2006). In contrast to this study is a self-report of dietary type by Japanese Americans in Hawaii. Those reporting a Western dietary pattern had a 9.6% diabetes prevalence rate as compared to 5.2% with an Oriental dietary pattern (Huang et al. 1996).

A multicenter study in the United States identified six dietary patterns. The "dark bread" and "wine" patterns were more positively associated with lower fasting insulin and higher insulin sensitivity than were "white bread," "fries," "low-frequency eaters," or "fruit" dietary patterns in both unadjusted models and models adjusted for demographics. When also adjusted for energy intake, differences in insulin sensitivity among dietary patterns were lost. However, the "wine" pattern represented only 3.8% of the sample and the "dark bread" pattern represented only 0.9% (Liese et al. 2004). A modest decrease in relative risk for type 2 diabetes was found in men in the Health Professionals Follow-Up Study as dairy intake increased from less than one serving per day up to three or more servings per day after adjustment for total energy intake, demographics, presence or absence of hypertension or hypercholesterolemia, as well as other dietary factors (Choi et al. 2005).

Data from the Nurses Health Study were analyzed for contribution of sugar-sweetened drinks to the relative risk of type 2 diabetes. Although an increase in relative risk was reported, data were not corrected for energy intake or BMI. Increased intake of sugar-sweetened beverages was associated with weight gain (Schulze et al. 2004). Dietary fat has also been investigated for a possible role in the development of type 2 diabetes. Although the ratio of polyunsaturated to saturated fat was inversely associated with the risk of type 2 diabetes, statistical significance was lost when data were corrected for BMI (Harding et al. 2004).

Researchers continue to investigate the role of dietary patterns as well as individual foods for associations with increased insulin sensitivity and glucose tolerance. Because weight maintenance and energy balance are so influential in diabetes development, research for optimal foods and patterns of food intake for diabetes prevention should adjust for differences in calories and BMI.

Inactivity

Physical activity can increase insulin sensitivity and glucose tolerance (Mayer-Davis et al. 2002). Insulin-mediated glucose uptake occurs primarily in skeletal muscle. Therefore, using those muscles in mild-to-moderate physical activity can decrease elevated blood glucose and improve glucose tolerance, especially in the obese. Glucose uptake is positively associated with lean mass and negatively with adipose tissue (Albright et al. 2000). Muscle contractions also stimulate glucose transport independently of insulin (Geiger et al. 2006). Differing results of studies focusing on insulin and glucose metabolism in response to exercise may be due in part to differences in intensity, frequency, and intensity of exercise training. While exercise alone can provide short-term improvements in glucose metabolism, weight loss through exercise and dietary modification provide longer-term impact (Coker et al. 2006).

Two large studies including a lifestyle intervention with diet and exercise counseling resulted in body weight loss and decreased relative risk in the incidence of diabetes (Diabetes Prevention Research Group 2006; Tuomilehto et al. 2001).

A review of studies of physical activity and risk of type 2 diabetes concluded that the role of physical activity independent of diet or weight loss requires additional study (Yates et al. 2007). The ADA recommends that those with impaired fasting glucose or impaired glucose tolerance receive counseling on exercise as a strategy to prevent or delay the onset of diabetes (American Diabetes Association and National Institute of Diabetes and Digestive and Kidney Disease 2002).

How We Can Slow Down and Possibly Reverse the Trend

Both the U.S. Diabetes Prevention Program and the Finnish Diabetes Prevention Study demonstrated that sustained changes in lifestyle with modification of eating habits and physical activity could achieve a relative risk reduction in progression from glucose intolerance to type 2 diabetes by 58% over approximately 3 years (Knowler et al. 2002; Tuomilehto et al. 2001). Other studies demonstrating a significant impact of lifestyle interventions include the Da Qing study in China (Pan et al. 1997), the Malmo study in Sweden (Eriksson and Lindgarde 1991), the United Kingdom Prospective Diabetes Study (United Kingdom Prospective Diabetes Study Group 1998), and the Indian Diabetes Prevention Programme (Ramachandran et al. 2006).

Because the effects of exercise on blood glucose levels are transient, to maintain an effect, exercise should be performed consistently at least every 2–3 days (Albright et al. 2000). Several studies have shown a positive effect of aerobic exercise on glucose control, glucose tolerance, and insulin sensitivity in subjects with type 2 diabetes (Tessier et al. 2000; Tokmakidis et al. 2004). A moderate physical activity of 45 min of aerobic activity at 60–70% of maximal heart rate for 3–4 days per week has been recommended (Frank et al. 2005). Both the ADA (American Diabetes Association 1999) and the American College of Sports Medicine (Pollock et al. 2000) recommend a combination of both strength training and aerobic exercise for those with diabetes. However, before starting an exercise program, it is recommended that persons with type 2 diabetes consult their physician for an evaluation of their glycemic control both at rest and after exercise, cardiovascular status, physical limitations, and any symptoms of neuropathy, retinopathy, and nephropathy (Albright et al. 2000).

A meta-analysis of nine lifestyle intervention studies examined the efficacy of education in reducing the risk of type 2 diabetes development. While lifestyle interventions varied widely among the studies, the risk of diabetes was decreased by ~50% (Yamaoka and Tango 2005). A follow-up to the Da Qing study suggests that those with less insulin resistance as measured by HOMA have a better response to lifestyle interventions than those with higher levels of insulin resistance. Authors suggest that those with higher degrees of insulin resistance may benefit from more restrictive lifestyle intervention or pharmacological treatment (Li et al. 2002). The Finnish Diabetes Prevention Study follow-up focused on the sustainability of risk reduction after the intervention had been completed. At a median of 7 years follow-up, the absolute difference in risk between control and intervention group remained the same as at the completion of the intervention, demonstrating a long-term benefit to lifestyle interventions (Lindstrom et al. 2006).

Clearly, type 2 diabetes is a chronic disease whose impact is growing worldwide. The interaction between genetics and environment cannot be ignored, but emphasis on changing environmental factors can attenuate the influence of genetics. The environmental factors reflect lifestyle habits that overlap, such as diet and physical activity relating to obesity. Additional studies examining specific foods, patterns of intake, intensity and duration of physical activity, and degree of urbanization across ethnicity will bring new insights into effective strategies to curb the rising prevalence of type 2 diabetes.

References

Albright A, Franz M, Hornsby G, Kriska A, Marrero D, Ullrich I, Verity LS. 2000. American College of Sports Medicine position stand. Exercise and Type 2 diabetes. Med Sci Sports Exerc 32(7):1345–1360.

Alberti KG, Zimmet PZ. 1998. Definition, diagnosis and classification of diabetes mellitus and its complications. Part 1: Diagnosis and classification of diabetes mellitus provisional report of a WHO consultation. Diabet Med 15(7):539–553.

American Diabetes Association. 2005. Position statement: Diagnosis and classification of diabetes mellitus. Diabetes Care 28(Suppl 1):S37–S42.

American Diabetes Association and National Institute of Diabetes and Digestive and Kidney Diseases. 2002. The prevention or delay of type 2 diabetes. Diabetes Care 20:109–116.

Barcelo A, Aedo C, Rajpathak S, Robles S. 2003. The cost of diabetes in Latin America and the Caribbean. Bull World Health Organ 81(1):19–27.

Benjamin SM, Valdez R, Geiss LS, Rolka DB, Narayan KMV. 2003. Estimated number of adults with prediabetes in the US in 2000. Diabetes Care 26:645–649.

Botas P, Delgado E, Castano G, Diaz de Grenu C, Prieto J, Diaz-Cadorniga FJ. 2003. Comparison of the diagnostic criteria for diabetes mellitus, WHO-1985, ADA-1997, and WHO-1999 in the adult population of Asturias (Spain). Diabet Med 20(11):904–998.

Carter EA, MacCluer JW, Dyke B, Howard BV, Devereux RB, Ebbesson SO, Resnick HE. 2005. Diabetes mellitus and impaired fasting glucose in Alaska Eskimos: The Genetics of Coronary Artery Disease in Alaska Natives (GOCADAN) Study. Diabetologia 49(1):29–35.

Choi HK, Willett WC, Stampfer MJ, Rimm E, Hu FB. 2005. Dairy consumption and risk of Type 2 diabetes mellitus in men: A prospective study. Arch Intern Med 165(9):997–1003.

Cockram CS. 2000. Diabetes mellitus: Perspective from the Asia-Pacific region. Diabetes Res Clin Pract 50(Suppl 2):S3–S7.

Coker RH, Hays NP, Williams RH, Brown AD, Freeling SA, Kortebein PM, Sullivan DH, Starling RD, Evans WJ. 2006. Exercise-induced changes in insulin action and glycogen metabolism in elderly adults. Med Sci Sports Exerc 38(3):433–438.

Cowie CC, Rust KF, Byrd-Holt DD, Eberhardt MS, Flegal KM, Engelgau MM, Saydah SH, Williams DE, Geiss LS, Gregg EW. 2006. Prevalence of diabetes and impaired fasting glucose in adults in the U.S. population: National Health and Nutrition Examination Survey 1999–2002. Diabetes Care 29(6):1263–1268.

Davis TM, Cull CA, Holman RR; U.K. Prospective Diabetes Study (UKPDS) Group. 2001. Relationship between ethnicity and glycemic control, lipid profiles, and blood pressure during the first 9 years of type 2 diabetes: U.K. Prospective Diabetes Study (UKPDS 55). Diabetes Care 24(7):1167–1174.

Diabetes Prevention Program Research Group. 2006. Relationship of body size and shape to the development of diabetes in the Diabetes Prevention Program. Obesity 14(11):2107–2117.

Dong Y, Gao W, Nan H, Yu H, Li F, Duan W, Wang Y, Sun B, Qian R, Tuomilehto J, Qiao Q. 2005. Prevalence of Type 2 diabetes in urban and rural Chinese populations in Qingdao, China. Diabet Med 22(10):1427–1433.

Eriksson KF, Lindgarde F. 1991. Prevention of type 2 (non-insulin-dependent) diabetes mellitus by diet and physical exercise: The 6-year Malmo feasibility study. Diabetologia 34(12):891–898.

Fagot-Campagna A, Bourdel-Marchasson I, Simon D. 2005. Burden of diabetes in an aging population: Prevalence, incidence, mortality, characteristics, and quality of care. Diabetes Metab 31(2):5S35–5S52.

Folsom AR, Kushi LH, Anderson KE, Mink PJ, Olson JE, Hong CP, Sellers TA, Lazovich D, Prineas RJ. 2000. Associations of general and abdominal obesity with multiple health outcomes in older women: The Iowa Women's Health Study. Arch Intern Med 160(14):2117–2128.

Forouhi NG, Merrick D, Goyder E, Ferguson BA, Abbas J, Lachowycz K, Wild SH. 2006. Diabetes prevalence in England, 2001—estimates from an epidemiological model. Diabet Med 23(2):189–197.

Frank LL, Sorensen BE, Yasui Y, Tworoger SS, Schwartz RS, Ulrich CM, Irwin ML, Rudolph RE, Rajan KB, Stanczyk F, Bowen D, Weigle DS, Potter JD, McTiernan A. 2005. Effects of exercise on metabolic risk variables in overweight postmenopausal women: A randomized clinical trial. Obes Res 13(3):615–625.

Garcia-Garcia G, Aviles-Gomez R, Luquin-Arellano VH, Padilla-Ochoa R, Lepe-Murillo L, Ibarra-Hernandez M, Briseño-Renteria G. 2006. Cardiovascular risk factors in the Mexican population. Ren Fail 28(8):677–687.

Geiger PC, Han DH, Wright DC, Holloszy JO. 2006. How muscle insulin sensitivity is regulated: Testing of a hypothesis. Am J Physiol Endocrinol Metab 291(6):E1258–E1263.

Gimeno SG, Ferreira SR, Franco LJ, Iunes M, for the Japanese–Brazilian Diabetes Study Group. 1998. Comparison of glucose tolerance categories according to World Health Organization and American Diabetes Association diagnostic criteria in a population-based study in Brazil. Diabetes Care 21(11):1889–1892.

Harding AH, Day NE, Khaw KT, Bingham S, Luben R, Welsh A, Wareham NJ. 2004. Dietary fat and the risk of clinical type 2 diabetes: the European prospective investigation of Cancer-Norfolk study. Am J Epidemiol. 159(1):73–82.

Harris MI, Eastman RC, Cowie CC, Flegal KM, Eberhardt MS. 1997. Comparison of diabetes diagnostic categories in the U.S. population according to the 1997 American

Diabetes Association and 1980–1985 World Health Organization diagnostic criteria. Diabetes Care 20(12):1859–1862.

Hotamisligil GS. 2006. Inflammation and metabolic disorders. Nature 444(7121):860–867.

Huang B, Rodriguez BL, Burchfiel CM, Chyou PH, Curb JD, Yano K. 1996. Acculturation and prevalence of diabetes among Japanese-American men in Hawaii. Am J Epidemiol 144(7):674–681.

International Diabetes Federation. 2003. Diabetes Atlas, 2nd ed. Brussels, Belgium: International Diabetes Federation.

International Diabetes Federation. 2006. Diabetes Atlas, 3rd ed. Brussels, Belgium: International Diabetes Federation.

Jorgensen LG, Brandslund I, Petersen PH, Olivarius Nde F, Stahl M. 2003. The effect of the new ADA and WHO guidelines on the number of diagnosed cases of diabetes mellitus. Clin Chem Lab Med 41(9):1246–1250.

Kershaw EE, Flier JS. 2004. Adipose tissue as an endocrine organ. J Clin Endocrinol Metab 89(6):2548–2556.

King H, Aubert RE, Herman WH. 1998. Global burden of diabetes, 1995–2025: Prevalence, numerical estimates, and projections. Diabetes Care 21:1414–1431.

Knowler WC, Barrett-Connor E, Fowler SE, Hamman RF, Lachin JM, Walker EA, Nathan DM; Diabetes Prevention Program Research Group. 2002. Reduction in the incidence of type 2 diabetes with lifestyle intervention or metformin. N Eng J Med 346(6):393–403.

Li G, Hu Y, Yang W, Jiang Y, Wang J, Xiao J, Hu Z, Pan X, Howard BV, Bennett PH; Da Qing IGT and Diabetes Study. 2002. Effects of insulin resistance and insulin secretion on the efficacy of interventions to retard development of type 2 diabetes mellitus: The Da Qing IGT and Diabetes Study. Diabetes Res Clin Pract 58(3):193–200.

Liese AD, Schulz M, Moore CG, Mayer-Davis EJ. 2004. Dietary patterns, insulin sensitivity and adiposity in the multi-ethnic Insulin Resistance Atherosclerosis Study population. Br J Nutr 92(6):973–984.

Lindstrom J, Ilanne-Parikka P, Peltonen M, Aunola S, Eriksson JG, Hemio K, Hamalainen H, Harkonen P, Keinanen-Kiukaanniemi S, Laakso M, Louheranta A, Mannelin M, Paturi M, Sundvall J, Valle TT, Uusitupa M, Tuomilehto J; Finnish Diabetes Prevention Study Group. 2006. Sustained reduction in the incidence of type 2 diabetes by lifestyle intervention: Follow-up of the Finnish Diabetes Prevention Study. Lancet 368(9548):1673–1679.

Mainous AG III, Baker R, Koopman RJ, Saxena S, Diaz VA, Everett CJ, Majeed A. 2007. Impact of the population at risk of diabetes on projections of diabetes burden in the United States: An epidemic on the way. Diabetologia 50(5):934–940.

Mainous AG III, Diaz VA, Saxena S, Baker R, Everett CJ, Koopman RJ, Majeed A. 2006. Diabetes management in the USA and England: Comparative analysis of national surveys. J R Soc Med 99(9):463–469.

Mannucci E, Bardini G, Ognibene A, Rotella CM; American Diabetes Association. 1999. Comparison of ADA and WHO screening methods for diabetes mellitus in obese patients. Diabet Med 16(7):579–585.

Mattout C, Bourgeois D, Bouchard P. 2006. Type 2 diabetes and periodontal indicators: Epidemiology in France 2002–2003. J Periodontal Res 41(4):253–258.

Mayer-Davis EJ, D'Agostino R Jr, Karter AJ, Haffner SM, Rewers MJ, Saad M, Bergman RN. 2002. Intensity and amount of physical activity in relation to insulin sensitivity: The Insulin Resistance Atherosclerosis Study. JAMA 279(9):669–674.

Meisinger C, Doring A, Thorand B, Heier M, Lowel H. 2006. Body fat distribution and risk of type 2 diabetes in the general population: Are there differences between men and women? The MONICA/KORA Augsburg cohort study. Am J Clin Nutr 84(3):483–489.

Menon VU, Kumar KV, Gilchrist A, Sugathan TN, Sundaram KR, Nair V, Kumar H. 2006. Prevalence of known and undetected diabetes and associated risk factors in central Kerala—ADEPS. Diabetes Res Clin Pract 74(3):289–294.

Ministry of Health, Mexico. 2003. Estadisticas de mortalidad en Mexico: muertes registradas en el ano 2001. Salud Publica Mex 44:565–576.

Mizoue T, Yamaji T, Tabata S, Yamaguchi K, Ogawa S, Mineshita M, Kono S. 2006. Dietary patterns and glucose tolerance abnormalities in Japanese men. J Nutr 136(5):1352–1358.

Montonen J, Knekt P, Harkanen T, Jarvinen R, Heliovaara M, Aromaa A, Reunanen A. 2005. Dietary patterns and the incidence of type 2 diabetes. Am J Epidemiol 161(3):219–227.

Muhammad S. 2004. Epidemiology of diabetes and obesity in the United States. Compend Contin Educ Dent 25(3):195–198, 200, 202.

Narayan KMV, Zhang P, Kanaya AM, Williams DE, Engelgau MM, Imperatore G, Ramachandran A. 2006. Diabetes: The pandemic and potential solutions. In Disease Control Priorities in Developing Countries, 2nd ed., edited by Jamison DT, Breman JG, Measham AR, Alleyne G, Claeson M, Evans DB, Jha P, Mills A, Musgrove P. Washington, DC: Oxford University Press and the World Bank.

Oda E. 2006. Incidence of type 2 diabetes in individuals with central obesity in a rural Japanese population: The Tanno and Sobetssu study. Response to Ohnishi and others. Diabetes Care 29(8):1988–1989.

Ohnishi H, Saitoh S, Takagi S, Katoh N, Chiba Y, Akasaka H, Nakamura Y, Shimamoto K. 2006. Incidence of type 2 diabetes in individuals with central obesity in a rural Japanese population: The Tanno and Sobetsu study. Diabetes Care 29(5):1128–1129.

Pan XR, Li GW, Hu YH, Wang JX, Yang WY, An ZX, Hu ZX, Lin J, Xiao JZ, Cao HB, Liu PA, Jiang XG, Jiang YY, Wang JP, Zheng H, Zhang H, Bennett PH, Howard BV. 1997. Effects of diet and exercise in preventing NIDDM in people with impaired glucose tolerance: The Da Qing IGT and Diabetes Study. Diabetes Care 20(4):537–544.

Pollock ML, Franklin BA, Balady GJ, Chaitman BL, Fleg JL, Fletcher B, Limacher M, Piña IL, Stein RA, Williams M, Bazzarre T, AHA Science Advisory. 2000. Resistance exercise in individuals with and without cardiovascular disease: Benefits, rationale, safety, and prescription. An advisory from the Committee on Exercise, Rehabilitation, and Prevention, Council on Clinical Cardiology, American Heart

Association; Position paper endorsed by the American College of Sports Medicine. Circulation 101(7):828–833.

Rajala MW, Scherer PE. 2003. Minireview: The adipocyte—at the crossroads of energy homeostasis, inflammation, and atherosclerosis. Endocrinology 144(9):3765–3773.

Ramachandran A, Snehalatha C, Mary S, Mukesh B, Bhaskar AD, Vijay V; Indian Diabetes Prevention Programme. 2006. The Indian Diabetes Prevention Programme shows that lifestyle modification and metformin prevent type 2 diabetes in Asian Indian subjects with impaired glucose tolerance (IDPP-1). Diabetologia 49(2):289–297.

Ramachandran A, Snehalatha C, Vijay V, Wareham NJ, Colagiuri S. 2005. Derivation and validation of diabetes risk score for urban Asian Indians. Diabetes Res Clin Pract 70(1):63–70.

Rull JA, Aguilar-Salinas CA, Rojas R, Rios-Torres JM, Gómez-Pérez FJ, Olaiz G. 2005. Epidemiology of type 2 diabetes in Mexico. Arch Med Res 36(3):188–196.

Sanchez-Castillo CP, Velasquez-Monroy O, Lara-Esqueda A, Berber A, Sepulveda J, Tapia-Conyer R, James WP. 2005. Diabetes and hypertension increases in a society with abdominal obesity: Results of the Mexican National Health Survey 2000. Public Health Nutr 8(1):53–60.

Sanghera DK, Bhatti JS, Bhatti GK, Ralhan SK, Wander GS, Singh JR, Bunker CH, Weeks DE, Kamboh MI, Ferrell RE. 2006. The Khatri Sikh Diabetes Study (SDS): Study design, methodology, sample collection, and initial results. Hum Biol 78(1):43–63.

Schulze MB, Manson JE, Ludwig DS, Colditz GA, Stampfer MJ, Willett WC, Hu FB. 2004. Sugar-sweetened beverages, weight gain, and incidence of type 2 diabetes in young and middle-aged women. JAMA 292(8):927–934.

Simmons D, Thompson CF, Volklander D. 2001. Polynesians: Prone to obesity and Type 2 diabetes mellitus but not hyperinsulinaemia. Diabet Med 18(3):193–198.

Stevens J, Couper D, Pankow J, Folsom AR, Duncan BB, Nieto FJ, Jones D, Tyroler HA. 2001. Sensitivity and specificity of anthropometrics for the prediction of diabetes in a biracial cohort. Obes Res 9(11):696–705.

Stock SA, Redaelli M, Wendland G, Civello D, Lauterbach KW. 2006. Diabetes—prevalence and cost of illness in Germany: A study evaluating data from the statutory health insurance in Germany. Diabet Med 23(3):299–305.

Stumvoll M, Goldstein BJ, van Haeften TW. 2005. Type 2 diabetes: Principles of pathogenesis and therapy. Lancet 365(9467):1333–1346.

Tessier D, Menard J, Fulop T, Ardilouze J, Roy M, Dubuc N, Dubois M, Gauthier P. 2000. Effects of aerobic physical exercise in the elderly with type 2 diabetes mellitus. Arch Gerontol Geriatr 31(2):121–132.

Tokmakidis SP, Zois CE, Volaklis KA, Kotsa K, Touvra AM. 2004. The effects of a combined strength and aerobic exercise program on glucose control and insulin action in women with Type 2 diabetes. Eur J Appl Physiol 92(4–5):437–442.

Torrens JI, Skurnick J, Davidow AL, Korenman SG, Santoro N, Soto-Greene M, Lasser N, Weiss G. 2004. Study of Women's Health Across the Nation (SWAN). Ethnic differences in insulin sensitivity and beta-cell function in premenopausal or early perimenopausal women without diabetes: The Study of Women's Health Across the Nation (SWAN). Diabetes Care 27(2):354–361.

Tuomilehto J, Lindstrom J, Eriksson JG, Valle TT, Hamalainen H, Ilanne-Parikka P, Keinanen-Kiukaanniemi S, Laakso M, Louheranta A, Rastas M, Salminen V, Uusitupa M; Finnish Diabetes Prevention Study Group. 2001. Prevention of type 2 diabetes mellitus by changes in lifestyle among subjects with impaired glucose tolerance. N Engl J Med 344(18):1343–1350.

United Kingdom Prospective Diabetes Study Group. 1998. Intensive blood glucose control with sulphonylureas or insulin compared with conventional treatment and risk of complications in patients with type 2 diabetes (UKPDS 33). Lancet 352:837–853.

van Winkel R, De Hert M, Van Eyck D, Hanssens L, Wampers M, Scheen A, Peuskens J. 2006. Screening for diabetes and other metabolic abnormalities in patients with schizophrenia and schizoaffective disorder: Evaluation of incidence and screening methods. J Clin Psychiatry 67(10):1493–1500.

Wang Y, Rimm EB, Stampfer MJ, Willett WC, Hu FB. 2005. Comparison of abdominal adiposity and overall obesity in predicting risk of type 2 diabetes among men. Am J Clin Nutr 81(3):555–563.

Wei M, Gaskill SP, Haffner SM, Stern MP. 1997. Waist circumference as the best predictor of noninsulin dependent diabetes mellitus (NIDDM) compared to body mass index, waist/hip ratio, and other anthropometric measurements in Mexican Americans—a 7-year prospective study. Obes Res 5(1):16–23.

Wild S, Roglic G, Green A, Sicree R, King H. 2004. Global prevalence of diabetes. Estimates for the year 2000 and projections for 2030. Diabetes Care 27:1047–1053.

Wong KC, Wang Z. 2006. Prevalence of type 2 diabetes mellitus of Chinese populations in Mainland China, Hong Kong, and Taiwan. Diabetes Res Clin Pract 73(2):126–134.

World Health Organization. 1985. Diabetes mellitus: Report of a WHO study group. Technical Report Series No. 727.

World Health Organization. 1999. Definition, diagnosis and classification of diabetes and its classifications: Report of WHO Consultation Part 1. Diagnosis and classification of diabetes mellitus. Geneva, World Health Organization.

World Health Organization. 2006. Definition and diagnosis of diabetes mellitus and intermediate hyperglycemia: Report of WHO Consultation. Geneva, World Health Organization.

Yamaoka K, Tango T. 2005. Efficacy of lifestyle education to prevent type 2 diabetes: A meta-analysis of randomized controlled trials. Diabetes Care 28(11):2780–2786.

Yates T, Khunti K, Bull F, Gorely T, Davis MJ. 2007. The role of physical activity in the management of impaired glucose tolerance: A systemic review. Diabetologia 50(6):1116–1126.

Zimmet PZ. 1992. Kelly West Lecture 1991. Challenges in diabetes epidemiology—from West to the rest. Diabetes Care 15(2):232–252.

Chapter 3

Preventing Type 2 Diabetes Mellitus

Frank Greenway, MD

Introduction

The prevalence of obesity in the United States was relatively level between 1960 and 1980, but since 1980 it has been rising steadily (Flegal 2005). Since the prevalence of diabetes follows obesity by approximately 10 years in a population, it is not surprising that the prevalence of diabetes began to rise by about 1990 (Bray 1998). Between 1976 and 1980, only 8.9% of the U.S. population between the ages of 40 and 74 years had diabetes, but this rose to 12.3% by 1988–1994 (Harris et al. 1998). It is predicted that the lifetime risk for developing diabetes for people born in the United States during the year 2000 will be 33% for men and 39% for women (Curtis and Wilson 2005). The increasing prevalence of diabetes is not just a problem for the United States. The global prevalence of diabetes is expected to rise from 118 million in 1995 to 220 million in the year 2010 (Younis et al. 2004).

Obesity is the major driver of the increasing prevalence of diabetes. Type 2 diabetes is 3–7 times more common in obese adults compared to their normal-weight counterparts, and those with a body mass index (BMI) >35 kg/m^2 are 20 times more likely to develop diabetes compared to those with a BMI <25 kg/m^2 (Klein et al. 2004). It has been estimated that the cause of 60–90% of type 2 diabetes can be attributed to obesity (Anderson et al. 2003).

An increase in weight by 5 kg increases the risk of cardiovascular disease by 30–50%. A person with a BMI of 33 kg/m^2 has a 320% greater risk of cardiovascular disease compared to a person with a BMI of 23 kg/m^2 (Anderson et al. 2003). Cardiovascular disease is responsible for 70–80% of the deaths in diabetic individuals, and weight loss is well documented to reduce cardiovascular risk factors. The increase in diabetes and its associated risk of cardiovascular disease represents a serious economic threat. The direct and indirect costs of diabetes to the United States are over $100 billion per year, a number that is sure to rise, and despite the extent of the problem in the United States, developing countries with far less resources are being hardest hit (American Diabetes Association 1998).

If the impending global catastrophe of diabetes and cardiovascular disease is to be averted, along with the suffering and economic challenges that it entails, type 2 diabetes must be prevented. The mainstay in the treatment of obesity has been diet and lifestyle treatment. This chapter reviews the efficacy of diet and lifestyle in preventing diabetes and the potential for pharmacological approaches to act as adjunctive measures.

Diet and Lifestyle

There have been several major studies evaluated diet and lifestyle change to prevent diabetes and all have come to similar conclusions. These important studies are reviewed in Table 3.1.

Swedish Diabetes Prevention Study—The Malmo Study

The Malmo study was a nonrandomized trial that demonstrated a diet and exercise program administered over 5 years reduced the incidence of diabetes (Eriksson and Lindgarde 1991). A unique aspect of this study is the 12-year follow-up report in which the mortality in the impaired glucose tolerance (IGT) intervention group ($n = 288$) was equivalent to the 6.5% mortality in the normal glucose tolerance group and less than the IGT group ($n = 135$) that received routine care (Eriksson and Lindgarde 1998). This is the only diabetes prevention study that has documented improvement in mortality.

Chinese Diabetes Prevention Study—The Da Qing IGT and Diabetes Study

The National Diabetes Data Group recognized the classification of IGT in 1979 due to evidence that it was associated with a higher incidence of diabetes and was associated with obesity, hypertension, hyperlipidemia, and an increased risk of cardiovascular disease (National Diabetes Data Group 1979). In 1986, the Da Qing IGT and Diabetes Study began in Da Qing, China. Baseline data revealed that those with IGT were twice as likely to have hypertension, obesity, and abnormal urinary albumin excretion as those with normal glucose tolerance. Plasma cholesterol and triglycerides were higher and high-density lipoprotein (HDL) cholesterol lower in subjects with IGT. Subjects with IGT had higher plasma insulin concentrations but lower insulin-to-glucose ratio at 1 h after a 75 g glucose load. The prevalence of electrocardiographically recognized coronary heart disease was 9.5-fold greater in IGT, and IGT remained an independent coronary heart disease risk factor after adjusting for confounders (Pan et al. 1993).

In the Da Qing IGT and Diabetes Study, subjects with IGT were randomized to one of four groups by clinic: (1) control, (2) diet alone, (3) exercise alone, and (4) diet plus exercise. Subjects were approximately 45 years of age, with a mean BMI

Table 3.1. Randomized studies of lifestyle and diet in subjects with IGT.

Group	Number	M/F	BMI	Age	Control/intervention		% Red	Reference
Chinese	577	1/1	26	45	6 yr 1/–0.5 kg	6 yr % 68/46	42	Pan et al. (1997)
Finnish	522	1/2	31	55	3 yr –.9/–3.5 kg	3 yr % 23/11	58	Tuomilehto et al. (2001)
American	3,234	1/2	34	51	3 yr –.1/–5.6 kg	3 yr % 29/14	58	Knowler et al. (2002)
Indian	531	4/1	26	46	3 yr 0 kg	3 yr % 55/40	28	Ramachandran et al. (2006)
Japanese	458	1/0	24	47	4 yr –0.4/–2.2 kg	4 yr % 3/9.3	67	Kosaka et al. (2005)

Abbreviations: M, male; F, female; BMI, body mass index; yr %, year percent; yr kg, year kilogram; % Red, percent reduction in diabetes.

of 26 kg/m^2 and both genders were equally represented. Oral glucose tolerance tests were conducted every 2 years, and at 6 years, the incidence of diabetes was approximately 68% in the control group, while the diet, exercise, and diet plus exercise groups were 44, 41, and 46%, respectively. This represented a 31, 46, and 42% reduction in the risk of developing diabetes in the diet, exercise, and diet plus exercise groups, respectively, in a proportional hazards analysis adjusted for differences in baseline BMI and fasting glucose. All groups were significantly different from control, but not from each other. Thus, it appeared that either diet or exercise reduced conversion of IGT to diabetes, but combining them was not more effective than using either one separately (Pan et al. 1997). The interventions were more effective in those with less insulin resistance (Li et al. 2002).

Finnish Diabetes Prevention Study

The Finnish Diabetes Prevention Study was preceded by a pilot study in which newly diagnosed diabetic patients either participated in a diet and exercise program or were given usual care. After 2 years, only 12.5% of the intervention group was receiving oral antidiabetic medication compared to 34.8% in the usual care group. The intervention group had a greater incidence of a 5-kg weight reduction compared to the usual care group, 42% vs. 12% ($P < 0.001$). It appeared that the main determinants of success in improving glucose tolerance were weight reduction and a reduction in saturated fats (Uusitupa 1996).

Based on the results of this pilot study, the Finnish Diabetes Prevention Study in subjects with IGT had several aims for the intervention group: increased physical activity, weight reduction, decreased dietary saturated fat, and increased dietary fiber intake (Uusitupa et al. 2000). A total of 523 overweight subjects with IGT were randomized to a diet and lifestyle intervention or usual care in a 1:1 ratio. At 1 year, there was a significant weight loss in the intervention group, 4.7 vs. 0.9 kg ($P < 0.001$) (Eriksson et al. 1999). At years 2 and 3, the weight loss difference maintained significance, 3.5 vs. 0.8 kg and 3.5 vs. 0.9 kg. At year 4, the incidence of diabetes was 11% in the intervention group and 23% in the usual care group, which represented a 58% reduction in the incidence of diabetes compared to control ($P < 0.001$) (Lindstrom 2003b; Tuomilehto et al. 2001). The goal to decrease dietary saturated fat, increase dietary fiber, and increase physical activity was also met (Lindstrom et al. 2003a).

Insulin sensitivity measured by frequently sampled insulin glucose tolerance testing improved proportional to the weight loss, and insulin secretion remained constant in those who were able to maintain a weight loss (Uusitupa et al. 2003). Subjects in the highest tertile of physical activity had a 65% reduction in the risk of developing diabetes, demonstrating that physical activity was an important aspect of the intervention program (Laaksonen et al. 2005). The intensive lifestyle group had beneficial effects on fibrinolysis, manifested by a 31% drop in plasminogen activator inhibitor-1 levels, but not fibrinogen (Hamalainen et al. 2005). Genetic analyses demonstrated polymorphisms in the study cohort that predicted impaired first-phase insulin secretion, development of diabetes, and weight loss (Laukkanen et al. 2004; Siitonen et al. 2004; Todorova et al. 2004). Seven-year follow-up of the Finnish Diabetes Prevention Study has recently been published, and a 43% relative risk reduction in developing diabetes in the intervention group has been sustained, despite only yearly visits for the past 3 years. As expected, the success in preventing diabetes continued to be strongly correlated with maintaining the original study goals of increased physical activity, lower dietary saturated fat, weight reduction, and increased dietary fiber, but weight reduction appeared to be the most important of these elements (Lindstrom et al. 2006).

U. S. Diabetes Prevention Program

The largest randomized trial of diabetes prevention in subjects with IGT is the Diabetes Prevention Program (DPP). The goal of this trial was to recruit 3,000 subjects with IGT of both genders, of which 50% would be minorities and 20% would be over 65 years of age, in 27 U.S. centers over a 2.5-year period (The Diabetes Prevention Program Research Group 1999). The DPP screened 133,683 individuals over 2.7 years, of whom 26,518 had an oral glucose tolerance test, resulting in 3,234 randomized participants. The study was originally designed to randomize subjects to four groups: (1) usual care, (2) intensive lifestyle, (3) metformin 850 mg twice daily, and (4) troglitazone 400 mg daily. During the recruitment process, one of the study participants in the troglitazone group died of liver failure. At that point, there had been 585 subjects randomized to the troglitazone group and they were taken off the drug but followed in the trial off medication (Fujimoto 2000).

Of the participants randomized to the trial, 55% were Caucasian, 20% were African American, 16% were Hispanic, 5% were American Indian, and 4% were Asian American. Women represented 67% of the participants and 48% of them were postmenopausal. The average age of the participants was 51 years, 16% were less than 40 years, and although only 10% were over 65 years, 20% were over 60 years (The Diabetes Prevention Program Research Group 2000). Psychological correlates at entry into the study showed that a higher BMI was associated with obesity earlier in life, and more attempts at weight loss with less confidence in the ability to successfully lose weight (Delahanty et al. 2002).

The DPP was planned to run for 5 years, but since the results became conclusive, the Data Safety Monitoring Board stopped the trial at 2.8 years. The lifestyle group had a 58% reduction in the incidence of diabetes compared to usual care, and the metformin group had a 31% reduction in conversion to diabetes; both of these reductions were statistically significant. It was estimated that 7 people would need to be treated with lifestyle intervention for 3 years to prevent 1 case of diabetes. The goals of the lifestyle program were a 7% weight loss and 150 min of physical activity per week (Knowler et al. 2002).

The 31% reduction in the conversion to diabetes in the metformin group was more difficult to interpret, since it was not clear whether the lower risk of diabetes was due to a transient pharmacological effect of metformin or to a more lasting prevention of diabetes. To answer this question, subjects in the metformin group had a repeat oral glucose tolerance test after a 1–2-week washout period off metformin. The repeat glucose tolerance tests revealed that the group previously treated with metformin had a 25% reduction in the conversion to diabetes. Thus, only 6% of the 31% reduction in conversion to diabetes in the metformin group seen in the DPP trial was due to a transient pharmacological effect (The Diabetes Prevention Program Research Group 2003a). Metformin and lifestyle seemed to have a differential effect in the DPP based on age. Participants over 60 years were leaner, had the lowest insulin secretion, and the best insulin sensitivity. These subjects had a more robust response to intensive lifestyle treatment with only 3.3 cases of diabetes per 100 person-years compared to 6.3 cases of diabetes per 100 patient-years in the 25–44-year-old age group. Those in the metformin group were more insulin resistant and had trends in the opposite direction. Those over 60 years of age had 9.3 cases of diabetes per 100 person-years compared to only 6.7 cases of diabetes per 100 patient-years in the 24–44-year-old age group. Thus, it appears that in preventing the conversion to diabetes, metformin is more effective in the younger subjects with more insulin resistance, and lifestyle was more effective in the older population with more of the older population reaching the goals of weight loss and exercise time of the lifestyle program. Another interesting difference based on age was that the youngest group was more likely to convert to diabetes on the basis of fasting blood sugar while the older patient group was more likely to convert based on the 2-h postprandial value (Crandall et al. 2006).

Unlike metformin that had a predominant effect on preventing diabetes, troglitazone seemed to have a predominantly transient pharmacological effect. Between 1996 and 1998, there were 587 subjects randomized to metformin, 585 randomized to troglitazone, 582 randomized to usual care, and 589 randomized to intensive lifestyle. During the 0.9 years of troglitazone treatment, the conversion to diabetes per 100 person-years was only 3 cases in the troglitazone group compared with 12, 6.7, and 5.1 cases per 100 person-years in the usual care, metformin, and intensive lifestyle groups, respectively. Three years after the discontinuation of troglitazone from the trial, the conversion to diabetes was essentially identical to the usual care group (Knowler et al. 2005).

After the results of the DPP were announced, the question of cost naturally arose. The cost of identifying one subject with IGT was $139, and over 3 years the estimated costs relative to usual care were an additional $2,412 for metformin and $3,540 for intensive lifestyle (Hernan et al. 2003). This converts to $24,400 and $34,500 per case of diabetes prevented in the metformin and intensive lifestyle arms, respectively. This is equivalent to $51,600 and $99,200 per quality-adjusted life year gained for metformin and intensive lifestyle, respectively, which was judged to be cost-effective (The Diabetes Prevention Program Research Group 2003b). Analysis of the goals of the intensive lifestyle arm was also undertaken. At 24 weeks, 49 and 74% attained the 7% weight loss goal and 150 min per week physical activity goal of the lifestyle program, respectively, but this fell to 37 and 67% at 3 years. As expected, self-monitoring was positively associated with meeting the lifestyle goals, and those who met the lifestyle goals at 24 weeks were 1.5–3 times more likely to meet these goals at 3 years (Wing et al. 2004). The intensive lifestyle group decreased dietary calories (−452 kcal/day) and dietary fat (−6.6%) to a greater extent than the metformin and usual care arms (Mayer-Davis et al. 2004).

Other correlates examined in the DPP were depression symptoms and antidepressant use. The DPP participants were not particularly depressed and participation in the program was not associated with changes in depression on questionnaires or with changes in antidepressant medication use (Rubin et al. 2005). Adherence to taking medications as instructed was associated with a risk reduction for conversion to diabetes (Walker et al. 2006). Women in the intensive lifestyle group experienced less stress or urge of urinary incontinence (38%) compared to metformin (48%) or usual care (46%), and this change was driven by weight loss (Brown et al. 2006). The lifestyle arm also had the greatest decrease in pain and increase in nerve regeneration using skin biopsies at baseline and 1 year (Smith et al. 2006). The lifestyle arm had an improvement of autonomic nervous system functioning. Subjects had a lower heart rate with greater heart rate variability, suggestive of improved fitness (Carnethon et al. 2006).

Subjects in the DPP were more active at baseline than a comparable cohort with impaired fasting glucose (IFG) in the NHANES (National Health and Nutrition Examination Survey) III (Kriska et al. 2006). Predictors of adherence to the physical activity program were being a man, having a lower BMI, being ready to increase activity levels, and having confidence that one could increase exercise (Delahanty et al. 2006). Metabolic syndrome is defined as having three or more of the following: increased waist circumference, increased blood pressure, low levels of HDL cholesterol, IFG, and elevated triglycerides. Metabolic syndrome was present in 53% of the DPP participants and decreased 41% in the lifestyle group, but only 17% in the metformin group compared to usual care (Orchard et al. 2005). There was also a decrease in C-reactive protein in the lifestyle and metformin groups compared to usual care (Haffner et al. 2005). Subjects entering the DPP study had a 30% prevalence of hypertension that increased in the usual care and metformin groups while decreasing in the intensive lifestyle arm. Triglycerides fell in the intensive lifestyle arm while HDL cholesterol rose. The intensive lifestyle arm had a reduction in the proatherogenic low-density lipoprotein B (LDL-B) particles, and at 3 years the use of medications for hypertension and hyperlipidemia decreased 28 and 25%, respectively, compared to the usual care and metformin groups (Ratner et al. 2005).

There were three parameters that predicted conversion to diabetes in the DPP: low insulin sensitivity, low beta-cell function, and increased weight. All these parameters improved most in the intensive lifestyle group, and the metformin group had more improvement than the usual care group (Kitabchi et al. 2005). Results of the DPP are felt to have global implications for reducing the progressive rise in diabetes, if the intensive lifestyle program were widely adopted (Molitch et al. 2003). In fact, the importance of the DPP was felt to be so encouraging that the U.S. National Institutes of Health is now supporting Look AHEAD, a long-term trial to determine if an intensive lifestyle program will reduce mortality and major cardiovascular events in subjects with diabetes compared to usual care (Bray et al. 2006).

Indian Diabetes Prevention Program

Asians from India develop diabetes at an earlier age, at a lower BMI, and have more insulin resistance than North American and Northern European populations. Chinese Asians have less insulin resistance and a lower incidence of diabetes compared to Indian Asians. A total of 531 subjects were recruited from an urban middle-class population working in service organizations in India. The subjects consisted of 421 men and 110 women with IGT who were randomized equally into four groups: (1) control, (2) lifestyle, (3) metformin 500 mg/day, and (4) lifestyle plus metformin. The mean age was approximately 46 years, and the BMI was approximately 26 kg/m^2. The prevalence of IGT in the population was 12.3%, and annual oral glucose tolerance tests (OGTTs) were performed over the 3-year study. The cumulative incidence of diabetes in 3 years in the control group was 55% and approximately 40% in the three other groups. This represented a 28% reduction in the incidence of diabetes in the three treatment groups compared to control. It is interesting that metformin and lifestyle decreased the conversion to diabetes to an equivalent degree while combining lifestyle with metformin did not further decrease the conversion to diabetes. It was estimated that one would need to treat approximately 7 people having IGT with metformin, lifestyle, or the combination to prevent 1 case of diabetes (Ramachandran et al. 2006).

Japanese Diabetes Prevention Program

This study randomized Japanese male subjects with IGT in a 4:1 ratio to a control standard intervention group (356 subjects) or an intensive lifestyle group (102 subjects). The control group was to maintain a BMI <24 kg/m^2 and the intensive lifestyle group was to maintain a BMI <22 kg/m^2. The intensive lifestyle group met for instructions on lifestyle every 3–4 months. The cumulative 4-year incidence of diabetes was 9.3% in the control group and 3% in the intensive lifestyle group, a 67% reduction in conversion to diabetes ($P < 0.001$). Weight loss was 0.4 kg in the control group, 2.2 kg in the intensive lifestyle group, and the incidence of diabetes was positively correlated with weight change. Subjects with a higher fasting plasma glucose developed diabetes at a higher rate, and BMI had a linear correlation with weight loss. Since the slope in the reduction of diabetes was steeper than the reduction in BMI, the effect of the lifestyle treatment cannot be solely ascribed to weight loss (Kosaka et al. 2005).

Adjunctive Pharmacological Approaches

The evidence for an intensive lifestyle program consisting of diet, physical activity, and weight loss in preventing diabetes clearly has the most compelling evidence. There are also pharmacologic treatments that have been shown to prevent diabetes. The use of metformin and troglitazone as control arms in randomized trials testing intensive lifestyle programs to prevent diabetes has already been reviewed. Since the prevalence of diabetes and obesity has been rising despite national goals promoting intensive lifestyle programs, intensive lifestyle alone may not be sufficient for many individuals in need of diabetes prevention. A variety of pharmacologic agents have examined the effects of treatment versus control on the development of diabetes as a primary or secondary objective. Medications with the potential to contribute to diabetes prevention along with an intensive lifestyle program will now be reviewed (Table 3.2).

Tolbutamide

Tolbutamide is a first-generation sulfonylurea medication used for the treatment of diabetes. It reduces blood sugar by stimulating insulin release (Muller 2000). Between 1962 and 1965, 2,477 (1.1%) of 228,883 subjects had clinistix-positive glycosuria after a carbohydrate-rich lunch. Of these 2,477 subjects, 578 subjects had IGT without diabetes. From these 578 subjects, 267 men were randomized to one of five groups: (1) tolbutamide 500 mg three times a day and diet with annual OGTT ($n = 49$), (2) placebo three times a day and diet with annual OGTT ($n = 48$), (3) diet only with annual OGTT ($n = 50$), (4) no treatment with annual OGTT ($n = 61$), or (5) no treatment with an OGTT at year 10 ($n = 59$). In addition, there was a control group of men with normal OGGT ($n + 52$). Twenty-nine percent of those without diet or tolbutamide converted to diabetes, 13% on diet alone converted to diabetes, and none of those on tolbutamide and diet converted to diabetes (Sartor et al. 1980). A follow-up of this trial in 1987 demonstrated that those subjects treated with tolbutamide had a lower risk of all-cause mortality and ischemic heart disease mortality (Knowler et al. 1997; Sartor et al. 1980).

Acarbose

Acarbose is an alpha-glucosidase inhibitor that slows the breakdown of disaccharides in the gut, resulting in decreased postprandial hyperglycemia, improved insulin sensitivity, and reduced stress on beta cells. A trial to prevent diabetes randomized 714 subjects with IGT to acarbose 100 mg three times a day or placebo for 3 years. Despite the lack of a diet or lifestyle program, the conversion to diabetes was reduced by 32% in the acarbose compared to the placebo control (Chiasson et al. 2002). The acarbose group lost 0.5 kg and the placebo group gained 0.3 kg over the course of this 3.3-year study. A subgroup of the subjects in this study—66 in the placebo group and 66 in the acarbose group—had serial intimal medial

Table 3.2. Studies of medications to prevent diabetes.

Medication	Number	Dose	% Control/drug Diabetes	Diet	Comment	% Red	Reference
Tolbutamide	267	0.5 g tid	29/0	Yes	Lower mortality	100	Knowler et al. (1997), Sartor et al. (1980)
Acarbose	714	100 mg tid	42/32	No	3 years	32	Chiasson et al. (2002), Hanefeld et al. (2004)
Orlistat	694	120 mg tid	29/19	Yes	Weight loss	35	Torgerson et al. (2004)
Pravastatin	6,595	40 mg qd	3.8/2.7	No	Not primary endpoint	30	Freeman et al. (2001)
Estrogen and progesterone	2,763	.625/2.5 mg qd	9.5/6.2	No	Not primary endpoint	35	Kanaya et al. (2003)
Rosiglitazone	5,269	8 mg qd	26/11	No	IGT subjects	55	Gerstein et al. (2006)
Ramipril	5,720	2.5 mg qd	5.4/3.6	No	4.5 years	33	Yusuf et al. (2001)
Candesartan	5,436	32 mg qd	7.4/6	No	2–4 years	19	Yusuf et al. (2005)
Bezafibrate	303	400 mg qd	54/42	No	IFG and CAD	22	Tenenbaum et al. (2004)
Benzafibrate	339	400 mg qd	37/27	No	Obese	27	Tenenbaum et al. (2005)

Abbreviations: % Red, percent reduction in diabetes; tid, three times a day; qd, daily.

thickness determined in the carotid artery as a measure of atherosclerosis. There was a significant 50% reduction in the progression of intimal medial thickness for the acarbose group, suggesting that acarbose treatment is vasoprotective (Hanefeld et al. 2004).

Orlistat

Orlistat is a lipase inhibitor that causes 30% of fat in the diet to pass undigested as oil in the stool. In a 4-year double-blind prospective study, 3,305 patients were randomized to a lifestyle change program with placebo or orlistat 120 mg three times a day. All subjects had a BMI > 30 kg/m^2, but 694 subjects had IGT. The lifestyle change program consisted of a 30% fat-balanced diet reduced 800 kcal below maintenance calories. The energy intake was readjusted every 6 months for any change in weight. Participants received dietary counseling every 2 weeks for the first 6 months and monthly thereafter. Subjects were encouraged to walk at least 1 extra kilometer per day in addition to their usual activity, and all subjects kept activity diaries. OGTTs were performed every 6 months. The conversion to diabetes was 29% in the lifestyle alone group, but decreased to 19% in the orlistat plus lifestyle group. Weight loss in the orlistat group was 5.8 kg, which was significantly greater than the 3-kg weight loss in the placebo group (Torgerson et al. 2004).

Pravastatin

Pravastatin is an HMG Co-A reductase inhibitor that reduces LDL cholesterol, a known risk factor for developing coronary artery disease. The Scotland Coronary Prevention Study was a study of 6,595 men between 45 and 64 years of age who were randomized to pravastatin 40 mg/day or a placebo. Although the primary objective of this trial was prevention of coronary heart disease, a secondary analysis evaluated the effect of pravastatin on the development of diabetes. Those subject who reported diabetes at baseline or who had a baseline glucose >7 mmol/L were excluded from the secondary analysis. There were 5,974 subjects included in the secondary analysis, of which 139 became diabetic during the study. The pravastatin group had a 30% reduction in the conversion to diabetes compared to the placebo group (Freeman et al. 2001).

Estrogen–Progesterone

Estrogen with progesterone was a standard treatment for postmenopausal women to prevent menopausal symptoms and prevent coronary heart disease. Since the risk for coronary heart disease rises after menopause, it was postulated that estrogen and progesterone replacement would reverse this increased risk. The effect of estrogen and progesterone on the conversion to diabetes was analyzed as a secondary objective of this trial. The Heart and Estrogen/Progesterone Replacement

Study randomized 2,763 postmenopausal women with coronary heart disease to conjugated estrogen 0.625 mg/day and medroxyprogesterone acetate 2.5 mg/day, or a placebo for 4.1 years. Of the women participating, 734 had diabetes, 218 had IFG, and 1,811 were normoglycemic. The 2,029 women without diabetes were followed for development of diabetes by measuring fasting glucose at baseline, 1 year, and at the end of the 4.1-year study. Diabetes was defined as a fasting glucose >126 mg/dL. Fasting glucose levels increased in the placebo group but remained stable in the hormone replacement group. The group on estrogen and progesterone lost 0.8 kg over the course of the study while the placebo group gained 0.2 kg. The incidence of developing diabetes was 9.5% in the placebo group and 6.2% in the hormone replacement group ($P < 0.006$). The incidence of developing diabetes was reduced by 35%, and it was estimated that one would need to treat 30 women with hormone replacement therapy to prevent 1 case of diabetes (Kanaya et al. 2003).

Rosiglitazone

Rosiglitazone is a thiazoladinedione that increases insulin sensitivity by stimulating peroxisome proliferators-activated receptor (PPAR) gamma. PPAR gamma stimulation causes fat cells to divide, and small fat cells are more insulin sensitive. Thus, rosiglitazone was tested for its ability to prevent the development of diabetes in people with IGT who are at high risk of developing diabetes. A total of 5,269 subjects with IGT were randomized to rosiglitazone 8 mg/day or placebo for 3 years. At the end of 3 years, 26% of subjects in the placebo group converted to diabetes as compared to only 11.6% in the rosiglitazone group, a significant reduction in diabetes risk despite a 2.2-kg greater weight gain in the rosiglitazone group than the placebo group (Gerstein et al. 2006).

Antihypertensive Therapy

The Antihypertensive and Lipid-Lowering Treatment to Prevent Heart Attack Trial (ALLHAT) evaluated the conversion to diabetes based on fasting glucose levels above 126 mg/dL as a secondary endpoint. The ALLHAT compared the incidence of developing diabetes over 4.9 years in subjects treated for hypertension with chlorthalidone (8,419 subjects), amlodipine (4,958 subjects), or lisinopril (5,034 subjects). Both the amlodipine and lisinopril groups at 2 years had a lower conversion to diabetes than the chlorthalidone group, but due to the lack of a true control group, it is not clear if chlorthalidone increased diabetes risk or the other two drugs decreased it (Barzilay et al. 2006).

Ramapril

Ramapril is an angiotensin-converting enzyme inhibitor used for the treatment of high blood pressure, and angiotensin-converting enzyme inhibitors may also increase insulin sensitivity in muscle (Dietze et al. 1996). The Heart Outcomes Prevention Evaluation (HOPE) trial had a primary objective to reduce myocardial infarction, strokes, death, and the development of diabetic nephropathy. The HOPE study randomized 9,541 subjects, of which 38% had diabetes and 62% did not. The conversion to diabetes was evaluated in the 5,720 subjects without diabetes randomized to ramipril 10 mg/day or a placebo for 4.5 years as a secondary study endpoint. Only 3.6% of the ramipril group converted to diabetes as compared to 5.4% in the placebo group, a 33% reduction (Yusuf et al. 2001).

Candesartan

Candesartan is an angiotensin receptor blocker used for the treatment of high blood pressure. The Candesartan in Heart Failure Assessment of Reduction in Mortality and Morbidity Program (CHARM) evaluated the reduction of cardiovascular mortality and heart failure hospitalizations in patients with heart failure. The development of diabetes in heart failure patients taking candesartan or placebo was a stated secondary endpoint for this trial. CHARM randomized 5,436 subjects with heart failure, but not diabetes, to candesartan 32 mg/day or a placebo for 2–4 years. Six percent of the candesartan group converted to diabetes compared to 7.4% in the placebo group, which was a significant reduction (Yusuf et al. 2005). Thus, both an angiotensin-converting enzyme inhibitor and an angiotensin receptor blocker have been shown to decrease the conversion to diabetes.

Bezafibrate

Bezafibrate is a PPAR alpha agonist used to treat high triglycerides and low HDL cholesterol. Bezafibrate also increases insulin sensitivity (Taniguchi et al. 2001). Two studies have shown that bezafibrate decreases the conversion to diabetes. One study randomized 303 subjects with coronary artery disease and IFG to bezafibrate 400 mg/day or a placebo for 6.2 years. The conversion to diabetes was 42% in the bezafibrate group and 54% in the placebo group ($P = 0.04$) (Tenenbaum et al. 2004). This result was confirmed in a second study with 339 obese subjects randomized to bezafibrate 400 mg/day or placebo for 6.3 years. The conversion to diabetes was 37% in the placebo group and 27% in the bezafibrate group ($P = 0.01$) (Tenenbaum et al. 2005).

Discussion

Clearly, the intervention with the strongest evidence for prevention of diabetes is weight loss and increased physical activity administered as part of a lifestyle change program. There has been interest in determining the relative importance of the components of the lifestyle change program in reducing the conversion to diabetes. A 5-year study randomized 176 subjects with IGT to a reduced fat diet compared to a usual diet. At 1 year, the low-fat diet group lost 3.3 kg and had a 47% rate of conversion to diabetes compared to a 67% conversion rate in the usual diet group (Swinburn et al. 2001). Thus, diet clearly plays a role in preventing diabetes, but this study did not address whether the weight loss induced by the diet or the low-fat nature of the diet was responsible for the decreased rate of conversion. The Hoorn study assessed 2,393 subjects by questionnaire about alcohol intake and divided them into four groups: nondrinkers, 0–10 g/day, 10–30 g/day, and >30 g/day. Subjects who drank up to 10 g/day had the lowest conversion rate to diabetes, 8.0% compared to 12.9% for nondrinkers ($P < 0.05$) (de Vegt et al. 2002).

Studies suggest that greatest of the protection from conversion to diabetes is driven by an improvement in insulin resistance (Chiasson and Rabasa-Lhoret 2004). A systematic review concluded that weight loss was the most protective feature of the comprehensive lifestyle program and that a weight loss of 9–13 kg was most protective (Aucott et al. 2004). The Swedish Obese Subjects Study (SOS) divided 1,702 subjects into obesity surgery or a medically treated control group. At 10 years of follow-up, the medically treated group gained 1.6% of body weight and the surgically treated group maintained a 16.5% weight loss. The conversion to diabetes was 24% in the control group and only 7% in the surgery group (Sjostrom et al. 2004). Thus, the surgical experience seems to bear out the importance of weight loss to the reduction in the rate of conversion to diabetes. In fact, gastric bypass, probably due to its bypass of the foregut and increased delivery of indigested chyme to the distal gut, seems to be the most effective obesity surgery in protecting from conversion to diabetes (Greenway et al. 2002).

Conclusions

Although weight loss and increased physical activity in the context of a comprehensive lifestyle program should be the foundation of any DPP, sustaining such an effort is difficult for most people. The increasing prevalence of obesity and diabetes in the United States, despite public health policy encouraging overweight and obese people to reduce food intake and increase physical activity, is a testimony to difficulty experienced by most people. Obesity and IGT were associated with higher rates of diabetes, hypertension, hypercholesterolemia, and hypertriglyceridemia. The medications approved for these conditions that reduce the conversion to diabetes may be worth considering first, when these conditions require medication therapy in people with obesity and a predisposition to diabetes. Since weight loss improves all these conditions, safe treatments with greater efficacy are clearly needed for the treatment of obesity. Until these new safe and effective therapies arrive, lifestyle programs emphasizing weight loss and increased physical activity will remain the lynchpin of any DPP.

References

American Diabetes Association. 1998. Economic consequences of diabetes mellitus in the U.S. in 1997. Diabetes Care 21(2):296–309.

Anderson JW, Kendall CW, Jenkins DJ. 2003. Importance of weight management in type 2 diabetes: Review with meta-analysis of clinical studies. J Am Coll Nutr 22(5):331–339.

Aucott L, Poobalan A, Smith WC, Avenell A, Jung R, Broom J, Grant AM. 2004. Weight loss in obese diabetic and non-diabetic individuals and long-term diabetes outcomes—a systematic review. Diabetes Obes Metab 6(2):85–94.

Barzilay JI, Davis BR, Cutler JA, Pressel SL, Whelton PK, Basile J, Margolis KL, Ong ST, Sadler LS, Summerson J. 2006. Fasting glucose levels and incident diabetes mellitus in older nondiabetic adults randomized to receive 3 different classes of antihypertensive treatment: A report from the Antihypertensive and Lipid-Lowering Treatment to Prevent Heart Attack Trial (ALLHAT). Arch Intern Med 166(20):2191–2201.

Bray GA. 1998. Obesity: A time bomb to be defused. Lancet 352(9123):160–161.

Bray G, Gregg E, Haffner S, Pi-Sunyer XF, WagenKnecht LE, Walkup M, Wing R. 2006. Baseline characteristics of the randomised cohort from the Look AHEAD (Action for Health in Diabetes) study. Diab Vasc Dis Res 3(3):202–215.

Brown JS, Wing R, Barrett-Connor E, Nyberg LM, Kusek JW, Orchard TJ, Ma Y, Vittinghoff E, Kanaya AM. 2006. Lifestyle intervention is associated with lower prevalence of urinary incontinence: The Diabetes Prevention Program. Diabetes Care 29 (2):385–390.

Carnethon MR, Prineas RJ, Temprosa M, Zhang ZM, Uwaifo G, Molitch ME. 2006. The association among autonomic nervous system function, incident diabetes, and intervention arm in the Diabetes Prevention Program. Diabetes Care 29(4):914–919.

Chiasson JL, Josse RG, Gomis R, Hanefeld M, Karasik A, Laakso M. 2002. Acarbose for prevention of type 2 diabetes mellitus: The STOP-NIDDM randomised trial. Lancet 359(9323):2072–2077.

Chiasson JL, Rabasa-Lhoret R. 2004. Prevention of type 2 diabetes: Insulin resistance and beta-cell function. Diabetes 53(Suppl 3):S34–S38.

Crandall J, Schade D, Ma Y, Fujimoto WY, Barrett-Connor E, Fowler S, Dagogo-Jack S, Andres R. 2006. The influence of age on the effects of lifestyle modification and metformin in prevention of diabetes. J Gerontol A Biol Sci Med Sci 61(10):1075–1081.

Curtis J, Wilson C. 2005. Preventing type 2 diabetes mellitus. J Am Board Fam Pract 18(1):37–43.

de Vegt F, Dekker JM, Groeneveld WJ, Nijpels G, Stehouwer CD, Bouter LM, Heine RJ. 2002. Moderate alcohol consumption is associated with lower risk for incident diabetes and mortality: The Hoorn Study. Diabetes Res Clin Pract 57(1):53–60.

Delahanty LM, Conroy MB, Nathan DM. 2006. Psychological predictors of physical activity in the diabetes prevention program. J Am Diet Assoc 106(5):698–705.

Delahanty LM, Meigs JB, Hayden D, Williamson DA, Nathan DM. 2002. Psychological and behavioral correlates of baseline BMI in the diabetes prevention program (DPP). Diabetes Care 25(11):1992–1998.

The Diabetes Prevention Program Research Group. 1999. Design and methods for a clinical trial in the prevention of type 2 diabetes. Diabetes Care 22(4):623–634.

The Diabetes Prevention Program Research Group. 2000. The Diabetes Prevention Program: Baseline characteristics of the randomized cohort. Diabetes Care 23(11):1619–1629.

The Diabetes Prevention Program Research Group. 2003. Effects of withdrawal from metformin on the development of diabetes in the diabetes prevention program. Diabetes Care 26(4):977–980.

The Diabetes Prevention Program Research Group. 2003. Within-trial cost-effectiveness of lifestyle intervention or metformin for the primary prevention of type 2 diabetes. Diabetes Care 26(9):2518–2523.

Dietze GJ, Wicklmayr M, Rett K, Jacob S, Henriksen EJ. 1996. Potential role of bradykinin in forearm muscle metabolism in humans. Diabetes 45(Suppl 1):S110–S114.

Eriksson J, Lindstrom J, Valle T, Aunola S, Hamalainen H, Ilanne-Parikka P, Keinanen-Kiukaanniemi S, Laakso M, Lauhkonen M, Lehto P, Lehtonen A, Louheranta A, Mannelin M, Martikkala V, Rastas M, Sundvall J, Turpeinen A, Viljanen T, Uusitupa M, Tuomilehto J. 1999. Prevention of Type II diabetes in subjects with impaired glucose tolerance: The Diabetes Prevention Study (DPS) in Finland. Study design and 1-year interim report on the feasibility of the lifestyle intervention programme. Diabetologia 42(7):793–801.

Eriksson KF, Lindgarde F. 1991. Prevention of type 2 (non-insulin-dependent) diabetes mellitus by diet and physical exercise. The 6-year Malmo feasibility study. Diabetologia 34(12):891–898.

Eriksson KF, Lindgarde F. 1998. No excess 12-year mortality in men with impaired glucose tolerance who participated in the Malmo Preventive Trial with diet and exercise. Diabetologia 41(9):1010–1016.

Flegal KM. 2005. Epidemiologic aspects of overweight and obesity in the United States. Physiol Behav 86(5):599–602.

Freeman DJ, Norrie J, Sattar N, Neely RD, Cobbe SM, Ford I, Isles C, Lorimer AR, Macfarlane PW, McKillop JH, Packard CJ, Shepherd J, Gaw A. 2001. Pravastatin and the development of diabetes mellitus: Evidence for a protective treatment effect in the West of Scotland Coronary Prevention Study. Circulation 103(3):357–362.

Fujimoto WY. 2000. Background and recruitment data for the U.S. Diabetes Prevention Program. Diabetes Care 23(Suppl 2):B11–B13.

Gerstein HC, Yusuf S, Bosch J, Pogue J, Sheridan P, Dinccag N, Hanefeld M, Hoogwerf B, Laakso M, Mohan V, Shaw J, Zinman B, Holman RR. 2006. Effect of rosiglitazone on the frequency of diabetes in patients with impaired glucose tolerance or impaired fasting glucose: A randomised controlled trial. Lancet 368(9541):1096–1105.

Greenway SE, Greenway FL, III, Klein S. 2002. Effects of obesity surgery on non-insulin-dependent diabetes mellitus. Arch Surg 137(10):1109–1117.

Haffner S, Temprosa M, Crandall J, Fowler S, Goldberg R, Horton E, Marcovina S, Mather K, Orchard T, Ratner R, Barrett-Connor E. 2005. Intensive lifestyle intervention or metformin on inflammation and coagulation in participants with impaired glucose tolerance. Diabetes 54(5):1566–1572.

Hamalainen H, Ronnemaa T, Virtanen A, Lindstrom J, Eriksson JG, Valle TT, Ilanne-Parikka P, Keinanen-Kiukaanniemi S, Rastas M, Aunola S, Uusitupa M, Tuomilehto J. 2005. Improved fibrinolysis by an intensive lifestyle intervention in subjects with impaired glucose tolerance: The Finnish Diabetes Prevention Study. Diabetologia 48(11):2248–2253.

Hanefeld M, Chiasson JL, Koehler C, Henkel E, Schaper F, Temelkova-Kurktschiev T. 2004. Acarbose slows progression of intima-media thickness of the carotid arteries in subjects with impaired glucose tolerance. Stroke 35(5):1073–1078.

Harris MI, Flegal KM, Cowie CC, Eberhardt MS, Goldstein DE, Little RR, Wiedmeyer HM, Byrd-Holt DD. 1998. Prevalence of diabetes, impaired fasting glucose, and impaired glucose tolerance in U.S. adults. The Third National Health and Nutrition Examination Survey, 1988–1994. Diabetes Care 21(4):518–524.

Hernan WH, Brandle M, Zhang P, Williamson DF, Matulik MJ, Ratner RE, Lachin JM, Engelgau MM. 2003. Costs associated with the primary prevention of type 2 diabetes mellitus in the diabetes prevention program. Diabetes Care 26(1):36–47.

Kanaya AM, Herrington D, Vittinghoff E, Lin F, Grady D, Bittner V, Cauley JA, Barrett-Connor E. 2003. Glycemic effects of postmenopausal hormone therapy: The Heart and Estrogen/Progestin Replacement Study. A randomized, double-blind, placebo-controlled trial. Ann Intern Med 138(1):1–9.

Kitabchi AE, Temprosa M, Knowler WC, Kahn SE, Fowler SE, Haffner SM, Andres R, Saudek C, Edelstein SL, Arakaki R, Murphy MB, Shamoon H. 2005. Role of insulin secretion and sensitivity in the evolution of type 2 diabetes in the diabetes prevention program: Effects of lifestyle intervention and metformin. Diabetes 54(8):2404–2414.

Klein S, Sheard NF, Pi-Sunyer X, Daly A, Wylie-Rosett J, Kulkarni K, Clark NG. 2004. Weight management through lifestyle modification for the prevention and management of type 2 diabetes: rationale and strategies. A statement of the American Diabetes Association, the North American Association for the Study of Obesity, and the American Society for Clinical Nutrition. Am J Clin Nutr 80(2):257–263.

Knowler WC, Barrett-Connor E, Fowler SE, Hamman RF, Lachin JM, Walker EA, Nathan DM. 2002. Reduction in the incidence of type 2 diabetes with lifestyle intervention or metformin. N Engl J Med 346(6):393–403.

Knowler WC, Hamman RF, Edelstein SL, Barrett-Connor E, Ehrmann DA, Walker EA, Fowler SE, Nathan DM, Kahn SE. 2005. Prevention of type 2 diabetes with troglitazone in the Diabetes Prevention Program. Diabetes 54(4):1150–1156.

Knowler WC, Sartor G, Melander A, Schersten B. 1997. Glucose tolerance and mortality, including a substudy of tolbutamide treatment. Diabetologia 40(6):680–686.

Kosaka K, Noda M, Kuzuya T. 2005. Prevention of type 2 diabetes by lifestyle intervention: A Japanese trial in IGT males. Diabetes Res Clin Pract 67(2):152–162.

Kriska AM, Edelstein SL, Hamman RF, Otto A, Bray GA, Mayer-Davis EJ, Wing RR, Horton ES, Haffner SM, Regensteiner JG. 2006. Physical activity in individuals at risk for diabetes: The Diabetes Prevention Program. Med Sci Sports Exerc 38(5):826–832.

Laaksonen DE, Lindstrom J, Lakka TA, Eriksson JG, Niskanen L, Wikstrom K, Aunola S, Keinanen-Kiukaanniemi S, Laakso M, Valle TT, Ilanne-Parikka P, Louheranta A, Hamalainen H, Rastas M, Salminen V, Cepaitis Z, Hakumaki M, Kaikkonen H, Harkonen P, Sundvall J, Tuomilehto J, Uusitupa M. 2005. Physical activity in the prevention of type 2 diabetes: The Finnish Diabetes Prevention Study. Diabetes 54(1):158–165.

Laukkanen O, Pihlajamaki J, Lindstrom J, Eriksson J, Valle TT, Hamalainen H, Ilanne-Parikka P, Keinanen-Kiukaanniemi S, Tuomilehto J, Uusitupa M, Laakso M. 2004. Common polymorphisms in the genes regulating the early insulin signalling pathway: Effects on weight change and the conversion from impaired glucose tolerance to Type 2 diabetes. The Finnish Diabetes Prevention Study. Diabetologia 47(5):871–877.

Li G, Hu Y, Yang W, Jiang Y, Wang J, Xiao J, Hu Z, Pan X, Howard BV, Bennett PH. 2002. Effects of insulin resistance and insulin secretion on the efficacy of interventions to retard development of type 2 diabetes mellitus: The Da Qing IGT and Diabetes Study. Diabetes Res Clin Pract 58(3):193–200.

Lindstrom J, Eriksson JG, Valle TT, Aunola S, Cepaitis Z, Hakumaki M, Hamalainen H, Ilanne-Parikka P, Keinanen-Kiukaanniemi S, Laakso M, Louheranta A, Mannelin M, Martikkala V, Moltchanov V, Rastas M, Salminen V, Sundvall J, Uusitupa M, Tuomilehto J. 2003a. Prevention of diabetes mellitus in subjects with impaired glucose tolerance in the Finnish Diabetes Prevention Study: Results from a randomized clinical trial. J Am Soc Nephrol 14(7, Suppl 2):S108–S113.

Lindstrom J, Ilanne-Parikka P, Peltonen M, Aunola S, Eriksson JG, Hemio K, Hamalainen H, Harkonen P, Keinanen-Kiukaanniemi S, Laakso M, Louheranta A, Mannelin M, Paturi M, Sundvall J, Valle TT, Uusitupa M, Tuomilehto J. 2006. Sustained reduction in the incidence of type 2 diabetes by lifestyle intervention: Follow-up of the Finnish Diabetes Prevention Study. Lancet 368(9548):1673–1679.

Lindstrom J, Louheranta A, Mannelin M, Rastas M, Salminen V, Eriksson J, Uusitupa M, Tuomilehto J. 2003b. The Finnish Diabetes Prevention Study (DPS): Lifestyle intervention and 3-year results on diet and physical activity. Diabetes Care 26(12):3230–3236.

Mayer-Davis EJ, Sparks KC, Hirst K, Costacou T, Lovejoy JC, Regensteiner JG, Hoskin MA, Kriska AM, Bray GA. 2004. Dietary intake in the diabetes prevention program cohort: Baseline and 1-year post randomization. Ann Epidemiol 14(10):763–772.

Molitch ME, Fujimoto W, Hamman RF, Knowler WC. 2003. The diabetes prevention program and its global implications. J Am Soc Nephrol 14(7, Suppl 2):S103–S107.

Muller G. 2000. The molecular mechanism of the insulin-mimetic/sensitizing activity of the antidiabetic sulfonylurea drug Amaryl. Mol Med 6(11):907–933.

National Diabetes Data Group. 1979. Classification and diagnosis of diabetes mellitus and other categories of glucose intolerance. Diabetes 28(12):1039–1057.

Orchard TJ, Temprosa M, Goldberg R, Haffner S, Ratner R, Marcovina S, Fowler S. 2005. The effect of metformin and intensive lifestyle intervention on the metabolic syndrome: The Diabetes Prevention Program randomized trial. Ann Intern Med 142(8):611–619.

Pan XR, Hu YH, Li GW, Liu PA, Bennett PH, Howard BV. 1993. Impaired glucose tolerance and its relationship to ECG-indicated coronary heart disease and risk factors among Chinese: The Da Qing IGT and Diabetes Study. Diabetes Care 16(1):150–156.

Pan XR, Li GW, Hu YH, Wang JX, Yang WY, An ZX, Hu ZX, Lin J, Xiao JZ, Cao HB, Liu PA, Jiang XG, Jiang YY, Wang JP, Zheng H, Zhang H, Bennett PH, Howard BV. 1997. Effects of diet and exercise in preventing NIDDM in people with impaired glucose tolerance: The Da Qing IGT and Diabetes Study. Diabetes Care 20(4):537–544.

Ramachandran A, Snehalatha C, Mary S, Mukesh B, Bhaskar AD, Vijay V. 2006. The Indian Diabetes Prevention Programme shows that lifestyle modification and metformin prevent type 2 diabetes in Asian Indian subjects with impaired glucose tolerance (IDPP-1). Diabetologia 49(2):289–297.

Ratner R, Goldberg R, Haffner S, Marcovina S, Orchard T, Fowler S, Temprosa M. 2005. Impact of intensive lifestyle and metformin therapy on cardiovascular disease risk factors in the diabetes prevention program. Diabetes Care 28(4):888–894.

Rubin RR, Knowler WC, Ma Y, Marrero DG, Edelstein SL, Walker EA, Garfield SA, Fisher EB. 2005. Depression symptoms and antidepressant medicine use in Diabetes Prevention Program participants. Diabetes Care 28(4):830–837.

Sartor G, Schersten B, Carlstrom S, Melander A, Norden A, Persson G. 1980. Ten-year follow-up of subjects with impaired glucose tolerance: Prevention of diabetes by tolbutamide and diet regulation. Diabetes 29(1):41–49.

Siitonen N, Lindstrom J, Eriksson J, Valle TT, Hamalainen H, Ilanne-Parikka P, Keinanen-Kiukaanniemi S, Tuomilehto J, Laakso M, Uusitupa M. 2004. Association between a deletion/insertion polymorphism in the alpha2B-adrenergic receptor gene and insulin secretion and Type 2 diabetes: The Finnish Diabetes Prevention Study. Diabetologia 47(8):1416–1424.

Sjostrom L, Lindroos AK, Peltonen M, Torgerson J, Bouchard C, Carlsson B, Dahlgren S, Larsson B, Narbro K, Sjostrom CD, Sullivan M, Wedel H. 2004. Lifestyle, diabetes, and cardiovascular risk factors 10 years after bariatric surgery. N Engl J Med 351(26):2683–2693.

Smith AG, Russell J, Feldman EL, Goldstein J, Peltier A, Smith S, Hamwi J, Pollari D, Bixby B, Howard J, Singleton JR. 2006. Lifestyle intervention for pre-diabetic neuropathy. Diabetes Care 29(6):1294–1299.

Swinburn BA, Metcalf PA, Ley SJ. 2001. Long-term (5-year) effects of a reduced-fat diet intervention in individuals with glucose intolerance. Diabetes Care 24(4):619–624.

Taniguchi A, Fukushima M, Sakai M, Tokuyama K, Nagata I, Fukunaga A, Kishimoto H, Doi K, Yamashita Y, Matsuura T, Kitatani N, Okumura T, Nagasaka S, Nakaishi S, Nakai Y. 2001. Effects of bezafibrate on insulin sensitivity and insulin secretion in non-obese Japanese type 2 diabetic patients. Metabolism 50(4):477–480.

Tenenbaum A, Motro M, Fisman EZ, Adler Y, Shemesh J, Tanne D, Leor J, Boyko V, Schwammenthal E, Behar S. 2005. Effect of bezafibrate on incidence of type 2 diabetes mellitus in obese patients. Eur Heart J 26(19):2032–2038.

Tenenbaum A, Motro M, Fisman EZ, Schwammenthal E, Adler Y, Goldenberg I, Leor J, Boyko V, Mandelzweig L, Behar S. 2004. Peroxisome proliferator-activated receptor ligand bezafibrate for prevention of type 2 diabetes mellitus in patients with coronary artery disease. Circulation 109(18):2197–2202.

Todorova B, Kubaszek A, Pihlajamaki J, Lindstrom J, Eriksson J, Valle TT, Hamalainen H, Ilanne-Parikka P, Keinanen-Kiukaanniemi S, Tuomilehto J, Uusitupa M, Laakso M. 2004. The G-250A promoter polymorphism of the hepatic lipase gene predicts the conversion from impaired glucose tolerance to type 2 diabetes mellitus: The Finnish Diabetes Prevention Study. J Clin Endocrinol Metab 89(5):2019–2023.

Torgerson JS, Hauptman J, Boldrin MN, Sjostrom L. 2004. XENical in the prevention of diabetes in obese subjects (XENDOS) study: A randomized study of orlistat as an adjunct to lifestyle changes for the prevention of type 2 diabetes in obese patients. Diabetes Care 27(1):155–161.

Tuomilehto J, Lindstrom J, Eriksson JG, Valle TT, Hamalainen H, Ilanne-Parikka P, Keinanen-Kiukaanniemi S, Laakso M, Louheranta A, Rastas M, Salminen V, Uusitupa M. 2001. Prevention of type 2 diabetes mellitus by changes in lifestyle among subjects with impaired glucose tolerance. N Engl J Med 344(18):1343–1350.

Uusitupa M, Lindi V, Louheranta A, Salopuro T, Lindstrom J, Tuomilehto J. 2003. Long-term improvement in insulin sensitivity by changing lifestyles of people with impaired glucose tolerance: 4-year results from the Finnish Diabetes Prevention Study. Diabetes 52(10):2532–2538.

Uusitupa M, Louheranta A, Lindstrom J, Valle T, Sundvall J, Eriksson J, Tuomilehto J. 2000. The Finnish Diabetes Prevention Study. Br J Nutr 83(Suppl 1):S137–S142.

Uusitupa MI. 1996. Early lifestyle intervention in patients with non-insulin-dependent diabetes mellitus and impaired glucose tolerance. Ann Med 28(5):445–449.

Walker EA, Molitch M, Kramer MK, Kahn S, Ma Y, Edelstein S, Smith K, Johnson MK, Kitabchi A, Crandall J. 2006. Adherence to preventive medications: Predictors and outcomes in the Diabetes Prevention Program. Diabetes Care 29(9):1997–2002.

Wing RR, Hamman RF, Bray GA, Delahanty L, Edelstein SL, Hill JO, Horton ES, Hoskin MA, Kriska A, Lachin J, Mayer-Davis EJ, Pi-Sunyer X, Regensteiner JG, Venditti B, Wylie-Rosett J. 2004. Achieving weight and activity goals among diabetes prevention program lifestyle participants. Obes Res 12(9):1426–1434.

Younis N, Soran H, Farook S. 2004. The prevention of type 2 diabetes mellitus: Recent advances. Quart J Med 97(7):451–455.

Yusuf S, Gerstein H, Hoogwerf B, Pogue J, Bosch J, Wolffenbuttel BH, Zinman B. 2001. Ramipril and the development of diabetes. JAMA 286(15):1882–1885.

Yusuf S, Ostergren JB, Gerstein HC, Pfeffer MA, Swedberg K, Granger CB, Olofsson B, Probstfield J, McMurray JV. 2005. Effects of candesartan on the development of a new diagnosis of diabetes mellitus in patients with heart failure. Circulation 112(1):48–53.

Chapter 4

Glycemic Index and Glycemic Load: Effects on Glucose, Insulin, and Lipid Regulation

Julia MW Wong, RD, Andrea R Josse, MSc, Livia Augustin, PhD, Amin Esfahani, BSc, Monica S Banach, BSc, Cyril WC Kendall, PhD, and David JA Jenkins, MD

Overview

Different types of carbohydrates produce different metabolic responses, which may relate to factors associated with chronic disease. The glycemic index (GI) is a classification of carbohydrate containing foods based on their effect on postprandial blood glucose response. Low GI diets have been reported to improve glycemic control in patients with diabetes and to reduce serum lipids in hypercholesterolemic individuals. Low GI or glycemic load diets have also been associated with improvements in high-density lipoprotein cholesterol, C-reactive protein concentrations, and decreased risk of developing obesity, type 2 diabetes, cardiovascular disease, and certain cancers. As new research in the area of GI continues to expand, the clinical use of the GI may have important implications in promoting health.

Introduction

The prevalence of chronic disease and associated metabolic disorders continues to be a growing concern worldwide. This has resulted in the renewed interest in both diet and lifestyle modifications as the first line of prevention and treatment. Particularly, the health benefits of high carbohydrate/low fat diets, in the presence of increasing rates of obesity, cardiovascular disease (CVD), and diabetes, have been challenged. At the center of this debate is the metabolic effect of carbohydrates, and it may be that the nature of the carbohydrate is as important, if not more, than the total carbohydrate content. This review will summarize some of the metabolic advantages of slow release carbohydrates or low glycemic index (GI) foods in preventing and treating certain chronic diseases.

49

Glycemic Index

Carbohydrates have been a prime target of debate in determining the optimal diet for overall health, especially related to obesity, diabetes, and CVD. Both the quantity and quality of the carbohydrate are seen to be important factors in determining the metabolic response. Specifically, postprandial blood glucose responses were once believed to be attributed to the carbohydrate chain length, often referred to as "simple" or "complex." However, with emerging research in this area, this simple classification was questioned. Thus, the concept of the GI was created—which links the concepts of the dietary fiber hypothesis, first proposed by Burkitt and Trowell (1977), the insulin-resistance syndrome (Reaven 2003), and the proposal that postprandial events increase free radical generation and oxidative damage (Brownlee 2001; Ceriello and Motz 2004)—in an attempt to find a dietary strategy to prevent and treat chronic diseases such as diabetes and coronary heart disease (CHD).

The GI was originally defined as the incremental area under the blood glucose response curve (IAUC) elicited by a 50 g available carbohydrate portion of food expressed as a percentage of the response after 50 g anhydrous glucose taken by the same subject (Wolever 2006). Later, 50 g of carbohydrate from bread was used since this seemed an appropriate standard with which to compare other study foods (Jenkins et al. 1983). Nevertheless, these two standards were interchangeable using the conversion factor of 1.42 glucose to bread (Foster-Powell and Miller 1995). This quantitative classification of carbohydrate foods is determined primarily by the rate of carbohydrate absorption as reflected in the glycemic response (Jenkins et al. 1981, 1984). The GI emphasizes differences in the nature of the carbohydrate that result in different physiological responses (Table 4.1). In other words, the GI is considered to be a specific property of the food itself and is different from the "glycemic response," which is an individual's change in blood glucose after ingestion of the food (Wolever 2006). Foods with lower glycemic indices tend to be the starchy staples of traditional cultures such as pasta, whole grain pumpernickel breads, cracked wheat or barley, rice, dried peas, beans, and lentils (Jenkins et al. 1980, 1986; Thorne et al. 1983). It is noteworthy that in groups such as the Pima Indians and the Australian aborigines, the relatively recent shift in the traditional use of low GI foods to much greater intakes of high GI foods may partially explain the appearance of greater rates of diabetes among these populations (Boyce and Swinburn 1993; O'Dea 1991; Thorburn et al. 1987).

Extensive research in the area of GI has resulted in the creation of comprehensive international GI food tables that encompass >750 different types of foods tested with use of standard GI methods (Foster-Powell et al. 2002). This has greatly facilitated clinical research and other general applications of the GI concept. Furthermore, the concept of glycemic load (GL), which is the product of dietary GI and the amount of available dietary carbohydrate in a food or diet, has been developed to assess the total glycemic impact of foods or dietary patterns (Salmeron et al. 1997a). Current research on the GI appears to address the five underlying diagnostic components of the metabolic syndrome: increased glucose, triglycerides,

Table 4.1.   Possible effects of prolonged carbohydrate absorption time.

- Lower postprandial glucose rise (Bertelsen et al. 1993; Jenkins et al. 1990, 1992; Jones et al. 1993)
- Reduced daily mean insulin levels (Bertelsen et al. 1993; Jenkins et al. 1990, 1992; Jones et al. 1993)
- Flatter gastric inhibitory polypeptide response (Bertelsen et al. 1993; Jenkins et al. 1990, 1992)
- Decreased 24-h urinary C-peptide output (Jenkins et al. 1989, 1992)
- Prolonged suppression of plasma free fatty acids (Jenkins et al. 1990)
- Reduced urinary catecholamine output (Jenkins et al. 1989, 1990)
- Lower total and low-density lipoprotein cholesterol levels (Arnold et al. 1993; Cohn 1964; Jenkins et al. 1989, 1995)
- Reduced hepatic cholesterol synthesis (Jones et al. 1993)
- Decreased serum apolipoprotein B levels (Jenkins et al. 1989)
- Decreased serum uric acid levels (Jenkins et al. 1995)
- Increased urinary uric acid excretion (Jenkins et al. 1995)

Adapted from Jenkins and Jenkins (1995).

blood pressure, body weight, and decreased HDL cholesterol, as defined by the National Cholesterol Education Program Adult Treatment Panel III (Expert Panel on Detection 2001).

Mechanisms of Action

The metabolic advantages observed from consuming low GI foods are suggested to be primarily due to the reduced rate of carbohydrate absorption from the small intestine (Figure 4.1). Slower absorption rates result in a lower rise in blood glucose and insulin levels. Many intrinsic and extrinsic factors exist that may affect the rate of carbohydrate absorption and, in turn, the GI value of the food. These include the rate of digestion (Englyst et al. 1999; Jenkins et al. 1981), food form (physical form, particle size), type of preparation (cooking method and processing) (Haber et al. 1977; Jenkins et al. 1982a; O'Dea et al. 1980; Sheard et al. 2004), ripeness (Englyst and Cummings et al. 1986), nature of the starch (amylose or amylopectin) (Sheard et al. 2004; Wursch et al. 1986), monosaccharide components, presence of antinutrients such as α-amylase inhibitors (phytic acid and phenolic compounds) (Isaksson et al. 1982; Yoon et al. 1983), transit time (Englyst et al. 1992), and amount and type of fiber, fat, and protein (Krezowski et al. 1986; Thorne et al. 1983).

The metabolic benefits associated with a reduced rate of carbohydrate absorption have been demonstrated in studies of healthy subjects and those with diabetes. For example, when a glucose solution was sipped at an even rate over 180 min (sipping) compared to the same amount of glucose taken over 5–10 min (bolus) (Jenkins

(a)   (b)

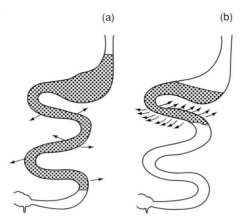

Figure 4.1.   Hypothetical effect of feeding diets with a low (a) or high (b) glycemic index on gastrointestinal glucose absorption. (Adapted from Jenkins et al. 2002.)

et al. 1990), marked reductions in both postprandial insulin secretion and serum free fatty acid (FFA) levels were seen with sipping. Similar improvements are also observed when low GI foods were consumed, where the rate of glucose absorption reduces the postprandial rise in gut hormones (e.g., incretins) and insulin demand. In addition, the prolonged absorption of carbohydrate will suppress the postmeal appearance of FFA in the blood and other counterregulatory responses (Jenkins et al. 1990; Ludwig et al. 1999; Wolever et al. 1995). Over time, the lower FFA concentrations and sustained tissue insulinization may increase the rate of glucose removal from the circulation. As a result, blood glucose concentrations will return to baseline levels despite continued absorption in the small intestine. Attenuation of the peak rise in postprandial and the incremental area under the blood glucose curve is observed. A "second meal" effect has also been seen such that after intravenous (IV) glucose, there was a more rapid uptake of glucose (increased K_G) after glucose sipping than after a bolus glucose drink (Jenkins et al. 1990), possibly related to a prolonged suppression of FFA levels (Jenkins et al. 1982b).

Obesity and GI

Obesity, especially abdominal obesity, contributes to insulin resistance which is a metabolic disorder associated with increased risk of type 2 diabetes and CVD (Rader 2007). Low GI diets may have a potential role in body weight regulation and, in turn, insulin resistance. This hypothesis has been supported by large epidemiological studies (Ma et al. 2005; Toeller et al. 2001). It has been observed that a higher body mass index (BMI) was positively associated with the consumption of high GI foods which suggests that the type of carbohydrate is an important determinant of body weight (Ma et al. 2005). This observation was consistent with the EURODIAB Complications Study which assessed over 3,000 individuals with type 1 diabetes and indicated that a lower GI diet was associated with lower waist-to-hip ratio and waist circumference measurements (Toeller et al. 2001).

A number of clinical trials have also investigated the effects of GI in the context of weight reduction on body weight and body composition. In a study of healthy obese females, the effects of varying GI levels in an energy restricted diet were evaluated in a 12-week parallel design ($n = 30$), which was followed by some subjects crossing over to the alternative treatment for another 12 weeks ($n = 16$) (Slabber et al. 1994). Weight reductions on the high and low GI diets were similar after the parallel study (7.4 kg low GI vs. 9.3 kg high GI, respectively). However, a greater weight reduction was observed after the low GI diet compared to the high GI diet after the crossover study (7.4 kg vs. 4.5 kg, respectively, $P = 0.04$) (Slabber et al. 1994). Low GI diets have also been shown to improve body composition. Despite body weight remaining comparable after a 5 week intervention on a low or high GI diet, a greater reduction in body fat mass ($\sim$700 g reduction) and a tendency to increase lean body mass, as measured by dual-energy X-ray absorptiometry (DEXA), was observed on the low GI diet (Bouche et al. 2002). The reduction in body fat mass was mainly attributed to a decline in trunk fat. Similarly, a study with 14 obese adolescents who received either a conventional diet or an energy restricted lower GI and GL diet for 6 months followed by a 6-month follow-up demonstrated a significant reduction in both body weight and fat mass (measured by DEXA) at the end of the intervention (Ebbeling et al. 2003). Although these studies suggest that low GI diets are beneficial in reducing body weight or body fat mass, the results have not been consistent (Ebbeling et al. 2005; Frost et al. 2004; Sloth et al. 2004). Some of the variation may be explained by the postprandial serum insulin levels. In a study by Ebbeling et al (Ebbeling et al. 2007), it was shown that the low GI bread was most effective in reducing body weight in those whose 30 min postprandial insulin concentrations were highest. Obesity control combined with positive metabolic effects relating to risk factors for CHD and diabetes may warrant the use of low GI diets as a nutrition strategy aimed to target those with a cluster of disease risk factors, including insulin resistance, known as the metabolic syndrome (Grundy et al. 2004). Further well-controlled studies of longer duration and greater sample size are needed before public health policies are likely to change significantly. More importantly, care should be taken to ensure that the study designs match the two comparison diets for palatability, energy density, fiber content, and macronutrient composition (Sloth and Astrup 2006), and probably more importantly, motivation to adhere to the diets equally.

Diabetes and GI

Epidemiological and clinical studies have shown that diets of lower GI may be beneficial in relation to the development and management of type 2 diabetes. Large prospective cohort studies have found a positive association between GI and diabetes risk, where the higher the dietary GI, the greater the risk (Hodge et al. 2004; Salmeron et al. 1997a, b; Schulze et al. 2004). However, this was not observed in the Iowa Women's Health Study and the Atherosclerosis Risk in Communities (ARIC) Study (Meyer et al. 2000; Stevens et al. 2002). The lack of association between GI and GL with type 2 diabetes in the Iowa Women's Health Study is possibly related to the inclusion of an elderly cohort which could introduce a selection bias (Meyer et al. 2000), and the dietary assessment tool used in the ARIC study was not specifically designed to assess GI (Meyer et al. 2000; Stevens et al. 2002).

The positive results observed in epidemiological studies have been further supported by two recent meta-analyses summarizing the effects of low GI diets on risk factors for diabetes. Significant reductions in fructosamine and haemoglobin A1c (HbA1c) were observed in those receiving low GI diets (Kelly et al. 2004; Opperman et al. 2004), but no significant changes were observed in blood glucose and insulin (Kelly et al. 2004). A meta-analysis of randomized controlled trials comparing low GI diets to conventional or high GI diets on glycemic control in individuals with diabetes demonstrated that low GI diets reduced glycated proteins by 7.4% and HbA1c by 0.43% compared to high GI diets (Brand-Miller et al. 2003). This meta-analysis included 14 clinical trials of either randomized crossover or parallel designs that lasted between 12 days and 12 months (mean 10 weeks) and included over 350 subjects. Subsequent studies are required to confirm the results of these meta-analyses (Jimenez-Cruz et al. 2003; Rizkalla et al. 2004).

Drug therapies that are used to manage diabetes and its complications, and are associated with altering the rate of glucose absorption, have been shown to have similar metabolic benefits as low GI diets. In the UK Prospective Diabetes Study (UKPDS), acarbose (α-glucosidase enzyme inhibitor) was given at a dosage of 100 mg three times daily in subjects with type 2 diabetes and resulted in a significant lowering of HbA1c compared to placebo at 3 years (Holman et al. 1999). This improvement in glycemic control was comparable to that achieved by monotherapy with either sulfonylureas, metformin, or insulin, but without the weight gain seen with sulfonylureas and insulin (UKPDS et al. 1998). Similarly, in the STOP-NIDDM trial, subjects with impaired glucose tolerance who were given a similar dosage of acarbose demonstrated a 25% reduction in the risk of progression to diabetes, in addition to a significant increase in reversion of impaired glucose tolerance to normal glucose tolerance (Chiasson et al. 2002).

Cardiovascular Disease and GI

Low GI foods have been suggested to play a protective role in the prevention and development of CVD. In a cohort of over 75,000 nurses (Nurses' Health Study), a direct relation between fatal and nonfatal myocardial infarction and GI as well as GL was observed (Liu et al. 2000). Furthermore, in those with a BMI >23 kg/m^2, the association between dietary GI and GL with CHD risk was more evident, suggesting that GI may be more important in those with a higher BMI and thus possibly with greater insulin resistance (Liu et al. 2000). This was also observed in a case-control study where a significant positive association between GI and myocardial infarction was observed in overweight individuals (BMI > 25) (Tavani et al. 2003). Similarly, a high carbohydrate intake in the context of a high GI diet tended to be positively associated with atherosclerotic progression in postmenopausal women (Mozaffarian et al. 2004). However, no significant association of GI or GL and CHD was observed in the Zutphen study of older men (van Dam et al. 2000). This may be due to the smaller sample size and age at baseline.

As previously mentioned, drug therapies used to reduce the rate of glucose absorption have not only been shown to improve diabetes prevention and management, but are also effective in reducing CVD risk. In the STOP-NIDDM trial, it was found that decreasing postprandial hyperglycemia with acarbose was associated with a 49% relative risk reduction in the development of cardiovascular events and a 34% relative risk reduction in new cases of hypertension (Chiasson et al. 2003).

The effect of GI on risk factors associated with CHD has also been explored in a number of studies. Epidemiological studies have indicated that a low GI diet is associated with lower serum triglycerides and/or higher serum HDL cholesterol, suggesting that low GI diets may preserve HDL-C (Ford and Liu 2001; Frost et al. 1999; Liu et al. 2001; Slyper et al. 2005). Furthermore, GI was positively associated with CRP in the Women's Health Study (Liu et al. 2002). Meta-analyses of the effect of low GI diets on major risk factors for CVD have been recently summarized (Kelly et al. 2004; Opperman et al. 2004). In these studies, 15 or 16 clinical trials were included in each analysis with the majority of the subjects having types 2 diabetes. Low GI diets demonstrated no change in HDL cholesterol, triglycerides, and LDL cholesterol levels compared to high GI diets. However, improvements in total cholesterol were observed (Kelly et al. 2004; Opperman et al. 2004) with greater reductions in those with a higher baseline value (Opperman et al. 2004). Interestingly, the observed improvement in HDL cholesterol levels in epidemiological studies is not reflected in the results from clinical trials. Despite the appearance of only a weak effect of low GI diets on CHD risk factors, it was concluded that the studies conducted to date have been too short-term and of small sample size to allow advice to be formulated on this issue. Therefore, there is a need for more well-designed randomized controlled trials that are adequately powered and of greater duration to determine the effects of low GI diets on CHD risk (Kelly et al. 2004).

Other surrogate markers of CHD risk have also been investigated. In subjects with type 2 diabetes, plasminogen activator inhibitor-1 (PAI-1) levels were reduced on a low GI diet (Jarvi et al. 1999; Rizkalla et al. 2004). Moreover, a low GL diet compared to a low fat diet, in the context of weight reduction, found marked improvements in heart disease risk factors such as insulin resistance, TG, CRP, and blood pressure (Pereira et al. 2004).

Cancer and GI

McKeown-Eyssen (McKeown-Eyssen et al. 1994) and Giovannucci (1995) were among the first to note similarities in risk factors for insulin resistance and development of colorectal cancer and possibly other types of cancer such as breast and prostate (Boyd 2003). In this context, a sedentary lifestyle and dietary risk factors result in hyperinsulinemia, hyperglycemia, and hypertriglyceridemia, all of which are factors associated with insulin resistance (Giovannucci 1995; McKeown-Eyssen et al. 1994). Increased promotion of tumor development may result as a consequence of increased insulin levels and increased insulin-like growth factors and increased glucose and triglycerides (i.e., increased available energy). Studies looking at the link between insulin resistance and/or diabetes and cancer risk have supported these initial observations. A link between cancer and diabetes mellitus had been suspected for more than 100 years and until the 1920s, hyperglycemia was used as a marker for cancer screening (Ellinger and Landsman 1944; Freund et al. 1885; Marble 1934). In a meta-analysis of diabetes and incidence and mortality of colorectal cancer, a positive association was observed in both men and women (Larsson et al. 2005). Furthermore, a similar association was observed in women with insulin resistance (Bruning et al. 1992) and type 2 diabetes (Michels et al. 2003) and incidence of breast cancer. Hyperglycemia has also been linked to increased cancer risk in large nondiabetic cohorts (Stattin et al. 2007) and in diabetic individuals in a dose-dependent fashion (Jee et al. 2005). Therefore, low GI and GL diets may be a possible strategy for cancer prevention.

Direct associations between GI or GL and various types of cancer have been observed in epidemiological studies. In a case-control study, it was observed that colorectal cancer risk increased with an increase in dietary GI (odd ratio for highest vs. lowest quintile: 1.7; 95% CI: 1.4–2.0) and GL (OR = 1.8; 95% CI: 1.5–2.2) (Franceschi et al. 2001). Similarly, an increased risk in developing cancer in the breast (Augustin et al. 2001), upper aero-digestive tract (Augustin et al. 2003b), prostate (Augustin et al. 2004a), ovaries (Augustin et al. 2003c), and endometrium (Augustin et al. 2003a; Silvera et al. 2005b) was associated with increases in dietary GI and GL, whereas gastric cancer was only positively associated with GL (Augustin et al. 2004b). GI and GL tend to be less strongly associated with cancer risk in prospective studies than in case-control studies (Higginbotham et al. 2004b). A few prospective studies found no association between GI and/or GL and cancer risk (Larsson et al. 2007b; Silvera et al. 2005a; Terry et al. 2003), while others found significant positive associations only in overweight/obese and physically inactive individuals, suggesting that those with insulin resistance may be more sensitive to dietary carbohydrate quality (Higginbotham et al. 2004a; Larsson et al. 2007a; Michaud et al. 2002, 2005; McCarl et al. 2006). To date, there is evidence supporting the concept of GI and GL in decreasing cancer risk; however, long-term intervention studies are required.

Conclusions

The concept of GI suggests that slowly digested or absorbed carbohydrates may help in the prevention and treatment of chronic diseases, such as obesity, cancer, CVD, and diabetes through amelioration of their associated metabolic disorders including insulin resistance and the metabolic syndrome. This dietary approach provides similar health benefits to those observed with pharmacological therapies that slow carbohydrate absorption, notably the α-glucosidase enzyme inhibitors, which are now routinely used in the management of diabetes (Chiasson et al. 2002; Holman et al. 1999). Therefore, despite some encouraging data, additional well-powered clinical studies focusing on palatable low GI diets are needed to determine efficacy and effectiveness before low GI diets can become part of general health recommendations and guidelines. In the meantime, in the absence of harm, it may still be prudent to advise more slowly absorbed traditional carbohydrate foods rather than those which are highly processed and rapidly absorbed.

References

Arnold LM, Ball MJ, Duncan AW, Mann J. 1993. Effect of isoenergetic intake of three or nine meals on plasma lipoproteins and glucose metabolism. Am J Clin Nutr 57:446–451.

Augustin LS, Dal Maso L, La Vecchia C, Parpinel M, Negri E, Vaccarella S, Kendall CW, Jenkins DJ, Francesch S. 2001. Dietary glycemic index and glycemic load, and breast cancer risk: A case-control study. Ann Oncol 12:1533–1538.

Augustin LS, Galeone C, Dal Maso L, Pelucchi C, Ramazzotti V, Jenkins DJ, Montella M, Talamini R, Negri E, Franceschi S, La Vecchia C. 2004a. Glycemic index, glycemic load and risk of prostate cancer. Int J Cancer 112:446–450.

Augustin LS, Gallus S, Bosetti C, Levi F, Negri E, Franceschi S, Dal Maso L, Jenkins DJ, Kendall CW, La Vecchia C. 2003a. Glycemic index and glycemic load in endometrial cancer. Int J Cancer 105:404–407.

Augustin LS, Gallus S, Franceschi S, Negri E, Jenkins DJ, Kendall CW, Dal Maso L, Talamini R, La Vecchia C. 2003b. Glycemic index and load and risk of upper aero-digestive tract neoplasms (Italy). Cancer Causes Control 14:657–662.

Augustin LS, Gallus S, Negri E, La Vecchia C. 2004b. Glycemic index, glycemic load and risk of gastric cancer. Ann Oncol 15:581–584.

Augustin LS, Polesel J, Bosetti C, Kendall CW, La Vecchia C, Parpinel M, Conti E, Montella M, Franceschi S, Jenkins DJ, Dal Maso L. 2003c. Dietary glycemic index, glycemic load and ovarian cancer risk: A case-control study in Italy. Ann Oncol 14:78–84.

Bertelsen J, Christiansen C, Thomsen C, Poulsen PL, Vestergaard S, Steinov A, Rasmussen LH, Rasmussen O, Hermansen K. 1993. Effect of meal frequency on blood glucose, insulin, and free fatty acids in NIDDM subjects. Diabetes Care 16:4–7.

Bouche C, Rizkalla SW, Luo J, Vidal H, Veronese A, Pacher N, Fouquet C, Lang V, Slama G. 2002. Five-week, low-glycemic index diet decreases total fat mass and improves plasma lipid profile in moderately overweight nondiabetic men. Diabetes Care 25:822–828.

Boyce VL, Swinburn BA. 1993. The traditional Pima Indian diet. Composition and adaptation for use in a dietary intervention study. Diabetes Care 16:369–371.

Boyd DB. 2003. Insulin and cancer. Integr Cancer Ther 2:315–329.

Brand-Miller J, Hayne S, Petocz P, Colagiuri S. 2003. Low-glycemic index diets in the management of diabetes: A meta-analysis of randomized controlled trials. Diabetes Care 26:2261–2267.

Brownlee M. 2001. Biochemistry and molecular cell biology of diabetic complications. Nature 414:813–820.

Bruning PF, Bonfrer JM, van Noord PA, Hart AA, de Jong-Bakker M, Nooijen WJ. 1992. Insulin resistance and breast-cancer risk. Int J Cancer 52:511–516.

Burkitt DP, Trowell HC. 1977. Dietary fibre and western diseases. Ir Med J 70:272–277.

Ceriello A, Motz E. 2004. Is oxidative stress the pathogenic mechanism underlying insulin resistance, diabetes, and cardiovascular disease? The common soil hypothesis revisited. Arterioscler Thromb Vasc Biol 24:816–823.

Chiasson JL, Josse RG, Gomis R, Hanefeld M, Karasik A, Laakso M. 2002. Acarbose for prevention of type 2 diabetes mellitus: the STOP-NIDDM randomised trial. Lancet 359:2072–2077.

Chiasson JL, Josse RG, Gomis R, Hanefeld M, Karasik A, Laakso M. 2003. Acarbose treatment and the risk of cardiovascular disease and hypertension in patients with impaired glucose tolerance: The STOP-NIDDM trial. JAMA 290:486–494.

Cohn C. 1964. Feeding patterns and some aspects of cholesterol metabolism. Fed Proc 23:76–81.

Ebbeling CB, Leidig MM, Feldman HA, Lovesky MM, Ludwig DS. 2007. Effects of a low-glycemic load vs low-fat diet in obese young adults: a randomized trial. JAMA 297:2092–2102.

Ebbeling CB, Leidig MM, Sinclair KB, Hangen JP, Ludwig DS. 2003. A reduced-glycemic load diet in the treatment of adolescent obesity. Arch Pediatr Adolesc Med 157:773–779.

Ebbeling CB, Leidig MM, Sinclair KB, Seger-Shippee LG, Feldman HA, Ludwig DS. 2005. Effects of an ad libitum low-glycemic load diet on cardiovascular disease risk factors in obese young adults. Am J Clin Nutr 81:976–982.

Ellinger F, Landsman H. 1944. Frequency and course of cancer in diabetics. NY State J Med 44:259–265.

Englyst HN, Cummings JH. 1986. Digestion of the carbohydrates of banana (Musa paradisiaca sapientum) in the human small intestine. Am J Clin Nutr 44:42–50.

Englyst HN, Kingman SM, Cummings JH. 1992. Classification and measurement of nutritionally important starch fractions. Eur J Clin Nutr 46(Suppl 2):S33–S50.

Englyst KN, Englyst HN, Hudson GJ, Cole TJ, Cummings JH. 1999. Rapidly available glucose in foods: An in vitro measurement that reflects the glycemic response. Am J Clin Nutr 69:448–454.

Expert Panel on Detection, Evaluation, and Treatment of High Blood Cholesterol in Adults. Executive Summary of The Third Report of The National Cholesterol Education Program (NCEP) Expert Panel on Detection, Evaluation, And Treatment of High Blood Cholesterol In Adults (Adult Treatment Panel III). 2001. JAMA 285:2486–2497.

Ford ES, Liu S. 2001. Glycemic index and serum high-density lipoprotein cholesterol concentration among us adults. Arch Intern Med 161:572–576.

Foster-Powell K, Holt SH, Brand-Miller JC. 2002. International table of glycemic index and glycemic load values: 2002. Am J Clin Nutr 76:5–56.

Foster-Powell K, Miller JB. 1995. International tables of glycemic index. Am J Clin Nutr 62:871S–890S.

Franceschi S, Dal Maso L, Augustin L, Negri E, Parpinel M, Boyle P, Jenkins DJ, La Vecchia C. 2001. Dietary glycemic load and colorectal cancer risk. Ann Oncol 12:173–178.

Freund E. 1885. Zur Diagnose des Carcinoms. Wien Med. Blat 8:268–269.

Frost G, Leeds AA, Dore CJ, Madeiros S, Brading S, Dornhorst A. 1999. Glycaemic index as a determinant of serum HDL-cholesterol concentration. Lancet 353:1045–1048.

Frost GS, Brynes AE, Bovill-Taylor C, Dornhorst A. 2004. A prospective randomised trial to determine the efficacy of a low glycaemic index diet given in addition to healthy eating and weight loss advice in patients with coronary heart disease. Eur J Clin Nutr 58:121–127.

Giovannucci E. 1995. Insulin and colon cancer. Cancer Causes Control 6:164–179.

Grundy SM, Brewer HB, Jr, Cleeman JI, Smith SC, Jr, Lenfant C. 2004. Definition of metabolic syndrome: Report of the National Heart, Lung, and Blood Institute/American Heart Association conference on scientific issues related to definition. Circulation 109:433–438.

Haber GB, Heaton KW, Murphy D, Burroughs LF. 1977. Depletion and disruption of dietary fibre. Effects on satiety, plasma-glucose, and serum-insulin. Lancet 2:679–682.

Higginbotham S, Zhang ZF, Lee IM, Cook NR, Buring JE, Liu S. 2004a. Dietary glycemic load and breast cancer risk in the Women's Health Study. Cancer Epidemiol Biomarkers Prev 13:65–70.

Higginbotham S, Zhang ZF, Lee IM, Cook NR, Giovannucci E, Buring JE, Liu S. 2004b. Dietary glycemic load and risk of colorectal cancer in the Women's Health Study. J Natl Cancer Inst 96:229–233.

Hodge AM, English DR, O'Dea K, Giles GG. 2004. Glycemic index and dietary fiber and the risk of type 2 diabetes. Diabetes Care 27:2701–2706.

Holman RR, Cull CA, Turner RC. 1999. A randomized double-blind trial of acarbose in type 2 diabetes shows improved glycemic control over 3 years (U.K. Prospective Diabetes Study 44). Diabetes Care 22:960–964.

Isaksson G, Lundquist I, Ihse I. 1982. Effect of dietary fiber on pancreatic enzyme activity in vitro. Gastroenterology 82:918–924.

Jarvi AE, Karlstrom BE, Granfeldt YE, Bjorck IE, Asp NG, Vessby BO. 1999. Improved glycemic control and lipid profile and normalized fibrinolytic activity on a low-glycemic index diet in type 2 diabetic patients. Diabetes Care 22:10–18.

Jee SH, Ohrr H, Sull JW, Yun JE, Ji M, Samet JM. 2005. Fasting serum glucose level and cancer risk in Korean men and women. JAMA 293:194–202.

Jenkins DJ, Jenkins AL. 1995. Nutrition principles and diabetes. A role for "lente carbohydrate"? Diabetes Care 18:1491–1498.

Jenkins DJ, Kendall CW, Augustin LS, Franceschi S, Hamidi M, Marchie A, Jenkins AL, Axelsen M. 2002. Glycemic index: Overview of implications in health and disease. Am J Clin Nutr 76:266S–273S.

Jenkins DJ, Khan A, Jenkins AL, Illingworth R, Pappu AS, Wolever TM, Vuksan V, Buckley G, Rao AV, Cunnane SC, Brighenti F, Hawkins M, Abdolell M, Corey P, Patten R, Josse RG. 1995. Effect of nibbling versus gorging on cardiovascular risk factors: Serum uric acid and blood lipids. Metabolism 44:549–555.

Jenkins DJ, Ocana A, Jenkins AL, Wolever TM, Vuksan V, Katzman L, Hollands M, Greenberg G, Corey P, Patten R, Wong G, Josse RG. 1992. Metabolic advantages of spreading the nutrient load: Effects of increased meal frequency in non-insulin-dependent diabetes. Am J Clin Nutr 55:461–467.

Jenkins DJ, Thorne MJ, Camelon K, Jenkins A, Rao AV, Taylor RH, Thompson LU, Kalmusky J, Reichert R, Francis T. 1982a. Effect of processing on digestibility and the blood glucose response: A study of lentils. Am J Clin Nutr 36:1093–1101.

Jenkins DJ, Wolever TM, Jenkins AL, Giordano C, Giudici S, Thompson LU, Kalmusky J, Josse RG, Wong GS. 1986. Low glycemic response to traditionally processed wheat and rye products: Bulgur and pumpernickel bread. Am J Clin Nutr 43:516–520.

Jenkins DJ, Wolever TM, Jenkins AL, Josse RG, Wong GS. 1984. The glycaemic response to carbohydrate foods. Lancet 2:388–391.

Jenkins DJ, Wolever TM, Jenkins AL, Thorne MJ, Lee R, Kalmusky J, Reichert R, Wong GS. 1983. The glycaemic index of foods tested in diabetic patients: A new basis for carbohydrate exchange favouring the use of legumes. Diabetologia 24:257–264.

Jenkins DJ, Wolever TM, Ocana AM, Vuksan V, Cunnane SC, Jenkins M, Wong GS, Singer W, Bloom SR, Blendis LM, Josse RG. 1990. Metabolic effects of reducing rate of glucose ingestion by single bolus versus continuous sipping. Diabetes 39:775–781.

Jenkins DJ, Wolever TM, Taylor RH, Barker H, Fielden H, Baldwin JM, Bowling AC, Newman HC, Jenkins AL, Goff DV. 1981. Glycemic index of foods: A physiological basis for carbohydrate exchange. Am J Clin Nutr 34:362–366.

Jenkins DJ, Wolever TM, Taylor RH, Barker HM, Fielden H. 1980. Exceptionally low blood glucose response to dried beans: Comparison with other carbohydrate foods. Br Med J 281:578–580.

Jenkins DJ, Wolever TM, Taylor RH, Griffiths C, Krzeminska K, Lawrie JA, Bennett CM, Goff DV, Sarson DL, Bloom SR. 1982b. Slow release dietary carbohydrate improves second meal tolerance. Am J Clin Nutr 35:1339–1346.

Jenkins DJ, Wolever TM, Vuksan V, Brighenti F, Cunnane SC, Rao AV, Jenkins AL, Buckley G, Patten R, Singer W, Corey P, Josse RG. 1989. Nibbling versus gorging: Metabolic advantages of increased meal frequency. N Engl J Med 321:929–934.

Jimenez-Cruz A, Bacardi-Gascon M, Turnbull WH, Rosales-Garay P, Severino-Lugo I. 2003. A flexible, low-glycemic index mexican-style diet in overweight and obese subjects with type 2 diabetes improves metabolic parameters during a 6-week treatment period. Diabetes Care 26:1967–1970.

Jones PJ, Leitch CA, Pederson RA. 1993. Meal-frequency effects on plasma hormone concentrations and cholesterol synthesis in humans. Am J Clin Nutr 57:868–874.

Kelly S, Frost G, Whittaker V, Summerbell C. 2004. Low glycaemic index diets for coronary heart disease. Cochrane Database Syst Rev, CD004467.

Krezowski PA, Nuttall FQ, Gannon MC, Bartosh NH. 1986. The effect of protein ingestion on the metabolic response to oral glucose in normal individuals. Am J Clin Nutr 44:847–856.

Larsson SC, Friberg E, Wolk A. 2007a. Carbohydrate intake, glycemic index and glycemic load in relation to risk of endometrial cancer: A prospective study of Swedish women. Int J Cancer 120:1103–1107.

Larsson SC, Giovannucci E, Wolk A. 2007b. Dietary carbohydrate, glycemic index, and glycemic load in relation to risk of colorectal cancer in women. Am J Epidemiol 165:256–261.

Larsson SC, Orsini N, Wolk A. 2005. Diabetes mellitus and risk of colorectal cancer: A meta-analysis. J Natl Cancer Inst 97:1679–1687.

Liu S, Manson JE, Buring JE, Stampfer MJ, Willett WC, Ridker PM. 2002. Relation between a diet with a high glycemic load and plasma concentrations of high-sensitivity C-reactive protein in middle-aged women. Am J Clin Nutr 75:492–498.

Liu S, Manson JE, Stampfer MJ, Holmes MD, Hu FB, Hankinson SE, Willett WC. 2001. Dietary glycemic load assessed by food-frequency questionnaire in relation to plasma high-density-lipoprotein cholesterol and fasting plasma triacylglycerols in postmenopausal women. Am J Clin Nutr 73:560–566.

Liu S, Willett WC, Stampfer MJ, Hu FB, Franz M, Sampson L, Hennekens CH, Manson JE. 2000. A prospective study of dietary glycemic load, carbohydrate intake, and risk of coronary heart disease in US women. Am J Clin Nutr 71:1455–1461.

Ludwig DS, Majzoub JA, Al-Zahrani A, Dallal GE, Blanco I, Roberts SB. 1999. High glycemic index foods, overeating, and obesity. Pediatrics 103:E26.

Ma Y, Olendzki B, Chiriboga D, Hebert JR, Li Y, Li W, Campbell M, Gendreau K, Ockene IS. 2005. Association between dietary carbohydrates and body weight. Am J Epidemiol 161:359–367.

Marble A. Diabetes and cancer. New Engl J Med 211:339–349.

McCarl M, Harnack L, Limburg PJ, Anderson KE, Folsom AR. 2006. Incidence of colorectal cancer in relation to glycemic index and load in a cohort of women. Cancer Epidemiol Biomarkers Prev 15:892–896.

McKeown-Eyssen G. 1994. Epidemiology of colorectal cancer revisited: Are serum triglycerides and/or plasma glucose associated with risk? Cancer Epidemiol Biomarkers Prev 3:687–695.

Meyer KA, Kushi LH, Jacobs DR, Jr, Slavin J, Sellers TA, Folsom AR. 2000. Carbohydrates, dietary fiber, and incident type 2 diabetes in older women. Am J Clin Nutr 71:921–930.

Michaud DS, Fuchs CS, Liu S, Willett WC, Colditz GA, Giovannucci E. 2005. Dietary glycemic load, carbohydrate, sugar, and colorectal cancer risk in men and women. Cancer Epidemiol Biomarkers Prev 14:138–147.

Michaud DS, Liu S, Giovannucci E, Willett WC, Colditz GA, Fuchs CS. 2002. Dietary sugar, glycemic load, and pancreatic cancer risk in a prospective study. J Natl Cancer Inst 94:1293–1300.

Michels KB, Solomon CG, Hu FB, Rosner BA, Hankinson SE, Colditz GA, Manson JE. 2003. Type 2 diabetes and subsequent incidence of breast cancer in the Nurses' Health Study. Diabetes Care 26:1752–1758.

Mozaffarian D, Rimm EB, Herrington DM. 2004. Dietary fats, carbohydrate, and progression of coronary atherosclerosis in postmenopausal women. Am J Clin Nutr 80:1175–1184.

O'Dea K. 1991. Westernisation, insulin resistance and diabetes in Australian aborigines. Med J Aust 155:258–264.

O'Dea K, Nestel PJ, Antonoff L. 1980. Physical factors influencing postprandial glucose and insulin responses to starch. Am J Clin Nutr 33:760–765.

Opperman AM, Venter CS, Oosthuizen W, Thompson RL, Vorster HH. 2004. Meta-analysis of the health effects of using the glycaemic index in meal-planning. Br J Nutr 92:367–381.

Pereira MA, Swain J, Goldfine AB, Rifai N, Ludwig DS. 2004. Effects of a low-glycemic load diet on resting energy expenditure and heart disease risk factors during weight loss. JAMA 292:2482–2490.

Rader DJ. 2007. Effect of insulin resistance, dyslipidemia, and intra-abdominal adiposity on the development of cardiovascular disease and diabetes mellitus. Am J Med 120:S12–S18.

Reaven GM. 2003. The insulin resistance syndrome. Curr Atheroscler Rep 5:364–371.

Rizkalla SW, Taghrid L, Laromiguiere M, Huet D, Boillot J, Rigoir A, Elgrably F, Slama G. 2004. Improved plasma glucose control, whole-body glucose utilization, and lipid profile on a low-glycemic index diet in type 2 diabetic men: A randomized controlled trial. Diabetes Care 27:1866–1872.

Salmeron J, Ascherio A, Rimm EB, Colditz GA, Spiegelman D, Jenkins DJ, Stampfer MJ, Wing AL, Willett WC. 1997a. Dietary fiber, glycemic load, and risk of NIDDM in men. Diabetes Care 20:545–550.

Salmeron J, Manson JE, Stampfer MJ, Colditz GA, Wing AL, Willett WC. 1997b. Dietary fiber, glycemic load, and risk of non-insulin-dependent diabetes mellitus in women. JAMA 277:472–477.

Schulze MB, Liu S, Rimm EB, Manson JE, Willett WC, Hu FB. 2004. Glycemic index, glycemic load, and dietary fiber intake and incidence of type 2 diabetes in younger and middle-aged women. Am J Clin Nutr 80:348–356.

Sheard NF, Clark NG, Brand-Miller JC, Franz MJ, Pi-Sunyer FX, Mayer-Davis E, Kulkarni K, Geil P. 2004. Dietary carbohydrate (amount and type) in the prevention and management of diabetes: A statement by the american diabetes association. Diabetes Care 27:2266–2271.

Silvera SA, Rohan TE, Jain M, Terry PD, Howe GR, Miller AB. 2005a. Glycemic index, glycemic load, and pancreatic cancer risk (Canada). Cancer Causes Control 16:431–436.

Silvera SA, Rohan TE, Jain M, Terry PD, Howe GR, Miller AB. 2005b. Glycaemic index, glycaemic load and risk of endometrial cancer: A prospective cohort study. Public Health Nutr 8:912–919.

Slabber M, Barnard HC, Kuyl JM, Dannhauser A, Schall R. 1994. Effects of a low-insulin-response, energy-restricted diet on weight loss and plasma insulin concentrations in hyperinsulinemic obese females. Am J Clin Nutr 60:48–53.

Sloth B, Astrup A. 2006. Low glycemic index diets and body weight. Int J Obes (Lond) 30(Suppl 3):S47–S51.

Sloth B, Krog-Mikkelsen I, Flint A, Tetens I, Bjorck I, Vinoy S, Elmstahl H, Astrup A, Lang V, Raben A. 2004. No difference in body weight decrease between a low-glycemic-index and a high-glycemic-index diet but reduced LDL cholesterol after 10-wk ad libitum intake of the low-glycemic-index diet. Am J Clin Nutr 80:337–347.

Slyper A, Jurva J, Pleuss J, Hoffmann R, Gutterman D. 2005. Influence of glycemic load on HDL cholesterol in youth. Am J Clin Nutr 81:376–379.

Stattin P, Bjor O, Ferrari P, Lukanova A, Lenner P, Lindahl B, Hallmans G, Kaaks R. 2007. Prospective study of hyperglycemia and cancer risk. Diabetes Care 30:561–567.

Stevens J, Ahn K, Juhaeri Houston D, Steffan L, Couper D. 2002. Dietary fiber intake and glycemic index and incidence of diabetes in African-American and white adults: The ARIC study. Diabetes Care 25:1715–1721.

Tavani A, Bosetti C, Negri E, Augustin LS, Jenkins DJ, La Vecchia C. 2003. Carbohydrates, dietary glycaemic load and glycaemic index, and risk of acute myocardial infarction. Heart 89:722–726.

Terry PD, Jain M, Miller AB, Howe GR, Rohan TE. 2003. Glycemic load, carbohydrate intake, and risk of colorectal cancer in women: A prospective cohort study. J Natl Cancer Inst 95:914–916.

Thorburn AW, Brand JC, Truswell AS. 1987. Slowly digested and absorbed carbohydrate in traditional bushfoods: A protective factor against diabetes? Am J Clin Nutr 45:98–106.

Thorne MJ, Thompson LU, Jenkins DJ. 1983. Factors affecting starch digestibility and the glycemic response with special reference to legumes. Am J Clin Nutr 38:481–488.

Toeller M, Buyken AE, Heitkamp G, Cathelineau G, Ferriss B, Michel G. 2001. Nutrient intakes as predictors of body weight in European people with type 1 diabetes. Int J Obes Relat Metab Disord 25:1815–1822.

UKPDS. 1998. Intensive blood-glucose control with sulphonylureas or insulin compared with conventional treatment and risk of complications in patients with type 2 diabetes (UKPDS 33). UK Prospective Diabetes Study (UKPDS) Group. Lancet 352:837–853.

van Dam RM, Visscher AW, Feskens EJ, Verhoef P, Kromhout D. 2000. Dietary glycemic index in relation to metabolic risk factors and incidence of coronary heart disease: The Zutphen Elderly Study. Eur J Clin Nutr 54:726–731.

Wolever TM. 2006. The Glycaemic Index: A Physiological Classification of Dietary Carbohydrate Oxfordshire, UK: CABI Publishing.

Wolever TM, Bentum-Williams A, Jenkins DJ. 1995. Physiological modulation of plasma free fatty acid concentrations by diet. Metabolic implications in nondiabetic subjects. Diabetes Care 18:962–970.

Wursch P, Del Vedovo S, Koellreutter B. 1986. Cell structure and starch nature as key determinants of the digestion rate of starch in legume. Am J Clin Nutr 43:25–29.

Yoon JH, Thompson LU, Jenkins DJ. 1983. The effect of phytic acid on in vitro rate of starch digestibility and blood glucose response. Am J Clin Nutr 38:835–842.

Chapter 5

Glycemia: Health Implications

L Raymond Reynolds, MD, FACP, FACE

Introduction

The rapid increase in prevalence of diabetes mellitus in the past two decades is considered a worldwide pandemic and is expected to continue with the global burden of diabetes predicted to more than double by the year 2025 to approximately 300 million people (King et al. 1998). The prevalence of diabetes in the United States in 2002 was 9.3% among adults ≥20 years of age according to the National Health and Nutrition Examination Survey (NHANES), but an additional 26% had impaired fasting glucose (IFG), now considered "prediabetes" (Cowie et al. 2006). The American Diabetes Association recently lowered the threshold for diagnosis of IFG from 110 to 100 mg/dL to enable earlier intervention and to delay progression to diabetes (Genuth et al. 2003). This has increased the estimated number of Americans with IFG or prediabetes to 41 million and now the total U.S. population with diabetes or IFG is estimated at 35.3% or 73.3 million (Centers for Disease Control 2005). In 2005, 1.5 million new cases of diabetes were diagnosed in adults, with obesity accompanying >80% of cases. Diabetes was the sixth leading cause of death on death certificates in 2002, but studies have estimated that in only 10–15% of people dying with diabetes is it listed as the primary cause of death (Centers for Disease Control 2005).

Cardiovascular disease (CVD) accounts for 60–70% of the deaths in people with diabetes including a two–fourfold increased risk of myocardial infarction and stroke. Diabetes is the leading cause of new cases of blindness, kidney failure, and nontraumatic lower limb amputations in the United States The risk of such complications is clearly linked to the degree and duration of hyperglycemia in a continuum, likely beginning before people reach the diabetes threshold for hyperglycemia. However, the risks for CVD are not explained entirely by hyperglycemia or other accompanying traditional risk factors such as obesity, hypertension, and dyslipidemia. Insulin resistance or resistance to insulin-mediated glucose uptake, resulting in hyperinsulinemia and eventual hyperglycemia, is proposed as the core defect leading to a clustering of metabolic risk factors resulting in CVD. Considerable

65

research is now focusing on the inflammatory process underlying atherogenesis and nontraditional cardiovascular risk factors including biomarkers of inflammation, endothelial dysfunction, and impaired fibrinolysis. This discussion will now examine the evidence linking insulin resistance, metabolic syndrome, and diabetes to CVD and other chronic complications associated with hyperglycemia.

Prediabetes and the Metabolic Syndrome: Definitions and Implications

In 1988, Gerald Reaven first introduced the term "syndrome X" in his Banting lecture discussing the role of insulin resistance in disease. (Reaven 1988). However, Kylin was the first to describe the clustering of hypertension, hyperglycemia, and gout as a syndrome in 1923 (Kylin 1923). Vague in 1947 noted the common association of abdominal obesity with type 2 diabetes and CVD (Vague 1947). In addition to the term syndrome X, other descriptive terms have emerged including metabolic syndrome, dysmetabolic syndrome, the deadly quartet, and the "insulin resistance syndrome," as recommended by the American College of Endocrinology in 2003. Although the term "prediabetes" was formally introduced into modern usage in 2002 at a Health and Human Services press conference, investigators recognized for years the existence of an extended period of minimal hyperglycemia co-existing with hypertension, dyslipidemia, and obesity. The acknowledgment of the association of mild hyperglycemia with accelerated atherosclerosis led in 1979 to the establishment of a classification system developed by the National Diabetes Data Group that included the category of impaired glucose tolerance. This category identified individuals with a greater risk of developing macrovascular disease who did not have the usual risks for diabetic microvascular complications. Prediabetic individuals are at an increased risk for progression to type 2 diabetes, with a conversion of 10–15% of such individuals over a 3 year period, as demonstrated in the Diabetes Prevention Program (Rubin et al. 2002). The more recently published DREAM Trial, comparing rosiglitazone to placebo in prediabetic individuals showed that 25% of those who were placebo-treated converted to overt diabetes in 3 years (Gerstein et al. 2006). The observation that the prediabetic state is associated with an increased risk of CVD years before the onset of diabetes led to the "ticking clock" hypothesis (Haffner et al. 1990). Unlike microvascular complications that define the glycemic threshold for the diabetic state, in the prediabetic state the "clock starts ticking" on cardiovascular complications 10–15 years before the onset of diabetes and is magnified closer to diagnosis (see Figure 5.1).

The growing recognition and acceptance of the existence of the insulin resistance syndrome characterized by a clustering of cardiovascular risk factors including hypertension, hyperglycemia, dyslipidemia, and obesity finally led to the development of the first internationally accepted criteria by the World Health Organization in 1998. Subsequently, the European Group for the Study of Insulin Resistance (EGSIR) and an expert panel for the National Cholesterol Education Program (NCEP) also developed separate but similar definitions. (See Table 5.1.) These

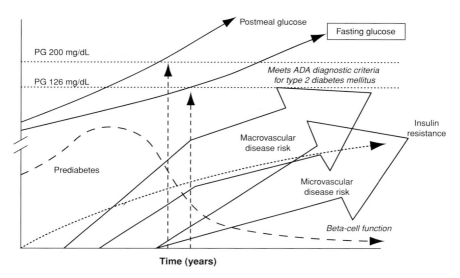

Postmeal glucose

Fasting glucose

PG 200 mg/dL

*Meets ADA diagnostic criteria
for type 2 diabetes mellitus*

PG 126 mg/dL

Insulin
resistance

Macrovascular
disease risk

Prediabetes

Microvascular
disease risk

Beta-cell function

Time (years)

Figure 5.1. Natural history of prediabetes and type 2 diabetes. (Adapted from R. Bergenstall, International Diabetes Center, Minneapolis, MN.)

definitions are in general agreement regarding the key components of glucose intolerance, hypertension, dyslipidemia, and obesity, but details of the criteria are somewhat different. The NCEP criteria are simpler and therefore better suited for clinical practice, whereas the WHO criteria include an assessment of insulin resistance generally more suitable for research settings. The National Heart, Lung, and Blood Institute (NHLIB) and the American Heart Association further revised the NCEP criteria (Grundy et al. 2004). More recently the International Diabetes Federation (IDF) proposed an updated set of criteria for the metabolic syndrome that are nearly identical to the NHLIB–AHA criteria, but include central obesity as a required element with lower thresholds for waist circumference for those ethnic groups particularly prone to insulin resistance such as Asians (Alberti et al. 2005). The IDF definition therefore recognizes the essential contribution of central obesity to the metabolic syndrome, underscoring the importance of waist circumference measurements in clinical practice.

Since Reaven's original description of syndrome X, thousands of studies have appeared in the literature detailing the relationship of obesity, lipid disorders, hypertension, and CVD. In 2005 a joint statement from the ADA and EASD appeared casting doubt on the existence of a metabolic syndrome due to imprecise definitions, uncertain pathogenesis, and insufficient scientific data (Kahn et al. 2005). Subsequently a "call to action" was published by the ADA and AHA indicating that the debate about the metabolic syndrome existence was merely a scientific discourse, not a disagreement about the importance of identifying and treating a core set risk factors for diabetes (Eckel et al. 2006). The clustering of the five major risk factors occurs more commonly than can be explained by chance, and

Table 5.1.   Metabolic syndrome definitions.

WHO (1998)	EGSIR (1999)	NCEP (2001)	NHLIB–AHA (2005)
		Three or more of below:	Three or more of below:
1. Diabetes, IFG, IGT, or insulin resistance (hyperinsuinemic) plus 2 or more of below:	1. Hyperinsuline-mic plus two or more of below:	1. Obesity: waist >102 cm (male), >88 cm (female)	1. Obesity: waist >102 cm (male), >88 cm (female)[a]
2. Obesity: BMI >30 or waist/hip >0.9 male, >0.85 female	2. Obesity: waist >94 cm (male) >80 cm (female)	2. Triglycerides ≥1.7 mmol/L	2. Triglycerides ≥1.7 mmol/L or drug treatment
3. Dyslipidemia: triglycerides >1.7 mmol/L or HDL chol <0.9 male, <1 female mmol/L	3. Dyslipidemia: triglycerides >2 mmol/L or HDL chol <1 mmol/L	3. HDL chol <1 mmol/L (males) <1.3 mmol/L (females)	3. HDLchol <1 mmol/L(males) <1.3 mmol/L (females) or drug treatment
4. Hypertension: >140/90	4. Hypertension: ≥140/90 or medication	4. Hypertension: ≥135/85 or medication	4. Hypertension: ≥130/85 or medication
5. Microalbuminuria; albumin excretion >20 μg/min	5. Fasting plasma glucose ≥6.1 mmol/L	5. Fasting plasma glucose ≥6.1 mmol/L	5. Fasting plasma glucose ≥5.6 mmol/L

[a]Lower waist circumference (≥90 cm in men and ≥80 cm in women is appropriate for Asians.

there may indeed be more than a single underlying pathogenetic mechanism. For purposes of this discussion we will continue to use the term metabolic syndrome due to its universal acceptance as a predictor of diabetes and cardiovascular risk (Table 5.2).

Association of Prediabetic Hyperglycemia and Cardiovascular Disease

Since 1979 multiple observational studies have been published documenting the relationship between blood glucose levels in the nondiabetic range and all-cause mortality and CVD. During this time the standards for defining impaired glucose tolerance have remained unchanged at 140–199 mg/dL (7.8–11.0 mmol/L) after a

Table 5.2. Summary of risks of insulin resistance and prediabetes.

All-cause mortality
Cardiovascular disease
Stroke
Cancer
Cognitive decline
Nonalcoholic fatty liver disease
Hypertension
Chronic kidney disease
Obstructive sleep apnea
Polycystic ovarian syndrome

glucose challenge. The definition of IFG, however, has been more recently revised from 110–126 mg/dL (6.1–6.9 mmol/L) to 100–125 mg/dL (5.6–6.9 mmol/L) by the ADA. The data from these population studies reported over 25 years are somewhat difficult to compare due to differences in methodology and definitions, but a common emerging theme is the clear association of mortality and CVD with increasing levels of glucose, even in prediabetic individuals.

In the Whitehall study of approximately 18,000 men aged 40–64 years, blood glucose was measured 2 h after a 50 g glucose load (Fuller et al. 1983). The 10-year mortality rates from coronary heart disease and stroke showed a nonlinear relation to blood glucose values, with a significantly increased risk for subjects with concentrations above the 95th percentile (5.3 mmol/L) and for diabetics. The excess CVD mortality was approximately 40% above age-adjusted control levels.

In the Paris Prospective Study, over 7,000 middle-aged policemen were followed for 11 years with measurements of glucose and insulin levels while fasting and again after an oral glucose challenge and annual coronary heart disease (CHD) events were noted (Eschwege et al. 1985). CHD mortality rates were highest in the upper 20% of the distribution of both glucose and insulin concentrations. Fasting plasma insulin levels were positively associated with risk independent of other common risk factors, such as smoking, hypertension, weight, and lipids. A similar study followed 1,000 Helsinki policemen over 10 years after baseline blood glucose and insulin levels were obtained 1 h after an oral glucose challenge (Pyorala et al. 1979). CVD mortality rates were significantly higher for those in the upper 20% of both glucose and insulin levels, both fasting and postglucose challenge.

The Chicago Heart Study used 9-year follow-up data for 11,000 men and 8,000 women aged 35–64 years at entry. This epidemiologic study explored the sex differential in risk of death from CHD in persons with or without clinically diagnosed diabetes or asymptomatic hyperglycemia (Pan et al. 1986). Blood glucose levels measured 1 h after an oral glucose load with a level >11.1 mmol/L were designated as hyperglycemia. CHD mortality rates were nearly doubled for hyperglycemic men for both clinically diagnosed diabetes and asymptomatic hyperglycemia. The extent of this association was greater in women than in men in regard to relative risk (RR). However, absolute excess risk for both diabetics and those with asymptomatic hyperglycemia was larger for men than for women. This study indicated an independent association of diabetes and possibly asymptomatic hyperglycemia with CHD mortality, with greater relative significance in women than in men. A subsequent 22-year follow-up analysis of white men confirmed an association of hyperglycemia with CHD in a linear fashion with increasing hyperglycemia conferring an increased risk (Lowe et al. 1997).

The Honolulu Heart Program investigated the association of glucose levels after a 50 g challenge on the development of CVD in approximately 7,000 nondiabetic men aged 45–70 at study entry, followed for 12 years (Donahue et al. 1987). The rate of fatal CHD increased linearly with glucose response. Men in the fourth quintile of postchallenge glucose (157–189 mg/dL) had twice the age-adjusted risk of fatal CHD of those in the lowest quintile. RR increased to threefold among those in the top quintile and remained statistically significant after adjustment for other risk factors including body mass, total cholesterol, hypertension, and left ventricular hypertrophy.

In the Rancho Bernardo study, 1,800 older women and men (aged 50–89 years) without a history of CVD or diabetes and with normal fasting glucose levels were evaluated with glucose tolerance testing at baseline and followed for 7 years. Women in the study with postchallenge hyperglycemia (2 h postglucose challenge >11.1 mm/L) had a 2.6 times increased risk of CVD that was independent of other risk factors (Barrett-Connor and Ferrara 1998).

The Nurses' Health Study provided a unique opportunity to examine whether the risk of CVD is elevated before clinical diagnosis of type 2 diabetes in women (Hu et al. 2002). A total of 117,629 female nurses aged 30–55 years without CVD at baseline were followed for 20 years. During follow-up, 1,556 new cases of myocardial infarction (MI), 1,405 strokes, 815 fatal CHD events, and 300 fatal strokes occurred. Among women who developed type 2 diabetes during follow-up, the risks of MI were nearly fourfold for the period before the diagnosis compared with women who remained free of diabetes. The risk of stroke was also significantly elevated more than twofold before diagnosis of diabetes. Adjustment for history of hypertension or hypercholesterolemia did not appreciably alter the results. This study confirmed a substantially elevated risk of CVD before clinical diagnosis of type 2 diabetes in women.

The Hoorn study evaluated the predictive value of fasting plasma glucose, 2 h postload glucose and HbA1c in a cohort of approximately 2,400 older (50–75 years) subjects, without known diabetes over 8 years (de Vegt et al. 1999). Postload glucose and HbA1c values were, even within the nondiabetic range, associated with an increased risk of CVD deaths in a linear fashion. After additional adjustment for known cardiovascular risk factors, the relative risks for all-cause and cardiovascular mortality were still statistically significant, with values of 2.20 and 3.00, respectively ($P < 0.05$).

A metaregression analysis of published data from 20 studies evaluated the relationship between nondiabetic glucose levels and cardiovascular risk (Coutinho et al. 1999). The study included an analysis of 95,783 people who had cardiovascular events over 12.4 years. Subjects with fasting and 2 h postprandial glucose levels exceeding 110 mg/dL and 140 mg/dL, respectively, had a 1.33 cardiovascular event risk compared to subjects with normal glucose levels of 75 mg/dL postprandial glucose values <140 mg/dL, further demonstrating the progressive relationship between glucose levels and cardiovascular risk.

The EPIC–Norfolk study evaluated the value of HbA1c, an indicator of average blood glucose levels over 3 months, as a predictor of death from CVD in men (Khaw et al. 2001). HbA1c levels were predictive in a continuous fashion of all-cause and cardiovascular mortality rates with the lowest rates in men with levels <5%. Among 3,200 nondiabetic men every 1% increase in HbA1c above 5% was associated with a 28% increased risk of death independent of other variables such as age, blood pressure, smoking, cholesterol levels, and body mass index.

The DECODE (Diabetes Epidemiology: Collaborative Analysis of Diagnostic Criteria in Europe) study group compared fasting and the 2 h glucose (2h-BG) criteria with regard to the prediction of mortality (Balkau et al. 1999). Data on glucose levels from 10 prospective European cohorts including approximately 15,000 men and 7,000 women with a median follow-up of 8.8 years were analyzed. Hazard ratios for death from all causes, CVD, and stroke were estimated. Hazard ratios (95% confidence intervals) in subjects with diabetes on 2h-BG were 1.73 (1.45–2.06) for all causes, 1.40 (1.02–1.92) for CVD, 1.56 (1.03–2.36) for coronary heart disease, and 1.29 (0.66–2.54) for stroke mortality, compared with the normal 2h-BG group. The largest number of excess deaths was observed in subjects who had impaired glucose tolerance but normal FBG levels, indicating that the postchallenge BG is a better predictor of deaths from all causes and CVD than is FBG (see Figure 5.2).

The Cardiovascular Health Study is a prospective study of CV risk factors in a population of approximately 4,000 adults aged 65 years and older (Smith et al. 2002). Included subjects were without treated diabetes or previous myocardial infarction or stroke. Baseline fasting and 2 h postchallenge glucose levels were analyzed in relationship to subsequent myocardial infarction, coronary death, or stroke over 8.5 years follow-up. Both an elevated fasting and 2 h postchallenge glucose were associated with an increased cardiovascular risk: hazard ratio 1.66 for fasting glucose ≥115 mg/dL; hazard ratio 1.29 or higher increased in linear fashion for postchallenge glucose ≥154 mg/dL or higher. The postchallenge glucose level was better than the fasting glucose at predicting CVD in older adults (Table 5.3).

Metabolic Syndrome and Cardiovascular Disease

The importance of the metabolic syndrome as a major healthcare problem relates not only to its pervasiveness in the United States, approximately one-third of

Table 5.3.   Summary of studies of glucose intolerance and CVD risk

Study	No. of subjects	Years of follow-up	Relative risks of CV disease	Reference
Nurses HS	5,894	20	3.17	Hu (2002)
Rancho Bernardo	1,858	7	2.6	Barrett-Connor and Ferrara (1998)
Honolulu	6,394	12	1.5	Donahue et al. (1987)
Paris	7,164	11.2	1.9	Eschwege et al. (1985)
CardioHS	4,014	8.5	1.66	Smith et al. (2002)
Chicago	11,554	22	1.2	Pan et al. (1986)
Helsinki	631	20	2.59	Pyorala et al. (1979)
Whitehall	10,025	20	1.23	Fuller et al. (1983)
Hoorn	2,400	8	3	de Vegt et al. (1999)

middle-aged adults, and now globally, but to the well-established risks of type 2 diabetes and CVD (Ford et al. 2004; Grundy et al. 2004). The metabolic syndrome can be precipitated by several risk factors; the most important are insulin resistance and central obesity. Insulin resistance, or decreased insulin-mediated glucose disposal, results in hyperinsulinemia, which has been associated with coronary heart disease in a number of studies (Despres et al. 1996; Eschwege et al. 1985; Folsom et al. 1997; Perry et al. 1996; Pyorala et al. 1979). Contributing to insulin resistance are increased circulating free fatty acids and decreased glucose transport due to

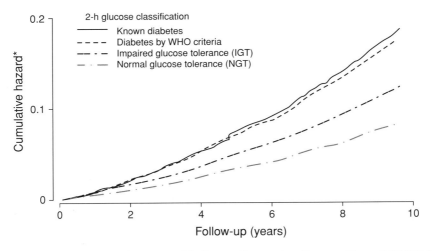

Figure 5.2.   Glycemia increases risk for cardiovascular disease. (From DECODE Study Group. Lancet 1999; 354(9179):617–621.)

changes in insulin-mediated signaling pathways. Central obesity is clearly a major pathogenetic factor, contributing to hepatic insulin resistance through an influx of free fatty acids and production of cytokines, such as tumor necrosis factor or TNF-α and interleukin-6 (Eckel et al. 2005).

The gold standard for measuring insulin resistance is the glucose clamp method as described by Defronzo (DeFronzo et al. 1979). This technique is laborious and not suitable for large population studies. Mathews proposed a simple method for estimating insulin sensitivity, the homeostasis model assessment of insulin sensitivity (HOMA), based on the mathematical modeling of fasting glucose and insulin concentrations (Matthews et al. 1985). The HOMA is defined as

$$\frac{\text{Fasting insulin } (\mu U/mL) \times \text{ fasting glucose } (mmol/L)}{22.5}.$$

To assess whether the HOMA method is a reliable surrogate measure of insulin sensitivity, Bonora et al. performed glucose clamp studies on 115 subjects and compared the results with the estimates of insulin sensitivity obtained by HOMA (Bonora et al. 2007). A strong correlation was found between glucose disposal as measured by the clamp technique and estimates of insulin sensitivity in men, women, obese, nonobese, diabetic, and nondiabetic subjects.

The DECODE registry of a Northern European population was also used to evaluate the association between the metabolic syndrome, CVD, and mortality (Hu et al. 2004). The prevalence of the metabolic syndrome was slightly higher in men (15.7%) than in women (14.2%). The hazard ratios for all-cause and cardiovascular mortality in persons with the metabolic syndrome were 1.44 and 2.26 in men and 1.38 and 2.78 in women after adjustment for age, blood cholesterol levels, and smoking, demonstrating that nondiabetic people with the metabolic syndrome have an increased risk of death from all causes as well as CVD. The Botnia study of nearly 5,000 subjects in Finland and Sweden demonstrated a threefold increased risk of coronary heart disease and stroke among those with the metabolic syndrome (Isomaa et al. 2001). In the Kupio study of 1,200 Finnish men, those with the metabolic syndrome were three times as likely to die of CVD (Lakka et al. 2002).

In a joint statement from the American Heart Association and National Heart, Lung, Blood Institute, it was noted the risk for atherosclerotic disease events is increased at least twofold and the risk for the development of diabetes increased fivefold with the metabolic syndrome (Grundy et al. 2004).

The Insulin Resistance Atherosclerosis Study (IRAS) was the first study that directly measured insulin sensitivity and the atherosclerosis in a large cohort (Howard et al. 1996). IRAS evaluated insulin sensitivity by the intravenous glucose tolerance test and intimal–medial thickness (IMT) of the carotid artery by ultra-sonography in 1,400 multiethnic subjects. Carotid IMT is considered a surrogate marker of atherosclerotic progression and predictive of future coronary events. There was a significant negative association between insulin sensitivity and the IMT of the carotid artery both in Hispanics and in non-Hispanic whites. This effect was reduced but not totally explained by adjustment for traditional CVD risk factors including dyslipidemia, hyperglycemia, and obesity. This suggests that the relationship of insulin sensitivity is partly mediated by an association with these traditional risk factors. Since the publication of the original IRAS report, two additional studies have confirmed an association of insulin resistance with carotid artery atherosclerosis (Watarai et al. 1999; Wohlin et al. 2003).

More recently the IRAS investigators also studied the relationship of insulin sensitivity, based on the intravenous glucose tolerance test to CVD incidence, including myocardial infarction, coronary bypass grafts, and coronary angioplasty (Rewers et al. 2004). Low insulin sensitivity was strongly associated with coronary heart disease, largely independent of traditional risk factors including dyslipidemia, hypertension, diabetes, smoking, and obesity. The odds ratio for coronary disease was greatest among those in the lower quartiles of insulin sensitivity and was stronger than the association with fasting or postprandial insulin levels, commonly used as a surrogate marker of insulin sensitivity.

The San Antonio Heart Study is a population-based study of CVD and diabetes among Mexican-Americans and Caucasians selected from low-, medium-, and high-income households. The investigators studied the relationship of the homeostasis model assessment of insulin resistance (HOMA-IR), insulin levels, and CVD incidence with an average 8 year follow-up (Hanley et al. 2002). A significant trend of increased risk of CVD was noted across quintiles of HOMA-IR. Subjects in the highest quintile had greater than a twofold increased risk of CVD compared to those in the lowest quintile. Adjustments for related variables such as lipids, smoking, blood pressure, waist circumference, and exercise did not alter the significance of these findings. The findings were similar for the measurements of insulin levels and CVD.

The Bruneck study of nearly 1,000 people living in northeastern Italy recently reported an analysis of the association of insulin resistance measured by the HOMA method and CVD (Bonora et al. 2007). The purpose of this study was to evaluate whether insulin resistance is associated with CVD and whether this association can be explained by traditional and novel CVD risk factors. Insulin-resistant subjects were considered those with a HOMA-IR score in the top quartile. Subjects were followed up for 15 years for cardiovascular incidents, including fatal and nonfatal MI, stroke, transient ischemic attack, and any revascularization procedure. Levels of HOMA-IR were higher at baseline among subjects who developed CVD (2.8) compared with those remaining free of CVD (2.5) ($P < 0.05$). Insulin-resistant subjects had 2.1-fold increased risk of incident symptomatic CVD. After adjustment for classic and novel risk factors such as hsCRP, oxidized LDL cholesterol, and adiponectin, the association between HOMA-IR and CVD remained significant and virtually unchanged (hazard ratio 2.2). HOMA-IR in this population was therefore associated with subsequent CVD independent of classic and novel risk factors.

Gami and colleagues recently reported a large meta-analysis evaluating the association between the metabolic syndrome and cardiovascular events and mortality that included 37 cohorts with 172,000 individuals (Gami et al. 2007). Studies were included that used either WHO or NCEP criteria for defining the metabolic syndrome, with some studies using modified criteria, substituting body mass index for waist circumference. Separate meta-analyses were completed for each outcome including cardiovascular events, CHD events, cardiovascular death, CHD death, and all-cause death. The risk for the different outcomes was quite similar, ranging from a RR of 1.6 for death to 2.18 for cardiovascular events. The pooled RR of cardiovascular events and death was 1.78. The association was stronger in women (RR 2.63 vs. 1.98, $P = 0.09$) than in men. The association remained after adjusting for traditional cardiovascular risk factors (RR 1.5), again indicating that the increased risk of CVD among people with the metabolic syndrome cannot be explained by the associated traditional risk factors. These multiple studies suggest that insulin resistance may be an important target to reduce CVD risk in the large numbers of adults worldwide with metabolic syndrome.

Metabolic Syndrome and Chronic Kidney Disease

Microalbuminuria is a well-known risk factor for the development of chronic kidney disease (CKD) in people with type 1 and type 2 diabetes, but is also recognized as a risk factor, predictive of CVD in the type 2 diabetic population (Deckert et al. 1989; de Zeeuw et al. 2006; Kuusisto et al. 1995). Several observational studies have now also linked metabolic syndrome to microalbuminuria and the risk of CKD. Investigators associated with the Insulin Resistance in Atherosclerosis Study (IRAS) studied the relationship of insulin resistance, estimated by the intravenous glucose tolerance test and fasting plasma insulin concentrations, to microalbuminuria (albumin-to-creatinine ratio > or = 2 mg/mmol) in nearly 1,000 nondiabetic subjects (Palaniappan et al. 2003). Subjects with microalbuminuria had a lower degree of insulin sensitivity and higher fasting insulin concentrations compared with subjects without microalbuminuria. An increasing degree of insulin sensitivity was associated with a decreasing prevalence of microalbuminuria. The presence (or absence) of hypertension did not affect the association between insulin sensitivity and microalbuminuria.

Chen et al. examined the association between the metabolic syndrome and risk for CKD and microalbuminuria using data from the third NHANES (Chen et al. 2004). CKD was defined as a glomerular filtration rate less than 60 mL/min per 1.73 m^2, and microalbuminuria was defined as a urinary albumin–creatinine ratio of 30–300 mg/g. The odds ratios of CKD and microalbuminuria in participants with the metabolic syndrome compared with participants without the metabolic syndrome were 2.60 and 1.89, respectively. The risk for CKD increased progressively in the presence of more components of the metabolic syndrome, including a graded relationship for the risk for CKD and the number of elements of metabolic syndrome present.

In the largest U.S. prospective study of adults to date, the Atherosclerosis Risk in Communities Study (ARIC) examined the relationship of CKD to the metabolic syndrome in a nondiabetic cohort of >10,000 subjects (Kurella et al. 2005). Incident CKD was defined as an estimated GFR <60 mL/min per 1.73 m^2 at the end of the study. The risk of developing CKD in participants with the metabolic syndrome was 1.43. Compared with participants without traits of the metabolic syndrome, those with an increasing number of criteria of the metabolic syndrome had a progressively greater risk of developing CKD. The risk of acquiring CKD was independent of hypertension or diabetes. Recently, studies of Asian cohorts have also demonstrated an increased risk of CKD in association with the metabolic syndrome (Kitiyakara et al. 2007; Ninomiya et al. 2006).

In addition to hypertension and diabetes as major risks for CKD, there is increasing evidence that obesity is an independent risk factor for CKD (Iseki et al. 2004; Kambham et al. 2001). In the NHANES cohort previously mentioned, abdominal obesity was associated with twofold increased risk for CKD (Chen et al. 2004). Possible mechanisms include proinflammatory cytokines associated with metabolic syndrome mediating renal damage, renal compression, small birth weight, and alterations in glucocorticoid and uric acid metabolism (Schelling and Sedor 2004). Therefore, of the major components of the metabolic syndrome, hyperglycemia, hypertension, and obesity, all represent modifiable risk factors for the development of CKD.

Metabolic Syndrome and Nonalcoholic Fatty Liver Disease

Nonalcoholic fatty liver disease (NAFLD) is a common disorder with a prevalence of 20–30% in the general population, but is found in nearly 75% of obese people and approximately 50% of those with the metabolic syndrome (Utzschneider and Kahn 2006). NAFLD is generally considered a benign condition characterized by fat accumulation in the liver, but the spectrum includes steatohepatitis (NASH), and in 5% of cases it can progress to cirrhosis and death from end stage liver disease (Sheth et al. 1997). Additionally, NAFLD subjects have higher all-cause mortality not only due to liver disease, but also related to the premature vascular disease of the metabolic syndrome. Insulin resistance is thought to be the underlying mechanism linking obesity, dyslipidemia, type 2 diabetes, and the metabolic syndrome. Insulin resistance increases the influx of free fatty acids into the liver and promotes triglyceride accumulation in the liver (Adams and Angulo 2006). Intra-abdominal fat may be a key determinant of hepatic fat, providing a source of free fatty acids and contributing to hepatic insulin resistance.

The diagnosis of NAFLD is usually considered in the setting of persistent elevation of alanine aminotransferase levels in the absence of other causes such as significant alcohol ingestion (<20 g/day), inflammatory bowel disease, malnutrition, and use of drugs such as glucocorticoids, amiodarone, tamoxifen, and antiviral agents (Angulo 2002). Imaging studies including ultrasound and computed tomographic scanning demonstrate distinctive images of fatty infiltration. Liver biopsy is necessary to diagnose steatohepatitis and provides precise information for staging and prognosis (American Gastroenterological Association 2002).

Treatment of NAFLD centers on weight loss to reduce insulin resistance and ameliorate the features of the metabolic syndrome. Most investigators have utilized a diet with both calorie and saturated fat restriction and encouraged exercise to improve insulin sensitivity (Dixon et al. 2004; Hickman et al. 2004). In diabetic individuals attempts to maximize glycemic control and achieve hemoglobin A1c levels <7% are also considered important in management (Utzschneider and Kahn 2006).

Metabolic Syndrome and Cognitive Decline

Cognitive decline, dementia, and type 2 diabetes are being recognized with increased frequency in the elderly population. The relationship of diabetes to dementia in a large cohort was investigated in the Rotterdam Study of over 6,000 subjects in 1996 (Ott et al. 1996). Of the dementia patients, 22.3% had diabetes with a positive association between diabetes and dementia (odds ratio: 1.3). Insulin-treated diabetic subjects had an even stronger association (odds ratio: 3.2), perhaps reflecting a longer duration of diabetes. The association was strongest for vascular dementia but was also present for Alzheimer's disease.

Kalmijn et al. studied the association of cognitive function with hyperinsulinemia, impaired glucose tolerance, and diabetes mellitus in a cohort of 462 men aged 69–89 years in the Netherlands (Kalmijn et al. 1995). Compared to subjects with normal glucose tolerance, diabetic subjects fared the worst on a minimental status exam with impaired glucose tolerance subjects having the next lowest number of accurate answers. Nondiabetic subjects with the highest serum insulin values also performed more poorly on the mental exams compared to those with lower insulin levels. Exclusion of other risk factors including smoking, hypertension, and CVD did not alter the results. The results did not change appreciably when potentially mediating factors, including CVD and risk factors associated with the insulin resistance syndrome, were taken into account. The results suggested that diabetes, prediabetes, and insulin resistance, demonstrated by hyperinsulinemia, are all associated with cognitive decline.

Further evidence regarding the association of insulin resistance and cognitive decline, as manifested by Alzheimer's disease, was demonstrated in a Finnish cohort of nearly 1,000 people (Kuusisto et al. 1997). A total of 4.7% of subjects were classified as having probable or possible Alzheimer's disease. Both hyperglycemia and hyperinsulinemia were associated with an increased risk of Alzheimer's disease: prevalence 7.5% versus 1.4% in those with normal insulin levels. Vanhanen evaluated cognitive function in relation to glucose and insulin levels at baseline and 3.5 years later in 980 subjects (Vanhanen et al. 1998). At follow-up, subjects with persistent impaired glucose tolerance consistently scored more poorly on cognitive function testing.

More recently, investigators studied fasting insulin levels as a surrogate marker of insulin resistance and the association with dementia in a sample of elderly subjects from the United States (Luchsinger et al. 2004). A total of 680 subjects were followed for 3,691 person years and the risk of Alzheimer's disease doubled in those with hyperinsulinemia. Hyperinsulinemia was also related to a significant decline in memory-related function. In a similar study of older Italians without diabetes, increased insulin resistance assessed by HOMA correlated with decreased frontal cortex function in those without dementia (Abbatecola et al. 2004). A study of older postmenopausal women demonstrated that women with prediabetes had poorer baseline cognitive function and a greater risk for cognitive decline than those with normal glucose tolerance (Yaffe et al. 2004a, b).

Investigators have also evaluated the possible association of the metabolic syndrome, using NCEP criteria, and inflammatory markers (CRP and IL-6) with cognitive decline among 2,600 people over 5 years (Yaffe et al. 2004a, b). Those with the metabolic syndrome were more likely to have cognitive impairment: RR 1.20 and those with increased levels of inflammatory markers showed the greatest risk, 1.66.

More recently middle-aged (aged 45–64 years at baseline) adults were studied with regard to the possible association of hyperinsulinemia with cognitive decline (Young et al. 2006). Tests of cognitive function were completed at baseline and 6 years later in addition to fasting insulin levels and HOMA-IR. Results: both hyperinsulinemia and HOMA-IR at baseline were associated with a significantly lower baseline cognitive function and greater declines at 6 years.

In conclusion, insulin resistance and the metabolic syndrome, in addition to diagnosed diabetes, are definite risk factors for cognitive decline and dementia and are potentially modifiable risk factors in midlife.

Chronic Diabetes Complications

The prevalence and impact of complications in people with type 2 diabetes were analyzed from two large national surveys and reported at the American Association of Clinical Endocrinologists (AACE) meeting in Seattle, WA in April 2007 (AACE 2007). Health economists used the National Health and Nutrition Survey (NHANES) of 5,000 Americans combined with data from the Medical Expenditure Panel Survey (MPES). The study demonstrated that 58% of people with type 2 diabetes have one or more of the complications associated with diabetes. The specific complications included heart disease, stroke, kidney disease, eye disease, and foot problems (Table 5.4).

The AACE report (2007) also demonstrated that the United States spends $22.9 billion annually on the complications of diabetes, and a total of $57 billion annually on diabetes care and complications. The annual healthcare costs for people with diabetes are threefold higher than people without diabetes.

Table 5.4. Prevalence of complications (AACE 2007).

	Macrovascular			Microvascular	
Complication	Prevalence in diabetes (%)	Prevalence in nondiabetes (%)	Complication	Prevalence in diabetes (%)	Prevalence in nondiabetes (%)
CHD	9.1	2.1	Kidney disease	27.8	6.1
CHF	7.9	1.1	Eye disease	18.9	N/A
Stroke	6.6	1.8	Foot problems	22.9	10

As indicated earlier in this chapter, diabetes is associated with a marked increase in the risk of CHD, accounting for 60–80% of the deaths in people with diabetes. The clinical paradigm of managing risk factors for CVD in diabetic patients as aggressively as previous myocardial infarction patients is based on multiple studies comparing the incidence of myocardial infarction among nondiabetic subjects with the incidence among diabetic subjects. An analysis of a Finnish cohort demonstrated incidence rates of myocardial infarction in nondiabetic subjects with and without prior myocardial infarction at baseline of 18.8 and 3.5%, respectively, compared to the rates of myocardial infarction in diabetic subjects with and without prior myocardial infarction at baseline of 45.0 and 20.2%, respectively ($P < 0.001$) (Haffner et al. 1998), see Figure 5.3. The hazard ratio for death from coronary heart disease for diabetic subjects without prior myocardial infarction as compared with nondiabetic subjects with prior myocardial infarction was not significantly

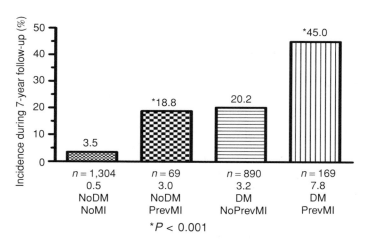

Figure 5.3. Incidence of coronary heart disease events in subjects with and without diabetes. (From Haffner SM et al. New Eng J Med 1998; 339:229)

different from 1.0 (hazard ratio, 1.4; 95% confidence interval, 0.7–2.6) after adjustment for age and sex, suggesting similar risks of infarction in the two groups. After further adjustment for total cholesterol, hypertension, and smoking, this hazard ratio remained close to 1.0 (hazard ratio, 1.2; 95% confidence interval, 0.6–2.4). These data provide the basis for the "coronary equivalence" paradigm that diabetic patients without previous myocardial infarction have the same risk of myocardial infarction as nondiabetic patients with previous myocardial infarction, underpinning the rationale for treating cardiovascular risk factors in diabetic patients as aggressively as in nondiabetic patients with prior myocardial infarction.

A common observation in medicine is that diabetic patients who experience an acute myocardial infarction (AMI) have poorer outcomes than nondiabetics. An analysis of the Munich myocardial infarction registry data evaluated the outcomes of diabetic patients following an AMI (Schnell et al. 2004). In 1999, coronary angiography, percutaneous transluminal coronary angioplasty (PTCA), and stenting were performed less frequently in diabetic than in nondiabetic patients. During this period, total hospital mortality (29 vs. 16%, $P < 0.01$) and mortality within 24 h after admission (14 vs. 5%, $P = 0.01$) were higher in diabetic than in nondiabetic patients. In 2001, frequencies of coronary angiography, PTCA, and stenting were increased in diabetic patients, and the interventions were comparable with those performed in nondiabetic patients with the addition of insulin infusions in 46% of diabetic subjects. In 2001, total hospital mortality decreased to 17% in diabetic subjects ($P = 0.028$ vs. 1999) and mortality within 24 h after admission declined to 4% ($P = 0.027$ vs. 1999). This analysis therefore revealed that intensification of therapeutic approaches is associated with a reduction in mortality of diabetic patients with AMI and that outcomes in diabetic patients with AMI can be markedly improved with advanced therapies.

In addition to the well-recognized microvascular complications such as retinopathy, nephropathy, and peripheral neuropathy, 20–40% of diabetic individuals experience some form of autonomic neuropathy (Vinik et al. 2003). The gastrointestinal tract can be affected from the esophagus through the colon, ranging from clinically silent to disabling motility disorders. The most clinically important autonomic neuropathy is cardiac autonomic neuropathy, which may cause orthostatic hypotension, tachycardia, silent ischemia, and sudden death (Maser and Lenhard 2005). Less commonly seen are focal and multifocal neuropathies such as cranial neuropathies and the amotrophy of the lower extremities seen in older patients with type 2 diabetes. The 2005 ADA statement on diabetic neuropathies provides an excellent overview of this subject (Boulton et al. 2005).

Diabetes and Cancer Risk

Type 2 diabetes is also associated with an increased risk of malignancy, particularly cancers of the pancreas, liver, colon, endometrium, breast, and melanoma (Czyzyk and Szczepanik 2000; Saydah et al. 2003; Stattin et al. 2007; Strickler et al. 2001).

Adjustment for obesity in some of these studies did not alter the association. The risk in the prospective studies of the United States, European, and Asian cohorts is greatest in type 2 diabetes, but is also increased in those with milder degrees of glucose intolerance. It is suggested that hyperinsulinemia and activation of insulin-like growth factor receptors result in increased mitogenesis and the development of tumors (Strickler et al. 2001).

Reducing Complications Through Improved Glycemic Control

The completion of two landmark glycemic intervention trials, the Diabetes Control and Complications Trial (DCCT) and the United Kingdom Prospective Diabetes Study (UKPDS) firmly established that improved glycemic control reduces the risk of diabetic complications in both type 1 and type 2 diabetes (DCCT Group 1993; UKPDS Group 1998). Both studies showed an approximate 35% reduction in risk of microvascular complications for each 1% reduction in HbA1c. In the UKPDS intensively managed group, myocardial infarction risk was reduced by 14%, which was of borderline statistical significance. However, in a long-term 17-year follow-up of DCCT participants, intensive glycemic control reduced the risk of any cardiovascular event by 42% ($P = 0.02$), correlated with microvascular risk reduction, indicating that strict glycemic control in the initial decade of hyperglycemia is critically important to reducing the probability of microvascular and macrovascular complications (DCCT/EDIC Group 2005).

Conclusions

In summary, the entire continuum of insulin resistance, metabolic syndrome, and all degrees of hyperglycemia is associated with wide-ranging deleterious effects beyond the well-known increased CVD including cognitive decline, liver, and renal disease. Efforts need to be focused at the earliest stage possible in this continuum to reduce the risks of premature disability and death and to reduce the global burden on healthcare systems.

References

Abbatecola AM, Paolisso G, Lamponi M, Bandinelli S, Lauretani F, Launer L, Ferrucci L. 2004. Insulin resistance and executive dysfunction in older persons. J Am Geriatr Soc 52:1713–1718.

Adams LA, Angulo P. 2006. Treatment of non-alcoholic fatty liver disease. Postgrad Med J 82:315–322.

Alberti KG, Zimmet P, Shaw J. 2005. The metabolic syndrome—a new worldwide definition. Lancet 366:1059–1062.

American Association of Clinical Endocrinologists Annual Meeting. 2007. State of Diabetes Complications in America. Seattle, WA.

American Gastroenterological Association. 2002. AGA Technical review on nonalcoholic fatty liver disease. Gastroenterology 123:1705–1725.

Angulo, P. 2002. Nonalcolholic fatty liver disease. N Engl J Med 346:1221–1231.

Barrett-Connor E, Ferrara A. 1998. Isolated postchallenge hyperglycemia and the risk of fatal cardiovascular disease in older women and men. The Rancho Bernardo Study. Diabetes Care 21:1236–1239.

Balkau B, Jarrett RJ, Pyorala K, Eschwege E. 1999. Fasting blood glucose and risk of cardiovascular disease. Diabetes Care 22:1385–1387.

Bonora E, Kiechl S, Willeit J, Oberhollenzer F, Egger G, Meigs JB, Bonadonna RC, Muggeo M. 2007. Insulin resistance as estimated by homeostasis model assessment predicts incident symptomatic cardiovascular disease in caucasian subjects from the general population: The bruneck study. Diabetes Care 30:318–324.

Boulton AJ, Vinik AI, Arezzo JC. 2005. Diabetic neuropathies: A statement by the American Diabetes Association. Diabetes Care 28:956–962.

Centers for Disease Control. 2005. National Diabetes Fact Sheet. Accessed at www.cdc.org.

Chen J, Muntner P, Hamm LL. 2004. The metabolic syndrome and chronic kidney disease in U.S. adults. Ann Intern Med 140:167–174.

Coutinho M, Gerstein HC, Wang Y, Yusuf S. 1999. The relationship between glucose and incident cardiovascular events. A metaregression analysis of published data from 20 studies of 95,783 individuals followed for 12.4 years. Diabetes Care 22:233–240.

Cowie CC, Rust KF, Byrd-Holt DD, Eberhardt MS, Flegal KM, Engelgau MM, Saydah SH, Williams DE, Geiss LS, Gregg EW. 2006. Prevalence of diabetes and impaired fasting glucose in adults in the U.S. population: National Health and Nutrition Examination Survey 1999–2002. Diabetes Care 29:1263–1268.

Czyzyk A, Szczepanik Z. 2000. Diabetes mellitus and cancer. Eur J Intern Med 11:245–252.

DCCT Group. 1993. The effect of intensive treatment of diabetes on the development and progression of long-term complications in insulin-dependant diabetes mellitus. N Engl J Med 329:977–986.

DCCT Group. 2005. Intensive diabetes treatment and cardiovascular disease in patients with type 1 diabetes. N Engl J Med 353:2643–2653.

Deckert T, Feldt-Rasmussen B, Borch-Johnsen K, Jensen T, Kofoed-Enevoldsen A. 1989.Albuminuria reflects widespread vascular damage. The Steno hypothesis. Diabetologia 32:219–226.

DeFronzo RA, Tobin JD, Andres R. 1979. Glucose clamp technique: a method for quantifying insulin secretion and resistance. Am J Physiol 237:E214-E223.

Despres JP, Lamarche B, Mauriege P, Cantin B, Dagenais GR, Moorjani S, Lupien PJ. 1996. Hyperinsulinemia as an independent risk factor for ischemic heart disease. N Engl J Med 334:952–957.

de Vegt F, Dekker JM, Ruhe HG, Stehouwer CD, Nijpels G, Bouter LM, Heine RJ. 1999. Hyperglycaemia is associated with all-cause and cardiovascular mortality in the Hoorn population: The Hoorn Study. Diabetologia 42:926–931.

de Zeeuw D, Parving HH, Henning RH. 2006. Microalbuminuria as an early marker for cardiovascular disease. J Am Soc Nephrol 17:2100–2105.

Dixon JB, Bhathal PS, Hughes NR, O'Brien PE. 2004. Nonalcoholic fatty liver disease: Improvement in liver histological analysis with weight loss. Hepatology 39:1647–1654.

Donahue RP, Abbott RD, Reed DM, Yano K.1987. Postchallenge glucose concentration and coronary heart disease in men of Japanese ancestry. Honolulu Heart Program. Diabetes 36:689–692.

Eckel RH, Grundy SM, Zimmet PZ. 2005. The metabolic syndrome. Lancet 365:1415–1428.

Eckel RH, Kahn R, Robertson RM, Rizza RA. 2006. Preventing cardiovascular disease and diabetes: A call to action from the American Diabetes Association and the American Heart Association. Diabetes Care 29:1697–1699.

Eschwege E, Richard JL, Thibult N, Ducimetiere P, Warnet JM, Claude JR, Rosselin GE. 1985. Coronary heart disease mortality in relation with diabetes, blood glucose and plasma insulin levels. The Paris Prospective Study, ten years later. Horm Metab Res Suppl 15:41–46.

Folsom AR, Szklo M, Stevens J, Liao F, Smith R, Eckfeldt JH. 1997. A prospective study of coronary heart disease in relation to fasting insulin, glucose, and diabetes. The Atherosclerosis Risk in Communities (ARIC) Study. Diabetes Care 20:935–942.

Ford ES, Giles WM, Mokdad AH. 2004. Increasing prevalence of the metabolic syndrome among U.S. adults. Diabetes Care 27:2444–2449.

Fuller JH, Shipley MJ, Rose G, Jarrett RJ, Keen H. 1983. Mortality from coronary heart disease and stroke in relation to degree of glycaemia: The Whitehall study. Br Med J (Clin Res Ed) 287:867–870.

Gami AS, Witt BJ, Howard DE, Erwin PJ, Gami LA, Somers VK, Montori VM. 2007. Metabolic syndrome and risk of incident cardiovascular events and death: A systematic review and meta-analysis of longitudinal studies. J Am Coll Cardiol 49:403–414.

Genuth S, Alberti KG, Bennett P, Buse J, Defronzo R, Kahn R, Kitzmiller J. 2003. Follow-up report on the diagnosis of diabetes mellitus. Diabetes Care 26:3160–3167.

Gerstein HC, Yusuf S, Bosch J, Pogue J, Sheridan P, Dinccag N, Hanefeld M, Hoogwerf B, Laakso M, Mohan V, Shaw J, Zinman B, Holman RR. 2006. Effect of rosiglitazone on the frequency of diabetes in patients with impaired glucose tolerance or impaired fasting glucose: A randomised controlled trial. Lancet 368:1096–1105.

Grundy SM, Brewer HB, Jr, Cleeman JI, Smith SC, Jr, Lenfant C. 2004. Definition of metabolic syndrome: Report of the National Heart, Lung, and Blood Institute/American Heart Association conference on scientific issues related to definition. Circulation 109:433–438.

Haffner SM, Stern MP, Hazuda HP, Mitchell BD, Patterson JK. 1990. Cardiovascular risk factors in confirmed prediabetic individuals. Does the clock for coronary heart disease start ticking before the onset of clinical diabetes? JAMA 263:2893–2898.

Haffner SM, Lehto S, Ronnemaa T, Pyorala K, Laakso M. 1998. Mortality from coronary heart disease in subjects with type 2 diabetes and in nondiabetic subjects with and without prior myocardial infarction. N Engl J Med 339:229–234.

Hanley AJ, Williams K, Stern MP, Haffner SM. 2002. Homeostasis model assessment of insulin resistance in relation to the incidence of cardiovascular disease: The San Antonio Heart Study. Diabetes Care 25:1177–1184.

Hickman IJ, Jonsson JR, Prins JB. 2004. Modest weight loss and physical activity in overweight patients with chronic liver disease results in sustained improvements in alanine aminotransferase, fasting insulin, and quality of life. Gut 53:413–419.

Howard G, O'Leary DH, Zaccaro D, Haffner S, Rewers M, Hamman R, Selby JV, Saad MF, Savage P, Bergman R, for The Insulin Resistance Atherosclerosis Study (IRAS) Investigators. 1996. Insulin sensitivity and atherosclerosis. Circulation 93:1809–1817.

Hu FB, Stampfer MJ, Haffner SM, Solomon CG, Willett WC, Manson JE. 2002. Elevated risk of cardiovascular disease prior to clinical diagnosis of type 2 diabetes. Diabetes Care 25:1129–1134.

Hu G, Qiao Q, Tuomilehto J, Balkau B, Borch-Johnsen K, Pyorala K. 2004. Prevalence of the metabolic syndrome and its relation to all-cause and cardiovascular mortality in nondiabetic European men and women. Arch Intern Med 164:1066–1076.

Iseki K, Ikemiya Y, Kinjo K, Inoue T, Iseki C, Takishita S. 2004. Body mass index and the risk of development of end-stage renal disease in a screened cohort. Kidney Int 65:1870–1876.

Isomaa B, Almgren P, Tuomi T, Forsen B, Lahti K, Nissen M, Taskinen MR, Groop L. 2001. Cardiovascular morbidity and mortality associated with the metabolic syndrome. Diabetes Care 24:683–689.

Kahn R, Buse J, Ferrannini E, Stern M. 2005. The metabolic syndrome: Time for a critical appraisal: Joint statement from the American Diabetes Association and the European Association for the Study of Diabetes. Diabetes Care 28:2289–2304.

Kalmijn S, Feskens EJ, Launer LJ, Stijnen T, Kromhout D. 1995. Glucose intolerance, hyperinsulinaemia and cognitive function in a general population of elderly men. Diabetologia 38:1096–1102.

Kambham N, Markowitz GS, Valeri AM, Lin J, D'Agati VD. 2001. Obesity-related glomerulopathy: An emerging epidemic. Kidney Int 59:1498–1509.

Khaw KT, Wareham N, Luben R, Bingham S, Oakes S, Welch A, Day N. 2001. Glycated haemoglobin, diabetes, and mortality in men in Norfolk cohort of european prospective investigation of cancer and nutrition (EPIC–Norfolk). BMJ 322:15–18.

King H, Aubert RE, Herman WH. 1998. Global burden of diabetes, 1995–2025: Prevalence, numerical estimates, and projections. Diabetes Care 21:1414–1431.

Kitiyakara C, Yamwong S, Cheepudomwit S, Domrongkitchaiporn S, Unkurapinun N, Pakpeankitvatana V, Sritara P. 2007. The metabolic syndrome and chronic kidney disease in a Southeast Asian cohort. Kidney Int 71:693–700.

Kurella M, Lo JC, Chertow GM. 2005. Metabolic syndrome and the risk for chronic kidney disease among nondiabetic adults. J Am Soc Nephrol 16:2134–2140.

Kuusisto J, Mykkanen L, Pyorala K, Laakso M. 1995. Hyperinsulinemic microalbuminuria. A new risk indicator for coronary heart disease. Circulation 91:831–837.

Kuusisto J, Koivisto K, Mykkanen L, Helkala EL, Vanhanen M, Hanninen T. 1997. Association between features of the insulin resistance syndrome and Alzheimer's disease independently of apolipoprotein E4 phenotype: Cross sectional population based study. BMJ 315:1045–1049.

Kylin E. 1923. Studien ueber das Hypertonie-Hyperglkamie-Hyperuikamiesyndrom. Zentralbl fur Inn Med 44:105–127.

Lakka HM, Laaksonen DE, Lakka TA, Niskanen LK, Kumpusalo E, Tuomilehto J, Salonen JT. 2002. The metabolic syndrome and total and cardiovascular disease mortality in middle-aged men. JAMA 288:2709–2716.

Lowe LP, Liu K, Greenland P, Metzger BE, Dyer AR, Stamler J. 1997. Diabetes, asymptomatic hyperglycemia, and 22-year mortality in black and white men. The Chicago Heart Association Detection Project in Industry Study. Diabetes Care 20:163–169.

Luchsinger JA, Tang MX, Shea S, Mayeux R. 2004. Hyperinsulinemia and risk of Alzheimer's disease. Neurology 63:1187–1192.

Maser RE, Lenhard MJ. 2005. Cardiovascular autonomic neuropathy due to diabetes mellitus: Clinical manifestations, consequences, and treatment. J Clin Endocrinol Metab 90:5896–5903.

Matthews DR, Hosker JP, Rudenski AS, Naylor BA, Treacher DF, Turner RC. 1985. Homeostasis model assessment: Insulin resistance and beta-cell function from fasting plasma glucose and insulin concentrations in man. Diabetologia 28:412–419.

Palaniappan L, Carnethon M, Fortmann SP. 2003. Association between microalbuminuria and the metabolic syndrome: NHANES III. Am J Hypertens 16:952–958.

Ninomiya T, Kiyohara Y, Kubo M, Yonemoto K, Tanizaki Y, Doi Y, Hirakata H, Iida M. 2006. Metabolic syndrome and CKD in a general Japanese population: The Hisayama Study. Am J Kidney Dis 48:383–391.

Ott A, Stolk RP, Hofman A, van Harskamp F, Grobbee DE, Breteler MM. 1996. Association of diabetes mellitus and dementia: The Rotterdam Study. Diabetologia 39:1392–1397.

Pan WH, Cedres LB, Liu K, Dyer A, Schoenberger JA, Shekelle, RB, Stamler R, Smith D, Collette P, Stamler J. 1986. Relationship of clinical diabetes and asymptomatic hyperglycemia to risk of coronary heart disease mortality in men and women. Am J Epidemiol 123:504–516.

Perry IJ, Wannamethee SG, Whincup PH, Shaper AG, Walker MK, Alberti KG. 1996. Serum insulin and incident coronary heart disease in middle-aged British men. Am J Epidemiol 144:224–234.

Pyorala K, Savolainen E, Lehtovirta E, Punsar S, Siltanen P. 1979. Glucose tolerance and coronary heart disease: Helsinki policemen study. J Chronic Dis 32:729–745.

Reaven GM. 1988. Banting lecture 1988. Role of insulin resistance in human disease. Diabetes 37:1595–1607.

Rewers M, Zaccaro D, D'Agostino R, Haffner S, Saad MF, Selby JV, Bergman R, Savage P. 2004. Insulin sensitivity, insulinemia, and coronary artery disease: The Insulin Resistance Atherosclerosis Study. Diabetes Care 27:781–787.

Rubin RR, Fujimoto WY, Marrero DG, Brenneman T, Charleston JB, Edelstein SL, Fisher EB, Jordan R, Knowler WC, Lichterman LC, Prince M, Rowe PM. 2002. The Diabetes Prevention Program: Recruitment methods and results. Control Clin Trials 23:157–171.

Saydah SH, Loria CM, Eberhardt MS, Brancati FL. 2003. Abnormal glucose tolerance and the risk of cancer death in the United States. Am J Epidemiol 157:1092–1100.

Schelling JR, Sedor JR. 2004. The metabolic syndrome as a risk factor for chronic kidney disease: More than a fat chance? J Am Soc Nephrol 15:2773–2774.

Schnell O, Schafer O, Kleybrink S, Doering W, Standl E, Otter W. 2004. Intensification of therapeutic approaches reduces mortality in diabetic patients with acute myocardial infarction: The Munich registry. Diabetes Care 27:455–460.

Sheth SG, Gordon FD, Chopra S. 1997. Nonalcoholic steatohepatitis. Ann Intern Med 126(2):137–145.

Smith NL, Barzilay JI, Shaffer D, Savage PJ, Heckbert SR, Kuller LH, Kronmal RA, Resnick HE, Psaty BM. 2002. Fasting and 2-hour postchallenge serum glucose measures and risk of incident cardiovascular events in the elderly: The Cardiovascular Health Study. Arch Intern Med 162:209–216.

Stattin P, Bjor O, Ferrari P, Lukanova A, Lenner P, Lindahl B, Hallmans G, Kaaks R. 2007. Prospective study of hyperglycemia and cancer risk. Diabetes Care 30:561–567.

Strickler HD, Wylie-Rosett J, Rohan T, Hoover DR, Smoller S, Burk RD, Yu H. 2001. The relation of type 2 diabetes and cancer. Diabetes Technol Ther 3:263–274.

UKPDS Group. 1998. Intensive blood-glucose control with sulphonylureas or insulin compared with conventional treatment and risk of complications in patients with type 2 diabetes (UKPDS 33). Lancet 352:837–853.

Utzschneider KM, Kahn SE. 2006. Review: The role of insulin resistance in nonalcoholic fatty liver disease. J Clin Endocrinol Metab 91:4753–4761.

Vague J. 1947. La differentiation sexuelle, facteur determinant des formes de l'obesite. Presse Med 30:339–340.

Vanhanen M, Koivisto K, Kuusisto J, Mykkanen L, Helkala EL, Hanninen T, Riekkinen P, Soininen H, Laakso M. 1998. Cognitive function in an elderly population with persistent impaired glucose tolerance. Diabetes Care 21:398–402.

Vinik AI, Maser RE, Mitchell BD, Freeman R. 2003. Diabetic autonomic neuropathy. Diabetes Care 26:1553–1579.

Watarai T, Yamasaki Y, Ikeda M, Kubota M, Kodama M, Tsujino T, Kishimoto M, Kawamori R, Hori M. 1999. Insulin resistance contributes to carotid arterial wall thickness in patients with non-insulin-dependent-diabetes mellitus. Endocr J 46:629–638.

Wohlin M, Sundstrom J, Arnlov J, Andren B, Zethelius B, Lind L. 2003. Impaired insulin sensitivity is an independent predictor of common carotid intima-media thickness in a population sample of elderly men. Atherosclerosis 170:181–185.

Yaffe K, Blackwell T, Kanaya AM, Davidowitz N, Barrett-Connor E, Krueger K. 2004a. Diabetes, impaired fasting glucose, and development of cognitive impairment in older women. Neurology 63:658–663.

Yaffe K, Kanaya A, Lindquist K, Simonsick EM, Harris T, Shorr RI, Tylavsky FA, Newman, AB. 2004b. The metabolic syndrome, inflammation, and risk of cognitive decline. JAMA 292:2237–2242.

Young SE, Mainous AG, III, Carnemolla M. 2006. Hyperinsulinemia and cognitive decline in a middle-aged cohort. Diabetes Care 29:2688–2693.

Chapter 6

Glycemic Health, Type 2 Diabetes, and Functional Foods

Kaye Foster-Powell, MNutrDiet, Alan Barclay, BSc, and Jennie Brand-Miller, PhD

Introduction

The goal of the clinical management of type 2 diabetes is optimal control of the common metabolic abnormalities without negatively impacting on quality of life. Ideally, blood glucose, blood lipids, and blood pressure should fall within the normal range to prevent or delay acute and long-term complications. Traditionally, a three-pronged approach has been applied: healthy eating, regular physical activity, and appropriate medication. Among these, none has been as controversial as diet.

Because diabetes is primarily a disorder of carbohydrate metabolism, dietary advice has always focused attention on the carbohydrate consumed. In principle, the carbohydrates that result in smaller glycemic excursions after consumption should be easier to metabolize by people with impairments in insulin action and/or insulin secretion. To identify the most desirable choices, David Jenkins and colleagues devised a system of comparing the blood glucose response to the same amount of carbohydrate in different foods—the glycemic index or GI (Jenkins et al. 1981). A quarter of a century later, our knowledge of the GI values of foods has grown from the original 50 foods tested to about 2,000 today. Over the same period of time, the evidence base supporting the benefits of low GI foods in the prevention and management of type 2 diabetes has accumulated. Around the world, the practical use of the GI by people with diabetes has been thwarted by lack of knowledge of the GI of local foods. Australian dietitians, however, with the benefit of a supportive food industry, have been at the forefront of change, incorporating the GI into day-to-day diabetes management over the past decade.

Definition

The GI is a standardized measurement of the glycemic potential of the carbohydrate in different foods. It ranks foods, gram for gram of carbohydrate, on the basis of their relative postprandial glucose response to a reference food (pure glucose or white bread). The carbohydrates in low GI foods and meals therefore lead to smaller fluctuations in blood glucose levels, compared to the same amount of carbohydrate in the high GI equivalent. An online database of GI values of common carbohydrate foods can be found at www.glycemicindex.com. Glycemic load (GL), the mathematical product of the GI of a food and its carbohydrate content (GL = GI% × total carbohydrate in grams), has been proposed as a global indicator of the glucose response and insulin demand induced by a serving of food (Salmeron et al. 1997). Although some scientists have suggested that GL provides a more complete picture of the total effect of a carbohydrate-containing food on blood glucose levels, it may be easily misconstrued to promote the consumption of fewer carbohydrates. This is because a low GL diet can be achieved by either reducing the GI of the carbohydrate foods or by reducing the amount of carbohydrate consumed. These two strategies have different effects beyond glycaemia that are not always consistent with optimal blood lipid levels or long-term health. To date, more evidence favors a moderately high carbohydrate intake (45–55% of energy) based on low GI sources, in preference to lower carbohydrate intake (<45% energy) (Anderson et al. 2004).

Contrary to popular belief, low GI foods are not the same as high fiber, whole-grain cereals, nor are high GI foods necessarily the ones high in sugars. In fact, the foods that produce the highest glycemic responses include many of the more common starchy foods consumed in Western countries including potato, bread, pancakes, and processed breakfast cereals, whether high or low in fiber. The reason relates to the fact that the starch in these foods is fully gelatinized and can be rapidly digested and absorbed (Jenkins et al. 2002).

A slower rate of starch digestion contributes to the lower GI of pasta, oats, and legumes, but it is the nature of the sugars in fruits, juices, and dairy products that accounts for their lower GI. Up to half of the weight of carbohydrate in these foods is fructose or galactose, which has little effect on glycemia. In fact, the overall GI of the diet has been shown to have an inverse correlation with total sugars (refined plus naturally occurring) expressed as a proportion of total carbohydrate (Wolever et al. 1994).

Evidence of Clinical Benefits

Evidence for the benefit of low GI diets can be found in three main areas: clinical interventions, epidemiology, and basic science.

Clinical Interventions

A dozen or more controlled intervention studies have compared diets containing conventional carbohydrate foods (with moderate to high GI) with low GI foods in the management of type 2 diabetes. (Brand-Miller et al. 2003) In most, there were favorable improvements in glucose and/or lipid metabolism with the low GI intervention and, importantly, none showed any disadvantage. There are important limitations of these studies: most had small numbers of subjects, they were of short duration (<12 months), and in some cases there were doubts surrounding dietary adherence. Longer and better studies are still needed. A meta-analysis showed that, on average, the low GI interventions decreased glycated proteins (HbA1c and/or fructosamines), markers of average blood glucose levels, by 0.43% points, compared to high GI (conventional) interventions. The benefits seen were in addition to those obtained by following standard dietary recommendations for the management of diabetes. Although the effect of a low GI diet on long-term markers of glycemic control may be considered modest, it is comparable or better than the effect of many pharmacological therapies. Any improvement in HbA1c, no matter how small, is important because there is no glycemic threshold for risk reduction: the lower the HbA1c, the less the risk of diabetic complications (kidney, eye, heart, and blood vessel disease, etc. . .) (Stratton et al. 2001).

There is additional evidence that conventional (high GI) diets may directly increase insulin resistance through their effect on blood glucose levels, free fatty acids, and counter-regulatory hormone secretion (e.g., glucagon). Conventional high carbohydrate diets may also increase fasting triglyceride levels and reduce HDL cholesterol levels (Garg and Jampol 2005). In human studies, it is difficult to control all dietary factors simultaneously and differences in fiber and energy intake may have contributed to the findings. However, these factors can be carefully matched in animal studies, so that the GI of the carbohydrate is the only difference. Pawlak and colleagues (2004) showed that, compared with diets based on low GI starch, the high GI starch had adverse effects on body composition, postprandial glycemia, and triglyceride concentration in two animal models.

Epidemiological Evidence

To date, seven studies have investigated the association between high GI or GL diets and the risk of developing type 2 diabetes. In a large cohort of 65,173 U.S. women aged 40–65 years, Salmeron et al. (1997) found the risk of developing diabetes was 37% higher in those consuming the diet with the highest GI versus lowest GI. Similarly, in 42,759 U.S. men aged 40–75 years, the relative risk in the highest versus lowest quintile of GI was 1.37 after adjustment for known confounders, including fiber intake (7). Schulze et al. (Schulze et al. 2004) studied a cohort of 91,249 U.S. women, aged 24–44, finding that a high GI diet increased the risk of developing type 2 diabetes by 59%. Hodge et al. (2004) studied a cohort of 36,787 Australian men and women, aged 40–69. The study found that the high GI diet increased the risk of developing type 2 diabetes by 32%. Zhang et al. (2006) studied a cohort of 13,110 U.S. women, aged 24–44. They found that the high GI diet increased the risk of developing gestational diabetes by 68%. Meyer et al. (2000) studied a cohort of 35,988 U.S. women, aged 55–69. This study did not find any positive associations between GI and risk of type 2 diabetes. Instead, it found that a diet high in dietary fiber (>23.6 g/day) decreased the risk of type 2 diabetes by 22%. Similarly, Stevens et al. (2002) studied a cohort of 9,529 U.S. men, aged 45–64. This study also did not find any positive associations between GI and risk of type 2 diabetes. Instead, it found that a diet high in total dietary fiber (26.1–27.5 g/day), particularly cereal fiber, (4–5.1 g/day) decreased the risk of type 2 diabetes by 14–25%. It is important to note that the tool used to assess the average GI of individuals diets in the studies of Meyer et al. (2000) and Stevens et al. (2002) did not appear to be valid for the measurement of total carbohydrate, and it was not validated for GI; therefore the null result found in these two studies is of questionable significance. Overall, the evidence supports the hypothesis that high GI diets increase the risk of developing type 2 diabetes, independent of other known risk factors. There is also evidence from a number of population studies that the average dietary GI is inversely correlated with HDL cholesterol levels.

The Practicalities of Using Low GI Foods in Diabetes

The glycemic potential of the diet can be lowered by changing the type of carbohydrate consumed, while keeping the habitual total carbohydrate intake the same. This has important implications for sustainability, because most studies have shown that altering the amount of total carbohydrate (e.g., lowering it) is difficult to sustain for more than on average 6 months. Altering the amount of carbohydrate by substituting more protein and/or fat for carbohydrate will also have differing metabolic consequences that may be unfavorable in the long term. As such, a diet with a low GI can still be high in carbohydrate. However, a diet with a lower GL can be achieved by reducing or avoiding sources of carbohydrate.

Research into the GI of foods has encouraged most practitioners to reassess the range of carbohydrate foods they recommended to the person with diabetes. After all, the traditional, low-fat, high-carbohydrate choices of Western diets such as bread, refined breakfast cereals, and potato have high GI values, and may be detrimental to long-term diabetes management.

One of the sticking points over the implementation of GI has been debate over the importance of carbohydrate quality versus carbohydrate quantity in diabetes medical nutrition therapy. Clearly, the amount of carbohydrate is the primary determinant of the glycemic response to meals, but reducing carbohydrate is not the only strategy to manage hyperglycaemia. Wolever and colleagues (2003) documented that together the quantity and quality (GI) of carbohydrate explained 90% of the variation in blood glucose response to mixed meals, with each component contributing almost equal amounts. In the face of accumulating evidence on the efficacy of low GI carbohydrate foods in promoting improved glycemia, the American Diabetes Association recently acknowledged that the GI of carbohydrate should be considered as well as the amount (American Diabetes Association 2004).

An individual who is newly diagnosed with diabetes will often ask the question "what to eat" before they ask "how much to eat." Incorporating carbohydrates with a low GI into meal planning for the person with diabetes is one simple, effective, dietary change which can improve glycemia and body weight.

Worldwide, most diabetes organizations now advocate the use of GI. Organizations include:

- Diabetes Nutrition Study Group of the European Association for the Study of Diabetes (EASD)
- Canadian Diabetes Association (CDA)
- Diabetes Australia
- Diabetes UK

However, a survey of Canadian dietitians found that the majority of responders were not applying GI in diabetes management partly because they believe that it is too complex for clients to understand and use (Kalafut et al. 2000). In our experience this has not been the case. In practice, applying the GI is a simple matter of substituting one source of carbohydrate for another. A system of "this for that" exchanging high GI foods with low GI foods within the same food categories makes lowering the GI of the diet easy. Breads, bakery products, and breakfast cereals make a large contribution to the GL of Western diets, and therefore a change to low GI alternatives is especially useful.

It is unnecessary to choose only low GI carbohydrates because exchanging around half of the carbohydrate from high to low GI will lower the GI of the whole diet by on average 15 units, sufficient to bring about clinical improvements in glucose metabolism in people with diabetes (Brand-Miller et al. 2003).

Dietary characteristics that will assist in lowering the GI of the diet include:

- Choosing coarsely ground flours instead of fine flours (whether white or wholemeal)
- Regular consumption of legumes, pasta, and foods with a high proportion of whole intact grains (kernels)
- Using high amylose varieties of rice (e.g., Basmati)
- Choosing low GI breakfast cereals and breads
- Fruits that have a low GI and are more acidic will help to lower the overall GI
- The use of salad dressings containing ingredients such as lemon juice or vinegar in a meal will lower the GI by slowing down the absorption of carbohydrate

Food Industry

The popularity of GI as a nutritional attribute of foods which consumers seek and their increasing awareness of its health consequences have presented a challenge to the food industry to produce palatable low GI foods. Some of the strategies that marketers and food formulators are using include:

- Modifying existing products to lower their GI (e.g., the addition of soluble fibers to white bread to produce "low GI white bread")
- Research and development of specific agricultural products which have a low GI such as higher amylose rice (e.g., Doongara)
- Development of GI lowering ingredients. For example, guar gum and fenugreek extract which can be added to foods to lower their GI
- Increased marketing of the naturally low GI of existing products. For example, food labels on some canned legumes, fruit juices, yogurts now espouse their low GI property.

Low GI foods have been developed for specific applications, some more soundly supported by scientific evidence than others, for example, low GI "sports waters" and low GI enteral nutrition supplements for patients with diabetes and low GI snack bars intended to lessen the risk of overnight hypos in people with diabetes.

Glycemic Index Tested Program

In 2001, the University of Sydney, Diabetes Australia, and the Juvenile Diabetes Research Foundation joined forces and developed the Glycemic Index Tested Program. The centerpiece is an easily recognized logo that represents a signpost for healthier food options that have been reliably GI tested (Figure 6.1).

Figure 6.1.   © & ™The University of Sydney.

Specifically, the logo on a food label is an indication that the GI has been measured by an approved GI testing facility that uses the Australian Standard for GI testing (http://www.saiglobal.com/shop/script/Details.asp?DocN= AS0733779662AT), and that the food also meets the program's strict nutrition criteria. The nutrition criteria require foods to contain at least 10 g of carbohydrate per serve, and must meet specified limits for

Energy
Total fat
Saturated and trans fat
Sodium
Dietary fiber, and
Calcium

The nutrient criteria ensure that the food has a nutrition profile that is consistent with the dietary guidelines for Australians and essentially the same as the Dietary Guidelines for Americans. It is also a requirement that the food lists its GI value near the nutrition panel on the pack so that people can confirm themselves that the claim "low GI" on the front of pack is accurate, and consistent with international recommendations.

The program was officially launched in Australia in July 2002 and over 100 separate food items now carry the logo. The number is expected to increase threefold over the next few years, corresponding with the introduction into one of Australia's largest supermarket chains on house-branded products. Several international food companies now produce foods that carry the Glycemic Index Tested logo, including some in the United States. More information about the program can be found on the website (www.gisymbol.com).

Conclusions

The incorporation of nutritious, low GI foods into the diet of people with diabetes is a safe and inexpensive strategy for improving glycemic management and reducing the risk of acute and long-term complications. Together with appropriate energy intake and regular physical activity, a diet based on low GI carbohydrate choices may also prevent or delay the development of type 2 diabetes. In practical terms, the GI is not complex, involving the simple substitution of low GI varieties of carbohydrate-containing foods for their high GI counterparts. About half of the carbohydrate foods that are eaten each day should be low GI. Because the GI of foods cannot be guessed from the nutrient composition or physical attributes of the food, GI testing of local foods is critical. The Glycemic Index Tested Program is a shopping tool that utilizes a certified trademark for identifying foods with a healthy nutritional profile that have been GI tested according to a standardized protocol.

Ideally, it will become an international symbol recognized by people with diabetes all around the world.

References

American Diabetes Association. 2004. Nutrition principles and recommendations in diabetes. Diabetes Care 27:S36–S46.

Anderson JW, Randles KM, Kendall CWC, Jenkins DJA. 2004. Carbohydrate and fiber recommendations for individuals with diabetes: A quantitative assessment and meta-analysis of the evidence. J Am Coll Nutr 23:5–17.

Brand-Miller J, Hayne S, Petocz P, Colagiuri S. 2003. Low-glycemic index diets in the management of diabetes: A meta-analysis of randomized controlled trials. Diabetes Care 26:2261–2267.

Garg S, Jampol LM. 2005. Systemic and intraocular manifestations of West Nile virus infection. Surv Ophthalmol 50:3–13.

Hodge AM, English DR, O'Dea K, Giles GG. 2004. Glycemic index and dietary fiber and the risk of type 2 diabetes. Diabetes Care 27:2701–2706.

Jenkins DJ, Kendall CW, Augustin LS, Franceschi S, Hamidi M, Marchie A, Jenkins AL, Axelsen M. 2002. Glycemic index: Overview of implications in health and disease. Am J Clin Nutr 76:266S–273S.

Jenkins DJ, Wolever TM, Taylor RH, Barker H, Fielden H, Baldwin JM, Bowling AC, Newman HC, Jenkins AL, Goff DV. 1981. Glycemic index of foods: A physiological basis for carbohydrate exchange. Am J Clin Nutr 34:362–366.

Kalafut MA, Schriger DL, Saver JL, Starkman S. 2000. Detection of early CT signs of >1/3 middle cerebral artery infarctions: Interrater reliability and sensitivity of CT interpretation by physicians involved in acute stroke care. Stroke 31:1667–1671.

Meyer KA, Kushi LH, Jacobs DR, Jr, Slavin J, Sellers TA, Folsom AR. 2000. Carbohydrates, dietary fiber, and incident type 2 diabetes in older women. Am J Clin Nutr 71:921–930.

Pawlak DB, Kushner JA, Ludwig DS. 2004. Effects of dietary glycaemic index on adiposity, glucose homoeostasis, and plasma lipids in animals. Lancet 364:778–785.

Salmeron J, Manson JE, Stampfer MJ, Colditz GA, Wing AL, Willett WC. 1997. Dietary fiber, glycemic load, and risk of non-insulin-dependent diabetes mellitus in women. JAMA 277:472–477.

Schulze MB, Liu S, Rimm EB, Manson JE, Willett WC, Hu FB. 2004. Glycemic index, glycemic load, and dietary fiber intake and incidence of type 2 diabetes in younger and middle-aged women. Am J Clin Nutr 80:348–356.

Stevens J, Ahn K, Juhaeri, Houston D, Steffan L, Couper D. 2002. Dietary fiber intake and glycemic index and incidence of diabetes in African-American and white adults: The ARIC study. Diabetes Care 25:1715–1721.

Stratton IM, Kohner EM, Aldington SJ, Turner RC, Holman RR, Manley SE, Matthews DR. 2001. UKPDS 50: Risk factors for incidence and progression of retinopathy in Type II diabetes over 6 years from diagnosis. Diabetologia 44:156–163.

Wolever TM, Nguyen PM, Chiasson JL, Hunt JA, Josse RG, Palmason C, Rodger NW, Ross SA, Ryan EA, Tan MH. 1994. Determinants of diet glycemic index calculated retrospectively from diet records of 342 individuals with non-insulin-dependent diabetes mellitus. Am J Clin Nutr 59:1265–1269.

Wolever TM, Vorster HH, Bjorck I, Brand-Miller J, Brighenti F, Mann JI, Ramdath DD, Granfeldt Y, Holt S, Perry TL, Venter C, Xiaomei W. 2003. Determination of the glycaemic index of foods: Interlaboratory study. Eur J Clin Nutr 57:475–482.

Zhang C, Liu S, Solomon CG, Hu FB. 2006. Dietary fiber intake, dietary glycemic load, and the risk for gestational diabetes mellitus. Diabetes Care 29:2223–2230.

Chapter 7

Dietary Fiber and Associated Phytochemicals in Prevention and Reversal of Diabetes

James W Anderson, MD

Introduction

Cereal fiber intake from whole grain sources appears to be one of the strongest preventive measures for type 2 diabetes mellitus (Lindstrom et al. 2006; Liu 2003; Liu et al. 2000; Weickert et al. 2006). While the explosive worldwide increase of diabetes can be most closely associated with the dramatic increase in the obesity prevalence (Anderson et al. 2003), distinct changes in food consumption patterns have also occurred over the past 20 years. As energy intake has increased, a higher percentage of the energy consumed has come from refined foods—laden with sugars and fats—and a lower percentage from whole grains and fiber-rich foods (Bhargava and Amialchuk 2007). The reduction in whole grains, dietary fiber, and associated phytonutrients may further enhance the risk for type 2 diabetes (Bo et al. 2006; Zhang et al. 2006).

Lifestyle measures are crucial for the prevention and reversal of type 2 diabetes. As reviewed by Greenway (Chapter 3), increased physical activity and weight losses of approximately 5% may decrease the emergence of diabetes in high-risk individuals by almost 60% over a 3–5 year period of follow-up (Nathan and Berkwits 2007). Obviously, these important lifestyle changes—because of their broad health-promoting effects, lack of side effects, and low expense—are preferable to pharmacotherapy.

The primary purpose of this chapter is to review the associations between intake of dietary fiber and associated phytochemicals as they relate to development of diabetes. As background, the consumption of specific nutrient and their association with prevalence of diabetes will be reviewed. Because of the close relationship between prevalence of obesity and diabetes, the potential role of dietary fiber intake and development of obesity will be assessed. The phytochemicals associated with high-fiber foods and their potential protective role for diabetes will be

Table 7.1.   Associations between consumption of different foods and nutrients and prevalence of diabetes.

Parameter	No. of studies	No. of subjects	RR	LCI	UCI
Total dietary fiber	5	239,485	0.81	0.70	0.93
Fruit fiber	5	239,485	0.95	0.86	1.04
Vegetable fiber	5	239,485	1.06	0.95	1.16
Total carbohydrate	4	235,169	0.91	0.82	1.00
Refined grain	3	175,416	1.02	0.90	1.13
Total sugars[a]	3	106,974	0.82	0.72	0.92
Fructose	3	166,582	1.30	1.19	1.42
Glycemic index	6	247,404	1.02	0.96	1.08
Glycemic load	6	247,404	1.02	0.96	1.08
Alcohol	13	369,862	0.69	0.58	0.81
Coffee	7	181,509	0.65	0.54	0.78
Magnesium	3	143,920	0.66	0.56	0.76
Red meat	3	171,059	1.36	1.19	1.52

Modified from a previous report by Anderson and Conley (2007).
Abbreviations: RR, relative risk; LCI, lower 95% confidence interval; UCI, upper 95% confidence interval.
[a] Total sugars include glucose, fructose, sucrose, and lactose.

discussed. Finally, the utility of high-fiber intakes for management of diabetes and its comorbid complications will be outlined.

Nutrients and Nutraceuticals Associated With Prevention of Diabetes

Nutrient and nutraceutical intakes in association with the prevalence of diabetes are reviewed extensively in this publication. Over the past decade persuasive data have emerged relating to diabetes risk and intake of dietary fiber, alcohol, coffee, magnesium, and red meat (Table 7.1) (Anderson and Conley 2007; Gross et al. 2004; Schulze and Hu 2005). An "unhealthy" diet pattern high in sugar-sweetened soft drinks, refined grains, and processed meats was associated with a significantly greater risk (relative risk (RR) of 2.56–2.93) for developing diabetes than a "healthy" diet pattern high in vegetables, coffee, wine, and other recommended food choices (Schulze et al. 2005). Individuals who consume the largest amount of dietary fiber have an estimated 19% reduction in risk for developing diabetes compared to those with the lowest level of intake. In a meta-analysis of five reports, including almost 250,000 persons, the individuals with the highest quintile of total dietary fiber intake had a relative risk for developing diabetes of

0.81 (95% confidence interval (CI), 0.70–0.93 (Meyer et al. 2000; Montonen et al. 2003; Salmeron et al. 1997a, b; Schulze et al. 2004b) (Table 7.1). The intakes of fruit, vegetables, fruit fiber, or vegetable fiber do not have a discernible effect on the risk for diabetes (Liu et al. 2004; Meyer et al. 2000; Montonen et al. 2003; Salmeron et al. 1997a, b; Schulze et al. 2004b).

While it has been widely assumed by consumers that high intake of carbohydrate (sugars and starches) increases risk for diabetes, the data indicate that those individuals with the highest quintile of total carbohydrate intake have a 9% lower risk for diabetes than those with the lowest quintile for carbohydrate consumption. In a meta-analysis of four studies, including almost 250,000 people, we noted that those individuals with total carbohydrate intake in the highest quintile had a relative risk for developing diabetes of 0.91 (95% CI, 0.82–1.00) compared to those with the lowest consumption of carbohydrate (Anderson and Conley 2007). Refined grain foods, compared to whole grain products, are not rich sources of dietary fiber and are associated with a slight, but insignificant, increased risk for coronary heart disease (CHD) (Jacobs et al. 1999). However, the available data indicate that refined grain intakes do not affect the risk for developing diabetes (Table 7.1).

The effects of sugar consumption on risk for diabetes are controversial and politically charged (Gross et al. 2004; Nestle 2006). Since the development of diabetes is closely linked to weight gain and obesity (Anderson et al. 2003), it is useful to review the evidence linking sugar intake to obesity. Most of the available evidence supports the hypothesis that higher levels of sugar consumption are associated with higher prevalence of obesity (Malik et al. 2006; Vartanian et al. 2007), but the data are not conclusive (Saris 2003; Vermunt et al. 2003). Emerging evidence links sugar intake, especially from sugar-sweetened colas and fruit juices, with weight gain or obesity in children (Berkey et al. 2004; Ludwig et al. 2001) and adults (Bes-Rastrollo et al. 2006b; Gross et al. 2004; Malik et al. 2006; Schulze et al. 2004b). High fructose corn syrup use has been specifically linked to the epidemic increase in obesity (Bray et al. 2004). The potential mechanisms linking sugar intake to obesity are reviewed by Bachman and colleagues (2006). While available evidence strongly suggests that increased intake of sugar-sweetened beverages and other foods sweetened with high fructose corn syrup contribute to the current worldwide epidemic of obesity, these data are from epidemiological studies. During the past two decades there have been many changes in industrial societies; adolescents have significantly decreased their physical activity (Kim et al. 2002) and, perhaps unrelated, cellular phone usage has increased dramatically. It is uncertain which of the many concurrent societal changes make a synergistic contribution to the epidemic of obesity (Jacobs 2006).

Diabetes prevalence has also been linked to sugar intake, but persuasive data are not yet available. Since the development of diabetes chronologically follows the development of obesity by approximately 10 years (Bray 1998), the impact of obesity and its dietary associations related to diabetes prevalence may not be detectable yet. The available data are summarized in Table 7.1. Two large studies have drawn conflicting conclusions (Janket et al. 2003; Schulze et al. 2004b). Analysis of 39,345 middle-aged women from the Women's Health Study showed no significant association with diabetes for women in the highest compared to the lowest quintile for consumption of total sugar, sucrose, fructose, glucose, or lactose (Janket et al. 2003). In contrast, study of 91,249 young and middle-aged women from the Nurses' Health Study II indicated that women with the highest intake of sugar-sweetened soft drinks had a significantly higher risk for diabetes than women with the lowest use. Rather than compare quintiles of intake, they compared women with ≥ one soft drink daily (9.3% of study group) to women with < one monthly (53% of study group). Even when one estimates the effects of consumption of ≥ two soft drinks weekly, representing 20.5% of the study group, the relative risk for diabetes appears significant and represents an increased risk of about 60% (Schulze et al. 2004b). Analyses of three studies (Hodge et al. 2004; Janket et al. 2003; Meyer et al. 2000) indicate that intake of total sugars (glucose, fructose, sucrose, and lactose) is associated with a 19% reduction in risk for diabetes; the risk ratio for highest compared to lowest quintile for intake of total sugars is 0.81 (95% CI, 0.71–0.91). However, analysis of three studies reporting fructose intake (Janket et al. 2003; Meyer et al. 2000) or high fructose sweetened soft drinks (Schulze et al. 2004b) suggests that those persons with the highest fructose intake have a 30% higher risk for developing diabetes than those with the lowest intake; the risk ratio for the highest versus lowest intake of fructose was 1.30 (95% CI, 1.19–1.42). Thus, more data are required to provide convincing evidence that increased fructose intake is associated with increased risk for diabetes.

Consuming high glycemic index (GI) foods—ones that increase blood glucose values substantially after intake—has been linked to increased risk for diabetes in some studies (Liu 2002). Because high GI foods stimulate more insulin release than low GI foods, it is postulated that they may exhaust the pancreas, especially in susceptible individuals. However, a meta-analysis of available data from five studies including 247,404 persons indicates that those individuals in the highest quintile have a risk for diabetes that is virtually identical to individuals in the lowest quintile—those with the lowest intake of high GI foods. The relative risk reported from these studied ranged from 0.89 to 1.62 with a variance-weighted (meta-analysis) value of 1.02 (95% CI, 0.96–1.02) (Meyer et al. 2000; Salmeron et al. 1997a, b; Schulze et al. 2004a; Stevens et al. 2002). The glycemic load (GL) is calculated as GI of the food multiplied by grams of carbohydrate in the food (Liu 2002). White bread has a high GL because of its high GI and large amount of carbohydrate, while most beans have a low GL because of their low GI and relatively less carbohydrate per serving. If high carbohydrate intake as well as high GI food intake produced a diathesis for diabetes, the GL would capture this outcome. However, individuals in the highest quintile for GL appear to have a similar risk for diabetes as do persons with the lowest quintile for GL food consumption (Table 7.1) (Meyer et al. 2000; Salmeron et al. 1997a, b; Schulze et al. 2004a; Stevens et al. 2002).

The protection from diabetes associated with alcohol and coffee consumption are reviewed elsewhere (Koppes et al. 2005; Van Dam and Hu 2005). Magnesium intake appears to protect from developing diabetes (Meyer et al. 2000). It is interesting to note that refining wheat flour removes about 85% of the magnesium (Liu 2002). Persons with the highest level of red meat consumption appear to have a 36% increased risk for developing diabetes compared to those with the lowest intake levels (Table 7.1).

Whole Grain and Cereal Fiber Intake and Prevalence of Diabetes

For almost 50 years keen observers have suggested that consumption of low-fiber foods contributed to development of diabetes, obesity, and other Western diseases (Burkitt and Trowell 1975b; Trowell 1960, 1972, 1975). The first scientific documentation of this hypothesis was provided when Morris and colleagues (1977) reported that cereal fiber intake was associated with a significantly lower risk for CHD among British men. Kromhout et al. (1982) provided further evidence in reporting that Dutch men in the highest quintile of fiber intake had a death rate from CHD that was only one-quarter that for men in the lowest quintile. In the past decade many studies have confirmed that dietary fiber intake—especially cereal fiber—or whole grain consumption are associated with lower prevalence rates for CHD, diabetes, and obesity (Jacobs and Gallaher 2004; Liu 2002; Slavin 2004).

Nine reports have evaluated the effects of whole grain consumption or cereal fiber intake on risk for diabetes (Table 7.2) (Barclay et al. 2006; Bhargava 2006; Bo et al. 2006; Esmaillzadeh and Azadbakht 2006; Helgeson et al. 2006; Lindstrom et al. 2006; Weickert et al. 2006; Zhang et al. 2006). Five studies have assessed the specific associations between whole grain consumption and development of diabetes. All five studies reported that whole grain consumption was associated with a reduction in risk for diabetes and, in four of the five studies, this association was significant. In the meta-analysis of these studies, whole grain consumption, for the highest quintile, was associated with a relative risk for diabetes of 0.74 (95% CI, 0.68–0.81) compared to the lowest quintile for whole grain intake. Thus, the consistent intake of whole grain foods is associated with a significant reduction in risk for diabetes that is approximately 26%. Eight studies have reported on the relative risk for diabetes related to cereal fiber intake. Seven of the eight studies indicate that consumption of cereal fiber is associated with a significant reduction in risk for diabetes. The aggregated values indicate that cereal fiber intake, from an analysis of 288,507 individuals, is associated with a relative risk for diabetes of 0.68 (95% CI, 0.63–0.74). When all 13 measurements are aggregated to give an estimate for 427,935 individuals, the relative risk for diabetes is 0.71 (95% CI, 0.67–0.75). This may provide the most accurate estimate for the available data indicating that individuals who consume the highest level of whole grains or cereal fiber have a risk for diabetes that is approximately 29% lower than persons who have the lowest intake of these foods.

Table 7.2. Relative risk (RR) for development of diabetes based on intake levels of whole grains or cereal fiber

Study	No.	WG (RR)	SE	Cereal fiber (RR)	SE
Salmeron et al. (1997a)	42,759	NA	NA	0.7	0.097
Salmeron et al. (1997b)	65,173	NA	NA	0.72	0.071
Liu (2002)	87,899	0.73	0.051	NA	NA
Meyer et al. (2000)	35,988	0.79	0.071	0.64	0.056
Stevens et al. (2002)	8,947[a]	NA	NA	0.75	0.079
	3,288[b]	NA	NA	0.86	0.107
Hodge et al. (2004)	31,641	0.86	0.117	0.97	0.122
Fung et al. (2002)	51,529	0.7	0.066	NA	NA
Montonen et al. (2003)	4,316	0.65	0.148	0.39	0.097
Schulze et al. (2004a)	91,249	NA	NA	0.63	0.082
	Total	RR	95% CI	RR	95% CI
All whole grain	216,519	0.74	0.68–0.81		
All cereal fiber	288,507			0.68	0.63–0.74
Whole grain and cereal fiber	427,935			0.71	0.67–0.75

Abbreviations: No., number of subjects followed; WG, whole grains; 95% CI, 95% confidence intervals; NA, not available.
[a] Whites.
[b] African-American.

Dietary Fiber and Obesity

The relationship between fiber depleted diets and obesity was recognized more than 50 years ago (Burkitt and Trowell 1975a; Cleave 1956) and an early clinical trial suggested that individuals lost more weight when fiber was included in the diet than those on low fiber diets (Yudkin 1959). An early clinical study clearly documented that greater satiety was seen after eating apples than after consuming the same energy intake from apple juice (Heaton et al. 1973). Subsequently, epidemiologic studies have shown a very strong inverse relationship between fiber intake and weight gain or obesity (Lairon 2007). Laboratory studies have documented the effects of different types of fiber intake on gastrointestinal function and recently many gut anorexigenic (satiating) and orexigenic (appetite inducing) hormones have been discovered (Konturek et al. 2004; Woods 2004). Finally, clinical trials using high-fiber diets or fiber supplements promote better weight loss than control diets that are low in fiber content (Howarth et al. 2001). The physiologic effects of fiber as it relates to eating behavior, the epidemiologic observations, and the clinical studies will be briefly reviewed.

Physiologic Effects of Fiber

Dietary fiber and fiber-rich foods have profound effects on the gastrointestinal tract and eating behavior while exerting distinct effects on energy balance and metabolic responses to food. These effects will be reviewed.

Gastric Emptying

Many soluble—also termed viscous or fermentable—fibers slow the emptying of food from the stomach. An early hypothesis was that these viscous fibers inhibited gastric emptying and this, in turn, resulted in slower glucose absorption and lower glycemic and insulin responses (Blackburn et al. 1984; Holt et al. 1979). This hypothesis was not supported by further research (Rainbird and Low 1986). Chronic intake of pectin results in ongoing inhibition of gastric emptying that persists for up to 3 weeks after pectin consumption is discontinued (Levine et al. 1982). Of interest, guar gum slows gastric emptying to a greater extent when given with a liquid meal than with a solid meal (Todd et al. 1990). Furthermore, the effects of one type of fiber cannot be extrapolated to other types of fibers (Bouin et al. 2000). Thus, the effects of dietary fiber intake on gastrointestinal function are complex, being affected by specific fibers, method of delivery, and timing of ingestion.

Decreasing Postprandial Glycemic and Insulin Responses

A large number of soluble fibers—pectin, guar, oat gum, psyllium, konjac mannan, carboxymethylcellulose, and others—diminish the increase in blood glucose and insulin levels after carbohydrate intake. Mechanisms appear to relate to a delay of gastric emptying (Doi 1995), slowing of intestinal motility (Bouin et al. 2000), slow glucose diffusion out of gut contents (Ebihara and Kiriyama 1982; Flourie et al. 1984), and gut hormones (Woods 2004). Different fibers with glycemic effects have different effects on gut function (Doi 1995). Fibers that do not have glycemic effects include gum Arabic (or acacia gum, a nonviscous soluble fiber) (Ross et al. 1983).

Fiber and Gut Hormones

The role of fiber related to these incretins, orexigens, and anorexigens requires much more study. Incretins are hormones that enhance the insulin response to food taken orally (Perfetti and Merkel 2000), orexigens (e.g., ghrelin) stimulate food intake, and anorexigens (e.g., cholecystokinin or CCK) act to decrease food intake (Konturek et al. 2004). Some of the effects of selected gut hormones are summarized in Table 7.3. Recently, four gut hormones have been involved in clinical trials to examine their role for weight loss and management. Based on available information, the efficacy of these agents for weight loss appears to as follows: Amylin (Pramlitide), modest; GLP-1 (Exenatide), modest; CCK (CCK-A agonist), modest; and PYY (PYY 3-36), moderate (Anderson 2007). At least 22 gut hormones associated with appetite-regulation properties have been identified (Woods 2004).

In reviewing the literature one is quickly struck by several paradoxes. The gut hormones incretins that stimulate insulin release (GLP-1, stimulating 80% of incremental insulin release, and GIP, 20%) also delay gastric emptying (Massimino et al. 1998). In contrast, other fibers that delay gastric emptying tend to decrease insulin release. CCK has received more attention than other anorexigens and its response to meals and fiber in humans has been reviewed (Bourdon et al. 1999). Liquid meals increase CCK more than solid meals. The effects of different fibers on CCK are mixed. In general, high-fiber meals produce higher CCK levels. Much more research is required to clarify our understanding of how foods, fibers, and eating patterns affect regulation of food intake.

Fiber and Satiety

High-fiber foods have long been considered to be "more filling" than low-fiber foods. Early clinical trials used fiber supplements as an aid to weight loss (Duncan et al. 1960; Yudkin 1959). The classical experiment of Haber and colleagues (1977) documented that apples providing 60 g of carbohydrate were significantly more satiating that fiber-free apple juice providing the same amount of carbohydrate. Careful laboratory experiments documented that high fiber intake decreased within meal food intake and, also, food intake at the next meal (Blundell and Burley 1987; Burley et al. 1987; Levine et al. 1989; Porikos and Hagamen 1986). Including pectin in the meal delayed gastric emptying and enhanced satiety (di Lorenzo C. et al. 1988). Subsequent studies have linked satiation to changes in orexigenic or anorexigenic hormones (Frost et al. 2003; Gruendel et al. 2006; Massimino et al. 1998). Unfortunately, as indicated in Table 7.3, the effects of different fibers on gut hormone secretion currently are unclear. However, determining the responses of key gut hormones to different types and formulations of fiber probably will greatly clarify our understanding of this area.

Epidemiological Association Between Fiber Intake and Obesity

The total daily intakes of dietary fiber, whole grains or cereal fiber, also have a positive effect on protection from obesity. Cross-sectional studies of two populations

Table 7.3. Pancreatic and gut hormones affecting metabolism and food intake.

Hormone	Source	Glucose effects	Gut effects	Appetite/food intake	Fiber effects on hormone release
Amylin (Fineman et al. 2002)	Pancreas: beta cells	Inhibits glucagon release	Delays gastric emptying	Decreases	
Apolipoprotein A-IV (Tso et al. 2001; Woods 2004)				Decreases	Review (Tso et al. 2001)
Bombesin family[a] (Woods 2004)				Decreases	
Cholecystokinin (Konturek et al. 2004; Woods 2004)	Duodenum and jejunum: I cells		Delays gastric emptying	Decreases	Increased by fiber; (Bourdon et al. 2001; Burton-Freeman et al. 2002) Not affected; (Boyd et al. 2003; di Lorenzo et al. 1988; Decreased (Geleva et al. 2003) Reviewed (Bourdon et al. 1999)
Enterostatin (Woods 2004)				Decreases	Fat intake required for the satiety effects (Levine et al. 2003)
Ghrelin (Cummings et al. 2002)	Stomach (Gr cells) and duodenum	Decreases fat catabolism		Increases	Fiber decreases (Gruendel et al. 2006)

(continued)

105

Table 7.3. (*continued*)

Hormone	Source	Glucose effects	Gut effects	Appetite/food intake	Fiber effects on hormone release
GIP (Drucker 2003)	Duodenum and jejunum: K cells	Enhances insulin release			Decreased by guar, Whole grain rye (Juntunen et al. 2002, 2003a), see (Morgan et al. 1990a, b)
Glucagon (Alberti et al. 1997)	Pancreas: alpha cells	Stimulates glucose release		Decreases	
GLP-1 (Perfetti and Merkel 2000)	Intestine and colon: L cells	Enhances insulin release, inhibits glucagon release	Delays gastric emptying	Decreases	Increased expression with fiber, (Gee et al. 1996; Reimer and McBurney 2004). Increased, (Massimino et al. 1998). Not affected; (Boyd et al. 2003). Decreased by whole grain rye, (Juntunen et al. 2002)
GLP-2 (Woods 2004)				Decreases	
Insulin (Alberti et al. 1997)	Pancreas: beta cells	Stimulates glucose metabolism		Decreases	Soluble fiber decreases response by slowing glucose absorption
Motilin (Asakawa et al. 2001)	Jejunum		Hastens gastric emptying	Increases	
Peptide YY (Batterham et al. 2003)	Intestine and colon: L cells			Decreases	Review (Onaga et al. 2002)
Somatostatin (Woods 2004)				Decreases	

Abbrndes: bombesin, gastrin-releasing peptide (GRP) and neuromedin B (Woods 2004).

106

with a total of over 100,000 persons as well as four prospective cohort studies including a total of over 100,000 persons indicate a strong negative association between fiber intake and presence or development of obesity. A fixed-effect meta-analysis technique was used to estimate pooled effect sizes (Anderson et al. 1995, 1999). The cross-sectional studies indicated that women and men with the highest fiber intake had a relative risk for obesity of 0.77 (95% CI, 0.68–0.87) compared to those with the lowest fiber consumption (Lairon et al. 2005; Maskarinec et al. 2006). Four prospective cohort studies also reported that men and women with the highest fiber intake showed lower rates of weight gain and less obesity than those with the lowest fiber consumption with relative risks of 0.70 (95% CI, 0.62–0.78) (Bes-Rastrollo et al. 2006a; Koh-Banerjee et al. 2004; Liu et al. 2003; Ludwig et al. 1999). In aggregate, these studies suggest that high fiber intakes reduce risk for weight gain or developing obesity by approximately 25%. Whole grain (Liu et al. 2003; McKeown et al. 2002) and cereal fiber intake (Koh-Banerjee et al. 2004) also appear to protect from weight gain, but the effects of fruit or vegetable intake are unclear (Bes-Rastrollo et al. 2006a; Koh-Banerjee et al. 2004). Because these studies include a wide diversity of ethnic/racial groups, they would appear to have wide applicability.

Clinical Trials Examining Fiber Effects on Body Weight Regulation

The physiological and epidemiological studies strongly support the role of dietary fiber in preventing and managing obesity in humans (Burton-Freeman 2000; Howarth et al. 2001; Koh-Banerjee et al. 2004; Liu et al. 2003). These observations have been supported by clinical trials that have used either fiber supplements and placebos or high-fiber diets compared to low-fiber diets. These studies will be reviewed.

Effects of fiber supplements on weight loss for individuals on weight-reducing diets have been examined in 17 placebo-controlled trials (Birketvedt et al. 2005; Cairella et al. 1995; Duncan et al. 1960; Ehmann and Ressin 1985; Rigaud et al. 1990; Rossner et al. 1987, 1988; Ryttig et al. 1984, 1989; Solum et al. 1987; Tuomilehto et al. 1980; Valle-Jones 1980; Walsh et al. 1984; Yudkin 1959). Some of the earlier studies were not as well controlled as more recent trials but 16 of the trials were randomized and placebo controlled. Large numbers of volunteers participated in these trials with 391 completing control interventions and 423 completing fiber-supplemented interventions. In most studies volunteers were instructed in the use of energy-restricted diets. Usually fiber was provided in tablets that were given three times daily. Fiber intake ranged from 4.5 to 20 g/day and averaged about 2.5 g three times daily with meals. Most studies used tablets containing predominantly insoluble fiber but guar gum or glucomannan were used in several studies. In only one of 17 studies was weight loss greater with placebo than fiber supplements (Rossner et al. 1988).

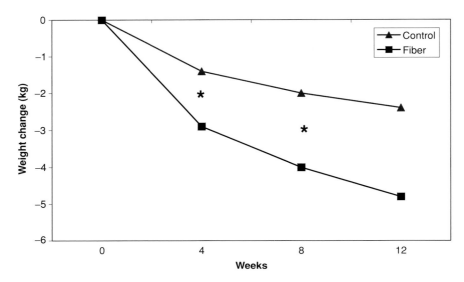

Figure 7.1. Weight losses comparing fiber-supplemented diets versus control diets. Values are means ± SEM. Data for 15 studies are available at 4 and 8 weeks and for 9 studies at 12 weeks. Significant differences are seen at 4 weeks ($P = 0.0063$) and 8 weeks ($P = 0.0088$).

Fiber supplements as an adjunct to an energy-restricted diet were associated with modestly greater weight loss than placebo. To estimate the effects of fiber supplements on weight loss we performed a variance-weighted (fixed-effect) analysis of weight losses at 4, 8, and 12 weeks (Anderson et al. 1995) (Figure 7.1). This analysis indicated that weight losses were as follows for placebo and fiber-supplemented groups, respectively: 4 weeks, -1.7 kg (95% CI, -1.3–2.0 kg) and -3.0 kg (95% CI, -2.6–3.4 kg) ($P = 0.0129$ vs. placebo); 8 weeks, -2.4 kg (95% CI, -1.9–2.9 kg) and -4.9 kg (95% CI, -3.5–4.5 kg) ($P = 0.0104$ vs. placebo); and 12 weeks, -2.7 kg (95% CI, -0.6–4.0) and -4.9 (95% CI, -0.1–8.0). Expressed as percentage of initial body weight, weight losses for placebo and fiber-supplemented diets, respectively, were: 4 weeks, 2.0% and 3.2%; 8 weeks, 2.9% and 4.9%; and 12 weeks, 2.7% and 4.9% (Birketvedt et al. 2005; Cairella et al. 1995; Duncan et al. 1960; Ehmann and Ressin 1985; Rigaud et al. 1990; Rossner et al. 1987, 1988; Ryttig et al. 1984, 1989; Solum et al. 1987; Tuomilehto et al. 1980; Valle-Jones 1980; Walsh et al. 1984; Yudkin 1959).

High-fiber foods or fiber-enhanced food products have also been evaluated by RCTs. Only five studies are available and the quality of some studies is suboptimal. Nevertheless, the average weight loss over an 8-week period with high-fiber diets was approximately 1 kg greater than with control diets (Henry et al. 1978; Krotkiewski 1985; Krotkiewski and Smith 1985; Mickelsen et al. 1979; Russ and

Atkinson 1985). More carefully designed clinical trials are required to clearly establish that high-fiber diets are associated with significantly more weight loss than control diets. One study comparing the effects of a diet high in complex carbohydrate (and, thus, higher in dietary fiber) provides persuasive data that this type of diet promotes more weight loss than a high simple carbohydrate (and thus lower fiber) diet (Poppitt et al. 2002). It seems very likely that well-designed studies using this model would document the weight-management benefits of higher fiber intakes.

Fiber and Obesity Conclusions

Clinical observations and epidemiological data clearly indicate that a higher fiber intake is associated with less weight gain than a lower fiber intake (Bes-Rastrollo et al. 2006a; Maskarinec et al. 2006). Studies in animals and humans demonstrate that intake of fiber tends to delay gastric emptying and create a sense of fullness (Haber et al. 1977; Rainbird and Low 1986). Animal and human studies also document that selected fiber intakes are associated with increases in satiating gut hormones (Bourdon et al. 2001; Cani et al. 2005; Frost et al. 2003; Massimino et al. 1998). Clinical trials with high-fiber foods have not provide consistent data that these diets are more efficacious for weight loss than low fiber control diets (Mickelsen et al. 1979; Russ and Atkinson 1985). However, a number of randomized, placebo-controlled, clinical trials have documented that fiber supplements are accompanied by more weight loss than are use of placebos (Birketvedt et al. 2005; Rossner et al. 1987; Solum et al. 1987). In aggregate, the available clinical evidence with the supporting mechanistic information provides strong indicators that dietary fiber consumption has beneficial effects on weight management.

Phytochemicals Associated with High Fiber Foods

Whole grains are crammed with carbohydrates, minerals, vitamins, and phytonutrients (Slavin 2004). These foods deliver antioxidants, regulators of gut function, hormone-response effectors, insulin sensitizers, satiating properties, and other effects (Table 7.4). Many of these effects have been attributed to the dietary fiber content but, as our understanding expands, it seems clear that many factors are working independently or synergistically to decrease risk for diabetes. As outlined earlier in this chapter, whole grain and cereal fiber intake have the strongest protective association related to diabetes, but magnesium also shows a strong beneficial association (Liese et al. 2003; Meyer et al. 2000). Many of these relationships are reviewed in detail in other chapters and will only be summarized here.

Whole grains have very high antioxidant levels that exceed those of most plant foods excepting certain berries (Miller et al. 2000). Multiple components of the whole grain contribute to these antioxidant properties. As reviewed by McCarty

Table 7.4. Components of whole grains potentially impacting risk for diabetes[a].

Component	Anti-oxidant	Gut modifier	Hormonal effects	Increase insulin sensitivity	Increase satiety	Other
Dietary fiber		X	X	X	X	
Oligosaccharides		X	X			Enhance immune function
Resistant starch		X	X	X	X	
Flavonoids	X					
Inositols	X					
Lignans	X		X	X		
Phenolics	X					
Phytates	X					
Amylase inhibitors		X				
Phytoestrogens	X		X	X		
Selenium	X					
Magnesium						Anti-diabetic properties
Zinc	X					

[a] Modified from Slavin (2004); lignan information from Hallmans et al. (2003); general information from Truswell (2002); and antioxidant information from Miller (Miller et al. 2000).

(Chapter 11) and elsewhere (Evans et al. 2003), experimental studies suggest that elevations of blood glucose and free fatty acid levels associated with prediabetes and early diabetes may increase production of reactive oxygen species and subsequent oxidative stress. Oxidative stress-related events may produce pancreatic B cell damage and aggravate insulin resistance in liver and muscle cells. Dietary antioxidants, such as those in whole grains, may attenuate this process and slow development of type 2 diabetes (McCarty and Inoguchi, Chapter 11).

The effects of dietary fiber, resistant starch, and possibly oligosaccharides in modifying the glucose and hormonal response to meals are discussed by Wong et al. (Chapter 4) and Foster-Powell et al. (Chapter 6). Simplistically the lower insulin response required to optimize glucose homeostasis creates less work for the beta cells and allows them to better preserve their function over time. The bacterial products of lignans—enterolactone and enterodiol—as well as isoflavones have important hormonal effects that may impact insulin sensitivity (Bhathena and Velasquez 2002; Juntunen et al. 2003b) (Anderson and Pasupuleti, Chapter 9).

The effects of whole grains on insulin sensitivity are affected by the type of grain and the method of preparation. Oat and barley products have lower glycemic indices than other grains because of their B glucan content (Foster-Powell et al. 2002). However, the following studies illustrate the limitations of our understanding of the effects of grain fibers on glucose homeostasis. High-fiber rye bread—rich in insoluble fiber—has more impact on insulin sensitivity than whole grain wheat bread (Juntunen et al. 2003b). Replacing refined rice with whole grains and legume powder enhances insulin sensitivity and lowers measures of lipid peroxidation in nondiabetic and diabetic subjects (Jang et al. 2001). Consumption of bread including arabinoxylan fiber extracted from wheat flour is followed by improved glucose and insulin responses compared to control meal tolerances using wheat bread (Lu et al. 2000). And, in addition, short-term intake of bread including a highly insoluble fiber-enhanced glucose and insulin responses and increased serum levels of satiating hormones when compared to the response to a low-fiber control bread (Weickert et al. 2005). Much more research is required to enhance our understanding of how different fibers affect short-term and long-term glucose and insulin homeostasis.

Most of the minerals are removed from whole grains as they are processed to refined flour. Since magnesium deficiency appears to contribute to development of diabetes, it is noteworthy that 85% of the magnesium is removed in the milling of white flour (Liu 2002). For selenium, the important antioxidant, 92% is removed in refining of whole grain to make white flour (Truswell 2002). Recent evidence suggests that whole grain consumption, compared to refined flour use, may enhance mineral absorption.

The fermentation of dietary fiber, oligosaccharides, and resistant starch in the colon changes the intestinal flora in a favorable manner and generates short-chain fatty acids. Some fibers and forms of resistant starch—as do most oligosaccharides—selectively promote development of heath-promoting bacteria that improve the immune system. Undigested carbohydrates that reach the colon are fermented by the microflora to produce short-chain fatty acids that nourish the colon and have favorable effects on lipid metabolism in the liver (Slavin 2004).

Some whole grain products have modest amylase inhibitory effects. By inhibiting amylase hydrolysis these foods have small effects that resemble those of acarbose, a prescription amylase inhibitor used for treatment of diabetes. This is of some interest because some data suggest that acarbose use may delay the onset of type 2 diabetes (Chiasson 2006; Maji et al. 2005).

Thus, whole grains provide a wide array of carbohydrates, minerals, vitamins, antioxidants, and other phytochemicals that appear independently to reduce risk for diabetes. Whether these effects are simply additive or are, more likely, synergistic needs further exploration. The consistent observations (Table 7.2) that whole grains significantly reduce risk for diabetes by almost 30% indicate that intake of whole grains with all its constituents is as potent for diabetes prevention as any other dietary intervention that has been identified. Only the combination of weight loss and increased physical activity appear more potent for diabetes prevention (Greenway, Chapter 3).

High Fiber Diets for Prevention and Reversal of Comorbid Complications of Diabetes

Metabolic Syndrome

The metabolic syndrome—a cluster of abnormalities including visceral adiposity, dyslipidemia, insulin resistance, and hypertension—can be slowed in development or reversed with high levels of fiber or whole grain consumption (Azadbakht et al. 2005; Esmaillzadeh and Azadbakht 2006; Laaksonen et al. 2005; McKeown et al. 2004; Pereira et al. 1998, 2002; Ylonen et al. 2003). High fiber or whole grain intakes specifically prevent or reverse these derangements associated with the metabolic syndrome: visceral adiposity (Lairon et al. 2005; Ludwig et al. 1999; McKeown et al. 2002), hypertriglyceridemia (Anderson 2000), low HDL-cholesterol values (Anderson and Gustafson 1988), hyperglycemia or insulin resistance (Weickert et al. 2006), systolic hypertension (Anderson 1983; Appel et al. 1997), or diastolic hypertension (Anderson 1983; Appel et al. 1997). The effects of whole grains on these "fellow travelers" are discussed in the specific paragraphs below. Of interest is the fact that whole grain intake has significant effects on all of these derangements except low HDL-cholesterol values. The nutrition management of low HDL-cholesterol values are discussed elsewhere but low glycemic index (GI) food intake or alcohol consumption appears to be the strongest nutritional effectors.

Coronary Heart Disease

The metabolic syndrome is associated with a significant increase in risk for CHD (Ford 2005; Maji et al. 2005) as is type 2 diabetes (Abbasi et al. 2002; Lakka et al. 2002, Reynolds, Chapter 5). The specific lipoprotein abnormalities associated with type 2 are as follows: hypertriglyceridemia (Ginsberg 2006); low HDL-cholesterol values (Ginsberg 2006), and dysfunctional HDL particles (Gowri et al. 1999; Hansel et al. 2004); glycosylated LDL particles, increases susceptibility of LDL to oxidation (Piarulli et al. 2005) and small dense LDL particles (Ginsberg 2006). High levels of whole grains and other dietary fibers improve many of these lipoprotein abnormalities and the better glycemic control accompanying appropriate high-fiber nutrition further improves these derangements, especially hypertriglyceridemia (Anderson et al. 2004). Hypertension is more prevalent in diabetic individuals, and increased fiber intake improves blood pressure (Anderson 1983; Appel et al. 1997). Other minor risk factors for CHD—such as elevated serum values for CRP, homocysteine, fibrinogen, and plasminogen activator inhibitor-1 (PAI-1)—also tend to be improved by a health-promoting, high-fiber diet (Bhargava 2006; Flight and Clifton 2006; Lee et al. 2007).

Glycemic Control

Expert panels around the world recommend high-carbohydrate, high-fiber (HCHF) diets as the treatment of choice for persons with type 1 or type 2 diabetes (Anderson 2005; Anderson et al. 2004). Using an artificial pancreas that provided an insulin infusion rate to maintain target blood glucose values, we documented that insulin requirements for persons with type 1 diabetes were significantly lower on a HCHF diet than on a low-carbohydrate, low-fiber diet (Anderson et al. 1991). To rigorously examine effects of diets on insulin sensitivity, we performed insulin clamp studies on healthy individuals who were consuming either HCHF or control diets using a random-sequence, crossover design. HCHF diets were associated with significantly greater sensitivity to insulin than were control diets (Fukagawa et al. 1990). While there have been many advocates for carbohydrate restriction and increased intake of monounsaturated fats (Garg 1998) for management of diabetes, the clinical trial evidence supporting these interventions is limited. Our meta-analysis of RCTs comparing various diets included 13 trials with 141 diabetic patients (Anderson et al. 2004). These studies indicate that HCHF diets, compared to control diets, are accompanied by these significant improvements: fasting plasma glucose, -14.3% (95% CI, $-19.2–9.4\%$); postprandial plasma glucose, -13.6%. (95% CI, $-24.6–2.6\%$); average daily plasma glucose, -12.5% (95% CI, $-22.0–3.0\%$); and hemoglobin A1c, -6.0% (95% CI, $-11.6–0.3\%$). In this meta-analysis, we also assessed the effects of low GI diets compared to control diet studied in 8 trials that included 104 diabetic subjects (Anderson et al. 2004). There were significant reductions in fasting plasma glucose (-7.5%, 95% CI, $-14.5–0.5\%$) and nonsignificant reductions in hemoglobin A1c (-3.3%) and fructosamine (-2.5%) comparing low GI to control diets.

Nutrition recommendations for persons with diabetes from nine influential international agencies were recently reviewed and aggregated as "Evidence-Based Recommendations" (Anderson et al. 2004). These recommendations are: attain and maintain desirable weight (BMI $\leq$ 25 kg/m^2); carbohydrate intake, 55–65% of energy; choose whole grains, legumes, and vegetables; use fruits and other sources of mono- and disaccharides in moderation; incorporate GI into exchanges and teaching material; fiber intake, 25–50 g/day (15–25 g/1,000 kcal); protein intake, 12–16% or energy; total fat intake, <30% of energy; saturated/trans fatty acids intake, <10% of energy; monounsaturated fat intake, 12–15% of energy; polyunsaturated fat intake, <10% of energy; and cholesterol intake, <200 mg/day.

Hypertension

Our early studies with HCHF diets documented a significant blood pressure reduction without weight loss (Anderson 1983). The DASH diet, generous in whole grains and fiber, significantly decreases blood pressure (Appel et al. 1997). Thus, incorporation of whole grains and dietary fiber significantly decreases blood pressure, a major CHD risk factor.

Dyslipidemia

Whole grain and fiber intake have favorable effects on virtually all derangements of lipid metabolism associated with diabetes (Anderson et al. 2004). The effects of HCHF diets on major lipoprotein fractions were assessed in our meta-analysis of 13 RCTs (Anderson et al. 2004). Compared to control diets, HCHC diets were associated with significant reductions in these fasting serum lipoproteins values: LDL-cholesterol, -16.4% (95% CI, -25.2–7.7%); triglycerides, -12.8% (95% CI -21.2–4.3%); and HDL-cholesterol, -6.3%, (95% CI, -10.9–1.7%). Despite concerns that high-carbohydrate diets would increase fasting serum triglyceride values, our early observations indicated that a diet rich in complex carbohydrates and fiber significantly decreases serum triglycerides (Kiehm et al. 1976), and this has been supported by a meta-analysis of effects of HCHF diets on serum triglycerides (Anderson 2000). The short-term effects of HCHF diets on serum HDL-cholesterol values may be misleading. First, when Tarahumara Indians switched from a low-fat, high-fiber native diet to an "affluent" typical American diet their serum HDL-cholesterol values decreased significantly (McMurry et al. 1991); whether this is an unfavorable change in their CHD risk or even their HDL-scavenger capacity is uncertain. Second, high-density lipoproteins—with more than a dozen distinct particles (Hansel et al. 2004)—may undergo dynamic compositional changes over time. Our observations with long-term dietary fiber use indicate that values decline initially and then return to above initial values (Anderson et al. 1992; Anderson and Gustafson 1988). With weight loss, we see a similar pattern with initial decreases of 25% and subsequent increases to 10% above initial values (Anderson et al. 2007). Third, low-GI diets, compared to higher GI diets, are with significantly higher HDL-cholesterol values (Anderson et al. 2004; Frost et al. 1999). Fourth, proinflammatory or proatherogenic HDL fractions (Navab et al. 2006) may be reduced by whole grains or high-fiber diets and increased by high-fat diets.

Conclusions

While pharmacotherapies with important diabetes-protective effects are emerging, nutritional approaches that include cereal fibers from whole grains and other fiber-rich foods with their associated phytochemicals remain the approach of choice for most individuals at moderate to high risk for developing type 2 diabetes. Epidemiologic studies strongly support the concept that high levels of dietary fiber intake associated with significant protection from development of diabetes and obesity. More specifically, cereal fiber or whole grains appear to convey the majority of these protective effects. Prospective cohort studies support these observations. Dietary fibers have many physiological actions that contribute to these protective effects. Most soluble fibers slow gastric emptying, decrease postprandial glycemia, and insulinemia, and have favorable effects on gut hormones. High-fiber food intakes are followed by greater satiation and satiety after intake and after the second meal. Randomized, controlled, clinical trials document that higher fiber intake is accompanied by more weight loss than are control diets. Whole grain and other fiber-rich foods, in addition to their fiber content, contain complex carbohydrates, vitamins, minerals, antioxidants, phytochemicals, and other compounds that may contribute synergistically to the protection from diabetes. High-fiber diets offer substantial therapeutic benefits for persons with the metabolic syndrome, prediabetes, or diabetes and form a cornerstone for the nutrition therapy for persons with diabetes.

References

Abbasi F, Brown BW, Lamendola C, McLaughlin T, Reaven GM. 2002. Relationship between obesity, insulin resistance, and coronary heart disease. J Am Coll Cardiol 40:943.

Alberti KGMM, Zimmet P, DeFronzo RA, Keen H. 1997. International Textbook of Diabetes, 2nd ed. New York: John Wiley & Sons.

Anderson JW. 1983. Plant fiber and blood pressure. Ann Intern Med 98:842–846.

Anderson JW. 2000. Dietary fiber prevents carbohydrate-induced hypertriglyceridemia. Curr Artheroscler Rep 2:536–541.

Anderson JW. 2005. In Diabetes Mellitus: Medical Nutrition Therapy, Modern Nutrition in Health and Disease, edited by Shils ME, Shike M, Ross AC, Caballero B, Cousins RJ, pp. 1043–1066. Philadelphia: Lea & Febiger.

Anderson JW. 2007. Orlistat for management of overweight and obesity: Review of potential for 60 mg over-the-counter dose. Expert Opin Pharmacother 8:1733–1742.

Anderson JW, Conley SB. 2007. In Whole Grains and Diabetes, Whole Grains and Health, edited by Marquart L, Jacobs DR, Jr, McIntosh GH, Poutanen K, Reicks M, pp. 29–45. Ames, IA: Blackwell Publishing Professional.

Anderson JW, Conley SB, Nicholas AS. 2007. One-hundred pound weight losses with intensive behavioral program: Report of 118 patients with risk factor changes and long-term follow-up. Am J Clin Nutr 86:301–307.

Anderson JW, Garrity TF, Wood CL, Whitis SE, Smith BM, Oeltgen PR. 1992. Prospective, randomized, controlled comparison of the effects of low-fat and low-fat plus high-fiber diets on serum lipid concentrations. Am J Clin Nutr 56:887–894.

Anderson JW, Gustafson NJ. 1988. Hypocholesterolemic effects of oat and bean products. Am J Clin Nutr 48:749–753.

Anderson JW, Johnstone BM, Cook-Newell ME. 1995. Meta-analysis of effects of soy protein intake on serum lipids in humans. New Engl J Med 333:276–282.

Anderson JW, Johnstone BM, Remley DT. 1999. Breast-feeding and cognitive function: A meta-analysis. Am J Clin Nutr 70:525–535.

Anderson JW, Kendall CWC, Jenkins DJA. 2003. Importance of weight management in type 2 diabetes: Review with meta-analysis of clinical studies. J Am Coll Nutr 22:331–339.

Anderson JW, Randles KM, Kendall CWC, Jenkins DJA. 2004. Carbohydrate and fiber recommendations for individuals with diabetes: A quantitative assessment and meta-analysis of the evidence. J Am Coll Nutr 23:5–17.

Anderson JW, Zeigler JA, Floore TL, Dillion DW, Wood CL, Oeltgen PR, Whitley RJ. 1991. Metabolic effects of high-carbohydrate, high-fiber diets for insulin-dependent diabetic individuals. Am J Clin Nutr 54:936–943.

Appel LJ, Moore TJ, Obarzanek E, Vollmer WM, Svetkey LP, Sacks FM, Bray GA, Vogt TM, Cutler JA, Windhauser MM, Lin PH, Karanja N. 1997. A clinical trial of the effects of dietary patterns on blood pressure. DASH Collaborative Research Group. New Engl J Med 336:1117–1124.

Asakawa A, Inui A, Kaga T, Yuzuriha H, Nagata N, Ueno N, Makino S, Fujimiya M, Nijima A, Fujino MA, Kasuga M. 2001. Ghrelin is an appetite-stimulating signal from the stomach with structural resemblance to motilin. Gastroenterology 120:337–345.

Azadbakht L, Mirmiran P, Esmaillzadeh A, Azizi T, Azizi F. 2005. Beneficial effects of a Dietary Approaches to Stop Hypertension eating plan on features of the metabolic syndrome. Diabetes Care 28:2823–2831.

Bachman CM, Baranowski T, Nicklas TA. 2006. Is there an association between sweetened beverages and adiposity? Nutr Rev 64:153–174.

Barclay AW, Brand-Miller JC, Mitchell P. 2006. Macronutrient intake, glycaemic index and glycaemic load of older Australian subjects with and without diabetes: Baseline data from the Blue Mountains Eye study. Br J Nutr 96:117–123.

Batterham RL, Cohen MA, Ellis SM, Le Roux CW, Withers DJ, Frost GS, Ghatei MA, Bloom SR. 2003. Inhibition of food intake in obese subjects by peptide YY3-36. New Engl J Med 349:941–948.

Berkey CS, Rockett HR, Field AE, Gillman MW, Colditz GA. 2004. Sugar-added beverages and adolescent weight change. Obes Res 12:778–788.

Bes-Rastrollo M, Martinez-Gonzalez MA, Sanchez-Villegas A, de la Fuente AC, Martinez JA. 2006a. Association of fiber intake and fruit/vegetable consumption with weight gain in a Mediterranean population. Nutrition 22:504–511.

Bes-Rastrollo M, Sanchez-Villegas A, Gomez-Gracia E, Martinez JA, Pajares RM, Martinez-Gonzalez MA. 2006b. Predictors of weight gain in a Mediterranean cohort: The Seguimiento Universidad de Navarra Study 1. Am J Clin Nutr 83:362–370.

Bhargava A. 2006. Fiber intakes and anthropometric measures are predictors of circulating hormone, triglyceride, and cholesterol concentrations in the women's health trial. J Nutr 136:2249–2254.

Bhargava A, Amialchuk A. 2007. Added sugars displaced the use of vital nutrients in the National Food Stamp Program Survey. J Nutr 137:453–460.

Bhathena SJ, Velasquez MT. 2002. Beneficial role of dietary phytoestrogens in obesity and diabetes. Am J Clin Nutr 76:1191–1201.

Birketvedt GS, Shimshi M, Erling T, Florholmen J. 2005. Experiences with three different fiber supplements in weight reduction. Med Sci Monit 11:I5–I8.

Blackburn NA, Redfern JS, Jarjis H, Holgate AM, Hanning I, Scarpello JHB, Johnson IT, Read NW. 1984. The mechanism of action of guar gum in improving glucose tolerance in men. Clin Sci 66:329–336.

Blundell JE, Burley VJ. 1987. Satiation, satiety and the action of fibre on food intake. Int J Obes 11(Suppl 1):9–25.

Bo S, Durazzo M, Guidi S, Carello M, Sacerdote C, Silli B, Rosato R, Cassader M, Gentile L, Pagano G. 2006. Dietary magnesium and fiber intakes and inflammatory and metabolic indicators in middle-aged subjects from a population-based cohort. Am J Clin Nutr 84:1062–1069.

Bouin M, Savoye G, Maillot C, Hellot M-F, Guedon C, Denis P, Ducrotte P. 2000. How do fiber-supplemented formulas affect antroduodenal motility during enteral feeding? A comparative study between mixed and insoluble fibers. Am J Clin Nutr 72:1040–1046.

Bourdon I, Olson B, Backus R, Davis PA, Schneeman BO. 2001. Beans as a source of dietary fiber increase cholecystokinin and apolipoprotein b-48 response to test meal in man. J Nutr 131:1485–1490.

Bourdon I, Yokoyama W, Davis P, Hudson C, Backus R, Richter D, Knuckles B, Schneeman BO. 1999. Postprandial lipid, glucose, insulin, and cholecystokinin responses in men fed barley pasta enriched with b-glucan. Am J Clin Nutr 59:55–63.

Boyd KA, O'Donovan DG, Doran S, Wishart J, Chapman IA, Horowitz M, Feine C. 2003. High-fat diet effects on gut motility, hormone and appetite responses to duodenal lipid in healthy man. Am J Physiol Gastrointest Liver Physiol 282:G188–G196.

Bray GA. 1998. Obesity: A time bomb to be defused. Lancet 352:160–161.

Bray GA, Nielsen SJ, Popkin BM. 2004. Consumption of high-fructose corn syrup in beverages may play a role in the epidemic of obesity. Am J Clin Nutr 79:537–543.

Burkitt DP, Trowell HC. 1975a. In Diabetes Mellitus and Obesity, Refined Carbohydrate Foods and Diseases: Some Implications of Dietary Fibre, edited by Burkitt DP, Trowell HC. London: Academic Press.

Burkitt DP, Trowell HC. 1975b. Refined Carbohydrate Foods and Diseases: Some Implications of Dietary Fibre. London: Academic Press.

Burley VJ, Leeds AR, Blundell JE. 1987. The effect of high and low-fibre breakfasts on hunger, satiety and food intake in a subsequent meal. Int J Obes 11(Suppl 1):87–93.

Burton-Freeman B. 2000. Dietary fiber and energy regulation. J Nutr 130:272S–275S.

Burton-Freeman B, Davis PA, Schneeman BO. 2002. Plasma cholecystokinin is associated with subjective measures of satiety in women. Am J Clin Nutr 76:659–667.

Cairella G, Cairella M, Marchini G. 1995. Effect of dietary fibre on weight correction after modified fasting. Eur J Clin Nutr 49(Suppl 3):S325–S327.

Cani PD, Neyrinck AM, Maton N, Delzenne NM. 2005. Oligofructose promotes satiety in rats fed a high-fat diet: Involvement of glucagon-like Peptide-1. Obes Res 13:1000–1007.

Chiasson JL. 2006. Acarbose for the prevention of diabetes, hypertension, and cardiovascular disease in subjects with impaired glucose tolerance: The Study to Prevent Non-Insulin-Dependent Diabetes Mellitus (STOP-NIDDM) Trial. Endocr Pract 12(Suppl 1):25–30.

Cleave TL. 1956. The neglect of the natural principles in current medical practice. J R Nav Med Serv 42:55–60.

Cummings DE, Weigle DS, Frayo RS, Breen PQ, Ma MK, Dellinger EP, Purnell JQ. 2002. Plasma ghrelin levels after diet-induced weight loss or gastric bypass surgery. New Engl J Med 346:1623–1630.

di Lorenzo C, Williams CM, Hajnal F, Valenzuela JE. 1988. Pectin delays gastric emptying and increases satiety in obese subjects. Gastroenterology 95:1211–1215.

Doi K. 1995. Effect of konjac fibre (glucomanna) on glucose and lipids. Eur J Clin Nutr 49:S190–S197.

Drucker DJ. 2003. Enhancing incretin action for treatment of type 2 diabetes. Diabetes Care 26:2929–2940.

Duncan LJP, Rose K, Meikeljohn AP. 1960. Phenmetrazine hydrochloride and methyl-cellulose in the treatment of "refractory" obesity. Lancet 1:1262–1265.

Ebihara K, Kiriyama S. 1982. Comparative effects of water-soluble and water-insoluble dietary fibers on various parameters relating to glucose tolerance in rats. Nutr Rep Int 26:193–201.

Ehmann D, Ressin W. 1985. About the significance of dietary fibre in the dietetic treatment of overweight individuals. Pharm Ztg 130:124–127.

Esmaillzadeh A, Azadbakht L. 2006. Whole-grain intake, metabolic syndrome, and mortality in older adults. Am J Clin Nutr 83:1439–1440.

Evans JL, Goldfine ID, Maddux BA, Grodsky GM. 2003. Are oxidative stress-activated signaling pathways mediators of insulin resistance and beta-cell dysfunction? Diabetes 52:1–8.

Fineman MS, Koda JE, Shen LZ, Strobel SA, Maggs DG, Weyer C, Kolterman OG. 2002. The human amylin analog, pramlitide, corrects postprandial hyperglucagone-mia in patients with type 1 diabetes. Metabolism 51:636–641.

Flight I, Clifton P. 2006. Cereal grains and legumes in the prevention of coronary heart disease and stroke: A review of the literature. Eur J Clin Nutr 60:1145–1159.

Flourie B, Vidon N, Florent CH, Bernier JJ. 1984. Effect of pectin on jejunal glucose absorption and unstirred layer thickness in normal man. Gut 25:936–941.

Ford ES. 2005. Risks for all-cause mortality, cardiovascular disease, and diabetes asso-ciated with the metabolic syndrome: A summary of the evidence. Diabetes Care 28:1769–1778.

Foster-Powell K, Holt SHA, Brand-Miller JC. 2002. International table of glycemic index and glycemic load values: 2002. Am J Clin Nutr 76:5–56.

Frost GS, Brynes AE, Dhillo WS, Bloom SR, McBurney MI. 2003. The effects of fiber enrichment fo pasta and fat content on gastric emptying, GLP-1, glucose, and insulin responses to a meal. Eur J Clin Nutr 57:293–298.

Frost G, Leeds AA, Dore CJ, Madeiros S, Brading S, Dornhorst A. 1999. Glycaemic index as a determinant of serum HDL-cholesterol concentration. Lancet 353:1045–1048.

Fukagawa NK, Anderson JW, Young VR, Minaker KL. 1990. High-carbohydrate, high-fiber diets increase peripheral insulin sensitivity in healthy young and old adults. Am J Clin Nutr 52:524–528.

Fung TT, Hu FB, Pereira MA, Liu S, Stampfer MJ, Colditz GA, Willett WC. 2002. Whole-grain intake and the risk of type 2 diabetes: A prospective study in men. Am J Clin Nutr 76:535–540.

Garg A. 1998. High-monounsaturated-fat diets for patients with diabetes mellitus: A meta-analysis. Am J Clin Nutr 67:577S–582S.

Gee JM, Lee-Finglas W, Wortley GW, Johnson IT. 1996. Fermentable carbohydrates elevate plasma enteroglucagon but high viscosity in also necessary to stimulate small bowel mucosal proliferation in rats. J Nutr 126:373–379.

Geleva D, Thomas W, Gannon MC. 2003. A solubilized cellulose fiber decreases peak postprandial cholecystokinin concentrations after a liquid mixed meal in hyperc-holesterolemic men and women. J Nutr 133:2194–2203.

Ginsberg HN. 2006. Review: Efficacy and mechanisms of action of statins in the treatment of diabetic dyslipidemia. J Clin Endocrinol Metab 91:383–392.

Gowri MS, Van Der Westhuyzen DR, Bridges SR, Anderson JW. 1999. Decreased protection by HDL from poorly controlled type 2 diabetic subjects against LDL oxidation may be due to the abnormal composition of HDL. Anterioscl Throm Vas 9:2226–2233.

Gross LS, Li L, Ford ES, Liu S. 2004. Increased consumption of refined carbohydrates and the epidemic of type 2 diabetes in the United States: An ecologic assessment. Am J Clin Nutr 79:774–779.

Gruendel S, Garcia AL, Otto B, Mueller C, Steiniger J, Weickert MO, Speth M, Katz N, Koebnick C. 2006. Carob pulp preparation rich in insoluble dietary fiber and polyphenols enhances lipid oxidation and lowers postprandial acylated ghrelin in humans. J Nutr 136:1533–1538.

Haber GB, Heaton KW, Murphy D, Burroughs LF. 1977. Depletion and disruption of dietary fibre. Effects on satiety, plasm-glucose, and serum-insulin. Lancet 2:679–682.

Hallmans G, Zhang JX, Lundin E, Stattin P, Johansson A, Johansson I, Hulten K, Winkvist A, Aman P, Lenner P, Adlercreutz H. 2003. Rye, lignans and human health. Proc Nutr Soc 62:193–199.

Hansel B, Giral P, Nobecourt E, Chantepie S, Bruckert E, Chapman MJ, Kontush A. 2004. Metabolic syndrome is associated with elevated oxidative stress and dysfunctional dense high-density lipoprotein particles displaying impaired antioxidative activity. J Clin Endocrinol Metab 89:4963–4971.

Heaton KW, Emmett PM, Henry CL, Thornton JR, Manhire A, Hartog M. 1973. Food fibre as an obstacle to energy intake. Lancet 2:1418–1421.

Helgeson VS, Viccaro L, Becker D, Escobar O, Siminerio L. 2006. Diet of adolescents with and without diabetes: Trading candy for potato chips? Diabetes Care 29:982–987.

Henry RW, Stout RW, Love AH. 1978. Lack of effect of bran enriched bread on plasma lipids, calcium, glucose and body weight. Ir J Med Sci 147:249–251.

Hodge AM, English DR, O'Dea K, Giles GG. 2004. Glycemic index and dietary fiber and the risk of type 2 diabetes. Diabetes Care 27:2701–2706.

Holt S, Heading RC, Carter DC, Prescott LF, Tothill P. 1979. Effect of gel fibre on gastric emptying and absorption of glucose and paracetomol. Lancet 1:636–639.

Howarth NC, Saltzman E, Roberts SB. 2001. Dietary fiber and weight regulation. Nutr Rev 59:129–139.

Jacobs DR, Jr. 2006. Fast food and sedentary lifestyle: A combination that leads to obesity. Am J Clin Nutr 83:189–190.

Jacobs DR, Jr., Gallaher DD. 2004. Whole grain intake and cardiovascular disease: A review. Curr Atheroscler Rep 6:415–423.

Jacobs DR, Meyer KA, Kushi LH, Folsom AR. 1999. Is whole grain intake associated with reduced total and cause-specific death rates in older women? The Iowa Women's Health Study. Am J Public Health 89:322–329.

Jang Y, Lee JH, Kim OY, Park HY, Lee SY. 2001. Consumption of whole grain and legume powder reduces insulin demand, lipid peroxidation, and plasma homocysteine concentrations in patients with coronary artery disease: Randomized controlled clinical trial. Arterioscler Thromb Vasc 21:2065–2071.

Janket SJ, Manson JE, Sesso H, Buring JE, Liu S. 2003. A prospective study of sugar intake and risk of type 2 diabetes in women. Diabetes Care 26:1008–1015.

Juntunen KS, Laaksonen DE, Autio K, Niskanen LK, Holst JJ, Savolainen KE, Liukkonen KH, Poutanen KS, Mykkanen HM. 2003a. Structural differences between rye and wheat breads but not total fiber content may explain the lower postprandial insulin response to rye bread. Am J Clin Nutr 78:957–964.

Juntunen KS, Laaksonen DE, Poutanen KS, Nikanen LK, Mykkanen HM. 2003b. High-fiber rye bread and insulin secretion and sensitivity in healthy postmenopausal women. Am J Clin Nutr 77:385–391.

Juntunen KS, Niskanen LK, Liukkonen KH, Poutanen KS, Holst JJ, Mykkanen HM. 2002. Postprandial glucose, insulin, and incretin responses to grain products in healthy subjects. Am J Clin Nutr 75:254–262.

Kiehm TG, Anderson JW, Ward K. 1976. Beneficial effects of a high carbohydrate, high fiber diet on hyperglycemic diabetic men. Am J Clin Nutr 29:895–899.

Kim SYS, Glynn NW, Kriska AM, Barton BA, Kronsberg SS, Daniels SR, Crawford PB, Sabry ZI, Liu K. 2002. Decline in physical activity in black girls and white girls during adolescence. New Engl J Med 347:709–715.

Koh-Banerjee P, Franz M, Sampson L, Jacobs DR, Spiegelman D, Willett W, Rimm E. 2004. Changes in whole-grain, bran, and cereal fiber consumption in relationship to 8-y weight gain among men. Am J Clin Nutr 80:1237–1245.

Konturek SJ, Konturek JW, Pawlik T, Brzozowki T. 2004. Brain-gut axis and its role in the control of food intake. J Physiol Pharmacol 55:137–154.

Koppes L, Dekker JM, Hendriks H, Bouter L, Heine RJ. 2005. Moderate alcohol consumption lowers risk of type 2 diabetes. Diabetes Care 28:719–725.

Kromhout D, Bosschieter EB, De Lezenne Coulander C. 1982. Dietary fiber and 10-year mortality from coronary hear disease, cancer and all causes. Lancet 2:518–522.

Krotkiewski M. 1985. In Use of Fibers in Different Weight Reduction Programs, Dietary Fiber and Obesity, edited by Bjorntorp P, Kritchevsky D, pp. 85–109. New York: Alan R. Liss.

Krotkiewski M, Smith U. 1985. In Dietary Fibre in Obesity, Dietary Fibre Perspectives. Reviews and Bibliography, edited by Leeds AA, Avenell A, pp. 61–67. London: John Libbey.

Laaksonen DE, Toppinen LK, Juntunen KS, Autio K, Liukkonen KH, Poutanen KS, Niskanen L, Mykkanen HM. 2005. Dietary carbohydrate modification enhances insulin secretion in persons with the metabolic syndrome. Am J Clin Nutr 82:1218–1227.

Lairon D. 2007. Dietary fiber and control of body weight. Nutr Metab Cardiovasc Dis 17:1–5.

Lairon D, Arnault N, Bertrais S, Planells R, Clero E, Hercberg S, Boutron-Ruault MC. 2005. Dietary fiber intake and risk factors for cardiovascular disease in French adults. Am J Clin Nutr 82:1185–1194.

Lakka H-M, Laaksonen DE, Lakka TA, Niskanen LK, Kumpusalo E, Tuomilehto J, Salonen JT. 2002. The metabolic syndrome and total and cardiovascular disease mortality in middle-aged men. J Am Med Assoc 288:2709–2716.

Lee S, Harnack L, Jacobs DR, Jr, Steffen LM, Luepker RV, Arnett DK. 2007. Trends in diet quality for coronary heart disease prevention between 1980–1982 and 2000–2002: The Minnesota Heart Survey. J Am Diet Assoc 107:213–222.

Levine AS, Kotz CM, Gosnell BA. 2003. Sugars: Hedonic aspects, neuroregulation, and energy balance. Am J Clin Nutr 78:834S–842S.

Levine RA, Schwartz SE, Singh A, Rogus JB, Track NS. 1982. In Chronic Pectin Ingestion Delays Gastric Emptying, Motility of the Digestive Tract, edited by Weinbeck M, pp. 379–385. New York: Raven.

Levine AS, Tallman JP, Grace MK, Parker SA, Billington CJ, Levitt MD. 1989. Effeect of breakfast cereals on short-term food intake. Am J Clin Nutr 50:1303–1307.

Liese AD, Roach AK, Sparks KC, Marquart L, D'Agostino RB, Mayer-Davis E. 2003. Whole-grain intake and insulin sensitvity: The Insulin Restistance Atherosclerosis Study. Am J Clin Nutr 78:965–971.

Lindstrom J, Peltonen M, Eriksson JG, Louheranta A, Fogelholm M, Uusitupa M, Tuomilehto J. 2006. High-fibre, low-fat diet predicts long-term weight loss and decreased type 2 diabetes risk: The Finnish Diabetes Prevention Study. Diabetologia 49:912–920.

Liu S. 2002. Intake of refined carbohydrates and whole grain foods in relation to risk of type 2 diabetes mellitus and coronary heart disease. J Am Coll Nutr 21:298–306.

Liu S. 2003. Whole-grain foods, dietary fiber, and type 2 diabetes: Searching for a kernel of truth. Am J Clin Nutr 77:527–529.

Liu S, Manson JE, Stampfer MJ, Hu FB, Giovannucci E, Colditz GA, Hennekens CH, Willett WC. 2000. A prospective study of whole grain intake and risk of type 2 diabetes mellitus in US women. Am J Public Health 90:1409–1415.

Liu S, Serdula M, Janket S-J, Cook NR, Sesso HD, Willet WC, Manson JE, Buring JE. 2004. A prospective study of fruit and vegetable intake and the risk of type 2 diabetes in women. Diabetes Care 27:2993–2996.

Liu S, Willett WC, Manson JE, Hu FB, Rosner B, Colditz G. 2003. Relation between changes in intakes of dietary fiber and grain products and changes in weight and development of obesity among middle-aged women. Am J Clin Nutr 78:920–927.

Lu ZX, Walker KZ, Muir JG, Mascara T, O'Dea K. 2000. Arabinoxylan fiber, a product of wheat flour processing, reduces the postprandial glucose response in normoglycemic subjects. Am J Clin Nutr 71:1123–1128.

Ludwig DS, Pereira MA, Kroenke CH, Hilner JE, Van HL, Slattery ML, Jacobs DR, Jr. 1999. Dietary fiber, weight gain, and cardiovascular disease risk factors in young adults. JAMA 282:1539–1546.

Ludwig DS, Peterson SL, Gortmaker SL. 2001. Relation between consumption of sugar-sweetened drinks and childhood obesity: A prospective, observational analysis. Lancet 357:505–508.

Maji D, Roy RU, Das S. 2005. Prevention of type 2 diabetes in the prediabetic population. J Indian Med Assoc 103:609–611.

Malik VS, Schulze MB, Hu FB. 2006. Intake of sugar-sweetened beverages and weight gain: A systematic review. Am J Clin Nutr 84:274–288.

Maskarinec G, Takata Y, Pagano I, Carlin L, Goodman MT, Le ML, Nomura AM, Wilkens LR, Kolonel LN. 2006. Trends and dietary determinants of overweight and obesity in a multiethnic population. Obesity (Silver. Spring) 14:717–726.

Massimino SP, McBurney MI, Field CJ. 1998. Fermentable dietary fiber increases GLP-1 secretion and improves glucose homeostasis despite increased intestinal glucose transport capacity in healthy dogs. J Nutr 128:1786–1793.

McKeown NM, Meigs JB, Liu S, Saltzman E, Wilson PW, Jacques PF. 2004. Carbohydrate nutrition, insulin resistance, and the prevalence of the metabolic syndrome in the Framingham Offspring Cohort. Diabetes Care 27:538–546.

McKeown NM, Meigs JB, Liu S, Wilson PWF, Jacques PF. 2002. Whole-grain intake is favorably associated with metabolic risk facotrs for type 2 diabetes and cardiovascular disease in the Framingham Offspring Study. Am J Clin Nutr 76:390–398.

McMurry MP, Cerqueira MT, Connor SL, Connor WE. 1991. Changes in lipid and lipoprotein levels and body weight in Tarahumara Indians after consumption of an affluent diet. New Engl J Med 325:1704–1708.

Meyer KA, Kushi LH, Jacobs DR, Slavin J, Sellers TA, Folsom AR. 2000. Carbohydrates, dietary fiber, and incidence of type 2 diabetes in older women. Am J Clin Nutr 71:921–930.

Mickelsen O, Makdani DD, Cotton RH, Titcomb ST, Colmey JC, Gatty R. 1979. Effects of a high fiber bread diet on weight loss in college-age males. Am J Clin Nutr 32:1703–1709.

Miller HE, Rigelhop F, Marquart L, Prakash A, Kanter M. 2000. Whole-grain products and antioxidants. Cereal Food World 45:59–63.

Montonen J, Knekt P, Jarvinen R, Aromaa A, Reunanen A. 2003. Whole-grain and fiber intake and the incidence of type 2 diabetes. Am J Clin Nutr 77:622–629.

Morgan LM, Goulder TJ, Tsiolakis D, Marks V. 1990a. The effect of unabsorble carbohydrate on gut hormones. Modification of post-prandial GIP secretion by guar. Diabetologia 17:85–89.

Morgan LM, Tredger JA, Wright J, Marks V. 1990b. The effect of soluble- and insoluble-fibre supplements on post-prandial glucose tolerance, insulin, and gastric inhibitory polypeptide in healthy subjects. Br J Nutr 64:103–110.

Morris JN, Marr JW, Clayton DG. 1977. Diet and heart: A postscript. Br Med J 2:1307–1314.

Nathan DM, Berkwits M. 2007. Trials that matter: Rosiglitazone, ramipril, and the prevention of type 2 diabetes. Ann Intern Med 146:461–463.

Navab M, Anantharamaiah GM, Reddy ST, Van Lenten BJ, Ansell BJ, Fogelman AM. 2006. Mechanisms of disease: Proatherogenic HDL—an evolving field. Nat Clin Pract Endocrinol Metab 2:504–511.

Nestle M. 2006. Food marketing and childhood obesity—a matter of policy. N Engl J Med 354:2527–2529.

Onaga T, Zabielski R, Kato S. 2002. Multiple regulation of peptide YY secretion in the digestive tract. Peptides 23:279–290.

Pereira MA, Jacobs DR, Jr., Pins JJ, Raatz SK, Gross MD, Slavin JL, Seaquist ER. 2002. Effect of whole grains on insulin sensitivity in overweight hyperinsulinemic adults. Am J Clin Nutr 75:848–855.

Pereira MA, Jacobs DR, Slattery ML, Ruth K, Van Horn L, Hilner J, Kushi LH. 1998. The association between whole grain intake and fasting insulin in a bi-racial cohort of young adults: The CARDIA study. CVD Prevention 1(3):231–242.

Perfetti R, Merkel P. 2000. Glucagon-like peptide-1: A major regularor of pancreatic B-cell function. Eur J Endocrinol 143:717–725.

Piarulli F, Lapolla A, Sartore G, Rossetti C, Bax G, Noale M, Minicuci N, Fiore C, Marchioro L, Manzato E, Fedele D. 2005. Autoantibodies against oxidized LDLs and atherosclerosis in type 2 diabetes. Diabetes Care 28:653–657.

Poppitt SD, Keogh GF, Prentice AM, Williams DE, Sonnemans HM, Valk EE, Robinson E, Wareham NJ. 2002. Long-term effects of ad libitum low-fat, high-carbohydrate diets on body weight and serum lipids in overweight subjects with metabolic syndrome. Am J Clin Nutr 75:11–20.

Porikos K, Hagamen S. 1986. Is fiber satiating? Effects of high fiber preload on subsequent food intake of normal-weight and obese young men. Appetite 7:153–162.

Rainbird AL, Low AG. 1986. Effects of various types of dietary fibre on gastric emptying in growing pigs. Br J Nutr 55:111–121.

Reimer RA, McBurney MI. 2004. Dietary fiber modulates intestinal proglucagon messenger ribonucleic acid and postprandial secretion of glucagon-like peptide-1 and insulin in rats. Endocrinology 137:3948–3956.

Rigaud D, Ryttig KR, Angel LA, Apfelbaum M. 1990. Overweight treated with energy restricition and a dietry fibre supplement: A 6-month randomized, double-blind, placebo-controlled trial. Int J Obes 14:763–769.

Ross AHM, Eastwood MA, Brydon WG, Anderson JR, Anderson DM. 1983. A study of the effects of dietary gum arabic in humans. Am J Clin Nutr 37:368–375.

Rossner S, Andersson IL, Ryttig K. 1988. Effects of a dietary fibre supplement to a weight reduction programme on blood pressure. A randomized, double-blind, placebo-controlled study. Acta Med Scand 223:353–357.

Rossner S, von ZD, Ohlin A, Ryttig K. 1987. Weight reduction with dietary fibre supplements. Results of two double-blind randomized studies. Acta Med Scand 222:83–88.

Russ CS, Atkinson RL. 1985. Use of high fiber diets in the outpatient treatment of obesity. Nutr Rep Int 32:193–199.

Ryttig KR, Larsen S, Haegh L. 1984. Treatment of slightly to moderately overweight persons. A double-blind placebo-controlled study with diet and fibre tablets (DumoVital). Tidsskr Nor Laegeforen 104:989–991.

Ryttig KR, Tellnes G, Haegh L, Boe E, Fagerthun H. 1989. A dietary fibre supplement and weight maintenance after weight reduction: A randomized, double-blind, placebo-controlled long-term trial. Int J Obes 13:165–171.

Salmeron J, Ascherio A, Rimm EB, Colditz GA, Spiegelman D, Jenkins DJ, Stampfer MJ, Wing AL, Willett WC. 1997a. Dietary fiber, glycemic load, and risk of NIDDM in men. Diabetes Care 20:545–550.

Salmeron J, Manson JE, Stampfer MJ, Coldtiz GA, Wing AL, Willet WC. 1997b. Dietary fiber, glycemic load, and risk of non-insulin-dependent diabetes mellitus in women. JAMA 277:472–477.

Saris WH. 2003. Sugars, energy metabolism, and body weight control. Am J Clin Nutr 78:850S–857S.

Schulze MB, Hoffmann K, Manson JE, Willett WC, Meigs JB, Weikert C, Heidemann C, Colditz GA, Hu FB. 2005. Dietary pattern, inflammation, and incidence of type 2 diabetes in women. Am J Clin Nutr 82:675–684.

Schulze MB, Hu FB. 2005. Primary prevention of diabetes: What can be done and how much can be prevented? Ann Rev Public Health 26:445–467.

Schulze MB, Liu S, Rimm EB, Manson JE, Willett WC, Hu FB. 2004a. Glycemic index, glycemic load, and dietary fiber intake and incidence of type 2 diabetes in younger and middle-aged women. Am J Clin Nutr 80:348–356.

Schulze MB, Manson JE, Ludwig DS, Colditz GA, Stampfer MJ, Willett WC, Hu FB. 2004b. Sugar-sweetened beverages, weight gain, and incidence of type 2 diabetes in young and middle-aged women. JAMA 292:927–934.

Slavin J. 2004. Whole grains and human health. Nutr Res Rev 17:99–110.

Solum TT, Ryttig KR, Sollum E, Larsen S. 1987. The influence of a high-fibre diet on body weight, serum lipids and blood pressure in slightly overweight persons. A randomized, double-blind, placebo-controlled investigation with diet and fibre tablets. Int J Obes 11:67–71.

Stevens J, Ahn K, Juhaer I, Houston D, Steffan L, Couper D. 2002. Dietary fiber intake and glycemic index and incidence of diabetes in African-American and white adults. Diabetes Care 25:1715–1721.

Todd PA, Benfield P, Goa KL. 1990. Guar gum a review of its pharmacological properties, and use as a dietary adjunct in hypercholesterolaemia. Drugs 39:917–928.

Trowell HC. 1960. Non-Infective Diseases. London: Edward Arnold.

Trowell H. 1972. Ischemic heart disease and dietary fiber. Am J Clin Nutr 25:926–932.

Trowell HC. 1975. Dietary-fiber hypothesis of the etiology of diabetes mellitus. Diabetes Care 24:762–766.

Truswell AS. 2002. Cereal grains and coronary heart disease. Eur J Clin Nutr 56:1–14.

Tso P, Liu M, Kalogeris TJ, Thomson AB. 2001. The role of apolipoprotein A-IV in the regulation of food intake. Ann Rev Nutr 21:231–254.

Tuomilehto J, Voutilainen E, Huttunen J, Vinni S, Homan K. 1980. Effect of guar gum on body weight and serum lipids in hypercholesterolemic females. Acta Med Scand 208:45–48.

Valle-Jones JC. 1980. The evaluation of a new appetite-reducing agent (Prefil) in the management of obesity. Br J Clin Pract 34:72–74.

Van Dam RM, Hu FB. 2005. Coffee consumption and risk of type 2 diabetes: A systematic review. JAMA 294:97–104.

Vartanian LR, Schwartz MB, Brownell KD. 2007. Effects of soft drink consumption on nutrition and health: A systematic review and meta-analysis. Am J Public Health 97:667–675.

Vermunt SH, Pasman WJ, Schaafsma G, Kardinaal AF. 2003. Effects of sugar intake on body weight: A review. Obes Rev 4:91–99.

Walsh DE, Yaghoubian V, Behforooz A. 1984. Effect of glucomannan on obese patients: A clinical study. Int J Obes 8:289–293.

Weickert MO, Mohlig M, Koebnick C, Holst JJ, Namsolleck P, Ristow M, Osterhoff M, Rochlitz H, Rudovich N, Spranger J, Pfeiffer AF. 2005. Impact of cereal fibre on glucose-regulating factors. Diabetologia 48:2343–2353.

Weickert MO, Mohlig M, Schofl C, Arafat AM, Otto B, Viehoff H, Koebnick C, Kohl A, Spranger J, Pfeiffer AF. 2006. Cereal fiber improves whole-body insulin sensitivity in overweight and obese women. Diabetes Care 29:775–780.

Woods SC. 2004. Gastrointestinal satiety signals: I. an overview of gastrointestinal signals that influence food intake. Am J Physiol Gastrointest Liver Physiol 286:G7–G13.

Ylonen K, Saloranta C, Kronberg-Kippila C, Groop L, Aro A, Virtanen SM. 2003. Associations of dietary fiber with glucose metabolism in nondiabetic relatives of subjects with type 2 diabetes: The Botnia Dietary Study. Diabetes Care 26:1979–1985.

Yudkin JS. 1959. The causes and cure of obesity. Lancet 2:1135–1138.

Zhang C, Liu S, Solomon CG, Hu FB. 2006. Dietary fiber intake, dietary glycemic load, and the risk for gestational diabetes mellitus. Diabetes Care 29:2223–2230.

Chapter 8

Cinnamon, Glucose, and Insulin Sensitivity

Richard A Anderson, PhD, and Anne-Marie Roussel, PhD

Overview

Compounds found in cinnamon not only improve the function of insulin but also function as antioxidants and may be anti-inflammatory. Human studies involving subjects with the metabolic syndrome, type 2 DM, and polycystic ovary syndrome (PCOS) show beneficial effects of whole cinnamon and aqueous extracts of cinnamon on glucose, insulin, lipids, and antioxidant status. There may also be effects of the cinnamon compounds on lean body mass, body composition, and inflammatory response. Type A procyanidin polymers were isolated from cinnamon that function both as antioxidants and also enhance insulin function; there are also other cinnamon compounds that may be beneficial. Cinnamon compounds activate insulin receptors by increasing their tyrosine phosphorylation activity and by decreasing phosphatase activity leading to increased insulin sensitivity. These compounds also increase the amount of insulin receptor-beta and GLUT4 protein, increase glycogen synthase activity and glycogen accumulation, and increase the amount of the early response anti-inflammatory proteins. All these activities are associated with increased insulin sensitivity that is associated with decreased incidence of diabetes, cardiovascular, and related diseases.

Introduction

Cinnamon is the most commonly consumed spice and has been known from antiquity. Cinnamon was considered more valuable than gold and a gift fit for kings and monarchs. Cinnamon was imported to Egypt from China as early as 2000 BC and is mentioned in the Bible. Emperor Nero is said to have burned a year's supply of cinnamon at the funeral for his wife in 65 AD to denote her importance to him (http://en.wikipedia.org/wiki/Cinnamon).

In 1990, we reported that compounds found in cinnamon have insulin-potentiating properties and may be involved in the alleviation of the signs and symptoms of diabetes and cardiovascular disease related to insulin resistance (Khan et al.

127

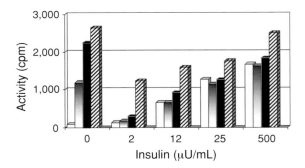

Figure 8.1.   Aqueous extracts of cinnamon increase insulin activity in vitro. The first open bar at each level of insulin denotes insulin only. The second, third, and fourth bar at each level of insulin denote the purified cinnamon fraction (7 mg/mL), diluted 1:20, 1:10 and 1:2, respectively; 25 μL was diluted into the 2 mL assay mixture. (From Anderson et al. 2004.)

1990). Following the demonstration that water soluble compounds found in cinnamon potentiate insulin activity, 49 herb, spice, and medicinal extracts were evaluated for insulin-like or insulin-potentiating activity in an in vitro model (Broadhurst et al. 2000). Aqueous extracts of cinnamon potentiated insulin activity more than 20-fold, higher than any other compound tested at comparable dilutions. The in vitro assay is a measure of the insulin-dependent utilization of glucose (Anderson et al. 1978). Figure 8.1 shows the effects of an aqueous extract of cinnamon and insulin on the insulin-dependent incorporation of glucose into lipids. The open bars are simply an insulin titration and obviously with increasing concentrations of insulin there are increases in the product formed. However, the most important point is to analyze the changes in the product formed at no added insulin, the first set of four bars. The effects of adding more of the aqueous extract of cinnamon are similar to adding more insulin. This is extremely important from a human health standpoint since it results in increased insulin sensitivity, and less insulin is required to have larger insulin effects.

Wortmanin, a potent inhibitor of phosphoinositol-3′kinase (PI-3-kinase), decreased the biological responses of the cinnamon extract and insulin similarly in vitro indicating that cinnamon is affecting elements upstream of PI-3 kinase. Water soluble cinnamon compounds also stimulate the autophosphorylation of the insulin receptor (Imparl-Radosevich et al. 1998). Since the assay used contained only the cinnamon extract, purified truncated version of the β-subunit of insulin receptor and radiolabelled ATP, it is assumed that there is a direct interaction between a component in the cinnamon extract and the kinase domain. It was hypothesized that the cinnamon compound enters the cells, interacts with intracellular kinase domain, and triggers an insulin-like response (Imparl-Radosevich et al. 1998).

Cinnamon compounds also inhibit phophotyrosine phosphatase (PTP-1B), an enzyme functioning in the dephosphorylation of the insulin receptor

(Imparl-Radosevich et al. 1998). This inhibition is specific since there was no inhibition of alkaline phosphatase. PTP-1B preferentially dephosphorylates a phosphopeptide of the insulin receptor-beta subunit (Imparl-Radosevich et al. 1998). The activation of the phosphorylation of the insulin receptor and the inhibition of the dephosphorylation of the insulin receptor lead to increased phosphorylation of the insulin receptor which is associated with increased insulin sensitivity. Subjects with type 2 diabetes mellitus (DM) have reduced phosphorylation of the insulin receptor (Cusi et al. 2000); water soluble products found in cinnamon reverse this effect leading to increased insulin sensitivity (Ziegenfuss et al. 2006).

Cinnamon Compounds

The main constituent of cinnamon bark is cinnamon oil that mainly contains cinnamic acid, cinnamaldehyde, and cinnamic alcohol (Duke 1992). Cinnamaldehyde is the most prevalent compound in cinnamon with concentrations ranging from 6,000 to 30,000 ppm. Its activities include anesthetic, antibacterial, antiinflammatory, antiulcer, antiviral, COX-2 inhibitor, hypotensive, tranquilizer, etc (Duke 1992). Cinnamaldehyde has also been shown to reduce blood glucose and lipids in rats made chemically diabetic, increase circulating insulin, decrease HbA1c, and restore the activities of plasma enzymes including aspartate aminotransferase, alanine aminotransferase, lactate dehydrogenase, and alkaline and acid phosphatases (Subash et al. 2007). However, cinnamaldehye functions as a prooxidant and, when given orally to rats at 73.5 mg/kg body weight for 90 days, leads to increased levels of thiobarbituric acid-reactive substances (Gowder and Devaraj 2006). As discussed later, the aqueous extracts of cinnamon function as antioxidants in experimental animals and humans.

Cinnamic acid and its derivatives have been reported to possess a variety of pharmacological properties including antioxidant and antihyperglycemic effects (Duke 1992). 3,4-Di(OH)-cinnamate and 3,4-di(OH)-hydroxycinnamate lowered cholesterol and triglycerides in rats as well as hepatic HMG-CoA reductase (Lee et al. 2001). In addition, these compounds displayed antioxidant activity with lowered thiobarbituric acid-reactive substances and elevated glutathione peroxidase activity (Lee et al. 2001). 2-Alkoxydihydrocinnamates function as peroxisome proliferator-activated receptor (PPAR) agonists, and in vivo studies in a genetic diabetic animal model (ZDF rat) demonstrated that these compounds decrease blood glucose and triglycerides (Martin et al. 2005). These compounds function similar to glitazone drugs used in the treatment of diabetes, which are also potent inhibitors of PPAR-gamma. The naphthalenemethyl ester derivative of dihydroxycinnamic acid lowered blood glucose to near normal in chemically and genetically diabetic mice, and increased glucose transport by increasing translocation of GLUT4 (Kim et al. 2006). P-Methoxycinnamic acid reduced the blood glucose of chemically diabetic rats and normalized hepatic glucose-6-phosphatase, hepatic hexokinase, glucokinase, phosphofructokinase, hepatic glycogen, and glucose-6-phosphate (Adisakwattana et al. 2005). Anderson et al. (2004) demonstrated that

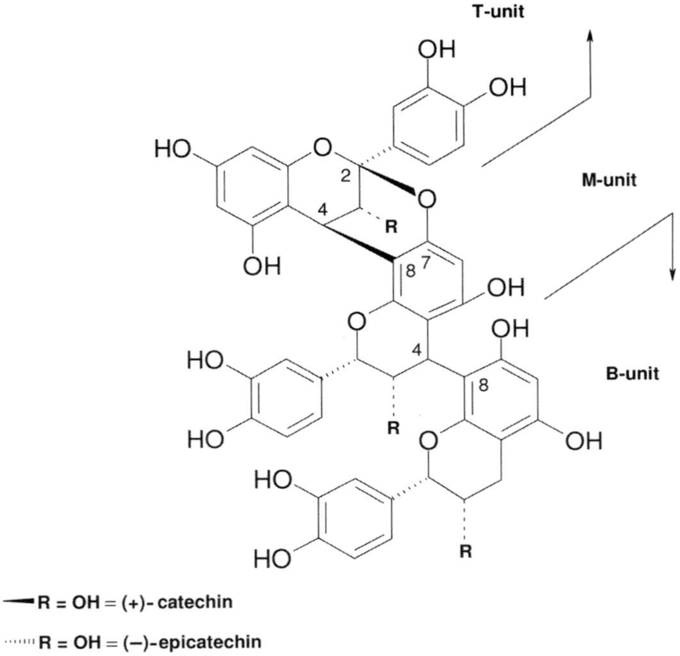

Figure 8.2. Structure of doubly linked procyanidin type-A polymers found in cinnamon that enhances insulin activity. (From Anderson et al. 2004.)

the in vitro insulin-potentiating activity found in cinnamon was present in the aqueous fraction. Aqueous extracts of "spent cinnamon," in which many of the organic components found in cinnamon, including cinnamaldehyde, are removed when cinnamon oil is extracted from whole cinnamon, has basically the same in vitro insulin-potentiating activity as extracts from the cinnamon before the cinnamon oil is removed. Since we were unable to identify known cinnamon compounds that acted similarly to the cinnamon extract and the activity was in aqueous fractions, we proceeded to isolate and characterize insulin-potentiating compounds present in the aqueous extracts of cinnamon. The structure of a class of water soluble cinnamon polyphenol compounds that were shown to display insulin-potentiating activity is shown in Figure 8.2. Two different trimers and a tetramer of the type A polymers have been isolated and all were shown to have in vitro insulin-potentiating activity (Anderson et al. 2004).

Animal Studies

In rats fed a control diet, the administration of aqueous extracts of cinnamon improves glucose metabolism and potentiates the action of insulin (Qin et al.

2003). After 3 weeks of oral administration of an aqueous cinnamon extract at 30 and 300 mg kg/body weight, there was significantly greater glucose utilization monitored during euglycemic clamp studies. Skeletal muscle insulin-stimulated insulin receptor-β and IRS-1 tyrosine phosphorylation levels and the IRS-1/PI 3-kinase ratio were also increased. These results suggest that increased glucose uptake is due to enhancing of the insulin-signaling pathway (Qin et al. 2003). In rats fed a high-fructose diet to simulate the metabolic syndrome, cinnamon extract at 300 mg/kg/day also improved in vivo glucose utilization monitored using euglycemic clamp studies compared with values for the control animals. Cinnamon extract fed to animals consuming the high fructose diet also prevented the development of the metabolic syndrome (Qin et al. 2004).

The aqueous extract of *Cinnamomon cassia* was slightly more efficacious on lowering blood glucose in rats in response to a glucose load than the extract of *Cinnamomon zeylanicum*. Effects on fasting glucose were not significant. Circulating insulin levels increased in the animals receiving cinnamon, and insulin release was also higher in an insulin secreting cell line (Verspohl et al. 2005).

Many commonly ingested nutrients or dietary elements known to augment insulin resistance are also associated with elevated blood pressure, for example, fatty acids and simple sugars. In contrast, dietary factors generally accepted to enhance insulin sensitivity such as soluble fibers and chromium are associated with lower blood pressure. Certain drugs, such as metformin and troglitazone, and exercise, which all augment insulin sensitivity, are also recognized to lower blood pressure (Preuss et al. 2006). This suggests that perturbed glucose/insulin metabolism is directly or indirectly involved in some forms of hypertension. Based on the correlation between insulin metabolism and blood pressure regulation, Preuss et al. (2006) examined the ability of cinnamon to influence systolic blood pressure of spontaneously hypertensive rats (SHR) with systolic blood pressure raised even higher by the presence of a moderate amount of sucrose in the diet. The amount of sucrose in the diet was 18%, which is similar to the average sugar content in the American diet. Cinnamon decreased hypertension, even in the starch group, indicating that cinnamon not only was improving the sucrose-induced hypertension but also the genetic component of the hypertension in these rats. The effects of the aqueous extracts were similar to those of the whole cinnamon, substantiating our previous studies demonstrating that the bioactivity relating to insulin function is present in the water-soluble material (Anderson et al. 2004; Broadhurst et al. 2000; Imparl-Radosevich et al. 1998; Jarvill-Taylor et al. 2001; Khan et al. 1990). There were significant effects at all levels of cinnamon tested, 10–100 g/kg of diet, indicating that lower levels of the spice may also be effective.

In a genetic mouse model of type 2 DM, the db/db mouse, an aqueous extract of cinnamon fed at 50–200 mg/kg body weight, also decreased blood glucose in a dose-dependent manner. In addition, serum insulin and HDL-cholesterol increased. Triglycerides, total cholesterol, and intestinal α-glucosidase activity were also significantly lower after 6 weeks of intake of the aqueous cinnamon extract (Kim et al. 2006). Cinnamon led to decreased sucrase, maltase, and lactase activities, suggesting that the mechanism of action of cinnamon may also be to inhibit the conversion of carbohydrates to simple sugars altering their subsequent absorption.

Human Studies

Following the observations that cinnamon potentiates insulin action in vitro, we conducted a human study involving 60 people with type 2 DM, 30 males and 30 females, who were taking sulfonylurea drugs (Khan et al. 2003). Subjects were divided randomly into six groups. Groups 1–3 received 1, 3, or 6 g of cinnamon as Cinnamomon cassia per day for 40 days. From 40 to 60 days, there was a washout period in which subjects did not receive capsules. Groups 4–6 received the same number of placebo capsules as the corresponding cinnamon groups. There were three placebo groups since the number of capsules in each of the groups was different. It was postulated that there would be a dose response to the three levels of cinnamon. After 40 days, all three levels of cinnamon reduced mean fasting serum glucose (18–29%) (3 groups of 10 people), triglycerides (23–30%), total cholesterol (12–26%), and LDL-cholesterol (7–27%) (Figure 8.3). Values in Figure 8.3 are only for the group of 10 people that received 1 g of cinnamon and the corresponding placebo group of 10 people. There were no significant changes in any of the three placebo groups. Values after the 20-day washout period were returning to baseline but were still significantly lower than the values at the onset of the study. This study basically confirms the effects of cinnamon since results of groups 2 and 3 repeated those in group 1 and all groups had their own respective placebo groups.

In a separate study involving 22 subjects with the metabolic syndrome, subjects were divided into two groups and given either 500 mg/day of a commercially available aqueous extract of cinnamon (Cinnulin PF®, Integrity Nutraceuticals, Sarasota, FL) or a placebo for 12 weeks. An estimated 41 million people in the United States are postulated to have the metabolic syndrome characterized by elevated blood sugar, hyperlipidemia, hypertension, and obesity. The National Cholesterol Education Program has established the current criteria for the metabolic syndrome as three or more than three of the following: (1) waist circumference >102 cm (40 inches) for men or >88 cm (35 inches) for women; (2) a fasting total serum triacylglycerol concentration of >1.7 mmol/L (150 mg/dL); (3) a fasting HDL-cholesterol concentration <1.0 mmol/L (40 mg/dL) for men or <1.2 mmol/L (50 mg/dL) for women; (4) a fasting glucose >6.1 mmol/L (110 mg/dL); or blood pressure >130 mm Hg (systolic) and/or 85 mm Hg (diastolic) and/or the use of antihypertensive medication.

Subjects in the group receiving the capsules containing the aqueous extract of cinnamon displayed decreases in fasting blood glucose (-8.4%), systolic blood pressure (-3.8%), and increases in lean mass ($+1.1\%$), compared with the placebo group. There were also significant decreases in body fat (-0.7%) in the cinnamon group (Figure 8.4) (Ziegenfuss et al. 2006).

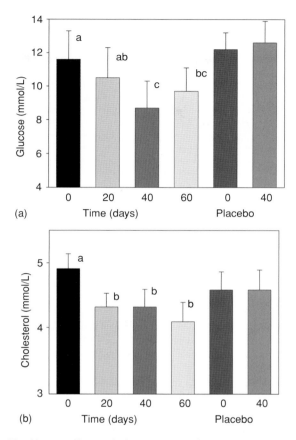

Figure 8.3. Significant effects of cinnamon on glucose (a) and cholesterol (b) of people with type 2 DM taking sulfonylurea drugs. Values in the panels are for subjects taking 1 g of cinnamon per day. Similar results were observed for subjects taking 3 and 6 g of cinnamon per day. Subjects ($n = 10$) consumed capsules for 40 days followed by a 20-day washout period. Bars with different superscripts are significantly different at $P < 0.05$. (From Khan et al. 2003.)

Oxidative stress, which is increased in obesity, plays an important role in the development of diabetes and cardiovascular diseases in obese people (Yu and Lyons 2005). Hyperglycemia causes the auto-oxidation of glucose, glycation of proteins, and the activation of polyol metabolism (Robertson 2004). These changes accelerate the generation of reactive oxygen species (ROS) and increase oxidative modifications of lipids and proteins (Osawa and Kato 2005). Roussel et al. (2008) found a significant positive correlation between plasma glucose levels and plasma malondyaldehyde (MDA), a measure of lipid peroxidation. In that study, cinnamon improved antioxidant status of people with the metabolic syndrome.

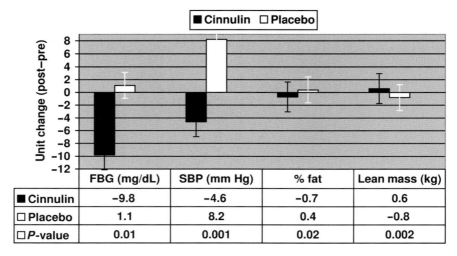

	FBG (mg/dL)	SBP (mm Hg)	% fat	Lean mass (kg)
■ Cinnulin	−9.8	−4.6	−0.7	0.6
□ Placebo	1.1	8.2	0.4	−0.8
□ P-value	0.01	0.001	0.02	0.002

Figure 8.4. Beneficial effects of aqueous extract of cinnamon on glucose, blood pressure, % body fat, and lean mass of subjects with metabolic syndrome. Data are expressed as change scores (week 12–week 0). Values represent group means ± SE. *P*-values were calculated from repeated measures (group × time) ANOVA, except for % fat where a dependent t-test was used following a marginally significant ($P < 0.06$) interaction term. (From Ziegenfuss et al. 2006.)

The improvement of impaired fasting glycemia due to cinnamon was correlated with the antioxidant effects of cinnamon supplementation assessed by plasma MDA, sulfyhydryl groups, and plasma antioxidant status evaluated using ferric reducing antioxidant power (FRAP). A significant positive correlation between plasma glucose levels and plasma MDA confirms a previous study showing that plasma glucose levels play a role in determining oxidative status (Hayden and Tyagi 2004). In subjects with metabolic syndrome, plasma MDA levels were reduced by the aqueous extract of cinnamon, indicating decreased lipid peroxidation, while plasma sulfhydryl groups were increased, indicating a protection of antioxidant sulfhydryl groups against oxidation (Roussel et al. 2008). The oxidation of lipids is thought to play a crucial role in the generation of atherosclerotic lesions in obese patients (Davi and Falco 2005). For patients suffering from obesity, cardiovascular diseases and diabetes, it is well known that decreasing lipid peroxidation is an important health challenge to avoid oxidative damage of the arterial walls and oxidative complications (Couillard et al. 2005). Functional consequences of SH group losses include protein misfolding, catalytic inactivation, and decreased antioxidative capacity (Balcerczyk and Bartosz 2003). In the group receiving cinnamon, plasma SH groups were increased after 12 weeks of supplementation, suggesting that cinnamon acts in protecting both lipids and proteins against oxidation. In parallel, the FRAP, which is a measure of the total antioxidant capacity of plasma, was increased thereby providing a contributory factor to the protective effects of cinnamon supplementation (Roussel et al. 2008). In contrast, cinnamon intake did not alter the activity of antioxidant enzymes in erythrocytes. Consistent with this observation, in humans, polyphenols in red wine do not alter the activities of renal antioxidant enzymes, while plasma antioxidant capacity is enhanced following red wine consumption (Rodrigo and Bosco 2006). However, in rats fed a high-fat diet with spices rich in cinnamon, antioxidant enzymes activities were enhanced (Dhuley 1999).

The antioxidant effects of cinnamon may favorably impact hormone signaling via adipokine cross-talk and may help explain the effects of cinnamon on body composition (Ziegenfuss et al. 2006). For example, Chrysohoou et al. (2006) recently reported an inverse relationship between body fat and antioxidant capacity, even after controlling for smoking, physical activity patterns, dietary habits, blood pressure, glucose levels, and lipid concentrations. It is therefore possible that the observed improvements in fasting blood glucose, as well as plasma antioxidant status, are responsible, at least in part, for cinnamon's beneficial effects on body composition and features of the metabolic syndrome (Ziegenfuss et al. 2006).

PCOS is one of the most common endocrinopathies among women of childbearing age, affecting 5–10% of the population (Wang et al. 2007). Insulin resistance and compensatory hyperinsulinemia are present in 50–70% of the women with PCOS and maybe as high as 95% in overweight women. Excess insulin secretion may also be implicated in the increased metabolic and cardiovascular risks reported in this disorder. Since insulin-sensitizing agents such as chromium and troglitazone have been shown to be beneficial in the treatment of PCOS, it was postulated that the insulin-potentiating water soluble polyphenol compounds found in cinnamon may also be beneficial for women with PCOS (Wang et al. 2007). During an 8-week treatment period, oral cinnamon extract, 1 g/day, resulted in a significant reduction in fasting glucose as well as in insulin resistance. Oral glucose tolerance tests (OGTT) also showed a 21% reduction in mean glucose and an increase in Matsuda's insulin sensitivity index. The cinnamon extract improved insulin resistance in the women with PCOS to that in the control group (Figure 8.5).

The beneficial effects of cinnamon are greater in the study of Khan et al. (2003) than those observed in the studies of Mang et al. (2006), Wang et al. (2007), and Ziegenfuss et al. (2006), but the subject populations were very different. Subjects in the Khan et al. study had type 2 DM and were taking sulfonylurea drugs that increase insulin secretion. Since compounds found in cinnamon increase insulin sensitivity, they are likely to have larger effects in subjects taking sulfonylurea drugs. Insulin resistance would also be larger in subjects with type 2 DM than in subjects who are still prediabetic. Mang et al. (2006) reported similar decreases in blood glucose of subjects with type 2 DM as those of Ziegenfuss et al. (2006) in a group of subjects with metabolic syndrome. Similar decreases in blood glucose were reported by Wang et al. (2007) in women with insulin resistance related to polycystic ovary syndrome. There were no significant differences in lipid profiles in these later studies. The duration of the supplementation seems important to

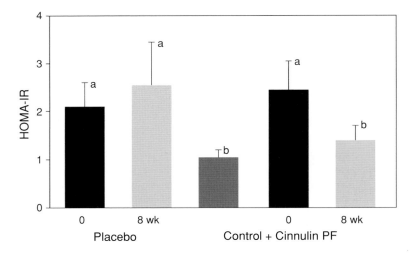

Figure 8.5.   Aqueous extract of cinnamon (Cinnulin PF) increases insulin sensitivity of women with polycystic ovary syndrome. Women (14) with polycystic ovary syndrome were divided into two groups and given either placebo or capsules containing 333 mg of aqueous cinnamon extract (Cinnulin PF) three times per day for 8 weeks. Homeostasis model insulin resistance HOMA-IR was calculated using the following formula: fasting glucose (mmol/L) × fasting insulin (μU/mL)/22.5. Control is for control women who do not have PCOS. Bars with different superscripts are significantly different at $P < 0.05$. (From Wang et al. 2007.)

consider, since, in the studies of Ziegenfuss et al. (2006) and Roussel et al. (2008), there were no effects on blood glucose after a 6-week intervention with supplementation of an aqueous cinnamon extract (500 mg/day) but only after 12 weeks. Similarly, there were no beneficial effects in postmenopausal women with type 2 DM after only 6 weeks (Vanschoonbeek et al. 2006). Antioxidant effects were also not significant after 6 weeks but were after 12 weeks in the study of Roussel et al. (2008).

Vanschoonbeek et al. (2006) reported no effect of 1.5 g/day of cinnamon powder for 6 weeks on indices of glycemic control (fasting blood glucose, insulin, OGTT, HBA$_{1c}$, insulin sensitivity, and insulin resistance) or blood lipids in 25 postmenopausal women from the Netherlands. In addition to duration, the study population of Vanschoonbeek et al. (2006) was different from those of Khan et al. (2003), Mang et al. (2006), Roussel et al. (2008), Wang et al. (2007), and Ziegenfuss et al. (2006) in that only postmenopausal females were included as subjects. Whether differences in hormonal milieu affect the potential interaction between cinnamon supplementation and glucose control is unknown at this time. In all the human studies involving cinnamon or aqueous extracts of cinnamon, there have

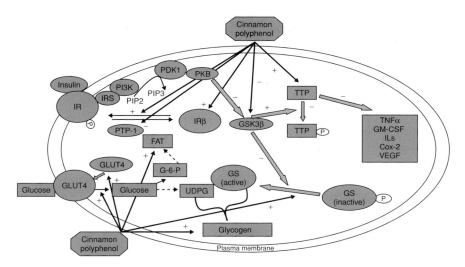

Figure 8.6. A model of actions of cinnamon polyphenols (CP) in the insulin signal transduction pathway leading to beneficial effects in people with insulin resistance: (1) CP activate insulin receptors by increasing their tyrosine phosphorylation and by decreasing phosphatase activity that inactivates the receptor; (2) CP increase the amount of insulin receptor-β and GLUT4 proteins; (3) CP increase glycogen synthase activity and glycogen accumulation; (4) CP decrease glycogen synthetase kinase-3 β (GSK3β) activity; (5) CP increase the amount of tristetraprolin (TTP) protein; and (6) CP may increase the activity of TTP by decreasing its phosphorylation through inhibition of GSK3β activity. (From Cao et al. 2007.)

been no reported adverse events and subjects with the poorest glycemic control appear to benefit the most.

Model of Cinnamon Effects

A model of the potential diverse effects of cinnamon is depicted in Figure 8.6 (Cao et al. 2007). Purified cinnamon polyphenols (CP) and an aqueous extract of cinnamon were tested in 3T3-L1 cells. CP affect multiple steps related to glucose and insulin function. CP activate insulin receptors by increasing their tyrosine phosphorylation activity and by decreasing phosphatase activity that inactivates the insulin receptor (Imparl-Radosevich et al. 1998). CP also increases the amount of IRβ and GLUT4 protein (Cao et al. 2007). CP increases glycogen synthase activity and glycogen accumulation (Jarvill-Taylor et al. 2001) with decreased GSK3β activity (Jarvill-Taylor et al. 2001). CP also increases the amount of the early response anti-inflammatory protein, tristetraprolin (TTP) (Cao et al. 2007). All these activities and other potential activities may eventually lead to more efficient glucose transport and utilization. In addition, CP-induced TTP accumulation may provide one of the molecular bases for the beneficial effects of cinnamon in improving the conditions of people with diabetes by down regulating the synthesis of proinflammatory cytokines.

Conclusions

Compounds found in cinnamon not only improve the function of insulin but also function as antioxidants and may be anti-inflammatory. This is very important since insulin function, antioxidant status, and inflammatory response are closely linked; with decreased insulin sensitivity there is also decreased antioxidant capacity. For example, people with metabolic syndrome or diabetes have both decreased insulin sensitivity and decreased antioxidant status. Type A procyanidin polymers were isolated from cinnamon that function both as antioxidants and also enhance insulin function but there are also other cinnamon compounds that may be beneficial. Animal and human studies involving subjects with the metabolic syndrome, type 2 DM, and polycystic ovary syndrome show beneficial effects of whole cinnamon and aqueous extracts of cinnamon on glucose, insulin, lipids, and antioxidant status. There may also be effects of the cinnamon compounds on lean body mass, body composition, and inflammatory response. All these effects would lead to decreased risk factors associated with diabetes and cardiovascular diseases.

References

Adisakwattana S, Roengsamran S, Hsu WH, Yibchok-anun S. 2005. Mechanisms of antihyperglycemic effect of p-methoxycinnamic acid in normal and streptozotocin-induced diabetic rats. Life Sci 78:406–412.

Anderson RA, Brantner JH, Polansky MM. 1978. An improved assay for biologically active chromium. J Agric Food Chem 26:1219–1221.

Anderson RA, Broadhurst CL, Polansky MM, Schmidt WF, Khan A, Flanagan VP, Schoene NW, Graves DJ. 2004. Isolation and characterization of polyphenol type-A polymers from cinnamon with insulin-like biological activity. J Agric Food Chem 52:65–70.

Balcerczyk A, Bartosz G. 2003. Thiols are main determinants of total antioxidant capacity of cellular homogenates. Free Radic Res 37:537–541.

Broadhurst CL, Polansky MM, Anderson RA. 2000. Insulin-like biological activity of culinary and medicinal plant aqueous extracts in vitro. J Agric Food Chem 48:849–852.

Cao H, Polansky MM, Anderson RA. 2007. Cinnamon extract and polyphenols affect the expression of tristetraprolin, insulin receptor, and glucose transporter 4 in mouse 3T3-L1 adipocytes. Arch Biochem Biophys 459:214–222.

Chrysohoou C, Panagiotakos DB, Pitsavos C, Skoumas I, Papademetriou L, Economou M, Stefanadis C. 2006. The implication of obesity on total antioxidant capacity in apparently healthy men and women: The ATTICA study. Nutr Metab Cardiovasc Dis 17:590–597.

Couillard C, Ruel G, Archer WR, Pomerleau S, Bergeron J, Couture P, Lamarche B, Bergeron N. 2005. Circulating levels of oxidative stress markers and endothelial adhesion molecules in men with abdominal obesity. J Clin Endocrinol Metab 90:6454–6459.

Cusi K, Maezono K, Osman A, Pendergrass M, Patti ME, Pratipanawatr T, DeFronzo RA, Kahn CR, Mandarino LJ. 2000. Insulin resistance differentially affects the PI 3-kinase- and MAP kinase-mediated signaling in human muscle. J Clin Invest 105:311–320.

Davi G, Falco A. 2005. Oxidant stress, inflammation and atherogenesis. Lupus 14:760–764.

Dhuley JN. 1999. Anti-oxidant effects of cinnamon (*Cinnamomum verum*) bark and greater cardamom (*Amomum subulatum*) seeds in rats fed high fat diet. Indian J Exp Biol 37:238–242.

Duke JA. 1992. Handbook of Phytomedical Constituents of GRAS Herbs and Other Economic Plants. Boca Raton, FL: CRC Press.

Gowder SJ, Devaraj H. 2006. Effect of the food flavour cinnamaldehyde on the antioxidant status of rat kidney. Basic Clin Pharmacol Toxicol 99:379–382.

Hayden MR, Tyagi SC. 2004. Neural redox stress and remodeling in metabolic syndrome, type 2 diabetes mellitus, and diabetic neuropathy. Med Sci Monit 10:RA291–RA307.

Imparl-Radosevich J, Deas S, Polansky MM, Baedke DA, Ingebritsen TS, Anderson RA, Graves DJ. 1998. Regulation of PTP-1 and insulin receptor kinase by fractions from cinnamon: Implications for cinnamon regulation of insulin signalling. Horm Res 50:177–182.

Jarvill-Taylor KJ, Anderson RA, Graves DJ. 2001. A hydroxychalcone derived from cinnamon functions as a mimetic for insulin in 3T3-L1 adipocytes. J Am Coll Nutr 20:327–336.

Khan A, Bryden NA, Polansky MM, Anderson RA. 1990. Insulin potentiating factor and chromium content of selected foods and spices. Biol Trace Elem Res 24:183–188.

Khan A, Safdar M, Ali Khan MM, Khattak KN, Anderson RA. 2003. Cinnamon improves glucose and lipids of people with type 2 diabetes. Diabetes Care 26:3215–3218.

Kim SH, Hyun SH, Choung SY. 2006. Anti-diabetic effect of cinnamon extract on blood glucose in db/db mice. J Ethnopharmacol 104:119–123.

Lee JS, Choi MS, Jeon SM, Jeong TS, Park YB, Lee MK, Bok SH. 2001. Lipid-lowering and antioxidative activities of 3,4-di(OH)-cinnamate and 3,4-di(OH)-hydrocinnamate in cholesterol-fed rats. Clin Chim Acta 314:221–229.

Mang B, Wolters M, Schmitt B, Kelb K, Lichtinghagen R, Stichtenoth DO, Hahn A. 2006. Effects of a cinnamon extract on plasma glucose, HbA, and serum lipids in diabetes mellitus type 2. Eur J Clin Invest 36:340–344.

Martin JA, Brooks DA, Prieto L, Gonzalez R, Torrado A, Rojo I, Lopez de UB, Lamas C, Ferritto R, Dolores Martin-Ortega M, Agejas J, Parra F, Rizzo JR, Rhodes GA,

Robey RL, Alt CA, Wendel SR, Zhang TY, Reifel-Miller A, Montrose-Rafizadeh C, Brozinick JT, Hawkins E, Misener EA, Briere DA, Ardecky R, Fraser JD, Warshawsky AM. 2005. 2-Alkoxydihydrocinnamates as PPAR agonists. Activity modulation by the incorporation of phenoxy substituents. Bioorg Med Chem Lett 15:51–55.

Osawa T, Kato Y. 2005. Protective role of antioxidative food factors in oxidative stress caused by hyperglycemia. Ann N Y Acad Sci 1043:440–451.

Preuss HG, Echard B, Polansky MM, Anderson R. 2006. Whole cinnamon and aqueous extracts ameliorate sucrose-induced blood pressure elevations in spontaneously hypertensive rats. J Am Coll Nutr 25:144–150.

Qin B, Nagasaki M, Ren M, Bajotto G, Oshida Y, Sato Y. 2003. Cinnamon extract (traditional herb) potentiates in vivo insulin-regulated glucose utilization via enhancing insulin signaling in rats. Diabetes Res Clin Pract 62:139–148.

Qin B, Nagasaki M, Ren M, Bajotto G, Oshida Y, Sato Y. 2004. Cinnamon extract prevents the insulin resistance induced by a high-fructose diet. Horm Metab Res 36:119–125.

Robertson RP. 2004. Chronic oxidative stress as a central mechanism for glucose toxicity in pancreatic islet beta cells in diabetes. J Biol Chem 279:42351–42354.

Rodrigo R, Bosco C. 2006. Oxidative stress and protective effects of polyphenols: Comparative studies in human and rodent kidney. A review. Comp Biochem Physiol C Toxicol Pharmacol 142:317–327.

Roussel AM, Hininger, I, Benaraba R, Ziegenfuss TN, Anderson RA. 2008. Antioxidant effects of cinnamon extract in people with impaired fasting glucose that are overweight or obese. J Am Coll Nutr (update).

Subash BP, Prabuseenivasan S, Ignacimuthu S. 2007. Cinnamaldehyde—a potential antidiabetic agent. Phytomedicine 14:15–22.

Vanschoonbeek K, Thomassen BJ, Senden JM, Wodzig WK, van Loon LJ. 2006. Cinnamon supplementation does not improve glycemic control in postmenopausal type 2 diabetes patients. J Nutr 136:977–980.

Verspohl EJ, Bauer K, Neddermann E. 2005. Antidiabetic effect of Cinnamomum cassia and Cinnamomum zeylanicum in vivo and in vitro. Phytother Res 19:203–206.

Wang JG, Anderson RA, Graham GM, III, Chu MC, Sauer MV, Guarnaccia MM, Lobo RA. 2007. The effect of cinnamon extract on insulin resistance parameters in polycystic ovary syndrome: A pilot study. Fertil Steril 88:240–243.

Yu Y, Lyons TJ. 2005. A lethal tetrad in diabetes: hyperglycemia, dyslipidemia, oxidative stress, and endothelial dysfunction. Am J Med Sci 330:227–232.

Ziegenfuss TN, Hofheins JE, Mendel RW, Landis J, Anderson RA. 2006. Effects of a water-soluble cinnamon extract on body composition and features of the metabolic syndrome in pre-diabetic men and women. J Int Soc Sport Nutr 3:45–53.

Chapter 9

Soybean and Soy Component Effects on Obesity and Diabetes

James W Anderson, MD, and Vijai K Pasupuleti, PhD

Introduction

Diabetes and obesity are very closely linked and the combination has been termed "diabesity." Both conditions are emerging at epidemic rates all over the world and preventive measures are desperately needed (Cowie et al. 2003; Flegal et al. 2002; James et al. 2001). Approximately 80% of diabetes can be attributed to obesity, and teenage obesity and weight gain can increase the risk for diabetes by more than 90-fold (Anderson et al. 2003). Insulin resistance precedes development of type 2 diabetes mellitus (T2DM) and is present in all persons with T2DM. Virtually all persons with obesity also have insulin resistance. The metabolic syndrome—with insulin resistance as the central component—is present in an estimated 24% of American adults, or ~47 million persons, and contributes to development of diabetes as well as obesity (Ford et al. 2002). Thus, interventions addressing insulin resistance have substantial promise in reducing emergence of insulin resistance and diabesity.

For more than 20 years clinicians have recognized the benefits of soy protein as a component of weight loss diets (Bosello et al. 1988; Fisler and Drenick 1984; Jenkins et al. 1989; Matzkies and Hubner 1981; Volgarev et al. 1989). Recent research has documented the satiating effects of increased protein intake (Eisenstein et al. 2002) and suggested that increasing protein intake will increase metabolism rates slightly (Mikkelsen et al. 2000). Studies in animals and humans suggest that soy protein intake selectively promotes loss of visceral adipose tissue (Anderson and Hoie 2003; Aoyama et al. 2000b). Soy protein intake by humans appears to have a preventative role for diabetes (Yang et al. 2004) and decreases insulin resistance in T2DM (Jayagopal et al. 2002). Recently, molecular biology techniques and animal studies have enhanced understanding of the mechanisms involved in these effects (Dang et al. 2003; Iqbal et al. 2002). Some of these relationships have recently been reviewed (Bhathena and Velasquez 2002; Lee and Kim 2000; McCarty 2003).

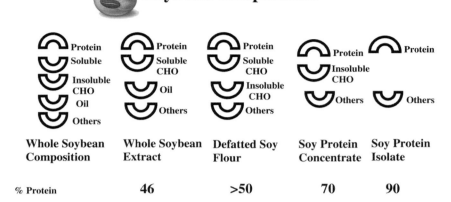

Figure 9.1.   Protein composition of commonly used soybean derived products.

This review will focus on the effects of soy foods and products on obesity and diabetes. Specifically, we will review these areas: the putative bioactive components contributing to the observed effects; *in vitro* studies examining the effects of specific components; animal studies examining the effects of these components; soy intake and the epidemiology of diabetes and obesity, clinical studies of soy components in obesity and diabetes; and safety considerations.

Bioactive Components of Soybeans

Soybeans are nutritious foods that provide high-quality protein, complex carbohydrates, polyunsaturated fats, dietary fiber, minerals, and many different phytonutrients and other phytochemicals. The protein composition of commonly used soybean derived products is depicted in Figure 9.1. It is beyond the scope of this chapter to discuss each and every bioactive component of soybeans. Based on the current literature, it appears that carbohydrate and lipid metabolism is influenced by these major bioactive components of soy: soy protein, soy peptides, and isoflavones that will be discussed in this chapter. Soy phytosterols (Sirtori et al. 2007), lipids, and fiber will not be discussed in this review. While other bioactive components of soy—pinitol (Kang et al. 2006), phospholipids, polyphenols (Apostolidis et al. 2007), saponins, phytates, etc. (see Table 9.1)—may have important biological effects of a minor or synergistic way on lipid or glucose metabolism, data are not available to carefully assess their contribution to weight management or insulin sensitivity. We will review the experimental data from *in vitro* and animal studies and clinical data from human studies related to soy protein, soy peptides, and isoflavones. An overview of these effects is presented in Table 9.2. While serum lipid changes are not a focus of this review, changes in serum cholesterol and

Table 9.1.   Typical composition of soybeans.

Soybean components
• Proteins
• Peptides
• Isoflavones
• Fiber
• Carbohydrates
• Phospholipids
• Phytic acid
• Saponins
• Phytochemicals
• Trypsin inhibitor
• Minerals
• Polyphenols

triglyceride values serve as biomarkers for the biological effects and are presented for comparative purposes.

All the biological effects of soy products related to carbohydrate and lipid metabolism have been documented with soy protein—such as isolated soy protein—providing the protein, peptides, and isoflavones. Soy peptides appear to have the most potent and consistent effects on carbohydrate and lipid metabolism

Table 9.2.   Summary of clinical data from human studies related to soy protein, soy peptides, and isoflavones.

Parameter	Protein		Peptides		Isoflavones	
	Humans	Animals	Humans	Animals	Humans	Animals
Serum cholesterol	+	+	+	+	0	0
Serum triglycerides	+	+	+	+	0	0
Body weight	+	+	Na	+	?	+
Visceral adipose tissue	+	+	+	+	Na	+
Food intake	+	+	Na	+	Na	?
Postprandial energy expenditure	+	+	+	+	Na	Na
Plasma glucose	+	+	Na	?	0	+
Insulin sensitivity	+	+	Na	+	0	+

Symbols and abbreviations: +, strong evidence; na, not available; 0, no supportive evidence; ?, evidence is not conclusive.

(Wang and de Mejia 2005). For example, consumption of 25–50 g of soy protein/day may decrease serum LDL-cholesterol in the range of 5–13% (Anderson et al. 1995), while intake of only 3–6 g/day of soy peptides from hydrolyzed soy protein may decrease serum LDL-cholesterol values by 27–43% (Hori et al. 2001; Wang et al. 1995). Recently, the role of soy peptides on lipid and carbohydrate metabolism has been extensively studied (Aoyama et al. 2000b; Ishihara et al. 2003b; Manzoni et al. 2003). Soy peptides are absorbed intact, circulate throughout the body, and maintain full biologic activity when extracted from blood or tissues (de Lumen et al. 2003; Kennedy 2003).

Earlier studies considered soy protein and its amino acid components to be the major mediators of soy effects on serum lipids (Kritchevsky 1979; Sanchez and Hubbard 1991; Terpstra et al. 1983). After the discovery of soy isoflavones (Setchell et al. 1984) and the effects of genistein as a potent inhibitor of tyrosine-specific protein kinase (Akiyama et al. 1987), the potential role of isoflavones gained credence (Anderson et al. 1995). This hypothesis was supported by studies indicating that ethanol-extracted soy protein, poor in isoflavones, had substantially less hypocholesterolemic effects than isoflavones-rich soy protein isolates (Anthony et al. 1996; Crouse et al. 1999). However, this suggestion was disputed by Sirtori (Gianazza et al. 2003; Sirtori et al. 1997), whose extensive work has substantially advanced our understanding of the mechanisms for soy-induced serum cholesterol reductions (Lovati et al. 1987, 1992, 1998, 2000; Sirtori et al. 1977). Soy isoflavones have estrogen-like and anti-estrogen-like activities and are best described as selective estrogen receptor modulators (SERMs) (Setchell 2001). Currently available data do not indicate that intake of isoflavone supplements without soy protein has cholesterol-lowering effects in animals or humans (Nestel et al. 1997). However, isoflavones in the absence of soy protein or peptides appear to have independent effects on body weight, feeding behavior, blood glucose values, and insulin sensitivity.

Soy Peptide Effects: Animal Studies

Soy peptides appear to have important effects on glucose and lipid metabolism and gene regulation based on studies using *in vitro* and animal models; representative studies are summarized in Table 9.3. The early and ongoing studies of the Sirtori group (Lovati et al. 1992, 1998, 2000; Manzoni et al. 1998) indicate that soy peptides upregulate LDL receptors in various model systems. The clinical importance of the studies are supported by documentation that soy protein intake increases LDL receptor activity in human monocytes (Baum et al. 1998; Lovati et al. 1987). Several studies indicate that feeding soy peptides to rodents decreases food intake and reduces weight gain (Aoyama et al. 2000b; Nishi et al. 2003b; Pupovac and Anderson 2002). Additional studies suggest that soy peptide administration to rodents suppresses appetite (Hara and Nishi 2003; Nishi et al. 2003a). Including a soy protein hydrolysate in the diet may increase postprandial energy

Table 9.3.   Effect of soy peptides *in vitro* and in animal models.

Area	Soy preparation	Species	Activity	References
Obesity	Soy protein hydrolylase	Obese mice and rats	Less weight gain	Aoyama et al. (2000a, b)
Obesity	Hydrolysed protein	Rats	Decreases food intake	Pupovac and Anderson (2002)
Obesity	Soy peptides	Rodents	Increased satiation	Hara and Nishi (2003)
Obesity	Soy protein hydrolylase	Diabetic mice	Increased postprandial energy expenditure	Ishihara et al. (2003a, b)
Obesity	Peptide B51-63	Rats	Appetite suppressant	Nishi et al. (2003a)
Obesity	B-conglycinin fraction	Rats	Decreases feeding	Nishi et al. (2003b)
Obesity	B-conglycinin fraction	Obese mice	Increased b-oxidation, decreased fatty acid synthetase	Moriyama et al. (2004)
Obesity	Soy protein hydrolysate	Adult male Sprague Dawley Rats	Increased satiation	Vaughn et al. (2007)
Obesity	Black soy peptid	Sprague Dawley rats	Weight loss and lower lipid levels	Rho et al. (2007)
Diabetes	Fermented soy, small peptides	*In vitro* adipocyte	Enhances glucose utilization	Kwon et al. (2006)
Diabetes	Fermented soy	Pancreatecto-mized diabetic rats	Enhanced insulin sensitivity	Kwon et al. (2007)

expenditure (Ishihara et al. 2003b). Direct injection of soy peptides via cannula in the brain of rats resulted in weight loss even when the rats have not changed their eating habits. This suggests that the weight loss is not by reducing food intake but by altering the metabolism (Vaughn et al. 2007). However, it remains to be seen that if their performance will be similar or not when fed directly to the rats and this will shed the light on peptide transport. Feeding the B-conglycinin component of soy protein also appears to increase fatty acid oxidation and decrease fatty acid synthesis (Moriyama et al. 2004). In another study rats receiving black soy peptide lost weight and had lower total cholesterol, lower LDL/HDL, and triglycerides suggesting that black soy peptide may be a bioactive source of anti-obesity properties through modulation of lipid composition which might contribute to the amelioration of obesity-related metabolic diseases such as atherosclerosis, cardiovascular disease, and diabetes (Rho et al. 2007). *In vitro* studies suggest that soy peptides enhance glucose utilization of adipose tissue cells and increase insulin sensitivity of diabetic rats (Kwon et al. 2006, 2007).

Soy Isoflavone Effects: *In vitro* and Animal Studies

The *in vitro* effects of genistein are complex and are incompletely explained by the protein tyrosine kinase inhibitory effects (Abler et al. 1992; Huang et al. 1999). Soy isoflavones have an array of *in vitro* effects ranging from affecting brain receptors relating to eating behavior, altering leptin secretion, and effects on lipid metabolism in fat cells; representative studies are presented in Table 9.4. The clinical relevance of these *in vitro* studies are unclear since most of the significant *in vitro* effects were usually achieved with genistein concentrations of 100–1,000 μM/L (median concentration, 225 μM/L, Table 9.3); free genistein concentrations in plasma with usual intakes of isoflavones-rich soy protein reach 2–10 μM/L (Dang et al. 2003; Setchell 2001; Sites et al. 2007). Genistein may affect food intake by increasing dopamine release (Bare et al. 1995a; Dong and Xie 2003) or inhibiting GABA receptors (Dunne et al. 1998; Huang et al. 1999) and, also, may decrease leptin secretion (Szkudelska et al. 2005). Isoflavones may affect adipocyte metabolism *in vitro* by decreasing rates of lipogenesis and increasing rates of lipolysis (Kim et al. 2005; Szkudelska et al. 2002) and by inducing apoptosis (Hwang et al. 2005; Kim et al. 2006). Dang and colleagues (2003) compared the effects of genistein on mesenchymal progenitor cells and adipocytes in culture and suggest that the biological effects of genistein are specific and distinct in different tissues; they observed that genistein reduced lipogenesis at low or moderate concentrations.

The effects on weight gain and fat accumulation seen with including isoflavones in the diet has also been examined in experimental animals. Using large amounts of isoflavones—50–5,000 mg/kg body weight compared to human consumption usually less than 1 mg/kg—the intake of isoflavones was associated with significantly less weight gain and fat accumulation in rodents (Ali et al. 2004; Naaz et al. 2003; Uesugi et al. 2001; Wu et al. 2004; Zhou et al. 1999).

In vitro studies indicate that genistein also has specific effects on glucose metabolism and insulin sensitivity. In isolated pancreatic islet cells, genistein enhanced glucose stimulated insulin release (Liu et al. 2006). Fermentation of isoflavones removes the carbohydrate side chain and convert them to aglycones; incubation of these fermented isoflavones with fat cells enhances glucose uptake

Table 9.4. Effect of soy isoflavones *in vitro* and in animal models.

Area	Isoflavone	Species	Amount	Procedure	Effect	References
Obesity	Genistein	*In vitro*	100–200 µM	Mouse striatal slices	Increased dopamine release	Bare et al. (1995a, b)
Obesity	Genistein	*In vitro*		Amygdala	Increased dopamine release	Dong and Xie (2003)
Obesity	Genistein	*In vitro*		HEK293 cells	Inhibited GABAAa receptors	Dunne et al. (1998)
Obesity	Genistein	*In vitro*	50–500 µM	HEK293 cells	Inhibited GABAAa receptors	Huang et al. (1999)
Obesity	Daidzein	*In vitro*	100–1000 µM	Rat adipocyte	Increased lipolysis at lower doses	Szkudelska et al. (2002)
Obesity	Genistein	*In vitro*	1–100 µM	Human hepG2 cells	Induced expression of genes promoting fat catabolism	Kim et al. (2005)
Obesity	Genistein	*In vitro*	20–200 µM	Cultured adipocytes	Inhibited differentiation, induced apoptosis	Hwang et al. (2005)
Obesity	Genistein	*In vitro*	250–1,000 µM	Rat adipocyte	Decreased leptin secretion	Szkudelska et al. (2005)
Obesity	Isoflavones	*In vitro*	1–400 µM	Adipocytes	Dose-related apoptosis	Kim et al. (2006)
Obesity	Genistein	Mice	8–200 mg/kg	Injection	Decreased adipose tissue weight	Naaz et al. (2003)

(continued)

Table 9.4. (*continued*)

Area	Isoflavone	Species	Amount	Procedure	Effect	References
Obesity	Isoflavones	Obese rats	0.1% isoflavones	Diet	Decreased weight gain, decreased fat	Ali et al. (2004)
Obesity	Isoflavones	Rat	50 mg/kg	Gavage	Decreased weight gain, decreased fat	Uesugi et al. (2001)
Obesity	Isoflavones	Mice	160 mg/kg	Diet	Decreased weight gain, decreased fat	Wu et al. (2004)
Obesity	Isoflavones fermented	Rat		Diet	Decreased weight gain	Zhou et al. (1999)
Diabetes	Genistein	*In vitro*	0.01–5 μM	Mouse pancreatic islets	Enhances glucose stim insulin release	Liu et al. (2006)
Diabetes	Fermented isoflavones	*In vitro*		Fat cells, *in vitro*	Enhanced glucose uptake by adipocytes	Kwon et al. (2006)
Diabetes	Fermented isoflavones	*In vitro*		Fat cells, *in vitro*	Enhances glucose utilization	Kwon et al. (2006)
Diabetes	Isoflavones	Lean and obese rats	0.10%	Diet	Decreased blood glucose	Ali et al. (2005)
Diabetes	Genistein	Stz-dm rats	600 mg/kg	Decreased glucose	Lee (2006)	
Diabetes	Fermented isoflavones	Pancr dm rats	Insulin clamp	Fermented soy but not cooked soy enhanced insulin sensitivity		Kwon et al. (2007)

and utilization (Kwon et al. 2006). In the perfused rat liver, genistein has effects on insulin receptor activity (Mackowiak et al. 1999). Feeding isoflavones in the diet in large amounts (600–5,000 mg/kg body weight) has favorable effects on glucose metabolism and insulin sensitivity. Two studies indicated that inclusion of isoflavones in the diet significantly decreases blood glucose values (Ali et al. 2005; Lee 2006), and another study reported that including fermented isoflavones in the diet enhanced insulin sensitivity (Kwon et al. 2007). Further studies are required to examine effects of dietary isoflavone on glucose metabolism and insulin sensitivity at doses in the human-usage range of 0.5–2 mg/kg body weight per day.

Soy Protein Effects: Animal Studies

Most studies of weight regulation or glucose metabolism have compared diets providing soy protein with diets containing casein or another animal protein. Representative studies are summarized in Table 9.5. In a number of different rodent models, soy protein consumption decreases weight gain (Banz et al. 2004; Bu et al. 2005; Lephart et al. 2001; Lund and Lephart 2001; Tovar et al. 2002) and decreases fat accumulation (Aoyama et al. 2000a; Nagasawa et al. 2003). Adiponectin—a protein secreted by fat cells that decreases risk for obesity—is also suppressed by soy protein intake (Matsuzawa 2005; Nagasawa et al. 2003).

Soy protein intake also has favorable effects on glucose metabolism and insulin sensitivity for rodents and monkeys (Table 9.5). Soy protein consumption is often accompanied by decreased blood glucose (Hurley et al. 1998; Lavigne et al. 2000; Lee 2006; Nagasawa et al. 2002; Noriega-Lopez et al. 2007) and insulin (Ascencio et al. 2004; Hurley et al. 1998) values as well as enhanced sensitivity to insulin (Lavigne et al. 2000; Wagner et al. 1997). Increased postprandial carbohydrate oxidation (Ishihara et al. 2003a) and increased insulin receptor gene expression (Iritani et al. 1997) have been observed in rodents fed soy protein. Soy protein intake may also protect from development of diabetes in selected rodent models (Lee and Park 2000; Reddy et al. 2001).

Epidemiology: Soy Foods and Obesity or Diabetes

Soy foods and their components have many beneficial effects on regulation of eating behavior, metabolism, body fat distribution, and insulin sensitivity that have been documented with *in vitro* and animal studies (Tables 9.3–9.5). Because soy food consumption is so low among Western people, it is difficult to assess the potential effects of soy intake for Western populations. The report of Goodman-Greun and colleagues (Goodman-Gruen and Kritz-Silverstein 2003) suggests that soy protein intake favorably alters regional fat distribution of menopausal women. Early results from the Shanghai Women's Health Study suggests that women who consume larger amounts of soy than women who consume smaller amounts are less likely to develop diabetes (Yang et al. 2004).

Table 9.5. Effect of soy protein diet in animal models.

Area	Soy source	Species	Soy effect	Reference
Obesity	SP diet	Mice	Decreased fat accumulation	Aoyama et al. (2000a, b)
Obesity	SP diet	Rats	Decreased weight gain	Lund and Lephart (2001)
Obesity	SP diet	Rats	Decreased weight gain	ephart et al. (2001)
Obesity	SP diet	Rats	Decreased weight gain	Tovar et al. (2002)
Obesity	SP diet	Wistar rats	Increased adiponectin, decreased visceral adiposity	Nagasawa et al. (2003)
Obesity	SP diet	Zucker obese	Decreased weight gain	Banz et al. (2004)
Diabetes	SP diet	Rats	Lower insulin levels on high-fat diet	Ascencio et al. (2004)
Obesity	SP diet	Rats	Decreased weight gain, decreased fat deposits	Bu et al. (2005)
Obesity	SP diet	Obese KK-A(y)	Increased adiponectin	Matsuzawa (2005)
Obesity/ Diabetes	SP diet	Rats	Decreased weight gain, lower glucose and insulin	Hurley et al. (1998)
Obesity/ Diabetes	SP diet	Obese KK A y mice	Decreased body fat, lower blood glucose	Nagasawa et al. (2002)
Obesity/ Diabetes	SP diet	Diabetic KK-A(y)	Increased postprandial carbohydrate oxidation	Ishihara et al. (2003a, b)
Diabetes	SP diet	Wistar fatty rats	Increased insulin receptor gene expression	Iritani et al. (1997)
Diabetes	SP diet	Monkeys	Increased insulin sensitivity	Wagner et al. (1997)

Table 9.5. (*continued*)

Area	Soy source	Species	Soy effect	Reference
Diabetes	SP diet	Rats	Improved blood glucose and insulin sensitivity	avigne et al. (2000)
Diabetes	SP diet	Diabetic rats	Lowered blood glucose and preserved islets	Lee and Kim (2000)
Diabetes	SP diet	NOD mice	Prevented cyclophosphamide-induced diabetes	Reddy et al. (2001)
Diabetes	SP diet	Diabetic rats	Decreased blood glucose	Lee (2006)
Diabetes	SP diet	Obese rats	Decreased blood glucose	Noriega-Lopez et al. (2007)

Human Clinical Trials of Soy Products for Obesity

Limited information related to isoflavone or peptide effects on body weight is available. Representative information from clinical trials is presented in Table 9.6. One report suggests that soy isoflavone intake is associated with weight loss in humans (Mori et al. 2004). An additional report suggests that soy peptides are associated with an increase in dietary thermogenesis (Claessens et al. 2007). Most of the clinical trials examining the effects of soy protein on body weight are not blinded, controlled studies. The study of Allison et al. (2003) compared use of a commercial weight loss product to nutrition counseling. The study of Deibert et al. (2004) compared use of a commercially available yogurt to nutrition counseling. In our first study (Anderson and Hoie 2005), we compared use of two commercially available products—one milk-based and the other soy-based—in a unblinded fashion and noted no differences in weight loss or adipose tissue loss. Our second study (Anderson et al. 2007) was a single-blind, carefully controlled study where all nutrition counseling and monitoring was identical for each group; over 16 weeks we observed no differences in weight loss or visceral adipose tissue changes. Other studies (Fontaine et al. 2003; Hermansen et al. 2005; Kok et al. 2006; Moeller et al. 2003; Sites et al. 2007; St-Onge et al. 2007) have reported that soy protein intake did not affect body weight significantly differently than animal protein intake. Some studies suggest that fatty tissue loss in specific portions of the body are greater with soy than control diets (Baba et al. 2004; Kohno et al. 2006; Moeller et al. 2003; Sites et al. 2007). One short-term metabolic study suggests

Table 9.6. Human clinical trials of soy products for obesity and diabetes.

Area	Study type	Soy preparation	Amount	Effect of soy	References
Obesity	RCT	Isoflavones	100 mg/day	Decreased body fat	Mori et al. (2004)
Diabetes	RCT	Isoflavones	114 mg/day	No effect observed	Nikander et al. (2004)
Diabetes	RCT	Isoflavones	100 mg/day	Decreased glucose and insulin	Cheng et al. (2004)
Diabetes	RCT	Isoflavone	50 mg/day	No blood glucose or insulin effects	Hall et al. (2006)
Obesity	Metabolic	SP hydrolysate	0.4 g/kg	Greater dietary thermogenesis	Claessens et al. (2007)
Obesity	Clinical trial	B-conglycinin	5 g/day	Decreased fat accumulation	Baba et al. (2004)
Obesity	RCT	B-conglycinin		Decreased visceral fat	Kohno et al. (2006)
Obesity	Metabolic	ISP	Test meal	Increased energy expenditure	Mikkelsen et al. (2000)
Obesity	RCT	ISP	90 g/day	Decreased body weight	Allison et al. (2003)
Obesity	RCT	ISP	40 g/day	Decreased thigh fat	Moeller et al. (2003)
Obesity	Clinical trial	ISP	Variable	Inconclusive	Fontaine et al. (2003)
Obesity	RCT	Soy yogurt	~ 30g/day	Decreased weight, body fat	Deibert et al. (2004)
Obesity	RCT	ISP	90 g/day	No effect observed	Anderson and Hoie (2005)
Obesity	RCT	ISP	30 g/day	Higher GIP levels	Hermansen et al. (2005)

Disease	Study type	Intervention	Dose	Result	Reference
Obesity	RCT	ISP	26 g/day	No effect observed	Kok et al. (2006)
Obesity	Metabolic	ISP	50 g meal	Soy and whey similarly decreased food intake	Bowen et al. (2006)
Obesity	RCT	ISP	62 g/day	No effect observed	Anderson et al. (2007)
Obesity	RCT	ISP	20 g/day	Decreased abdominal fat	Sites et al. (2007)
Obesity/diabetes	RCT	ISP	18 g/day	No body weight effect, decreased glucose and insulin	St-Onge et al. (2007)
Diabetes	RCT	ISP	50 g/day	Check	Hermansen et al. (2001)
Diabetes	RCT	ISP	30 g/day	Decreased insulin, HbA1c	Jayagopal et al. (2002)
Diabetes	Clinical trial	ISP???	30–60 g/day	Decreased glucose, insulin	Meshcheryakova et al. (2002)
Diabetes	RCT	Soy nuts	30 g/day	Lower blood glucose, C-peptide	Azadbakht et al. (2007)
Obesity	Review article	NA	NA	Increases adiponectin	Matsuzawa et al. (2005)
Obesity	Meta-analysis	NA	NA	Suggestion of faster weight loss	Anderson et al. (2004)

that soy protein intake is associated with greater 24 h energy expenditure than control diets (Mikkelsen et al. 2000). Thus, the eight clinical trials addressing the effects of soy protein intake on body weight are inconclusive. Some of the encouraging results observed in animal studies are suggested by these available trials but persuasive data are not available and more research is required.

Human Clinical Trials of Soy Products for Diabetes

Animal studies suggest that soy components—isoflavones, peptides, and soy protein—have important effects on regulation of glucose metabolism and insulin sensitivity. Because of the important effects of soy peptides on serum cholesterol (Hori et al. 2001; Wang et al. 1995) and dietary thermogenesis (Claessens et al. 2007) in humans, soy peptides may exert important effects on glucose metabolism in humans as seen in animals (Table 9.3); however, no clinical studies of soy peptide effects on glucose metabolism in humans are available. Three clinical trials have reported glucose and insulin values after soy isoflavone administration with one (Cheng et al. 2004) showing favorable effects on these values and two (Hall et al. 2006; Nikander et al. 2004) showing no effects. The limited number of clinical trials utilizing soy protein for diabetic subjects have reported favorable changes in blood glucose and insulin values (Hermansen et al. 2001; Jayagopal et al. 2002; Meshcheryakova et al. 2002). While soy peptides and protein appear to show substantial promise in preventing and ameliorating manifestations of diabetes in humans, much further research is required in this arena.

Discussion of Potential Mechanisms for Soy Regulation of Body Weight

Regulation of body weight is extraordinarily complex with many redundant pathways (Spiegelman and Flier 2001). This brief review will only cite a few studies indicating the potential interactions of soy components on these regulatory systems. Soy food intake may affect energy intake as well as energy expenditure— physical activity and metabolism rate. Recent research indicates that soy food intake may affect central nervous system regulation of feeding behavior. For example, dopamine release appears to decrease food intake (Wang et al. 2001), and *in vitro* studies indicate that genistein increases release of dopamine by inhibition of tyrosine kinase in mouse striatal slices (Bare et al. 1995b). Lifelong exposure of rats to genistein also affects regulation of dopamine release in such a manner as to reduce weight gain (Ferguson et al. 2002). Soy isoflavones may also affect weight management through brain gamma-aminobutyric acid (GABA) transport or receptor effects (Tsujii and Bray 1991). Overexpression of GABA transporter subtype I contributes to obesity in transgenic mice by causing significantly increased body weight and fat deposition (Ma et al. 2003). Altered GABAergic mechanisms may depress ventilation and decrease exercise capacity in obese Zucker rats, thus contributing to their weight gain (Lee et al. 2001). Genistein and daidzein appear to have direct inhibitory effects on $GABA_A$ receptors in HEK293 cells (Huang et al. 1999). These effects of genistein on $GABA_A$ receptors have been confirmed by Dunne and colleagues (1998). Other effects of soy isoflavones on CNS regulatory systems are being documented in this fertile field of research (Lephart et al. 2001).

Our understanding of the regulation of food intake by gut hormones continues to grow (Batterham et al. 2003; Gale et al. 2004; van der Lely et al. 2004; Woods 2004). The specific effects of soy food intake on these important hormones have not been well characterized. Cholecystokinin (CCK), produced in the brain and endocrine cells of the gut, acts to suppress food intake (Woods 2004). Nishi and colleagues (2003a, b) noted that the intraduodenal infusion of a soybean B-conglycinin suppresses food intake by stimulating CCK release in rats. Isoflavone supplement intake in humans may reduce serum ghrelin levels and thereby reduce hunger and weight gain (Nikander et al. 2004). Soy peptide intake may have effects on energy metabolism in obese animals (Ishihara et al. 2003b; Saito 1991).

Safety Concerns

Foods rich in protein may produce allergic reaction in humans. These foods listed in approximate rank for frequency are: cow's milk, fish, crustacea, eggs, peanuts, wheat, soybeans, and tree nuts (Taylor and Hefle 2002). However, the occurrence of allergy to soybeans in the United States is quite low (Anderson and Hanna 2001). Since soy foods are often rich in fiber, complex carbohydrate, and phytochemicals, they may provoke minor physiological effects such as gastrointestinal symptoms that are not medically significant. Soy isoflavones can have estrogen-agonist-like effects or estrogen-antagonist-like effects and probably are best described as natural SERMs (Setchell 2001). A rigorous review of their safety concluded that "the current literature supports the safety of isoflavones as typically consumed in diets based on soy or containing soy products" (Munro et al. 2003). The health consequences of early soy consumption have been carefully reviewed (Badger et al. 2002), and long-term follow-up of infants fed soy-based formula have been reported (Strom et al. 2001). These studies indicate that exposure to soy formula does not result in health or reproductive outcomes that differ from exposure to cow milk formula (Strom et al. 2001).

Observations in animals have suggested that soy protein intake might affect thyroid function (Forsythe 1995). However, careful observations in humans have not documented clinically meaningful effects of soy protein intake on a variety of serum assessments of thyroid function (Duncan et al. 1999a; Persky et al. 2002). The small changes in length of menstrual cycle and serum levels of female hormone (Duncan et al. 1999a, b; Persky et al. 2002; Setchell and Cassidy 1999) associated with soy protein intake do not appear to have clinical significance. The weight of evidence suggests that soy protein intake has protective effects with respect to risk for breast cancer, prostate cancer, and gastrointestinal cancer. The recent meta-analyses of available studies indicated the following relative risks for persons with high versus those with low soy protein intakes: for 22 studies, the relative risk for breast cancer was 0.78 (95% confidence intervals (CI), 0.68–0.91, $P < 0.001$); for 10 studies, the relative risk of prostate cancer was 0.66 (95% CI, 0.54–0.81, $P < 0.001$); and for 29 studies, the relative risk for gastrointestinal cancer was 0.70 (95% CI, 0.61–0.80, $P < 0.001$) (Solae 2004).

Conclusions

Soy foods and supplements provide many health benefits and the specific benefits in prevention and treatment of diabetes or obesity are rapidly being discovered. Soy foods provide a high-quality protein, isoflavones, dietary fiber, and other bioactive components. Soy peptides, resulting from hydrolysis of soy protein before it is consumed or from hydrolysis in the intestine, have a wide array of biological activities that are just being uncovered.

With *in vitro* systems, soy components—peptides or isoflavones—have effects on glucose and lipid metabolism as well as gene expression. Early studies demonstrated that soy peptides upregulated expression of LDL receptors contributing to the hypocholesterolemic effect of soy protein intake. Human consumption of 3–6 g of soy peptides daily decreases serum LDL-cholesterol values by 27–43%. Additional studies indicate that soy components have regulatory effects on a number of genes for glucose or lipid metabolism. Soy isoflavones also exert regulatory effects on genes for insulin receptors.

In experimental animals, intake of soy protein components—protein, peptides, or isoflavones—have well documented effects on body fat content, body fat distribution, serum glucose and insulin levels, and insulin sensitivity. Soy components in the diet also affect gene expression, especially those related to glucose and lipid metabolism as well as insulin receptors. Soy intake decreases risk for developing diabetes in several animal models. It is intriguing to note that a few studies have demonstrated that the diabetes is less prevalent in women especially post-menopausal women receiving hormone replacement therapy than men suggesting that estrogens might protect pancreatic beta cells (Le May et al. 2006; Rossi et al. 2004; Wilson 2003). Based on this, one can postulate that weak estrogen-like activity of soy isoflavones may be working independently or synergistically with other bioactive components of soy in the prevention or management of diabetes. Hopefully, in the future more definitive studies by way of double-blind placebo controlled studies will be performed to shed more light on this.

In humans, intake of soy foods increases insulin sensitivity and appear to decrease risk for developing diabetes. Soy protein or soy peptide intake significantly decreases serum cholesterol, LDL-cholesterol, and triglyceride values while increasing protective HDL-cholesterol values. Emerging data suggest that soy foods may affect regulation of food intake and energy expenditure in humans.

High protein intakes, including soy protein, decrease food intake. Increased intake of soy protein produces a small increase in basal metabolic rate. Preliminary data indicate that soy foods favorably affect distribution of body fat. While preliminary information suggests that obese individuals may lose weight more rapidly with soy protein compared to animal protein, much further research is required to support this hypothesis.

References

Abler A, Smith JA, Randazzo PA, Rothenberg PL, Jarett L. 1992. Genistein differentially inhibits postreceptor effects of insulin in rat adipocytes without inhibiting the insulin receptor kinase. J Biol Chem 267:3946–3951.

Akiyama T, Ishida J, Nakagawa S, Ogawana H, Watanabe S, Itoh NM, Shibuya M, Fukami Y. 1987. Genistein, a specific inhibitor of tyrosine-specific protein kinases. J Biol Chem 262:5592–5595.

Ali AA, Velasquez MT, Hansen CT, Mohamed AI, Bhathena SJ. 2004. Effects of soybean isoflavones, probiotics, and their interactions on lipid metabolism and endocrine system in an animal model of obesity and diabetes. J Nutr Biochem 15:583–590.

Ali AA, Velasquez MT, Hansen CT, Mohamed AI, Bhathena SJ. 2005. Modulation of carbohydrate metabolism and peptide hormones by soybean isoflavones and probiotics in obesity and diabetes. J Nutr Biochem 16:693–699.

Allison DB, Gadbury G, Schwartz LG, Murugesan R, Kraker JG, Heska S, Fontaine KR, Heymsfield SB. 2003. A novel soy-based meal replacement formula for weight loss among obese individuals: a randomized controlled clinical trial. Eur J Nutr 57:514–522.

Anderson JW, Fuller J, Patterson K, Blair R, Tabor A. 2007. Soy compared to casein meal replacement shakes with energy-restricted diets for obese women: Randomized controlled trial. Metabolism 56:280–288.

Anderson JW, Hanna TJ. 2001. In Soy Foods and Health Promotion, Vegetables, Fruits, and Herbs in Health Promotion, edited by Watson T, pp. 117–134. Boca Raton, FL: CRC Press.

Anderson JW, Hoie LH. 2003. Comparison of weight and lipid responses to soy and milk meal replacements. In Fifth International Symposium on the Role of Soy in Preventing and Treating Chronic Disease, Vol. 77, pp. 9–21.

Anderson JW, Hoie LH. 2005. Weight loss and lipid changes with low-energy diets: Comparison of milk-based versus soy-based liquid meal replacement interventions. J Am Coll Nutr 24:210–216.

Anderson JW, Johnstone BM, Cook-Newell ME. 1995. Meta-analysis of effects of soy protein intake on serum lipids in humans. New Engl J Med 333:276–282.

Anderson JW, Kendall CWC, Jenkins DJA. 2003. Importance of weight management in type 2 diabetes: Review with meta-analysis of clinical studies. J Am Coll Nutr 22:331–339.

Anderson JW, Luan J, Hoie LH. 2004. Structured weight-loss programs: A meta-analysis of weight loss at 24 weeks and assessment of effects of intensity of intervention. Adv Ther 21:61–75.

Anthony MS, Clarkson TB, Hughes CL, Morgan TM, Burke GL. 1996. Soybean isoflavones improve cardiovascular risk factors without affecting the reproductive system of peripubertal rhesus monkeys. J Nutr 126:43–50.

Aoyama T, Fukui K, Nakamori T, Hashimoto Y, Yamamoto T, Takamatsu K, Sugano M. 2000a. Effect of soy and milk whey protein isolates and their hydrolysates on weight reduction in genetically obese mice. Biosci Biotech Bioch 64:2594–2600.

Aoyama T, Fukui K, Takamatsu K, Hashimoto Y, Yamamoto T. 2000b. Soy protein isolate and its hydrolysate reduce body fat of dietary obese rats and genetically obese mice (yellow KK). Nutrition 16:349–354, 699–717.

Apostolidis E, Kwon YI, Shetty K. 2006. Potential of select yogurts for diabetes and hypertension management. J. Food Biochem 30:699–717.

Apostolidis E, Kwon YI, Ghaedian R, Shetty K. 2007. Fermentation of milk and soymilk by Lactobacillus bulgaricus and Lacto bacillus acidophilus enhances functionality for potential dietary management of hyperglycemia and hypertension. Food Biotechnol. 21:217–236.

Ascencio C, Torres N, Isoard-Acosta F, Gomwz-Perez FJ, Hernandez-Pando R, Tovar AR. 2004. Soy protein affects serum insulin and hepatic SREBP-1 mRNA and reduces fatty liver in rats. J Nutr 134:522–529.

Azadbakht L, Kimiagar M, Mehrabi Y, Esmaillzadeh A, Padyab M, Hu FB, Willett WC. 2007. Soy inclusion in the diet improves features of the metabolic syndrome: A randomized crossover study in postmenopausal women. Am J Clin Nutr 85:735–741.

Baba T, Ueda A, Kohno M, Fukui K, Miyazaki C, Hirotsuka M, Ishinaga M. 2004. Effects of soybean beta-conglycinin on body fat ratio and serum lipid levels in healthy volunteers of female university students. J Nutr Sci Vitaminol 50:26–31.

Badger TM, Ronis MJJ, Hakkak R, Rowlands JC, Korourian S. 2002. The health consequences of early soy consumption. J Nutr 132:559S–565S.

Banz WJ, Davis J, Peterson R, Iqbal MJ. 2004. Gene expression and adiposity are modified by soy protein in male Zucker diabetic fatty rats. Obes Res 12:1907–1913.

Bare DJ, Ghetti B, Richter JA. 1995a. The tyrosine kinase inhibitor genistein increases endogenous dopamine release from normal and weaver mutant mouse striatal slices. J Neurochem 65:2096–2104.

Bare DJ, Ghetti B, Richter JA. 1995b. The tyrosine kinase inhibitor genistein increases endogenous dopamine release from normal and weaver mutant mouse striatal slices. J Neurochem 65:2096–2104.

Batterham RL, Cohen MA, Ellis SM, Le Roux CW, Withers DJ, Frost GS, Ghatei MA, Bloom SR. 2003. Inhibition of food intake in obese subjects by peptide YY3-36. New Engl J Med 349:941–948.

Baum JA, Teng H, Erdman JW, Weigel RM, Klein BP, Persky VW, Freels S, Surya P, Bakhit RM, Ramos E, Shay NK, Potter SM. 1998. Long-term intake of soy protein improves blood lipid profiles and increases mononuclear cell low-density lipoprotein receptor messenger RNA in hypercholesterolemic, postmenopausal women. Am J Clin Nutr 68:545–551.

Bhathena SJ, Velasquez MT. 2002. Beneficial role of dietary phytoestrogens in obesity and diabetes. Am J Clin Nutr 76:1191–1201.

Bosello O, Cominacini L, Zocca I, Garbin U, Compri R, Davoli A, Brunetti L. 1988. Short and long-term effects of hypocaloric diets containing proteins of different sources on plasma lipids and apoproteins of obese subjects. Ann Nutr Metab 32:206–214.

Bowen J, Noakes M, Clifton PM. 2006. Appetite regulatory hormone responses to various dietary proteins differ by body mass index status despite similar reductions in ad libitum energy intake. J Clin Endocrinol Metab 91:2913–2919.

Bu L, Setchell KDR, Lephart ED. 2005. Influence of dietary soy isoflavones on metabolism but not nociception and stress hormone responses of ovariectomized female rats. Reprod Biol Endocrinol 3:58–65.

Cheng SY, Shaw NS, Tsai KS, Chen CY. 2004. The hypoglycemic effects of soy isoflavones on postmenopausal women. J Womens Health (Larchmt.) 13:1080–1086.

Claessens M, Calame W, Siemensma AD, Saris WH, Van Baak MA. 2007. The thermogenic and metabolic effects of protein hydrolysate with or without a carbohydrate load in healthy male subjects. Metabolis 56:1051–1059.

Cowie CC, Rust KF, Byrd-Holt D, Eberhardt MS, Geiss LS, Engelgau MM, Ford ES, Gregg EW. 2003. Prevalence of diabetes and impaired fasting glucose in adults—United States, 1999–2000. Morb Mortal Wkly Rep 52:833–837.

Crouse JR III, Morgan T, Terry JG, Ellis J, Vitolins M, Burke GL. 1999. A randomized trial comparing the effect of casein with that of soy protein containing varying amounts of isoflavones on plasma concentrations of lipids and lipoproteins. Arch Intern Med 159:2070–2076.

Dang Z-C, Audinot V, Papapoulos SE, Boutin JA, Löwik CWGM. 2003. Peroxisome proliferator-activated receptor g(PPAR g) as a molecular target for the soy phytoestrogen genistein. J Biol Chem 278:962–967.

Deibert P, Konig D, Schmidt-Trucksaess A, Kaenker KS, Frey I, Berg A. 2004. Weight loss without losing muscle mass in pre-obese and obese subjects induced by high-soy diet. Int J Obes 28:1349–1352.

deLumen BO, Lam Y, Galvez AF, Joeng HJ. 2003. The chemopreventive properties of the chromatin-binding soy peptide Lunasin[TM]. J Nutr 132:588S–619S.

Dong XL, Xie JX. 2003. Regulation of estrogen and phytoestrogen on the dopaminergic systems of amygdala in rats. Acta Physiol Sinica 55:589–593.

Duncan AM, Merz BE, Xu X, Nagel TC, Phipps WR, Kurzer MS. 1999a. Soy isoflavones exert modest hormonal effects in premenopausal women. J Clin Endocr Metab 84:192–197.

Duncan AM, Underhill KEW, Xu X, Lavalleur J, Phipps WR, Kurzer MS. 1999b. Modest hormonal effects of soy isoflavones in postmenopausal women. J Clin Endocr Metab 84:3479–3484.

Dunne EL, Moss SJ, Smart TG. 1998. Inhibition of GABA-A receptor function by tyrosine kinase inhibitors and their inactive analogues. Mol Cell Neurosci 12:300–310.

Eisenstein J, Roberts SB, Dallal G, Saltzman E. 2002. High-protein weight loss diets: Are they safe and do they work? a review of experimental and epidemiologic data. Nutr Rev 60:189–200.

Ferguson SA, Flynn KM, Delclos KB, Newbold RR, Gough BJ. 2002. Effects of lifelong dietary exposure to genistein or nonphenol on amphetamine-stimulated striatal dopamine release in male and female rats. Neurotoxicol Teratol 24:37–45.

Fisler JS, Drenick EJ. 1984. Calcium, magnesium, and phoshate balances during very low calorie diets of soy or collagen protein in obese men: Comparison to total fasting. Am J Clin Nutr 40:14–25.

Flegal KM, Carroll MD, Ogden CL, Johnson CL. 2002. Prevalence and trends in obesity among US adults, 1999–2000. J Am Med Assoc 288:1723–1727.

Fontaine KR, Yang D, Gadbury GL, Heshka S, Schwartz LG, Murugesan R, Kraker JL, Heo M, Heymsfield SB, Allison DB. 2003. Results of soy-based meal replacement formula on weight, anthropometry, serum lipids & blood pressure during a 40-week clinical weight loss trial. Nutr J 2:14.

Ford ES, Giles WH, Dietz WH. 2002. Prevalence of the metabolic syndrome among US adults. J Amer Med Assoc 287:356–359.

Forsythe WA. 1995. Soy protein, thyroid regulation and cholesterol metabolism. J Nutr 125:619S–623S.

Gale SM, Castracane VD, Mantzoros CS. 2004. Energy homeostasis, obesity and eating disorders: Recent advances in endocrinology. J Nutr 134:295–298.

Gianazza E, Eberini I, Arnoldi A, Wait R, Sirtori CR. 2003. A proteomic investigation of isolated soy proteins with variable effects in experimental and clinical studies. J Nutr 133:9–14.

Goodman-Gruen D, Kritz-Silverstein D. 2003. Usual dietary isoflavone intake and body composition in postmenopausal women. Menopause 10:427–432.

Hall WL, Vafeiadou K, Hallund J, Bugel S, Reimann M, Koebnick C, Zunft HJ, Ferrari M, Branca F, Dadd T, Talbot D, Powell J, Minihane AM, Cassidy A, Nilsson M, hlman-Wright K, Gustafsson JA, Williams CM. 2006. Soy-isoflavone-enriched foods and markers of lipid and glucose metabolism in postmenopausal women: Interactions with genotype and equol production. Am J Clin Nutr 83:592–600.

Hara H, Nishi T. 2003. Identification of amino acid sequence responsible for the satiety effect of soybean B-conglycinin. Soy Protein Res, Japan 6:94–99.

Hermansen K, Hansen B, Jacobsen R, Clausen P, Dalgaard M, Dinesen B, Holst JJ, Pedersen E, Astrup A. 2005. Effects of soy supplementation on blood lipids and arterial function in hypercholesterolaemic subjects. Eur J Clin Nutr 59:1843–1850.

Hermansen K, Soondergaard M, Hoie L, Carstensen M, Brock B. 2001. Beneficial effects of a soy-based dietary supplement on lipid levels and cardiovascular risk markers in type 2 diabetic subjects. Diabetes Care 24:228–233.

Hori G, Wang M-F, Chan Y-C, Komatsu T, Wong Y, Chen TH, Yamamoto K, Nagaoka S, Yamamoto S. 2001. Soy protein hydrolyzate with bound phospholipids reduces serum cholesterol levels in hypercholesterolemic adult male volunteers. Biosci Biotechnol Biochem 65:72–78.

Huang R-Q, Fang M-J, Dillon GH. 1999. The tyrosine kinase inhibitor genistein directly inhibits GABAA receptors. Mol Brain Res 67:177–183.

Hurley C, Richard D, Deshaires Y, Jacques H. 1998. Soy protein isolate in the presence of cornstarch reduces body fat gain in rats. Can J Physiol Pharmacol 76:1000–1007.

Hwang J-T, Park I-J, Shin J-I, Lee YK, Lee SK, Baik HW, Ha J, Park OJ. 2005. Genistein, EGCG, and capsaicin inhibit adipocyte differentiation process via activating AMP-activated protein kinase. Biochem Biophys Res Comm 338:694–699.

Iqbal MJ, Yaegashi S, Ahsan R, Lightfoot DA, Banz WJ. 2002. Differentially abundant mRNAs in rat liver in response to diets containing soy protein isolate. Physiol Genomics 11:219–226.

Iritani N, Sugimoto T, Fukuda H, Komiya M, Ikeda H. 1997. Dietary soybean protein increases insulin receptor gene expression in Wistar fatty rats when dietary polyunsaturated fatty acid level is low. J Nutr 127:1077–1083.

Ishihara K, Fukuchi Y, Mizunoya W, Mita Y, Fukuya Y, Fishiki T, Yasumoto K. 2003a. Amino acid composition of soybean protein increased carbohydrate oxidation in diabetic mice. Biosci Biotech Bioch 67:2505–2511.

Ishihara K, Oyaizu S, Fukuchi Y, Mizunoya W, Segawa K, Takahashi M, Mita Y, Fukuya Y, Fushiki T, Yasumoto K. 2003b. A soybean peptide isolate diet promotes postprandial carbohydrate oxidation and energy expenditure in type II diabetic mice. J Nutr 133:752–757.

James PT, Leach R, Kalamara E, Shayeghi M. 2001. The worldwide obesity epidemic. Obes Res 9:228S–233S.

Jayagopal V, Albertazzi P, Kilpatrick ES, Howarth EM, Jennings PE, Hepburn DA, Atkin SL. 2002. Beneficial effects of soy phytoestrogen intake in postmenopausal women with type 2 diabetes. Diabetes Care 25:1709–1714.

Jenkins DJA, Wolever TMS, Spiller GA, Buckley G, Lam Y, Jenkins AL, Josse RG. 1989. Hypocholesterolemic effects of vegetable protein in a hypocaloric diet. Atherosclerosis 78:99–107.

Kang MJ, Kim JI, Yoon SY, Kim JC, Cha IJ. 2006. Pinitol from soybeans reduces postprandial blood glucose in patients with type 2 diabetes mellitus. J Med Food 9:182–186.

Kennedy A. 2003. Absorption and health effects of the Bowman-Burke Inhibitor (BBI). J Nutr 134;1237–1238.

Kim H-K, Nelson-Dooley C, Della-Fera MA, Yang J-Y, Zhang W, Duan J, Hartzell DL, Hamrick MW, Baile CA. 2006. Genistein decreases food intake, body weight, and fat pad weight and causes adipose tissue apoptosis in ovariectomized female mice. J Nutr 136:409–414.

Kim S, Shin H-J, Kim SY, Kim JH, Lee YS, Kim D-H, Lee M-O. 2005. Genistein enhances expression of genes involved in fatty acid catabolism through activation of PPARa. Mol Cell Biochem 220:51–58.

Kohno M, Hirotsuka M, Kito M, Matsuzawa Y. 2006. Decreases in serum triacylglycerol and visceral fat mediated by dietary soybean beta-conglycinin. J Atheroscler Thromb 13:247–255.

Kok L, Kreijkamp-Kaspers S, Grobbee DE, Lampe JW, Van Der Schouw YT. 2006. Soy isoflavones, body composition, and physical performance. Maturitas 52:102–110.

Kritchevsky D. 1979. Vegetable protein and atherosclerosis. J Am Oil Chem Soc 56:135–140.

Kwon DY, Jang JS, Hong SM, Lee JE, Sung SR, Park HR, Park S. 2007. Long-term consumption of fermented soybean-derived Chungkookjang enhances insulinotropic action unlike soybeans in 90% pancreatectomized diabetic rats. Eur J Nutr 46:44–52.

Kwon DY, Jang JS, Lee JE, Kim YS, Shin DH, Park S. 2006. The isoflavonoid aglycone-rich fractions of Chungkookjang, fermented unsalted soybeans, enhance insulin signaling and peroxisome proliferator-activated receptor-gamma activity *in vitro*. Biofactors 26:245–258.

Lavigne C, Marette A, Jacques H. 2000. Cod and soy proteins compared to casein improve glucose tolerance and insulin sensitivity in rats. Am J Physiol Endocrinol Metab 278:E491–E500.

Le May C, Chu K, Hu M, Ortega CS, Simpson ER, Korach KS, Tsai M-J, Mauvais-Jarvis F. 2006. Estrogens protect pancreatic B-cells from apoptosis and prevent insulin-deficient diabetes mellitus in mice. Proc Natl Acad Sci USA 103:9232–9237.

Lee JS. 2006. Effects of soy protein and genistein on blood glucose, antioxidant enzyme activities, and lipid profile in streptozotocin-induced diabetic rats. Life Sci 79:1578–1584.

Lee M, Kim I. 2000. Soy protein and obesity. Nutrition 16:459–460.

Lee SD, Nakano H, Farkas GA. 2001. GABAergic modulation of ventilation and peak oxygen consumption in obese Zucker rats. J Appl Physiol 90:1707–1713.

Lee S-H, Park I-S. 2000. Effects of soybean diet on the B cells in the streptozotocin treated rats for the induction of diabetes. Diabetes Res Clin Pr 47:1–13.

Lephart ED, Adlercreutz H, Lund TD. 2001. Dietary soy phytoestrogen effects on brain structure and aromatase in Long–Evans rats. Neuroendocrinol 12:3451–3455.

Liu D, Zhen W, Yang Z, Carter JD, Si H, Reynolds KA. 2006. Genistein acutely stimulates insulin secretion in pancreatic beta-cells through a cAMP-dependent protein kinase pathway. Diabetes 55:1043–1050.

Lovati MR, Manzoni C, Canavesi A, Sirtori M, Vaccarino V, Marchi M, Gaddi G, Sirtori CR. 1987. Soybean protein diet increases low density lipoprotein receptor activity in mononuclear cells from hypercholesterolemic patients. J Clin Invest 80:1498–1502.

Lovati MR, Manzoni C, Corsini A, Granata A, Frattini R, Fumagalli R, Sirtori CR. 1992. Low density lipoprotein receptor activity is modulated by soybean globulins in cell culture. J Nutr 122:1971–1978.

Lovati MR, Manzoni C, Gianazza E, Arnoldi A, Kurowska E, Carroll KK, Sirtori CR. 2000. Soy protein peptides regulate cholesterol homeostasis in Hep G2 cells. J Nutr 130:2543–2549.

Lovati MR, Manzoni C, Gianazza E, Sirtori CR. 1998. Soybean protein products as regulators of liver low-density lipoprotein receptors. I. Identification of active beta-conglycinin subunits. J Agric Food Chem 46:2474–2480.

Lund TD, Lephart ED. 2001. Dietary soy phytoestrogens produce anxiolytic effects in the elevated plus-maze. Brain Res 21:180–184.

Ma YA, Hu JH, Zhou XG, Zeng RW, Mei ZT, Fei J, Guo LH. 2003. Trangenic mice overexpressing gamma-aminobutyric acid transporter subtype I develop obesity. Cell Res 10:303–310.

Mackowiak P, Nogowski L, Nowak KW. 1999. Effect of isoflavone genistein on insulin receptors in perfused liver of ovariectomized rats. J Recep Signal Transduct Res 19:283–292.

Manzoni C, Duranti M, Eberini I, Scharnag H, Marz W, Castiglioni S, Lovati MR. 2003. Subcellular localization of soybean 7S globulin in HepG2 cells and LDL receptor up-regulation by its a' constituent subunit. J Nutr 133:2149–2155.

Manzoni C, Lovati MR, Gianazza E, Morita Y, Sirtori CR. 1998. Soybean protein products as regulators of liver low-density lipoprotein receptors. II. Alpha-alpha prime rich commercial soy concentrate and alpha prime deficient mutant differently affect low-density lipoprotein receptor activity. J Agric Food Chem 46:2481–2484.

Matsuzawa Y. 2005. Adiponectin: Identification, physiology and clinical relevance in metabolic and vascular disease. Atheroscler Suppl 6:7–14.

Matzkies F, Hubner M. 1981. Cholesterol- and weight-reduction with a formula diet containing protein and legumes. Fortschr Med 99:195–199.

McCarty MF. 2003. The origins of western obesity: A role for animal protein. Med Hypotheses 54:488–494.

Meshcheryakova VA, Plotnikova OA, Sharafrtdinov KK, Iatsyshina TA. 2002. The use of the combined food products with soy protein in diet therapy for patients with diabetes mellitus type 2 (Russian). Vopr Pitan 71:19–24.

Mikkelsen PB, Toubro S, Astrup A. 2000. Effect of fat-reduced diets on 24-h energy expenditure: Comparisons between animal protein, vegetable protein, and carbohydrate. Am J Clin Nutr 72:1135–1141.

Moeller LE, Peterson CT, Hanson KB, Dent SB, Lewis DS, King DS, Alekel DL. 2003. Isoflavone-rich soy protein prevents loss of hip lean mass but does not prevent the shift in regional fat distribution in perimenopausal women. Menopause 10:322–331.

Mori M, Aizawa T, Tokoro M, Miki T, Yamori Y. 2004. Soy isoflavones tablets reduce osteoporosis risk factors and obesity in middle-aged Japanese women. Clin Exp Pharmacol Physiol 31:S44–S46.

Moriyama T, Kishimoto K, Nagai K, Urade R, Ogawa T, Utsumi S, Maruyama N, Maebuchi M. 2004. Soybean b-conglycinin diet suppresses serum triglyceride levels in normal and genetically obese mice by induction of b-oxidation, downregulation of fatty acid synthase, and inhibition of triglyceride absorption. Biosci Biotech Bioch 68:352–359.

Munro IC, Harwood M, Hlywka JJ, Stephen AM, Doull J, Flamm H, Adlercreutz H. 2003. Soy isoflavones: A safety review. Nutr Rev 61:1–33.

Naaz A, Yellayi S, Zakroczymski MA, Bunick D, Doerge DR, Lubahn DB, Helferich WG, Cooke PS. 2003. The soy isoflavone genistein decreases adipose deposition in mice. Endocrinology 144:3315–3320.

Nagasawa A, Fukui K, Funahashi T, Maeda N, Shimomura I, Kihara S, Waki M, Takamatsu K, Matsuzawa Y. 2002. Effects of soy protein diet on the expression of adipose genes and plasma adiponectin. Horm Metab Res 34:635–639.

Nagasawa A, Fukui K, Kojima M, Kishida K, Maeda N, Nagarentani H, Hibuse T, Nishizawa H, Kihara S, Waki M, Takamatsu K, Funahashi T, Matsuzawa Y. 2003. Divergent effects of soy protein diet on the expression of adipocytokines. Biochem Biophys Res Comm 311:909–914.

Nestel PJ, Yamashita T, Sasahara T, Pomeroy S, Dart A, Komesaroff P, Own A, Abbey M. 1997. Soy isoflavones improve systemic arterial compliance but not plasma lipids in menopausal and perimenopausal women. Anterioscl Throm Vas 17:3392–3398.

Nikander E, Tiitinen A, Laitinen K, Tikkanen M, Ylikorkala O. 2004. Effects of isolated isoflavones on lipids, lipoproteins, insulin sensitivity, and ghrelin in postmenopausal women. J Clin Endocr Metab 89:3567–3572.

Nishi T, Hara H, Asano K, Tomita F. 2003a. The soybean B-conglycinin B-51-63 fragment suppresses appetite by stimulating cholecystokinin release in rats. J Nutr 133:2537–2542.

Nishi T, Hara H, Tomita F. 2003b. Soybean B-conglycinin peptone suppresses food intake and gastric emptying by increasing plasma cholecystokinin levels in rats. J Nutr 133:352–357.

Noriega-Lopez L, Tovar AR, Gonzalez-Granillo M, Hernandez-Pando R, Escalante B, Santillan-Doherty P, Torres N. 2007. Pancreatic insulin secretion in rats fed soy protein-high fat diet depends on the interaction between amino acid pattern and isoflavones. J Biol Chem 282:20657–20666.

Persky VW, Turyk ME, Wang L, Freels S, Chatterton R, Barnes S, Erdman JW, Sepkovic DW, Bradlou HL, Potter S. 2002. Effect of soy protein on endogenous hormones in postmenopausal women. Am J Clin Nutr 75:145–153.

Pupovac J, Anderson GH. 2002. Dietary peptides induce satiety via cholecystokinin-A and peripheral opiod receptors in rats. J Nutr 132:2775–2780.

Reddy S, Karanam M, Robinson E. 2001. Prevention of cyclophosphamide-induced accelerated diabetes in the NOD mouse by nicotinamide or a soy protein-based infant formula. Int J Exp Diabetes Res 1:299–313.

Saito M. 1991. Effect of soy peptides on energy metabolism in obese animals. Nutr Sci 12:91.

Rho SJ, Park S, Ahn C-W, Shin J-K, Lee HG. 2007. Dietetic and hypocholesterolaemic action of black soy peptide in dietary obese rats. J Food Sci Agric 87:908–913.

Rossi R, Origliani G, Modena M. 2004. Transdermal 17-β-estradiol and risk of developing Type 2 diabetes in a population of healthy, nonobese postmenopausal women. Diabetes Care 27:645–649.

Sanchez A, Hubbard RW. 1991. Plasma amino acids and insulin/glucagon ratio as an explaination for the dietary protein modulation of atherosclerosis. Med Hypotheses 36:27–32.

Setchell KDR. 2001. Soy isoflavones—benefits and risks from nature's selective estrogen receptor modulators (SERMs). J Am Coll Nutr 20:354S–362S.

Setchell KDR, Borriello SP, Hulme P, Kirk DN, Axelson M. 1984. Nonsteroidal estrogens of dietary origin: Possible roles in hormone-dependent disease. Am J Clin Nutr 40:569–578.

Setchell KDR, Cassidy A. 1999. Dietary isoflavones: Biological effects and relevance to human health. J Nutr 129:758S–767S.

Sirtori CR, Agradi E, Conti F, Mantero O, Gatto E. 1977. Soybean-protein diet in the treatment of type-II hyperlipoproteinaemia. Lancet 1:275–277.

Sirtori CR, Anderson JW, Arnoldi A. 2007. Nutrition and nutraceutical considerations for dyslipidemia. Future Lipidol 3:313–339.

Sirtori CR, Gianazza E, Manzoni C, Lovati MR, Murphy PA. 1997. Role of isoflavones in the cholesterol reduction by soy proteins in the clinic. Am J Clin Nutr 65:166–167.

Sites CK, Cooper BC, Toth MJ, Gastaldelli A, Arabshahi A, Barnes S. 2007. Effect of a daily supplement of soy protein on body composition and insulin secretion in postmenopausal women. Fertil Steril 88:1609–1617.

Solae L. 2004. Health claim petition submission to FDA: Soy protein and the reduced risk of certain cancers. Available at http://www.fda.gov/ohrms/dockets/dockets/04q0151/04q0151.htm. Accessed on April 19, 2008.

Spiegelman BM, Flier JS. 2001. Obesity and the regulation of energy balance. Cell 104:531–543.

St-Onge MP, Claps N, Wolper C, Heymsfield SB. 2007. Supplementation with soy-protein-rich foods does not enhance weight loss. J Am Diet Assoc 107:500–505.

Strom BL, Schinnar R, Ziegler EE, Barnhart K, Sammel M, Macones G, Stallings V, Hanson SA, Nelson SE. 2001. Follow-up study of a cohort fed soy-based formula during infancy. J Amer Med Assoc 286:807–814.

Szkudelska T, Nogowski L, Pruszynska-Oszmalek E, Kaczmarek P, Szkudelska K. 2005. Genistein restricts leptin secretion from rat adipocytes. J Steroid Biochem Mol Biol 96:301–307.

Szkudelska K, Szkudelska T, Nogowski L. 2002. Daidzein, coumestrol and zearalenone affect lipogenesis and lipolysis in rat adipocytes. Phytomedicine 9:338–345.

Taylor SL, Hefle SL. 2002. Allergic reactions and food intolerances. In Nutritional Toxicology, edited by Kotsonis FN, Mackey MA. New York: Taylor & Francis.

Terpstra AHM, Hermus RJJ, West CE. 1983. The role of dietary protein in cholesterol metabolism. World Rev Nutr Diet 42:1–55.

Tovar AR, Merguia F, Cruz C, Hernandez-Pando R, Aguilar-Salinas CA, Pedraza-Chaverri J, Correa-Rotter R, Torres N. 2002. A soy protein diet alters hepatic lipid metabolism gene expression and reduces serum lipids and renal fibrogenic cytokines in rats with chronic nephrotic syndrome. J Nutr 132:2562–2569.

Tsujii S, Bray GA. 1991. GABA-related feeding control in genetically obese rats. Brain Res 540:48–54.

Uesugi T, Toda T, Tsuji K, Ishida H. 2001. Comparative study on reduction of bone loss and lipid metabolism abnormality in ovariectomized rats by soy isoflavones, daidzin, genistin, and glycitin. Biol Pharm Bull 24:368–372.

Van Der Lely AJ, Tschop M, Heiman ML, Ghigo E. 2004. Biological, physiological, pathophysiological, and pharmacological aspects of ghrelin. Endocrinol Rev 25:426–457.

Vaughn N, Beverly L, Rizzo A, Doane D, de Mejia EG. 2007. Intracerebro ventricular administration of soy hydrolysates reduced body weight without affecting food intake. FASEB J 21:A372–A372.

Volgarev MN, Vysotsky VG, Meshcheryakova VA, Yatsyshina TA, Steinke FH. 1989. Evaluation of isolated soy protein foods in weight reduction with obese hypercholesterolemic and normocholesterolemic obese individuals. Nutr Rep Int 39:61–72.

Wagner JD, Cefalu WT, Anthony MS, Litwak KN, Zhang L, Clarkson TB. 1997. Dietary soy protein and estrogen replacement therapy improve cardiovascular disease risk factors and decrease aortic cholesterol ester content in ovariectomized cynomolgus monkeys. Metabolis 46:698–705.

Wang W, de Mejia EG. 2005. A new frontier in soy bioactive peptides that may prevent age-related chronic diseases. Compr Rev Food Sci Food Saf 4:63–78.

Wang GJ, Volkow ND, Logan J, Pappas NR, Wong CT, Zhu W, Netusil N, Fowler JS. 2001. Brain dopamine and obesity. Lancet 357:354–357.

Wang M-F, Yamamoto S, Chung H-M, Chung S-Y, Miyatani S, Mori M, Okita T, Sugano M. 1995. Antihypocholesterolemic effect of undigested fraction of soybean protein in young female volunteers. J Nutr Sci Vitaminol 41:187–195.

Wilson PWF. 2003. Lower diabetes risk with hormone replacement therapy: An encore for estrogen. Ann Intern Med 138:69–70.

Woods SC. 2004. Gastrointestinal satiety signals: I. an overview of gastrointestinal signals that influence food intake. Am J Physiol Gastrointest Liver Physiol 286:G7–G13.

Wu J, Wang X, Chiba H, Higuchi M, Nakatoni T, Ezaki O, Cui H, Yamada K, Ishimi Y. 2004. Combined intervention of soy isoflavone and moderate exercise prevents body fat elevation and bone loss in ovariectomized mice. Metabolis 53:942–948.

Yang G, Shu X-O, Jin F, Elasy T, Li H-L, Li Q, Huang F, Gao Y-T, Zheng W. 2004. Soyfood consumption and risk of glycosuria: A cross-sectional study within the Shanghai Women's Health Study. Eur J Clin Nutr 58:615–620.

Zhou J-R, Pan W, Takebe M, Blackburn GL. 1999. Effects of isoflavone aglycone in the prevention of obesity in a diet-induced obesity mouse model. Atherosclerosis 143:81–90.

Chapter 10

Minerals and Insulin Health

Philip Domenico, PhD, and James R Komorowski, MS

Overview

Mineral deficiencies are often implicated in many chronic conditions, including those involving glycemic and insulin health. The best-supported minerals in this regard are chromium, magnesium, and zinc, yet vanadium, calcium, boron, and manganese may also play a role. The typical Western diet does not meet the requirements for these minerals. Chromium is a cofactor for insulin. Low chromium levels in various tissues are strongly correlated with diabetes. In 10 clinical trials treating DM2 with chromium picolinate (CrPic), the average HbA1c reduction was 0.95%, which represents substantial risk reduction. All 15 DM2 trials with CrPic intervention showed salutary effects in at least one parameter of diabetes management. Magnesium is a cofactor in hundreds of enzymes. Its effects on carbohydrate metabolism are multifactorial and are linked to insulin resistance and DM2. Several epidemiologic, animal, and human clinical studies support the benefits of magnesium in the prevention and treatment of diabetes and its complications. Zinc is also a cofactor in hundreds of enzymes, some of which are directly involved in the synthesis, storage, and secretion of insulin. Supplemental vanadium has been shown to enhance glucose oxidation, glycogen synthesis, and modulate hepatic glucose output. Calcium and vitamin D intake is associated with a lower prevalence of the metabolic syndrome and DM2. Boron influences many enzymes important in energy substrate metabolism, and is linked to insulin control. Manganese enzymes play important roles in carbohydrate, amino acid, and cholesterol metabolism. Chelated forms of these minerals are generally superior to inorganic forms, and their safety profiles are of little concern at recommended doses. The data suggest that certain minerals are altered in DM2 and that supplementation may be useful in preventing and treating this disorder.

Introduction

Mineral nutrition is beginning to take shape, as research in this area has grown exponentially over the past two decades. Minerals are facets of both structure and function in living systems. Beyond strong bones, minerals are often the active principle in enzymes, especially for energy generation, storage, and transport. Minerals also play a crucial role in cell signaling and regulation. Metalloproteins are pervasive in every aspect of biology, including replication, transcription, respiration, cellular fluid balance, electrochemical nerve and muscle activity, bone strength, and tissue elasticity. Minerals also contribute to pigmentation. In many respects, biology is an elaboration of mineral chemistry.

The list of missing minerals that impact metabolic function is growing. Mineral deficiencies result from several modern realities, to include soil depletion, deionizing water, food processing, cooking, and uneducated or undisciplined food choices. The medical establishment recommends meeting nutritional requirements with natural food sources, but few people are complying. People at high risk for diabetes are least likely to choose a healthy diet, or to have access to high-quality, mineral-rich foods (Schulze and Hu 2005). Though generally not in support of dietary supplements, the American Diabetes Association does concede their need in people at increased risk for micronutrient deficiencies, such as the elderly, vegans, dieters, and other special populations (O'Connell 2001; Yeh et al. 2003). The widespread use of dietary supplements is justified given the low micronutrient intake in the Western diet (Popkin 2001).

The likely result of a lifetime of poor eating habits and sedentary lifestyle is type 2 diabetes (DM2), a chronic condition characterized by high levels of blood glucose. It begins with insulin resistance (poor insulin function), followed by insulin deficiency as the disease progresses. DM2 is a leading cause of cardiovascular disease, blindness, kidney failure, gum disease, lower limb amputation, complications of pregnancy, and death (Centers for Disease Control and Prevention 2005).

Since DM2 is associated with older age and obesity, it is often responsive to dietary and lifestyle intervention. Managing disease can involve a number of options at different stages of progression. Many patients can control their blood glucose with diet, exercise, and supplementation, though some may require medication (Knowler et al. 2002). Insulin therapy may be necessary at later stages of disease.

Exactly what role can dietary supplements play, in conjunction with other modalities, to help prevent or treat disorders of insulin health and metabolism? Research has identified several nutritional supplements that could help control blood sugar and lipids, and prevent diabetic complications. These include antioxidants, antiinflammatory agents, B vitamins, minerals, herbs, essential oils, and amino acids. In the mineral category, several supplements belong in a preventive or therapeutic regimen for DM2. The evidence is strongly in support of chromium, magnesium, and zinc in diabetes prevention and therapy (Anderson 1998; Chausmer 1998; Tosiello 1996). Other minerals, such as vanadium, calcium, boron, and manganese affect metabolic health, and need to be addressed in any dietary platform for diabetes or prediabetes care. This chapter takes a closer look at each of these minerals and, in particular, chromium with regard to their role in insulin health and safety profile.

Minerals Affecting Metabolic Health

Chromium (Cr)

Certainly, the most popular and best-researched mineral for insulin function is Cr. This trace element is essential in carbohydrate, lipid, and protein metabolism (Anderson 1998; Evans 1989). As a cofactor for insulin, Cr increases insulin binding (Vincent 2000), the number of insulin receptors (Anderson 1995; Cefalu and Hu 2004), and insulin receptor phosphorylation (Wang et al. 2005), resulting in enhanced glucose transport into liver, muscle, and adipose tissue (Anderson 1998). Chromium also inhibits phosphotyrosine phosphatase (Davis et al. 1996), which cleaves phosphate from the insulin receptor and reduces insulin sensitivity. Since Cr is required for normal glucose and lipid metabolism, low Cr status can adversely affect blood glucose, insulin, total cholesterol, triglyceride, and high-density lipoprotein (HDL) cholesterol levels (Anderson 1998; Anderson et al. 1997; Cefalu and Hu 2004; Guallar et al. 2005; Heimbach and Anderson et al. 2005; Morris et al. 1999; Rajpathak et al. 2004).

Though the minimum estimated safe and adequate daily dietary intake for Cr is 50–200 μg/day for people 7 years or older, typical Western diets do not meet these requirements (Anderson et al. 1992; Anderson and Kozlovsky 1985; Lukaski 1999). Indeed, close to 90% of the U.S. population do not receive adequate Cr (Anderson and Kozlovsky 1985). A more recent report found that U.S. adults are consuming less than the established adequate intakes of 25–35 μg of Cr/day (Juturu and Komorowski 2003). An assessment in moderately obese, DM2 subjects revealed that the average Cr intake was 16.4 μg per 1,000 kcal, far lower than the RDI of 120 μg/day. The same study showed a significant correlation between dietary Cr intake and fasting insulin (FI), total cholesterol, LDL cholesterol, triglycerides, insulin sensitivity, and atherogenic index (Juturu et al. 2006). Similar findings documenting inadequate Cr intake were reported in Canada (Campbell 2001), Britain (Smart and Sherlock 1985), and Finland (Kumpulainen et al. 1980). One epidemiological study of over 2,000 subjects showed low Cr status in over 50% of the study population, based on hair analysis (Campbell 2001).

Dietary sources of Cr include brewer's yeast, broccoli, beer, whole grains, cheese, mushrooms, oysters, and liver. However, Cr content is not usually ample and can vary widely in these foods (Anderson et al. 1992; Offenbacher and Pi-Sunyer 1997). Some of the Cr measured in foods arises from food processing equipment and is not bioavailable (Offenbacher and Pi-Sunyer 1997). In modern times, natural Cr is not easily obtained, and more readily lost. Refining and processing of foods remove >90% of Cr (Davies et al. 1997). Most chromium is found in the germ of grains, which is removed to extend product shelf life. Diets high in sugar promote Cr excretion (Gross et al. 2004; Kozlovsky et al. 1986). Exercise also increases Cr losses (Rubin et al. 1998), as do common prescription medications, particularly corticosteroids (Ravina et al. 1999). Relative Cr deficiency is further exacerbated with age (Davies et al. 1997; Mertz 1990), illness (Anderson 1986), pregnancy (Anderson et al. 1993), burns (Anderson et al. 2005), and assorted stresses (Campbell et al. 1999).

In subjects with DM2, Cr metabolism is altered by inadequate intake, decreased absorption, and increased loss, exemplified by abnormal blood, tissue, and urine Cr levels (Anderson et al. 1997; Cefalu et al. 2002; Morris et al. 1999). Low Cr levels in serum (Davies et al. 1997; Ekmekcioglu et al. 2001), hair (Aharoni et al. 1992), or toenail tissue (Rajpathak et al. 2004) are strongly correlated with diabetes. High urine Cr levels in DM2 indicate that Cr is not reabsorbed well by the kidneys (Mita et al. 2005; Rajpathak et al. 2004). Higher doses of chromium (1,000 μg/day) may be required to provide significant clinical benefit in DM2 (Anderson 2000).

Inadequate Cr from the diet and increased loss of Cr with age increases the need for supplementation, especially in people who are insulin resistant. With a number of chromium products currently in the market, consumers and scientists alike are confused as to which Cr supplement is most effective. In general, organic or chelated forms of Cr are much better absorbed than inorganic forms. It has been shown conclusively that Cr is not well absorbed from $CrCl_3$ (Anderson et al. 2003; DiSilvestro and Dy 2005; Ghosh et al. 2002). Starch can further inhibit Cr absorption from $CrCl_3$, suggesting that a variety of foods can interfere with its bioavailability (Anderson et al. 2003).

Disparities in absorption also exist among organic Cr forms. CrPic is considered the Cr of choice, with regard to total sales and to its preference by the NIH. Again, it is about absorption. In two independent human studies, CrPic was significantly better absorbed than Cr polynicotinate (CrNic) or $CrCl_3$ (Anderson et al. 2003; DiSilvestro and Dy 2005). In animal studies, CrPic reached significantly higher tissue concentrations in muscle, liver, and heart than $CrCl_3$, CrNic, or Cr histidinate (Anderson et al. 1996). CrPic has also demonstrated higher absorption and insulin internalization rates compared to CrNic (Anderson et al. 2003; DiSilvestro and Dy 2005). One animal study (Olin et al. 1994) reported greater bioavailability of CrNic over CrPic, but based their analysis on percent Cr retained versus total amount absorbed. The available data indicate that CrNic is poorly absorbed. A new Cr supplement under development is Cr histidinate, which has shown superior absorption in humans (Anderson et al. 2004). However, to date, there is scant evidence supporting its efficacy or safety.

The absorption data also explain the difference in clinical health benefit among Cr supplements. Of 43 clinical studies to date that investigated Cr supplementation in DM2 or impaired glucose, 22 used CrPic, while 21 used other Cr forms (Table 10.1). Nineteen of 22 clinical studies using CrPic reported benefits in glucose or insulin parameters. In contrast, only 5 of 12 using $CrCl_3$, four of seven using Cr yeast, and none of two studies using CrNic reported significant glycemic

Table 10.1. Chromium intervention in subjects with diabetes/impaired glucose.

	Study	No. of patients	Results
	Chromium picolinate		
1	Evans (1989)	6	+
2	Lee and Reasner (1994)	28	−
3	Ravina et al. (1995)	162	+
4	Anderson et al. (1997)	105	+
5	Cefalu et al. (1999)	29	+
6	Jovanovic-Peterson et al. (1999)	20	+
7	Ravina et al. (1999)	44	+
8	Bahadori et al. (1999)	16	+
9	Cefalu et al. (1999)	15	+
10	Cheng et al. (1999)	833	+
11	Rabinowitz et al. (1983)	39	+
12	Morris et al. (2000)	5	+
13	Feng et al. (2002)	32	+
14	Ghosh et al. (2002)	43	+
15	Rabinovitz et al. (2004)	78	+
16	Vrtovec et al. (2005)	56	+
17	Lucidi et al. (2005)	6	+
18	Gunton et al. (2005)	20	−
19	Vladeva et al. (2005)	17	+
20	Kleefstra et al. (2006)	29	−
21	Lydic et al. (2006)	5	+
22	Martin et al. (2006)	17	+
	Total		19+/3−
	Chromium polynicotinate		
1	Thomas and Gropper (1996)	5	−
2	Juang et al. (20000	15	−
	Total		0+/2−
	Chromium yeast		
1	Offenbacher and Pi-Sunyer (1980)	4	+
2	Rabinowitz et al. (1983)	15	−
3	Elias et al. (1984)	6	+
4	Uusitupa et al. (1992)	26	−
5	Trow et al. (2000)	23	−
6	Bahijiri et al. (2000)	78	+
7	Racek et al. (2006)	36	+
	Total		4+/3−

(continued)

Table 10.1. (*continued*)

	Study	No. of patients	Results
	Chromium chloride		
1	Glinsmann and Mertz (19660	6	+
2	Sherman et al. (1968)	7	−
3	Levine (1968)	10	−
4	Rabinowitz et al. (1983)	15	−
5	Mossop (1983)	13	+
6	Uusitupa et al. (1983)	10	+
7	Anderson et al. (1983)	41	+
8	Potter et al. (1985)	5	−
9	Martinez et al. (1985)	17	−
10	Anderson et al. (1991)	8	−
11	Abraham et al. (1992)	27	−
12	Bahijiri et al. (2000)	78	+
	Total		5+/7−

+ = significant; − = not significant

benefits. The results are consistent with the absorption data: CrPic is consistently more effective and taken more seriously by scientists than other Cr forms.

At the appropriate dose and duration, supplementation with CrPic may affect a number of metabolic risk factors. Generally speaking, more consistent responses were observed with >200 μg Cr for a duration of ≥2 months (Cefalu and Hu 2004). Benefits observed from CrPic supplementation include reduced fasting and postprandial blood glucose and insulin levels (Anderson et al. 1997), reduced total cholesterol (total C), LDL cholesterol, triglycerides, body weight, body fat, and carbohydrate cravings (Docherty et al. 2004), and increased insulin sensitivity for subjects with prediabetes (Cefalu et al. 1999) or diabetes (Martin et al. 2005).

To date, there have been 15 clinical studies on CrPic supplementation for diabetes mellitus. Twelve of the studies were randomized controlled trials, and three were open label. Fourteen studies focused on DM2, and one each on DM1, corticosteroid-induced, and gestational diabetes. All together, 1,690 subjects, 1,505 of whom were assigned to CrPic, completed the trials. Dosages ranged from 200 to 1,000 μg Cr/day, with duration of supplementation from 1 week to 10 months. Measures of glycemic control—fasting glucose (FG), postprandial glucose (PPG), FI, postprandial insulin (PPI), glycated hemoglobin (HbA1c), or insulin sensitivity—differed among trials. Other metabolic risk factors (i.e., blood lipids, microalbuminuria, ApoA1, C reactive protein) and body composition (i.e., BMI, body fat, lean body mass) were also studied.

A randomized controlled trial (RCT) (Anderson et al. 1997) evaluated CrPic in 180 Chinese subjects with DM2. Subjects received either 200 or 1,000 μg Cr/day as CrPic for 4 months. Supplemental CrPic led to significant improvements in FG, PPG, FI, PPI ($P < 0.05$), and HbA1c ($P < 0.01$) levels. Significance was achieved as early as 2 months, especially at the higher dose. Subjects receiving the 1,000 μg Cr dose showed near 30% reductions in FG, PPG, FI, and HbA1c. A follow-up, open-label study was conducted in 833 Chinese subjects with DM2 on insulin or hypoglycemic drugs (Cheng et al. 1999). All patients received 500 μg Cr/day as CrPic for 10 months. Again, FG and PPG were significantly lowered (by 20.0 and 17.5%, respectively, $P < 0.05$) after the first month of therapy and remained so in the following 9 months. Close to 90% of subjects experienced marked relief from fatigue, thirst, and frequent urination, with no reported side effects. In an extension to this trial, over 1,000 DM2 subjects have received CrPic (500 μg Cr) daily for over 6 years, with progressive benefits in blood sugar control. FG dropped gradually from 12.1 to 7.4 mmol/L, and PPG went from 14.5 to 8.8 mmol/L in the 6-year trial period (Zhuang et al. 2005).

Another RCT (Feng et al. 2002) involving 136 Chinese subjects on insulin therapy taking 500 μg Cr/day as CrPic for 3 months showed significant reductions in FG and PPG (by 16.2 and 20.8%, respectively, $P < 0.01$). Three other studies, supplementing with 200–1,000 μg Cr/day as CrPic from 3 weeks to 6 months, reported significant improvement in both FG and HbA1c levels (Evans 1989; Martin et al. 2006; Rabinovitz et al. 2004). Reductions in FG in two of those studies were highly significant (21–24.3%, $P < 0.001$).

Another RCT on elderly subjects with DM2 recovering from stroke or hip fracture involved supplementation with 400 μg Cr as CrPic over 3 weeks in addition to their normal hypoglycemic and/or insulin medication (Rabinovitz et al. 2004). These patients showed a significant decrease in FG and HbA1c levels. Blood glucose levels decreased from 10.5 mmol/L at baseline to 8.3 mmol/L at the end of the study ($P < 0.001$), and HbA1c improved from 8.2 to 7.6% ($P < 0.01$).

A study on Asian Indian subjects taking CrPic (400 μg Cr/day for 12 weeks) showed highly significant improvements in most measures of glycemic control (-7.2% for FG, -16.4% for PPG, and -19.5% for FI) (Ghosh et al. 2002). Though the mean change in HbA1c was not different from baseline, it was significantly better than placebo ($P < 0.05$). Two studies on Caucasian subjects with DM2 either diet treated (Vrtovec et al. 2005) or on hypoglycemic drugs (Bahadori et al. 1999), reported that 1,000 μg Cr/day as CrPic for 3 or 4 months reduced FI significantly ($P < 0.05$). In the diet treated study, a significant decrease in FI was associated with a shortened QTc interval in 62% of subjects, especially those with high BMI (Vrtovec et al. 2005). QTc interval prolongation on a standard EKG is a powerful predictor of mortality, cardiac death, and stroke in patients with DM2.

Insulin sensitivity was also significantly increased in three studies (Martin et al. 2006; Morris et al. 2000; Ravina et al. 1995) after 3–6 months of CrPic supplementation.

Two of the 15 CrPic intervention studies for DM2 did not show significant improvement in glycemic markers. In one of these studies (Kleefstra et al. 2006), subject selection (advanced stage, poorly controlled DM2, high insulin intake) did

Table 10.2.   CrPic effect on glycemic control in DM2 subjects (% change from baseline).

Study	FG	PPG	FI	PPI	HbA1c
Anderson et al. 1997[a]	−10.8	−13.9	−25.5[c]	−18.2[c]	−24.5[c]
Anderson et al. 1997[b]	−27.6[c]	−30.4[c]	−33.1[c]	−15.6[c]	−34[d]
Bahadori et al. 1995	–	−5.0	−38.5[c]	−13.6	−2.4
Cheng et al. 1999	−20.0[c]	−17.5[c]	–	–	–
Evans 1989	−24.3[c]	–	–	–	−16.0[c]
Feng et al. 2002	−16.2[d]	−20.8[d]	–	–	–
Ghosh et al. 2002	−7.2[e]	−16.4[e]	−19.5[c]	–	−0.1[f]
Martin et al. 2006	–	–	–	–	−11.9[d]
Morris et al. 2000	−1.3	–	–	–	1.5
Rabinovitz et al. 2004	−21.0[e]	–	–	–	−7.3[d]
Vrtovec et al. 2005	−1.7	–	−28.4[c]	–	2.9
Mean ± SD	−15.3 ± 9.7	−18.9 ± 8.4	−29.8 ± 7.0	−15.0 ± 1.5	−9.0 ± 10.9

FG, fasting glucose; PPG, postprandial glucose; FI, fasting insulin; PPI, postprandial insulin; HbA1c, glycated hemoglobin; –, no data available.
[a] 200 mcg Cr/day.
[b] 1,000 mcg Cr/day.
[c] $P < 0.05$.
[d] $P < 0.01$.
[e] $P < 0.001$.
[f] $P < 0.05$ compared to control.

not favor a positive outcome. The other negative study (Lee and Reasner 1994) employed 200 μg Cr as CrPic for 2 months, a likely insufficient dose and duration. Nevertheless, both studies reported a significant impact on blood lipid risk factors. CrPic also improved glycemic control in gestational diabetes (Jovanovic-Peterson et al. 1999), corticosteroid-induced diabetes (Ravina et al. 1999), and reduced medication requirements (Ravina et al. 1995).

The combined effects of these changes attributed to CrPic intervention are summarized in Table 10.2. Pooled percent changes for FG, PPG, FI, PPI, and HbA1c, and their standard deviations are presented. Six of 10 studies reported significant improvement in FG from baseline, with a mean reduction of 1.5 mmol/L or −15.3%. This change is comparable to intensive control using sulfonylureas or insulin (Adler et al. 2000). Four of six studies measuring PPG showed significant results compared to baseline, with a mean reduction of 2.7 mmol/L or −18.9%. All four trials testing for FI reported significant improvements from baseline regardless of CrPic dose, with an average reduction of 45.2 pmol/L or −29.8%. Lowering the FI is associated with decreased risk of obesity, diabetes, and heart failure (Slabber et al. 1994). Two of three assessable studies reported improvements in PPI from the baseline, with an average reduction of 92.7 pmol/L or −15.0%. Of 10 studies that measured HbA1c, five were significant with respect to change from baseline, and one was significant compared to placebo. The average reduction in HbA1c for all 10 studies was −0.95%, or a 9.6% reduction from baseline. This represents substantial risk reduction, since a 1% drop in HbA1c equates to a 37% reduction in risk of microvascular complications and a 21% reduction in risk for diabetes-related mortality (Stratton et al. 2000).

Chromium deficiency is also associated with lipid abnormalities and an increased risk of atherosclerotic disease (Newman et al. 1978). Given the known predisposition for coronary heart disease in diabetes, improving blood lipid control with Cr may translate to reduced risk. Indeed, clinical trials show improved lipid profiles in DM2 subjects with CrPic supplementation. In one study, 200 μg/day Cr as CrPic for 1.5 months decreased total cholesterol and LDL cholesterol by 13 and 11%, respectively (Evans 1989). Supplementation with CrPic improved total cholesterol, HDL cholesterol, and triglycerides significantly in subjects with insulin-treated DM2 (Anderson et al. 1997; Cheng et al. 1999; Kleefstra et al. 2006). Rehabilitating, elderly DM2 patients showed significant improvement in total cholesterol ($P < 0.02$) and a trend toward reduction in triglycerides (Rabinovitz et al. 2004). Lee and Reasner (1994) reported a significant (17.4%) reduction in triglycerides in Hispanic subjects with DM2 after 2 months of CrPic supplementation (200 μg Cr/day). Martin et al. (2006) reported significant reductions in plasma free fatty acids after 6 months for DM2 subjects taking sulphonylurea and 1,000 μg Cr/day as CrPic. Kleefstra et al. (2006) showed improvement in blood lipid profile (LDL, total cholesterol, and total cholesterol to HDL cholesterol ratio) after 6 months of CrPic supplementation.

Improved insulin sensitivity and glucose control also resulted in improved body composition. This was supported in a recent study in which 27 subjects with DM2 on sulfonylurea received CrPic supplementation (1,000 μg Cr/day) or placebo for 6 months. Those on placebo showed a significant increase in body weight, percentage body fat, and total abdominal fat. Subjects randomized to sulfonylurea + CrPic experienced significant improvements in insulin sensitivity, HbA1c, and free fatty acids, which resulted in less body weight gain (2 vs. 5 lbs) and visceral fat accumulation compared to placebo (Martin et al. 2006).

Supplementation with CrPic may also reduce side effects (e.g., weight gain, elevated liver enzymes) associated with high sulphonylurea intake by reducing the requirement for this medication (Ravina et al. 1995, 1999). Shortening of QTc intervals further supports CrPic's efficacy in cardiovascular disease risk reduction.

Altogether, 13 of 15 clinical studies reported significant improvements in at least one outcome of glycemic control. All 15 studies showed significant benefits in at least one parameter of diabetes management, including blood lipid control. Other positive outcomes linked to CrPic therapy included improved EKGs, reduced need for hypoglycemic medications, and no reported adverse effects.

Other reviews (Althuis et al. 2002; Guerrero-Romero and Rodriguez-Moran 2005) were less positive about dietary Cr on glucose and insulin responses. However, they failed to differentiate by Cr form, which is crucial to the analysis. Other Cr complexes do not show the same consistent benefits (Abraham et al. 1992; Sherman et al. 1968; Thomas and Gropper 1996; Trow et al. 2000; Uusitupa et al. 1983). Secondly, subjects with DM2 may require much higher Cr intakes than do normal subjects to demonstrate significant benefits (Anderson et al. 1991, 1997). Thirdly, earlier reviews arbitrarily dismissed a number of CrPic clinical studies, which weakened their analysis. A more representative review of studies involving CrPic intervention was recently published (Broadhurst and Domenico 2006).

Treatment claims for diseases such as diabetes are not permissible with a dietary supplement. Thus, none of the above diabetes clinical trials using CrPic can support a petition for a qualified health claim (QHC). Nevertheless, the FDA issued a QHC specifically for CrPic in diabetes prevention (US Food and Drug Administration 2005). This QHC was the first for insulin resistance. It was based largely on one well-conducted study in nondiabetic, obese subjects that tested 1,000 μg Cr as CrPic for 8 months, using gold standards to measure insulin sensitivity (Cefalu et al. 1999). In their analysis, the FDA also affirmed the safety of CrPic at the doses used. Another recent study, using a reported daily dose of 800 μg Cr as CrPic for 3 months, did not show improved insulin sensitivity (Gunton et al. 2005). It was later determined that the scientists provided only 100 μg Cr per day (Komorowski and Juturu 2005). At best, it supports the need for higher CrPic doses in these trials.

A growing body of evidence suggests that a combination of CrPic and high-dose biotin act synergistically to reduce hyperglycemic insults. Biotin is a water-soluble B complex vitamin necessary in protein, fat, and carbohydrate metabolism. Biotin has been shown to elevate glucokinase levels, which can be depressed in subjects with DM2 (McCarty 1999). Biotin also enhanced insulin secretion in a rat model (Romero-Navarro et al. 1999). Animal studies and a human clinical trial of high-dose biotin supplementation (9 mg/day) showed reductions in FG without significant side effects (Maebashi et al. 1993).

In combination, CrPic and biotin improve both glucose and lipid metabolism by increasing the activation of IRS-1 insulin receptors, internalization of these activated receptors, GLUT-4 translocation, PI3-Kinase activity, liver glucokinase activity, and pancreatic beta-cell function (McCarty 2000). One study reported significant declines in HbA1c, PPG, fructosamine, or triglycerides with a combination of CrPic (600 mcg Cr) and biotin (2 mg) in subjects with poorly controlled DM2

(Geohas et al. 2004). Albaracin et al. (2004) reported significant improvements in FG, HbA1c, and TG/HDL ratios in the treatment group of 369 poorly controlled DM2 subjects receiving this CrPic and biotin combination for 90 days. An open-label extension phase to this study has confirmed the continuing statistically significant and clinically relevant declines in HbA1c in the additional 9 months of intervention (Albarracin et al. 2006). No adverse events related to the study product were reported. This nutrient combination could prove to be a cost-effective complement to existing pharmacological therapies for controlling DM2 (Fuhr et al. 2005).

Toxicity is a concern with any mineral supplement, with some minerals and some forms inherently more toxic than others. Toxicity can also result from contamination in poorly manufactured products. Toxicity increases when high levels of a mineral are ingested for prolonged periods, especially without supportive nutrients. Conversely, homeostatic mechanisms serve to keep mineral levels constant in tissues.

Chromium is considered one of the least toxic minerals in its dietary form (Cr3+). CrPic safety has been reviewed by the Institute of Medicine (IOM) and the Food Standards Agency (FSA) in the United Kingdom. No consistent, frequent side effects were evident from the human data, which spans over a decade of evidence (Institute of Medicine 2004). The recent FDA qualified health claim also confirms its safety (US Food and Drug Administration, 5 A.D. 1220/id).

In vivo and in vitro genotoxicity assays conducted by the National Toxicology Program have shown that CrPic does not produce chromosomal damage in rats (National Toxicology Program 2003). This and additional studies confirm that CrPic is not genotoxic to mammalian species (Berner et al. 2004). This refutes two negative studies that reported CrPic toxicity in cultured cells in the laboratory (Stearns et al. 1995, 2002). Two tests were recently duplicated to clarify these results, and showed CrPic to be safe at every test dose level (Gudi et al. 2005; Slesinski et al. 2005). Unlike the negative studies, these recent studies were conducted in compliance with the federal guidelines and applied Good Laboratory Practice (GLP) standards.

One of the most common misconceptions regarding CrPic is its alleged link to liver or kidney damage. No study has ever found dietary Cr to cause or contribute to liver or kidney damage in humans or animals, even when animals received massive daily doses for years (Rhodes et al. 2005). In fact, research shows that CrPic supplementation can lead to positive effects on renal function (Mita et al. 2005; Mozaffari et al. 2005; Shinde and Goyal 2003).

The safety of Cr supplementation has also been confirmed by numerous human studies that treated thousands of patients. Treatment with CrPic at doses up to 1,000 µg of chromium per day for long periods did not result in any substantiated or consistent adverse effects above control (Broadhurst and Domenico 2006). Moreover, rodents subjected to CrPic for 90 days showed no toxic or unusual effects at over 1,000 times the recommended Cr dose (Rhodes et al. 2005). Hopefully, these recent data have finally put the notion of CrPic toxicity to rest.

Magnesium

Magnesium is an essential mineral found in a variety of foods, including legumes, green leafy vegetables, whole grains, nuts, seeds, and fish (Sarubin 2000). Yet, this important mineral is continually being lost from the food supply, largely due to soil depletion, cooking, and food processing (Saris et al. 2000). Exposure to fluoride also reduces magnesium availability (Machoy-Mokrzynska 1995). Also, a diet high in sugar, fructose, protein intake, saturated fat, tannins, caffeine, alcohol, oxalate, and phytate may increase the need for magnesium (de Valk 1999; Eades and Eades 2001; Lemann et al. 1970; Sarubin 2000).

Though many experts consider the RDA for magnesium (400 mg) inadequate, most people do not achieve these levels in their daily intake (Institute of Medicine 1997). In fact, current magnesium intake in the United States is hundreds of milligrams less than a century ago (Altura 1994). Drugs (e.g., diuretics, insulin, antibiotics, corticosteroids, nicotine) and other stresses can further reduce magnesium status (Dean 2007). Up to 70% of the population does not receive adequate magnesium (Dean 2007).

Magnesium is second only to potassium as the most abundant element in human cells (Aikawa 1981). Magnesium regulates more than 325 enzymes in the body, many of which involve energy utilization and storage. Aspects of metabolism regulated by magnesium include nucleic acid synthesis, cell growth, and cell reproduction. The physiological roles for magnesium include energy production and transportation, protein synthesis, nerve signal transmission, and muscle relaxation. Magnesium deficiency may result in fatigue, muscle weakness, kidney disease, migraines, osteoporosis, anxiety, colitis, depression, diabetes, hypoglycemia, asthma, blood clotting, heart attacks, arrhythmia, and seizures (Aikawa 1981; Dean 2007).

The exact mechanisms by which magnesium affects insulin resistance, hypertension, and cardiovascular disease are unclear. However, with involvement of magnesium in macronutrient metabolism, cellular transport, intracellular signaling, platelet aggregation, vascular tone and contractility, electrolyte homeostasis, and phosphorylation and dephosphorylation reactions, the effects of magnesium on metabolic health are multifactorial.

Magnesium is involved on multiple levels in the secretion, binding, and activity of insulin, and is a critical cofactor of many enzymes in carbohydrate metabolism, especially those involved in phosphorylation reactions such as tyrosine kinase (Paolisso et al. 1990; Tosiello 1996). In animals, low magnesium triggers severe insulin resistance, and is reflected in deficient tyrosine kinase activity in the insulin pathway (Suarez et al. 1995). Magnesium deficiency negatively affects postreceptor insulin signaling (Dzurik et al. 1991; Suarez et al. 1995). Without magnesium, insulin is not properly secreted from the pancreas. In cells, magnesium is required for entrance of blood sugar.

Deficiency in magnesium has been linked repeatedly to DM2. While excessive urination in diabetes can induce mineral deficiencies, magnesium deficiency may also be a risk factor for developing DM2 (Tosiello 1996). People with diabetes, cardiovascular disease, or hypertension tend to have low magnesium levels (Ma et al. 1995; Paolisso et al. 1990), and supplementation with magnesium can rectify this problem (Eibl et al. 1995). Low magnesium levels occur in up to 40% of diabetes patients (de Lordes et al. 1998; Kao et al. 1999; Paolisso and Barbagallo 1997).

In humans, low magnesium diets can induce insulin resistance (Nadler et al. 1993). An inverse relationship between magnesium intake and FI levels, a marker for insulin resistance, has been established. Insulin production in older patients with or without DM2 improves with magnesium supplementation (Paolisso et al. 1989, 1992). One double-blind trial found benefit from 1,000 mg magnesium per day, but not from 500 mg per day, in people with DM2 (de Lordes et al. 1998). Another RCT study with 63 DM2 subjects reported improvements in insulin sensitivity and metabolic control with 2.5 g $MgCl_2$/day for 16 weeks (Rodriguez-Moran and Gurrero-Romero 2003). In contrast, a Dutch trial reported no improvement in FG in insulin-requiring DM2 patients taking magnesium (de Valk et al. 1998).

The metabolic syndrome contributes greatly to the current cardiovascular disease epidemic. Magnesium deficiency is unquestionably a major contributor to its signs and symptoms, including elevated triglycerides, obesity, disturbed insulin function (Ma et al. 1995), and insulin resistance (Alzaid et al. 1995; Barbagallo et al. 2001a; Dominguez et al. 1998; Humphries et al. 1999). Some have argued that a high calcium to magnesium ratio is a characteristic of the metabolic syndrome (Resnick 1992). A longitudinal study in 5,000 young adults showed that the likelihood of developing metabolic syndrome was related to magnesium intake (He et al. 2006). Reduced intake and increased magnesium losses due to elevated insulin contribute to metabolic syndrome in a vicious cycle.

Metabolic studies have shown a beneficial effect of magnesium supplementation on insulin action and blood glucose (de Valk 1999; Paolisso et al. 1992). A recent epidemiologic study of about 128,000 men and women showed that those in the highest quintile of magnesium intake were about 33% less likely to get DM2 than those in the lowest quintile over a 12–18 year period (Lopez-Ridaura et al. 2004). In some communities, magnesium-rich water helps confer protective benefits against diabetes (Yang et al. 1999; Zhao et al. 2001). Serum magnesium levels have been found to correlate inversely with HbA1c (Sjogren et al. 1986) and with FG (Fujii et al. 1994).

Benefits of magnesium supplementation on glucose metabolism or insulin sensitivity have been shown in some clinical studies (Paolisso et al. 1989, 1992, 1994), but not in others (de Lordes et al. 1998; de Valk et al. 1998; Eibl et al. 1995). Differences in magnesium form used, treatment regimens, and methods of evaluation explain some of the inconsistencies. One problem is that insulin resistance can interfere with magnesium uptake by cells (Alzaid et al. 1995). Moreover, copious urination in DM2 produces greater magnesium losses and more symptoms.

Diabetes complications are also linked to magnesium, the lack of which increases the risk of cardiovascular disease, eye symptoms, and nerve damage (de Lordes et al. 1998; Engelen et al. 2000; Shechter et al. 1999). Magnesium is important in the treatment of peripheral vascular disease associated with diabetes (Howard 1990). Diabetic retinopathy is also more prevalent with magnesium deficiency (McNair et al. 1978). All the complications of diabetes—to include neuropathy, impotence, atherosclerosis, blindness, infection, and nephropathy—are related to magnesium deficiency (Resnick et al. 1993).

Supplemental magnesium has also been shown to reduce cholesterol levels (Altura et al. 1990; Singh et al. 1991), including for those with diabetes (Corsica 1994). Indeed, modern diets that promote elevated cholesterol also cause magnesium deficiency. To be sure, heart disease is not associated with high cholesterol (e.g., meat, lard, cream, butter, and eggs) diets, but rather with refined and processed foods lacking in magnesium and other minerals (Gao et al. 1999; Liu et al. 2000).

Magnesium affects a number of enzymes involved in lipid metabolism. Magnesium acts by the same mechanism as statin drugs to lower cholesterol (Rosanoff and Seelig 2004) via the enzymatic reduction of HMG-CoA to mevalonate, which is the rate-limiting step in the cholesterol synthetic pathway. The controlling factor for this reaction is Mg-ATP, part of the enzyme that deactivates the reductase. Thus, magnesium slows down cholesterol production when in sufficient quantities. As with statins, this translates to a reduced incidence of heart attacks, angina, and other nonfatal cardiac events, as well as cardiac, stroke, and total mortality. The difference is that magnesium is the natural and safe intervention. Magnesium is also necessary for lecithin cholesterol acyl transferase (LCAT) activity, which lowers LDL C and triglycerides and raises HDL C levels. Desaturase is another Mg-dependent enzyme involved in lipid metabolism. Desaturase catalyzes the first step in conversion of linoleic acid and linolenic acid into prostaglandins, important in cardiovascular and overall health. Eighteen human studies support the role of magnesium supplementation for improved lipid metabolism (Rosanoff and Seelig 2004). Generally, low magnesium levels are associated with higher levels of "bad" cholesterol, and higher magnesium levels with "good" cholesterol.

Magnesium also plays a vital role in homocysteine control. High blood levels of homocysteine are the result of incomplete protein digestion, and promote cholesterol oxidation and blood vessel damage (McCully 1998). Elevated homocysteine is an even stronger marker than cholesterol for heart disease and blood clotting disorders (Boushey et al. 1995). When certain nutrients are present, protein digestion is more efficient and homocysteine levels are controlled. These nutrients include magnesium, vitamin B6, vitamin B12, and folic acid. Unfortunately, magnesium is commonly left out of this formula. Yet, elevated homocysteine typically means low magnesium. These nutrients are necessary to prevent blood vessel damage induced by homocysteine (Li et al. 1999; Rowley et al. 2001; Tice et al. 2001).

One overlooked major cause of primary hypertension is magnesium. Magnesium is a natural antihypertensive, muscle relaxant, antianxiety remedy, and sleep aid. Deficiency leads to blood vessels that are less relaxed and more susceptible to spasm and tension. It is widely known that diuretics lead to potassium loss, yet major magnesium loss also occurs with these medications (Shechter and Shechter 2005). Indeed, the common side effects of diuretics are weakness, muscle cramps, joint pain, and irregular heartbeat, which are the symptoms of magnesium deficiency (Dean 2007).

Diet should be the first consideration when treating hypertension. Epidemiologic studies show a direct relationship between magnesium intake and high blood pressure (Shechter 2003), and diets deficient in magnesium produce hypertension in animals (Shechter et al. 2003b). Increased dietary magnesium has a suppressive effect on calcium-regulating hormones, which influence blood pressure (Shechter et al. 2003a). Arterial blood pressure rises as magnesium levels drop (Shechter et al. 2000). However, some forms of severe hypertension are hereditary or are due to kidney disease, and require medication. Also, significant vessel damage may not respond to magnesium. The good news is that magnesium is safe, inexpensive, with many potential health benefits, and a little or no tendency toward drug interactions (Shechter et al. 2003a). Thus, it is worth exploring as an adjunct to existing therapy.

Oral supplements are available in numerous forms, but some research suggests that organic forms, such as magnesium citrate, glycinate, aspartate, malate, or taurate are more bioavailable (Sarubin 2000). Supplements up to 350 mg/day are appropriate; intakes >500 mg/day of elemental magnesium may cause diarrhea (Sarubin 2000). Experts have recommended supplementation with magnesium at up to 600 mg/day for DM2 patients with normal renal function. Patients with renal insufficiency must be monitored closely. Effects of supplementation on magnesium status are variable, but the data warrant relatively high doses of magnesium to start, followed by lower daily doses to restore and maintain magnesium in DM2 (Eibl et al. 1995; Lima et al. 1998).

Zinc

Zinc is a versatile mineral. It is a cofactor in hundreds of enzymes, with functions ranging from hormone action to cell growth. Sufficient levels of zinc are necessary for fertility, immunity, growth, and repair. Low zinc levels are associated with fatigue, arthritis, loss of taste, prolonged wound healing, retarded growth, sterility, and preeclampsia—a serious condition of pregnancy associated with elevations in blood pressure, fluid retention, and loss of protein in the urine (Frassinetti et al. 2006).

The best food sources for zinc come from meat, dairy, and seafood. Good vegetarian sources include pumpkin seeds, beans, yeast, nuts, seeds, and whole grain cereals. Zinc is best absorbed when taken with protein, but fiber can interfere with absorption. Organic zinc supplements include zinc acetate, citrate, gluconate, or picolinate. Inorganic zinc sulfate is the most popular, but least absorbed, zinc supplement. General intake is approximately 15 mg daily, but the amount of zinc needed for optimal health will differ among individuals (Food and Nutrition Board 2001)

There are several links between diabetes, insulin, and zinc. Zinc is an essential trace metal that is directly involved in the synthesis, storage, and secretion of insulin, as well as conformational integrity of insulin in the hexameric form (Chausmer 1998). Diabetes affects zinc homeostasis in many ways. Hyperglycemia is responsible for increased excretion and loss of total body zinc. Thus, people with DM2 tend to be low in zinc (Prasad 1988; Sasmita et al. 2004). However, zinc supplementation may not improve blood sugar control in DM2 (Niewoehner et al. 1986).

Several complications of diabetes may be related to increased intracellular oxidants and free radicals associated with decreases in zinc-dependent antioxidant enzymes (Sasmita et al. 2004). For animals fed a high-cholesterol diet, addition of zinc had an overall antiatherogenic effect, in part by reducing oxidative damage (Jenner et al. 2007). Zinc's critical antioxidant action provides significant prevention of oxidative damage to the heart (Song et al. 2005). In combination with other nutrients (i.e., magnesium, vitamin C + E), zinc may help reduce blood pressure in DM2 (Farvid et al. 2004).

Vanadium

Vanadium is a trace mineral that appears to be essential for humans, though it has not yet achieved essential status (Harland and Harden-Williams 1994; Sarubin 2000). It is ubiquitous in the environment but in extremely small quantities, making it difficult to measure accurately or to induce deficiencies. Natural sources include black pepper, spinach, shellfish, mushrooms, dill seed, parsley, soy, corn, olives, olive oil, wine, and gelatin (Harland and Harden-Williams 1994; Sarubin 2000).

There is no established RDA. American intake of vanadium is 10–60 μg/day, and general recommendations are 20–30 μg per day. Vanadium exists in several valence states. Vanadyl (+5) sulfate and sodium metavanadate (+4) are the most common supplemental forms. Absorption of these inorganic vanadium supplements is poor (<5%). Organic forms (e.g., amino acid chelates) are considered better absorbed, but the data are lacking here. Researchers are working to develop forms of vanadium that are better absorbed and have fewer side effects (O'Connell 2001).

Animals, including human, store vanadium primarily in bone. It is transported in the bloodstream by transferrin and cleared primarily through the kidney (Harland and Harden-Williams 1994). Female goats on a long-term vanadium-restricted diet gave birth to kids with serious birth defects. In humans, pharmacologic doses (i.e., 10–100 times normal intake) alter lipid and glucose metabolism by enhancing glucose oxidation, glycogen synthesis, and modulating hepatic glucose output (Harland and Harden-Williams 1994; O'Connell 2001). The evidence for efficacy is suggestive, but RCT trials are recommended (Yeh et al. 2003). High-dose vanadium has been reported to improve thyroid function and negatively affect bone and tooth development in animals (e.g., 15 mg/day).

The insulin-mimetic effect of vanadium is well established, and vanadate has been shown to improve insulin sensitivity in diabetic rats and humans (Mehdi et al. 2006). Improving insulin sensitivity may be mediated, in part, by the ability of vanadium to increase intracellular magnesium levels, which in turn, helps to determine cellular insulin action (Barbagallo et al. 2001b).

Vanadium appears to affect several aspects of the insulin signaling pathway, possibly leading to up regulation of the insulin receptor, insulin receptor autophosphorylation, increased protein tyrosine and serine threonine kinase activity, inhibition of phosphotyrosine phosphatase activity, increased adenylate cyclase activity, altered glucose-6-phosphatase activity, inhibition of hepatic gluconeogenesis, and increased glycogen synthesis (Cam et al. 2000; Poucheret et al. 1998). With a chemical structure similar to phosphorus, vanadium may act as a phosphate analog to alter the rate of activity of a number of adenosine triphosphatases, phosphatases, and phosphotransferases (Poucheret et al. 1998). Research suggests that insulin may be required for its effects (Cam et al. 2000; Poucheret et al. 1998).

Several small trials have evaluated the use of oral vanadium supplements in DM2. Vanadium increased insulin sensitivity in some (Cohen et al. 1995; Goldfine et al. 1995; Halberstam et al. 1996), but not all, trials (Boden et al. 1996). Glucose oxidation and glycogen synthesis increased, and hepatic glucose output decreased in two studies (Cohen et al. 1995; Halberstam et al. 1996). Supplementation decreased FG, HbA1c, and cholesterol levels, and stimulated kinase activity (Boden et al. 1996; Cohen et al. 1995; Goldfine et al. 1995; Halberstam et al. 1996). Benefits of vanadium were more pronounced among DM2 subjects than in obese or nondiabetic subjects (Halberstam et al. 1996). More significant effects were seen in animal models than in humans, and information on the long-term effects is lacking.

Since it is needed in such small quantities (50—500 ppb) and body stores are so low (100 μg), relatively small doses of supplemental vanadium are potentially toxic (Harland and Harden-Williams 1994). Patients using oral vanadium supplements most commonly report nausea, vomiting, cramping, flatulence, and diarrhea. These effects are transient and improve with a decrease in dose. Longer-term use has been associated with anorexia, decreased food and fluid intake, and weight loss. Animal studies indicate that long-term, high-dose supplementation (>10 mg/day of elemental vanadium) can be toxic, with neurological, hematological, nephrotoxic, hepatotoxic, and reproductive and developmental effects (Harland and Harden-Williams 1994; Sarubin 2000). Organic vanadium supplements appear safer than inorganic forms (Srivastava 2000).

Calcium

Evidence from both observational and interventional studies implicates calcium and vitamin D malnutrition as predisposing conditions for a number of chronic diseases, including hypertension and chronic inflammatory conditions (Peterlik and Cross 2005). Low serum vitamin D levels are also associated with impaired glucose tolerance and diabetes (Scragg et al. 2004). Data from the Nurses Health Study indicate that a combined daily intake of more than 1,200 mg of calcium and more than 800 international units (IU) of vitamin D was associated with a 33% lower risk of type 2 diabetes (Pittas et al. 2006). Calcium may play a role in normalizing glucose intolerance, and vitamin D insufficiency has been linked to insulin resistance and reduced function of pancreatic beta-cells (Scragg et al. 1995). Dairy consumption has been shown to be inversely associated with the metabolic syndrome (Liu et al. 2005), and calcium intake is associated with significant decreases in blood pressure (Jorde and Bonaa 2000). Calcium can also boost mitochondrial metabolism, which contributes to sustained second-phase insulin secretion (Wiederkehr and Wollheim 2006). Renal tubular reabsorption of magnesium and calcium is reduced in DM2, resulting in increased urinary losses of the two divalent cations in coordinated fashion (Olukoga et al. 1989).

Excess body fat is the major environmental cause of DM2, and even minor weight loss can prevent its development in high risk subjects. Low-fat dairy products are recommended for maintenance of a healthy body weight in susceptible individuals. Lean dairy products may enhance satiety, and facilitate weight control. It is possible that dairy calcium also promotes weight loss, although the mechanism of action remains unclear (Astrup 2006). Good sources of calcium beyond dairy foods include green leafy vegetables, sardines, nuts, and enriched beverages and foods.

Boron (B)

Boron appears to be an essential nutrient for animals and humans. Dietary boron influences the activity of many metabolic enzymes, steroid hormones, and several micronutrients, including calcium, magnesium, and vitamin D (Bakken and Hunt 2003; Hunt 1994). At least 26 enzymes are influenced by boron, many of which are important in energy substrate metabolism (Hunt 1998). Boron supplementation in animals has been shown to increase bone strength, and improve arthritis, plasma lipid profiles, and brain function (Devirian and Volpe 2003).

The U.S. Department of Agriculture found that within 8 days of boron supplementation at 3 mg/day, postmenopausal subjects showed 30–40% less loss of calcium and magnesium than before supplementation. Boron helps maintain magnesium levels, which decrease as estrogen levels decline (Nielsen 1990). Dietary boron deprivation has been shown to induce hyperinsulinemia in vitamin D deprived rats, and work in various animal models suggests that physiologic amounts of boron may help reduce the amount of insulin required to maintain plasma glucose (Bakken and Hunt 2003). Boron is known to inhibit the activity of two classes of enzymes directly involved in the inflammatory process, and reduced the incidence and severity of inflammation in animal models (Hunt 2003). Despite this evidence, boron has yet to be added to the "Essential Nutrient" list and given an RDA intake level (Hunt 1994).

Good food sources of boron include pears, prunes, pulses, raisins, tomatoes, and apples. Boron is best taken as part of a multivitamin and mineral supplement. Doses above 3 mg are considered unnecessary.

Manganese

Manganese is an essential mineral nutrient with diverse functions. Natural sources of manganese include legumes, cereals, green leafy vegetables, and tea. The RDA has been set at 2 mg/day. Manganese plays an essential role in bone and cartilage formation, as well as amino acid and energy metabolism (Leach and Harris 1997). Manganese is an important cofactor in the key enzymes of glucose metabolism (Nielsen 1999). Several manganese enzymes play important roles in carbohydrate, amino acids, and cholesterol metabolism (Food and Nutrition Board 2001). Pyruvate carboxylase, a manganese-containing enzyme, and phosphoenolpyruvate carboxykinase, a Mn-activated enzyme, play critical roles in gluconeogenesis. Arginase, another Mn-containing enzyme, is required to detoxify ammonia in the urea cycle (Leach and Harris 1997).

Manganese superoxide dismutase (MnSOD) is the principal antioxidant enzyme of mitochondria, which are especially vulnerable to oxidative stress. MnSOD catalyzes the conversion of superoxide radicals to hydrogen peroxide, which is reduced to water by other antioxidant enzymes (Leach and Harris 1997).

Manganese-deficient animals can be characterized by impaired insulin production, altered lipoprotein metabolism, impaired oxidant defense system, and perturbed growth factor metabolism (Keen et al. 1999). Manganese deficiency can result in diabetes in guinea pigs, and offspring who develop pancreatic abnormalities (Saner et al. 1985).

Many people are deficient in this mineral because as much as 75% of all manganese is lost in the refining of wheat to white flour. People with diabetes have been shown to have only one-half the manganese of normal individuals (Nielsen 1999). Lower manganese concentrations were detected in lymphocytes derived from patients with DM2 versus healthy subjects (Ekmekcioglu et al. 2001). Also, diabetics with liver disorders or those who were not treated with insulin excreted significantly more manganese than did their diabetic counterparts (el Yazigi et al. 1991). Young women fed a manganese-poor diet developed mildly abnormal glucose tolerance in response to an intravenous (IV) infusion of glucose (Johnson and Lykken 1991). Although manganese appears to play a role in glucose metabolism, there is little evidence that supplementation improves glucose tolerance in diabetic or nondiabetic individuals.

Summary

Healthcare providers need an unbiased resource on the numerous treatments available for people with DM2. The data suggest that certain minerals are altered in

Table 10.3. Essential minerals for the prevention or treatment of diabetes and its complications.

Mineral (best form)	Effective safe dose	Potential benefit/insulin health
Chromium (Cr picolinate)	200–1,000 μg/day	Insulin sensitivity Blood sugar/lipid control Improves mood, energy levels Reduces carb cravings Preserves lean muscle
Magnesium (Mg citrate, malate, taurate, glycinate)	100–350 mg/day	Multifactorial effects Insulin sensitivity Insulin secretion Blood sugar/lipid control Prevents complications, antihypertension effects
Zinc (Zn gluconate, picolinate)	15–30 mg/day	Multifactorial effects Insulin production, secretion Prevents complications Antiatherogenic Reduces oxidative damage
Vanadium (Vanadium amino acid chelate)	100–200 μg/day	Insulin sensitivity Insulin mimetic Blood sugar/lipid control Inhibits gluconeogenesis Enhances glycogen synthesis
Calcium (Ca citrate, ascorbate)	800 mg/day	Prevents inflammation Supports insulin function Maintains healthy body weight
Boron (B glycinate)	150 μg/day	Important in energy metabolism Improves blood lipid profiles Maintains Ca, Mg, vit D levels
Manganese (Mn ascorbate)	2–3 mg/day	Important in carbohydrate/energy metabolism May prevent oxidative stress

diabetes, and that this alteration affects glycemic control. The evidence is strong for the direct effect of some dietary minerals on preventing and treating diabetes, while with others it is at best supportive. A summary of possible benefits from minerals for DM2 is shown in Table 10.3. Obviously, the dose, duration, chemical form, and time of intervention can make a profound difference in the conferred benefits and safety profile. Generally speaking, at the prescribed dosages, dietary minerals have incredibly clean safety records. Essential minerals are also safer and work better in combination with other nutrients. With heated interest in this area of health, the next few years should be interesting and informative regarding the role of minerals in metabolic health.

References

Abraham AS, Brooks BA, Eylath U. 1992. The effects of chromium supplementation on serum glucose and lipids in patients with and without non-insulin-dependent diabetes. Metabolism 41:768–771.

Adler AI, Stratton IM, Neil HA, Yudkin JS, Matthews DR, Cull CA, Wright AD, Turner RC, Holman RR. 2000. Association of systolic blood pressure with macrovascular and microvascular complications of type 2 diabetes (UKPDS 36): Prospective observational study. BMJ 321(7258):412–419.

Aharoni A, Tesler B, Paltieli Y, Tal J, Dori Z, Sharf M. 1992. Hair chromium content of women with gestational diabetes compared with nondiabetic pregnant women. Am J Clin Nutr 55:104–107.

Aikawa JK. 1981. Magnesium: Its Biological Significance. Boca Raton, FL: CRC Press. pp 21–38.

Albaracin C, Fuqua B, Finch M, Juturu V, Komorowski J. 2004. Glycemic control is improved by the combination of chromium picolinate and biotin in type 2 diatetes mellitus [abstract 2483]. Diabetes 52(Suppl 2)A587.

Albarracin C, Fuqua B, Geohas J, Finch MR, Komorowski J. 2006. Effect of chromium picolinate and biotin combination on glycosylated hemoglobin and plasma glucose in subjects with type 2 diabetes mellitus with baseline HbA1c equal to or >10%. ENDO [abstract 1649]. Diabetes 55 (Suppl 1)A381.

Althuis MD, Jordan NE, Ludington EA, Wittes JT. 2002. Glucose and insulin responses to dietary chromium supplements: A meta- analysis. Am J Clin Nutr 76:148–155.

Altura BM. 1994. Introduction: Importance of Mg in physiology and medicine and the need for ion selective electrodes. Scand J Clin Lab Invest Suppl 217:5–9.

Altura BT, Brust M, Bloom S, Barbour RL, Stempak JG, Altura BM. 1990. Magnesium dietary intake modulates blood lipid levels and atherogenesis. Proc Natl Acad Sci USA 87:1840–1844.

Alzaid AA, Dinneen SF, Moyer TP, Rizza RA. 1995. Effects of insulin on plasma magnesium in noninsulin-dependent diabetes mellitus: Evidence for insulin resistance. J Clin Endocrinol Metab 80:1376–1381.

Anderson RA. 1986. Chromium metabolism and its role in disease processes in man. Clin Physiol Biochem 4:31–41.

Anderson RA. 1995. Chromium, glucose tolerance, diabetes and lipid metabolism. J Adv Med 8:37–50.

Anderson RA. 1998. Chromium, glucose intolerance and diabetes. J Am Coll Nutr 17:548–555.

Anderson RA. 2000. Chromium in the prevention and control of diabetes. Diabetes Metab 26:22–27.

Anderson RA, Bryden NA, Patterson KY, Veillon C, Andon MB, Moser-Veillon PB. 1993. Breast milk chromium and its association with chromium intake, chromium excretion, and serum chromium. Am J Clin Nutr 57:519–523.

Anderson RA, Bryden NA, Polansky MM. 1992. Dietary chromium intake. Freely chosen diets, institutional diet, and individual foods. Biol Trace Elem Res 32:117–121.

Anderson RA, Bryden NA, Polansky MM. 2003. Stability and absorption of chromium and absorption of chromium histidine by humans. J Trace Elem Med Biol 16:110.

Anderson RA, Bryden NA, Polansky MM, Gautschi K. 1996. Dietary chromium effects on tissue chromium concentrations and chromium absorption in rats. J Trace Elem Exp Med 9:11–25.

Anderson RA, Cheng N, Bryden NA, Polansky MM, Cheng N, Chi J, Feng J. 1997. Elevated intakes of supplemental chromium improve glucose and insulin variables in individuals with type 2 diabetes. Diabetes 46:1786–1791.

Anderson RA, Kozlovsky AS. 1985. Chromium intake, absorption and excretion of subjects consuming self- selected diets. Am J Clin Nutr 41(6):1177–1183.

Anderson RA, Polansky MM, Bryden NA. 2004. Stability and absorption of chromium and absorption of chromium histidinate complexes by humans. Biol Trace Elem Res 101:211–218.

Anderson RA, Polansky MM, Bryden NA, Canary JJ. 1991. Supplemental-chromium effects on glucose, insulin, glucagon, and urinary chromium losses in subjects consuming controlled low-chromium diets. Am J Clin Nutr 54:909–916.

Anderson RA, Polansky MM, Bryden NA, Patterson KY, Veillon C, Glinsmann WH. 1983. Effects of chromium supplementation on urinary Cr excretion of human subjects and correlation of Cr excretion with selected clinical parameters. J Nutr 113:276–281.

Anderson RA, Sandre C, Bryden NA, Agay D, Chancerelle Y, Polansky MM, Roussel AM. 2005. Burn-induced alterations of chromium and the glucose/insulin system in rats. Burns 32:46–51.

Astrup A. 2006. How to maintain a healthy body weight. Int J Vitam Nutr Res 76:208–215.

Bahadori B, Habersack S, Schneider H, Wascher TC, Toplak H. 1995. Treatment with chromium picolinate improves lean body mass in patients following weight reduction. Int J Obesity 19:38.

Bahadori B, Wallner S, Hacker C, Boes U, Komorowski JR, Wascher TC. 1999. Effects of chromium picolinate on insulin levels and glucose control in obese patients with Type-II diabetes mellitus. Diabetes 48(Suppl 1):A349.

Bahijiri SM, Mira SA, Mufti AM, Ajabnoor MA. 2000. The effects of inorganic chromium and brewer's yeast supplementation on glucose tolerance, serum lipids and drug dosage in individuals with type 2 diabetes. Saudi Med J 21:831–837.

Bakken NA, Hunt CD. 2003. Dietary boron decreases peak pancreatic in situ insulin release in chicks and plasma insulin concentrations in rats regardless of vitamin D or magnesium status. J Nutr 133:3577–3583.

Barbagallo M, Dominguez LJ, Bardicef O, Resnick LM. 2001a. Altered cellular magnesium responsiveness to hyperglycemia in hypertensive subjects. Hypertension 38:612–615.

Barbagallo M, Dominguez LJ, Resnick LM. 2001b. Insulin-mimetic action of vanadate: Role of intracellular magnesium. Hypertension 38:701–704.

Berner TO, Murphy MM, Slesinski R. 2004. Determining the safety of chromium tripicolinate for addition to foods as a nutrient supplement. Food Chem Toxicol 42:1029–1042.

Boden G, Chen X, Ruiz J, van Rossum GD, Turco S. 1996. Effects of vanadyl sulfate on carbohydrate and lipid metabolism in patients with non-insulin-dependent diabetes mellitus. Metabolism 45:1130–1135.

Boushey CJ, Beresford SA, Omenn GS, Motulsky AG. 1995. A quantitative assessment of plasma homocysteine as a risk factor for vascular disease. Probable benefits of increasing folic acid intakes. JAMA 274:1049–1057.

Broadhurst CL, Domenico P. 2006. Clinical studies on chromium picolinate supplementation in diabetes mellitus—a review. Diabetes Technol Ther 8:677–687.

Cam MC, Brownsey RW, McNeill JH. 2000. Mechanisms of vanadium action: insulin-mimetic or insulin-enhancing agent? Can J Physiol Pharmacol 78:829–847.

Campbell JD. 2001. Lifestyle, minerals and health. Med Hypotheses 57:521–531.

Campbell WW, Joseph LJ, Davey SL, Cyr-Campbell D, Anderson RA, Evans WJ. 1999. Effects of resistance training and chromium picolinate on body composition and skeletal muscle in older men. J Appl Physiol 86:29–39.

Cefalu WT, Bell-Farrow AD, Stegner J, Wand ZQ, King T, Morgan T, Terry JG. 1999. Effect of chromium picolinate on insulin sensitivity in vivo. J Trace Elem Exp Med 12:71–83.

Cefalu WT, Hu FB. 2004. Role of chromium in human health and in diabetes. Diabetes Care 27:2741–2751.

Cefalu WT, Wang ZQ, Zhang XH, Baldor LC, Russell JC. 2002. Oral chromium picolinate improves carbohydrate and lipid metabolism and enhances skeletal muscle Glut-4 translocation in obese, hyperinsulinemic (JCR-LA corpulent) rats. J Nutr 132:1107–1114.

Centers for Disease Control and Prevention. 2005. The 2005 National Diabetes Fact Sheet. www.cdc.gov/diabetes.

Chausmer AB. 1998. Zinc, insulin and diabetes. J Am Coll Nutr 17:109–115.

Cheng N, Zhu X, Hongli S, Wo W, Chi J, Cheng J, Anderson R. 1999. Follow-up survey of people in China with type 2 diabetes mellitus consuming supplemental chromium. J Trace Elem Med Biol 12:55–60.

Cohen N, Halberstam M, Shlimovich P, Chang CJ, Shamoon H, Rossetti L. 1995. Oral vanadyl sulfate improves hepatic and peripheral insulin sensitivity in patients with non-insulin-dependent diabetes mellitus. J Clin Invest 95:2501–2509.

Corsica F. 1994. Effects of oral magnesium supplementation on plasma lipid concentrations in patients with non-insulin-dependent diabetes mellitus. Magnes Res 7:43–46.

Davies S, Howard JM, Hunnisett A, Howard M. 1997. Age-related decreases in chromium levels in 51,665 hair, sweat, and serum samples from 40,872 patients—implications for the prevention of cardiovascular disease and type II diabetes mellitus. Metabolism 46:469–473.

Davis CM, Sumrall KH, Vincent JB. 1996. A biologically active form of chromium may activate a membrane phosphotyrosine phosphatase (PTP). Biochemistry 35:12963–12969.

de Lordes LM, Cruz T, Pousada JC, Rodrigues LE, Barbosa K, Cangucu V. 1998. The effect of magnesium supplementation in increasing doses on the control of type 2 diabetes. Diabetes Care 21:682–686.

de Valk HW. 1999. Magnesium in diabetes mellitus. Neth J Med 54:139–146.

de Valk HW, Verkaaik R, van Rijn HJ, Geerdink RA, Struyvenberg A. 1998. Oral magnesium supplementation in insulin-requiring Type 2 diabetic patients. Diabet Med 15:503–507.

Dean C. 2007. The Magnesium Miracle. New York: Ballantine Books. p. 309.

Devirian TA, Volpe SL. 2003. The physiological effects of dietary boron. Crit Rev Food Sci Nutr 43:219–231.

DiSilvestro RA, Dy E. 2005. Comparison of acute absorption of various types of chromium supplement complexes. FASEB 19:A92–A93.

Docherty J, Sack D, Roffman M, Finch M, Komorowski J. 2004. Chromium Picolinate: Atypical Depression, Carbohydrate Craving, and Clinical Response. NCDEU Poster 167:221.

Dominguez LJ, Barbagallo M, Sowers JR, Resnick LM. 1998. Magnesium responsiveness to insulin and insulin-like growth factor I in erythrocytes from normotensive and hypertensive subjects. J Clin Endocrinol Metab 83:4402–4407.

Dzurik R, Stefikova K, Spustova V, Fetkovska N. 1991. The role of magnesium deficiency in insulin resistance: an in vitro study. J Hypertens Suppl 9:S312–S313.

Eades MR, Eades MD. 2001. The Protein Power Lifeplan. New York: Grand Central Publishing. p. 429.

Eibl NL, Kopp HP, Nowak HR, Schnack CJ, Hopmeier PG, Schernthaner G. 1995. Hypomagnesemia in type II diabetes: Effect of a 3-month replacement therapy. Diabetes Care 18:188–192.

Ekmekcioglu C, Prohaska C, Pomazal K, Steffan I, Schernthaner G, Marktl W. 2001. Concentrations of seven trace elements in different hematological matrices in patients with type 2 diabetes as compared to healthy controls. Biol Trace Elem Res 79:205–219.

el Yazigi A, Hannan N, Raines DA. 1991. Urinary excretion of chromium, copper, and manganese in diabetes mellitus and associated disorders. Diabetes Res 18:129–134.

Elias AN, Grossman MK, Valenta LJ. 1984. Use of the artificial beta cell (ABC) in the assessment of peripheral insulin sensitivity: Effect of chromium supplementation in diabetic patients. Gen Pharmacol 15:535–539.

Engelen W, Bouten A, De LI, De Block C. 2000. Are low magnesium levels in type 1 diabetes associated with electro-myographical signs of polyneuropathy? Magnes Res 13:197–203.

Evans GW. 1989. The effect of chromium picolinate on insulin controlled parameters in humans. Intl J Biosoc Med Res 11:163–180.

Farvid MS, Jaali M, Siassi F, Saadat N, Hosseini M. 2004. The impact of vitamins and/or mineral supplementation on blood pressure in type 2 diabetes. J Am Coll Nutr 23:272–279.

Feng J, Lin D, Zheng A, Cheng N. 2002. Chromium picolinate reduces insulin requirement in people with type 2 diabetes mellitus. Diabetes 51(Suppl 2):A469.

Food and Nutrition Board, Institute of Medicine. 2001. Dietary Reference Intakes for Vitamin A, Vitamin K, Boron, Chromium, Copper, Iodine, Iron, Manganese, Molybdenum, Nickel, Silicon, Vanadium, and Zinc. Washington, DC: National Academy Press.

Frassinetti S, Bronzetti G, Caltavuturo L, Cini M, Della Croce C. 2006. The role of zinc in life: A review. J Environ Pathol Toxicol Oncol 25:597–610.

Fuhr JP, Jr, He H, Goldfarb N, Nash DB. 2005. Use of chromium picolinate and biotin in the management of type 2 diabetes: an economic analysis. Dis Manag 8:265–275.

Fujii S, Takemura T, Wada M, Akai T, Okuda K. 1994. Magnesium levels in plasma, erythrocyte and urine in patients with diabetes mellitus. J Assoc Phys India 42:720–721.

Gao M, Ikeda K, Hattori H, Miura A, Nara Y, Yamori Y. 1999. Cardiovascular risk factors emerging in Chinese populations undergoing urbanization. Hypertens Res 22:209–215.

Geohas J, Finch M, Juturu V, Greenberg D, Komorowski J. 2004. Improvement in fasting blood glucose with the combination of chromium picolinate and boitin in type 2 diatetes mellitus [abstract 191]. Diabetes 52 (Suppl)A45.

Ghosh D, Bhattacharya B, Mukherjee B, Manna B. 2002. Role of chromium supplementation in Indians with type 2 diabetes mellitus. J Nutr Biochem 13:690–697.

Glinsmann WH, Mertz W. 1966. Effect of trivalent chromium on glucose tolerance. Metabolism 15:510–520.

Goldfine AB, Simonson DC, Folli F, Patti ME, Kahn CR. 1995. Metabolic effects of sodium metavanadate in humans with insulin-dependent and noninsulin-dependent diabetes mellitus in vivo and in vitro studies. J Clin Endocrinol Metab 80:3311–3320.

Gross LS, Li L, Ford ES, Liu S. 2004. Increased consumption of refined carbohydrates and the epidemic of type 2 diabetes in the United States: An ecologic assessment. Am J Clin Nutr 79:774–779.

Guallar E, Jimenez FJ, van't Veer P, Bode P, Riemersma RA, Gomez-Aracena J, Kark JD, Arab L, Kok FJ, Martin-Moreno JM. 2005. Low toenail chromium concentration and increased risk of nonfatal myocardial infarction. Am J Epidemiol 162:157–164.

Gudi R, Slesinski RS, Clarke JJ, San RH. 2005. Chromium picolinate does not produce chromosome damage in CHO cells. Mutat Res 587:140–146.

Guerrero-Romero F, Rodriguez-Moran M. 2005. Complementary therapies for diabetes: The case for chromium, magnesium, and antioxidants. Arch Med Res 36:250–257.

Gunton JE, Cheung NW, Hitchman R, Hams G, O'Sullivan C, Foster-Powell K, McElduff A. 2005. Chromium supplementation does not improve glucose tolerance, insulin sensitivity, or lipid profile: A randomized, placebo-controlled, double-blind trial of supplementation in subjects with impaired glucose tolerance. Diabetes Care 28:712–713.

Halberstam M, Cohen N, Shlimovich P, Rossetti L, Shamoon H. 1996. Oral vanadyl sulfate improves insulin sensitivity in NIDDM but not in obese nondiabetic subjects. Diabetes 45:659–666.

Harland BF, Harden-Williams BA. 1994. Is vanadium of human nutritional importance yet? J Am Diet Assoc 94:891–894.

He K, Liu K, Daviglus ML, Morris SJ, Loria CM, Van Horn L, Jacobs DR, Savage PJ. 2006. Magnesium intake and incidence of metabolic syndrome among young adults. Circulation 113:1675–1682.

Heimbach JT, Anderson RA. 2005. Chromium: Recent studies regarding nutritional roles and safety. Nutr Today 40:2–8.

Howard JMH. 1990. Magnesium deficiency in peripheral vascular disease. J Nutr Med 1:39.

Humphries S, Kushner H, Falkner B. 1999. Low dietary magnesium is associated with insulin resistance in a sample of young, nondiabetic Black Americans. Am J Hypertens 12:747–756.

Hunt CD. 1994. The biochemical effects of physiologic amounts of dietary boron in animal nutrition models. Environ Health Perspect 102(Suppl 7):35–43.

Hunt CD. 1998. Regulation of enzymatic activity: One possible role of dietary boron in higher animals and humans. Biol Trace Elem Res 66:205–225.

Hunt CD. 2003. Dietary boron: An overview of the evidence for its role in immune function. J Trace Elem Exp Med 16:291–306.

Institute of Medicine, Food and Nutrition Board. 1997. Dietary Reference Intakes for Calcium, Phosphorus, Magnesium, Vitamin D, and Fluoride. Washington, DC: National Academy Press.

Institute of Medicine, Food and Nutrition Board. 2004. Chromium Picolinate—Prototype Monograph Summary. In: Dietary Supplements—A Framework for Evaluating Safety. Washington, DC: National Academy Press.

Jenner A, Ren M, Rajendran R, Ning P, Huat BT, Watt F, Halliwell B. 2007. Zinc supplementation inhibits lipid peroxidation and the development of atherosclerosis in rabbits fed a high cholesterol diet. Free Radic Biol Med 42:559–566.

Johnson PE, Lykken GI. 1991. Manganese and calcium absorption and balance in young women fed diets with varying amounts of manganese and calcium. J Trace Elem Exp Med 4:19–35.

Jorde R, Bonaa KH. 2000. Calcium from dairy products, vitamin D intake, and blood pressure: The Tromso Study. Am J Clin Nutr 71:1530–1535.

Jovanovic-Peterson L, Gutierrez M, Peterson CM. 1999. Chromium supplementation for women with gestational diabetes mellitus. J Trace Elem Med Biol 12:91–97.

Juang J-H, Lu W-T, Wu W-P. 2000. Effects of chromium on patients with impaired glucose tolerance. Diabetes 49(S1):A112–A113.

Juturu V, Daly A, Geohas J, Finch M, Komorowski J. 2006. Diabetes risk factors and chromium intake in moderately obese subjects with type 2 diabetes mellitus. Nutr Food Sci 36:390–398.

Juturu V, Komorowski JR. 2003. Consumption of selected food sources of chromium in the diets of American Adults. FASEB J 17:A1129.

Kao WH, Folsom AR, Nieto FJ, Mo JP, Watson RL, Brancati FL. 1999. Serum and dietary magnesium and the risk for type 2 diabetes mellitus: The Atherosclerosis Risk in Communities Study. Arch Intern Med 159:2151–2159.

Keen CL, Ensunsa JL, Watson MH, Baly DL, Donovan SM, Monaco MH, Clegg MS. 1999. Nutritional aspects of manganese from experimental studies. Neurotoxicology 20:213–223.

Kleefstra N, Houweling ST, Jansman FG, Groenier KH, Gans RO, Meyboom-de Jong B, Bakker SJ, Bilo HJ. 2006. Chromium treatment has no effect in patients with poorly controlled, insulin-treated type 2 diabetes in an obese Western population. Diabetes Care 29:521–525.

Knowler SE, Hamman RF, Lachin JM, Walker EA, Nathan DM; Diabetes Prevention Program Research Group. 2002. Reduction in the incidence of type 2 diabetes with lifestyle intervention or metformin. N Engl J Med 346:393–403.

Komorowski J, Juturu V. 2005. Chromium supplementation does not improve glucose tolerance, insulin sensitivity, or lipid profile: A randomized, placebo-controlled, double-blind trial of supplementation in subjects with impaired glucose tolerance: Response to Gunton et al. Diabetes Care 28:1841–1842.

Kozlovsky AS, Moser PB, Reiser S, Anderson RA. 1986. Effects of diets high in simple sugars on urinary chromium losses. Metabolism 35:515–518.

Kumpulainen J, Vuori E, Makinen S, Kara R. 1980. Dietary chromium intake of lactating Finnish mothers: Effect on the Cr content of their breast milk. Br J Nutr 44:257–263.

Leach RM, Harris ED. 1997. Manganese. In Handbook of Nutritionally Essential Mineral Elements, edited by O'Dell BL, Sunde RA, pp. 335–356. New York: Marcel Dekker.

Lee NA, Reasner CA. 1994. Beneficial effect of chromium supplementation on serum triglyceride levels in NIDDM. Diabetes Care 17:1449–1452.

Lemann J, Jr, Lennon EJ, Piering WR, Prien EL, Jr, Ricanati ES. 1970. Evidence that glucose ingestion inhibits net renal tubular reabsorption of calcium and magnesium in man. J Lab Clin Med 75:578–585.

Levine RA. 1968. Chromium and glucose tolerance. Nutr Rev 26:281–282.

Li W, Zheng T, Wang J, Altura BT, Altura BM. 1999. Extracellular magnesium regulates effects of vitamin B6, B12 and folate on homocysteinemia-induced depletion of intracellular free magnesium ions in canine cerebral vascular smooth muscle cells: possible relationship to [Ca2+], atherogenesis and stroke. Neurosci Lett 274:83–86.

Lima DLM, Cruz T, Pousada JC, Rodrigues LE, Barbarosa K, Cangucu V. 1998. The effect of magnesium supplementation in increasing doses on the control of type 2 diabetes. Diabetes Care 21:682–686.

Liu L, Mizushima S, Ikeda K, Hattori H, Miura A, Gao M, Nara Y, Yamori Y. 2000. Comparative studies of diet-related factors and blood pressure among Chinese and Japanese: Results from the China-Japan Cooperative Research of the WHO-CARDIAC Study. Cardiovascular Disease and Alimentary Comparison. Hypertens Res 23:413–420.

Liu S, Song Y, Ford ES, Manson JE, Buring JE, Ridker PM. 2005. Dietary calcium, vitamin D, and the prevalence of metabolic syndrome in middle-aged and older U.S. women. Diabetes Care 28:2926–2932.

Lopez-Ridaura R, Willett WC, Rimm EB, Liu S, Stampfer MJ, Manson JE, Hu FB. 2004. Magnesium intake and risk of type 2 diabetes in men and women. Diabetes Care 27:134–140.

Lucidi RS, Thyer AC, Easton CA, Holden AE, Schenken RS, Brzyski RG. 2005. Effect of chromium supplementation on insulin resistance and ovarian and menstrual cyclicity in women with polycystic ovary syndrome. Fertil Steril 84:1755–1757.

Lukaski HC. 1999. Chromium as a supplement. Annu Rev Nutr 19:279–302.

Lydic ML, McNurlan M, Bembo S, Mitchell L, Komaroff E, Gelato M. 2006. Chromium picolinate improves insulin sensitivity in obese subjects with polycystic ovary syndrome. Fertil Steril 86:243–246.

Ma J, Folsom AR, Melnick SL, Eckfeldt JH, Sharrett AR, Nabulsi AA, Hutchinson RG, Metcalf PA. 1995. Associations of serum and dietary magnesium with cardiovascular disease, hypertension, diabetes, insulin, and carotid arterial wall thickness: Atherosclerosis Risk in Communities Study. J Clin Epidemiol 48:927–940.

Machoy-Mokrzynska A. 1995. Fluoride magnesium interaction. Fluoride 28:175–177.

Maebashi M, Makino Y, Furukawa Y, Ohinata K, Kimura S, Sato T. 1993. Theraputic evaluation of the effect of biotin on typerglycemia in patients with non insulin dependent diabtes mellitus. J Clin Biochem Nutr 14:211–218.

Martin J, Matthews DE, Wang Z, Zhang X, Volaufova J, Cefalu WT. 2005. Effect of chromium picolinate on body composition, insulin sensitivity, and glycemic control in subjects with type 2 diabetes. Diabetes 54:A427.

Martin J, Wang ZQ, Zhang XH, Wachtel D, Volaufova J, Matthews DE, Cefalu WT. 2006. Chromium picolinate supplementation attenuates body weight gain and increases insulin sensitivity in subjects with type 2 diabetes. Diabetes Care 29:1826–1832.

Martinez OB, MacDonald AC, Gibson RS. 1985. Dietary chromium and effect of chromium supplementation on glucose tolerance of elderly canadian women. Nutr Res 5:609–620.

McCarty MF. 1999. High-dose biotin, an inducer of glucokinase expression, may synergize with chromium picolinate to enable a definitive nutritional therapy for type II diabetes. Med Hypotheses 52:401–406.

McCarty MF. 2000. Toward practical prevention of type 2 diabetes. Med Hypotheses 54:786–793.

McCully KS. 1998. Homocysteine, folate, vitamin B6, and cardiovascular disease. JAMA 279:392–393.

McNair P, Christiansen C, Madsbad S, Lauritzen E, Faber O, Binder C, Transbol I. 1978. Hypomagnesemia, a risk factor in diabetic retinopathy. Diabetes 27:1075–1077.

Mehdi MZ, Pandey SK, Theberge JF, Srivastava AK. 2006. Insulin signal mimicry as a mechanism for the insulin-like effects of vanadium. Cell Biochem Biophys 44:73–81.

Mertz W. 1990. The role of trace elements in the aging process. Prog Clin Biol Res 326:229–240.

Mita Y, Ishihara K, Fukuchi Y, Fukuya Y, Yasumoto K. 2005. Supplementation with chromium picolinate recovers renal Cr concentration and improves carbohydrate metabolism and renal function in type 2 diabetic mice. Biol Trace Elem Res 105:229–248.

Morris BW, Kouta S, Robinson R, MacNeil S, Heller S. 2000. Letters: Chromium supplementation improves insulin resistance in patients with Type 2 diabetes mellitus. Diabet Med 17:684–685.

Morris BW, MacNeil S, Hardisty CA, Heller S, Burgin C, Gray TA. 1999. Chromium homeostasis in patients with type II (NIDDM) diabetes. J Trace Elem Med Biol 13:57–61.

Mossop RT. 1983. Effects of chromium III on fasting blood glucose, cholesterol and cholesterol HDL levels in diabetics. Cent Afr J Med 29:80–82.

Mozaffari MS, Patel C, Ballas C, Schaffer SW. 2005. Effects of chronic chromium picolinate treatment in uninephrectomized rat. Metabolism 54:1243–1249.

Nadler JL, Buchanan T, Natarajan R, Antonipillai I, Bergman R, Rude R. 1993. Magnesium deficiency produces insulin resistance and increased thromboxane synthesis. Hypertension 21:1024–1029.

National Toxicology Program. 2003. Chromium picolinate. http://ntp-server.niehs.nih.gov.

Newman HA, Leighton RF, Lanese RR, Freedland NA. 1978. Serum chromium and angiographically determined coronary artery disease. Clin Chem 24:541–544.

Nielsen FH. 1990. Studies on the relationship between boron and magnesium which possibly affects the formation and maintenance of bones. Magnes Trace Elem 9:61–69.

Nielsen FH. 1999. Ultratrace minerals. In: Nutrition in Health and Disease, 9th ed., edited by Shils M, Olson JA, Shike M, Ross AC, pp. 283–303. Baltimore: Williams & Wilkins.

Niewoehner CB, Allen JI, Boosalis M, Levine AS, Morley JE. 1986. Role of zinc supplementation in type II diabetes mellitus. Am J Med 81:63–68.

O'Connell BS. 2001. Select vitamins and minerals in the management of diabetes. Diabetes Spectr 14:133–148.

Offenbacher EG, Pi-Sunyer FX. 1980. Beneficial effect of chromium-rich yeast on glucose tolerance and blood lipids in elderly subjects. Diabetes 29:919–925.

Offenbacher E, Pi-Sunyer FX. 1997. Chromium. In Handbook of Nutritionally Essential Mineral Elements, edited by O'Dell B, Sunde R, pp. 389–411. New York: Marcel Dekker.

Olin K, Stearn DM, Armstrong WH, Keen CL. 1994. Comparative retention/absorption of 51chromium (51Cr) from 51Cr chloride, 51Cr nicotinate and 51Cr picolinate in a rat model. Trace Elem Electrolytes 11:182–186.

Olukoga AO, Adewoye HO, Erasumus RT. 1989. Renal excretion of magnesium and calcium in diabetes mellitus. Cent Afr J Med 35:378–383.

Paolisso G, Barbagallo M. 1997. Hypertension, diabetes mellitus, and insulin resistance: The role of intracellular magnesium. Am J Hypertens 10:346–355.

Paolisso G, Scheen A, Cozzolino D, Di Maro G, Varricchio M, D'Onofrio F, Lefebvre PJ. 1994. Changes in glucose turnover parameters and improvement of glucose oxidation after 4-week magnesium administration in elderly noninsulin-dependent (type II) diabetic patients. J Clin Endocrinol Metab 78:1510–1514.

Paolisso G, Scheen A, D'Onofrio F, Lefebvre P. 1990. Magnesium and glucose homeostasis. Diabetologia 33:511–514.

Paolisso G, Sgambato S, Gambardella A, Pizza G, Tesauro P, Varricchio M, D'Onofrio F. 1992. Daily magnesium supplements improve glucose handling in elderly subjects. Am J Clin Nutr 55:1161–1167.

Paolisso G, Sgambato S, Pizza G, Passariello N, Varricchio M, D'Onofrio F. 1989. Improved insulin response and action by chronic magnesium administration in aged NIDDM subjects. Diabetes Care 12:265–269.

Peterlik M, Cross HS. 2005. Vitamin D and calcium deficits predispose for multiple chronic diseases. Eur J Clin Invest 35:290–304.

Pittas AG, Dawson-Hughes B, Li T, Van Dam RM, Willett WC, Manson JE, Hu FB. 2006. Vitamin D and calcium intake in relation to type 2 diabetes in women. Diabetes Care 29:650–656.

Popkin BM. 2001. The nutrition transition and obesity in the developing world. J Nutr 131:S871–S873.

Potter JF, Levin P, Anderson RA, Freiberg JM, Andres R, Elahi D. 1985. Glucose metabolism in glucose-intolerant older people during chromium supplementation. Metabolism 34:199–204.

Poucheret P, Verma S, Grynpas MD, McNeill JH. 1998. Vanadium and diabetes. Mol Cell Biochem 188:73–80.

Prasad AS. 1988. Zinc in growth and development and spectrum of human zinc deficiency. J Am Coll Nutr 7:377–384.

Rabinovitz H, Friedensohn A, Leibovitz A, Gabay G, Rocas C, Habot B. 2004. Effect of chromium supplementation on blood glucose and lipid levels in type 2 diabetes mellitus elderly patients. Int J Vitam Nutr Res 74:178–182.

Rabinowitz MB, Gonick HC, Levin SR, Davidson MB. 1983. Effects of chromium and yeast supplements on carbohydrate and lipid metabolism in diabetic men. Diabetes Care 6:319–327.

Racek J, Trefil L, Rajdl D, Mudrova V, Hunter D, Senft V. 2006. Influence of chromium-enriched yeast on blood glucose and insulin variables, blood lipids, and markers of oxidative stress in subjects with type 2 diabetes mellitus. Biol Trace Elem Res 109:215–230.

Rajpathak S, Rimm EB, Li T, Morris JS, Stampfer MJ, Willet WC, Hu F. 2004. Lower toenail chromium in men with diabetes and cardiovascular disease compared with healthy men. Diabetes Care 27:2211–2216.

Ravina A, Slezak L, Mirsky N, Bryden NA, Anderson RA. 1999. Reversal of corticosteroid-induced diabetes mellitus with supplemental chromium. Diabet Med 16:164–167.

Ravina A, Slezak L, Rubal A, Mirsky N. 1995. Clinical use of the trace element chromium (III) in the treatment of diabetes mellitus. J Trace Elem Med Biol 8:183–190.

Resnick LM. 1992. Cellular ions in hypertension, insulin resistance, obesity, and diabetes: A unifying theme. J Am Soc Nephrol 3:S78–S85.

Resnick LM, Altura BT, Gupta RK, Laragh JH, Alderman MH, Altura BM. 1993. Intracellular and extracellular magnesium depletion in type 2 (non-insulin-dependent) diabetes mellitus. Diabetologia 36:767–770.

Rhodes MC, Hebert CD, Herbert RA, Morinello EJ, Roycroft JH, Travlos GS, Abdo KM. 2005. Absence of toxic effects in F344/N rats and B6C3F1 mice following subchronic administration of chromium picolinate monohydrate. Food Chem Toxicol 43:21–29.

Rodriguez-Moran M, Gurrero-Romero F. 2003. Oral magnesium supplementation improves insulin sensitivity and metabolic control in type 2 diabetic subjects. Diabetes Care 26:1147–1152.

Romero-Navarro G, Cabrera-Valladares G, German MS, Matschinsky FM, Velazquez A, Wang J, Fernandez-Mejia C. 1999. Biotin regulation of pancreatic glucokinase and insulin in primary cultured rat islets and in biotin-deficient rats. Endocrinology 140:4595–4600.

Rosanoff A, Seelig MS. 2004. Comparison of mechanism and functional effects of magnesium and statin pharmaceuticals. J Am Coll Nutr 23:S501–S505.

Rowley KG, Su Q, Cincotta M, Skinner M, Skinner K, Pindan B, White GA, O'Dea K. 2001. Improvements in circulating cholesterol, antioxidants, and homocysteine after dietary intervention in an Australian Aboriginal community. Am J Clin Nutr 74:442–448.

Rubin MA, Miller JP, Ryan AS, Treuth MS, Patterson KY, Pratley RE, Hurley BF, Veillon C, Moser-Veillon PB, Anderson RA. 1998. Acute and chronic resistive exercise increase urinary chromium excretion in men as measured with an enriched chromium stable isotope. J Nutr 128:73–78.

Saner G, Dagoglu T, Ozden T. 1985. Hair manganese concentrations in newborns and their mothers. Am J Clin Nutr 41:1042–1044.

Saris NE, Mervaala E, Karppanen H, Khawaja JA, Lewenstam A. 2000. Magnesium. An update on physiological, clinical and analytical aspects. Clin Chim Acta 294:1–26.

Sarubin A. 2000. The Health Professional's Guide to Popular Dietary Supplements.Chicago:American Dietetic Association. p. 552.

Sasmita T, Sumathi S, Raj GB. 2004. Minerals nutritional status of Type 2 diabetic subjects. Int J Diab Dev Ctries 24:27–28.

Schulze MB, Hu FB. 2005. Primary prevention of diabetes: What can be done and how much can be prevented? Annu Rev Public Health 26:445–467.

Scragg R, Holdaway I, Singh V, Metcalf P, Baker J, Dryson E. 1995. Serum 25-hydroxyvitamin D3 levels decreased in impaired glucose tolerance and diabetes mellitus. Diabetes Res Clin Pract 27:181–188.

Scragg R, Sowers M, Bell C. 2004. Serum 25-hydroxyvitamin D, diabetes, and ethnicity in the Third National Health and Nutrition Examination Survey. Diabetes Care 27:2813–2818.

Shechter M. 2003. Does magnesium have a role in the treatment of patients with coronary artery disease? Am J Cardiovasc Drugs 3:231–239.

Shechter M, Bairey Merz CN, Stuehlinger HG, Slany J, Pachinger O, Rabinowitz B. 2003a. Effects of oral magnesium therapy on exercise tolerance, exercise-induced chest pain, and quality of life in patients with coronary artery disease. Am J Cardiol 91:517–521.

Shechter M, Hod H, Rabinowitz B, Boyko V, Chouraqui P. 2003b. Long-term outcome of intravenous magnesium therapy in thrombolysis-ineligible acute myocardial infarction patients. Cardiology 99:205–210.

Shechter M, Merz CN, Paul-Labrador M, Meisel SR, Rude RK, Molloy MD, Dwyer JH, Shah PK, Kaul S. 1999. Oral magnesium supplementation inhibits platelet-dependent thrombosis in patients with coronary artery disease. Am J Cardiol 84:152–156.

Shechter M, Merz CN, Rude RK, Paul Labrador MJ, Meisel SR, Shah PK, Kaul S. 2000. Low intracellular magnesium levels promote platelet-dependent thrombosis in patients with coronary artery disease. Am Heart J 140(2):212–218.

Shechter M, Shechter A. 2005. Magnesium and myocardial infarction. Clin Calcium 15:111–115.

Sherman L, Glennon JA, Brech WJ, Klomberg GH, Gordon ES. 1968. Failure of trivalent chromium to improve hyperglycemia in diabetes mellitus. Metabolism 17:439–442.

Shinde UA, Goyal RK. 2003. Effect of chromium picolinate on histopathological alterations in STZ and neonatal STZ diabetic rats. J Cell Mol Med 7:322–329.

Singh RB, Rastogi SS, Mani UV, Seth J, Devi L. 1991. Does dietary magnesium modulate blood lipids? Biol Trace Elem Res 30:59–64.

Sjogren A, Floren CH, Nilsson A. 1986. Magnesium deficiency in IDDM related to level of glycosylated hemoglobin. Diabetes 35:459–463.

Slabber M, Barnard HC, Kuyl JM, Dannhauser A, Schall R. 1994. Effects of a low-insulin-response, energy-restricted diet on weight loss and plasma insulin concentrations in hyperinsulinemic obese females. Am J Clin Nutr 60:48–53.

Slesinski RS, Clarke JJ, RH CS, Gudi R. 2005. Lack of mutagenicity of chromium picolinate in the hypoxanthine phosphoribosyltransferase gene mutation assay in Chinese hamster ovary cells. Mutat Res 585:86–95.

Smart GA, Sherlock JC. 1985. Chromium in foods and the diet. Food Addit Contam 2:139–147.

Song Y, Wang J, Li XK, Cai L. 2005. Zinc and the diabetic heart. Biometals 18:325–332.

Srivastava AK. 2000. Anti-diabetic and toxic effects of vanadium compounds. Mol Cell Biochem 206:177–182.

Stearns DM, Silveira SM, Wolf KK, Luke AM. 2002. Chromium(III) tris(picolinate) is mutagenic at the hypoxanthine (guanine) phosphoribosyltransferase locus in Chinese hamster ovary cells. Mutat Res 513:135–142.

Stearns DM, Wise JP, Sr, Patierno SR, Wetterhahn KE. 1995. Chromium(III) picolinate produces chromosome damage in Chinese hamster ovary cells. FASEB J 9:1643–1648.

Stratton IM, Adler AI, Neil HA, Matthews DR, Manley SE, Cull CA, Hadden D, Turner RC, Holman RR. 2000. Association of glycaemia with macrovascular and microvascular complications of type 2 diabetes (UKPDS 35): Prospective observational study. BMJ 321(7258):405–412.

Suarez A, Pulido N, Casla A, Casanova B, Arrieta FJ, Rovira A. 1995. Impaired tyrosine-kinase activity of muscle insulin receptors from hypomagnesaemic rats. Diabetologia 38:1262–1270.

Thomas VL, Gropper SS. 1996. Effect of chromium nicotinic acid supplementation on selected cardiovascular disease risk factors. Biol Trace Elem Res 55:297–305.

Tice JA, Ross E, Coxson PG, Rosenberg I, Weinstein MC, Hunink MG, Goldman PA, Williams L, Goldman L. 2001. Cost-effectiveness of vitamin therapy to lower plasma homocysteine levels for the prevention of coronary heart disease: Effect of grain fortification and beyond. JAMA 286:936–943.

Tosiello L. 1996. Hypomagnesemia and diabetes mellitus. A review of clinical implications. Arch Intern Med 156:1143–1148.

Trow LG, Lewis J, Greenwood RH, Sampson MJ, Self KA, Crews HM, Fairweather-Tait SJ. 2000. Lack of effect of dietary chromium supplementation on glucose tolerance, plasma insulin and lipoprotein levels in patients with type 2 diabetes. Int J Vitam Nutr Res 70:14–18.

US Food and Drug Administration. 2005. Qualified Health Claims: Letter of Enforcement Discretion—Chromium Picolinate and Insulin Resistance (Docket No. 2004Q-0144). http://www.cfsan.fda.gov/~dms/qhccr.html

Uusitupa MI, Kumpulainen JT, Voutilainen E, Hersio K, Sarlund H, Pyorala KP, Koivistoinen PE, Lehto JT. 1983. Effect of inorganic chromium supplementation on glucose tolerance, insulin response, and serum lipids in noninsulin-dependent diabetics. Am J Clin Nutr 38:404–410.

Uusitupa M, Mykkanen L, Siitonen O, Laakso M, Sarlund H, Kolehmainen P, Rasanen T, Kumpulainen J, Pyorala K. 1992. Chromium supplementation in impaired glucose tolerance of elderly: effects on blood glucose, plasma insulin, C-peptide and lipid levels. Br J Nutr 68:209–216.

Vincent JB. 2000. The biochemistry of chromium. J Nutr 130:715–718.

Vladeva S, Terzieva DD, Arabadjiiska DT. 2005. Effect of chromium on the insulin resitance in patients with type II diabetes mellitus. Folia Med XLVII:59–62.

Vrtovec M, Vrtovec B, Briski A, Kocijancic A, Anderson RA, Radovancevic B. 2005. Chromium supplementation shortens QTc interval duration in patients with type 2 diabetes mellitus. Am Heart J 149:632–636.

Wang H, Kruszewski A, Brautigan DL. 2005. Cellular chromium enhances activation of insulin receptor kinase. Biochemistry 44:8167–8175.

Wiederkehr A, Wollheim CB. 2006. Minireview: Implication of mitochondria in insulin secretion and action. Endocrinology 147:2643–2649.

Yang CY, Chiu HF, Cheng MF, Tsai SS, Hung CF, Tseng YT. 1999. Magnesium in drinking water and the risk of death from diabetes mellitus. Magnes Res 12:131–137.

Yeh GY, Eisenberg DM, Kaptchuk TJ, Phillips RS. 2003. Systematic review of herbs and dietary supplements for glycemic control in diabetes. Diabetes Care 26:1277–1294.

Zhao HX, Mold MD, Stenhouse EA, Bird SC, Wright DE, Demaine AG, Millward BA. 2001. Drinking water composition and childhood-onset Type 1 diabetes mellitus in Devon and Cornwall, England. Diabet Med 18:709–717.

Zhuang X, Yang Y, Zeng Y, Kung J, Zhou Z, Zhang D, Cheng N, Anderson RA. 2005. Chromium picolinate improves blood gucose: A six-year follow-up study of 1,056 patients with type 2 diabetes [abstract P310]. Diabet Med 23:61.

Chapter 11

Targeting Oxidant Stress as a Strategy for Preventing Vascular Complications of Diabetes and Metabolic Syndrome

Mark F McCarty, BA, and Toyoshi Inoguchi, MD, PhD

Overview

Oxidant stress plays a central role in mediating the macro- and microvascular complications of diabetes and metabolic syndrome. Radicals antagonize protective nitric oxide (NO) bioactivity, through direct quenching of NO and uncoupling of NO synthase, while promoting inflammation and fibrosis via activation of NF-kappaB and TGF-beta, respectively. Oxidants are key mediators of insulin resistance in hypertrophied adipocytes—which gives rise to systemic insulin resistance—and may also promote beta cell dysfunction. In diabetics, a major effect of peroxynitrite is to trigger PARP-mediated inhibition of glyceraldehyde-3-phosphate dehydrogenase (GAPDH); glycolytic intermediates pile up behind this bottleneck, boosting the activity of three key pathways known to mediate complications: diacylglycerol (DAG) synthesis (leading to protein kinase C activation), and the hexosamine and polyol pathways. The chief sources of excess oxidant stress in metabolic syndrome and diabetes are NADPH oxidase (activated by PKC, angiotensin II, and advanced glycation/lipoxidation endproducts), uncoupled NO synthase, and—in diabetics—mitochondria. Fortunately, it may be feasible to suppress the production and downstream effects of the radicals overproduced in these disorders, using safe nutraceuticals. Phycocyanobilin (PCB), a biliverdin-derived chromophore found in blue–green algae, has recently been shown to potently inhibit NADPH oxidase in a manner analogous to bilirubin; efforts to develop PCB-enriched algae extracts as antioxidant supplements are underway. High-dose folate can reconstitute the normal activity of uncoupled eNOS in oxidatively stressed endothelium, likely by scavenging the peroxynitrite-derived radicals that attack tetrahydrobiopterin. Lipoic acid—as well as the drug metformin—boosts the antioxidant defenses of endothelial mitochondria by activating AMP-activated kinase and PGC-1α; lipoic acid may also combat oxidant stress via phase II induction of glutathione and heme oxygenase-1. High doses of thiamine—or its better-absorbed analog benfotiamine—can increase transketolase activity, decreasing excess substrate in the upper glycolytic pathway by drawing it into the pentose phosphate pathway. Taurine has the potential to ameliorate the impact of myeloperoxidase-derived oxidants on diabetic complications, and suppresses development of neuropathy in diabetic rodents. The vitamer pyridoxamine can aid control of oxidative stress by inhibiting production of advanaced glycation/lipoxidation end products. Thus, a regimen of PCB, high-dose folate, lipoic acid, taurine, pyridoxamine, and benfotiamine may have considerable potential for preventing complications in patients who are diabetic and/or insulin resistant. Selenium, vitamin C, niacinamide, melatonin, and oligopeptide ACE inhibitors may also have some value in this regard. Clearly, there is a considerable scope for the development of rational, well-tolerated nutraceutical regimens which could substantially mitigate the health risks associated with metabolic syndrome and diabetes.

Oxidant Stress—Key Mediator of Diabetic Complications

Excessive production of superoxide—most notably in vascular endothelium—is believed to be a fundamental mediator of the macro- and microvascular complications of diabetes and metabolic syndrome (Du et al. 2003; Nishikawa et al. 2000). This increase in superoxide production reflects activation of NADPH oxidase, uncoupling of the endothelial isoform of nitric oxide synthase (eNOS), and—specifically in diabetes—short-circuiting of respiratory chain electron transport in mitochondria (Du et al. 2003; Inoguchi et al. 2000, 2003; Inoguchi and Nawata 2005; Milstien and Katusic 1999; Nishikawa et al. 2000; Shinozaki et al. 1999, 2004, 2005).

The increase in NADPH activity is induced by activation of certain isoforms of PKC that can phosphorylate p47phox, promoting its translocation to the plasma membrane (Fontayne et al. 2002); this increase in PKC activity is largely attributable to increased de novo synthesis of DAG, reflecting increased free fatty acid exposure and, particularly in diabetics, increased substrate in the upper portion of the glycolytic pathway within tissues such as endothelium that are highly glucose permeable (Du et al. 2003; Inoguchi et al. 2000; Nishikawa et al. 2000; Rask-Madsen and King 2005). DAG synthesis is further boosted in tissues expressing aldose reductase owing to an increase in the cytosolic $NADH/NAD^+$ ratio that promotes generation of glycerol-3-phosphate. Up regulation of the local renin–angiotensin system (Cooper 2004; Engeli et al. 2003; Gilbert et al. 2003; Ko et al. 2006; Sodhi et al. 2003), including a PKC-mediated increase in AT1 receptor expression (Holzmeister et al. 1997), further amplifies NADPH activity, in part by increasing the expression of NADPH subunits (Mollnau et al. 2002; Nakayama et al. 2005; Seshiah et al. 2002).

In diabetics, an additional increase in NADPH oxidase activity can be induced by advanced glycation and lipoxidation end products (AGEs and ALEs), via interaction with the RAGE receptor (Basta et al. 2005; Wautier et al. 2001; Wautier and Schmidt 2004).

Uncoupling of eNOS is secondary to increased oxidant production by NADPH oxidase and/or mitochondria; peroxynitrite readily oxidizes tetrahydrobiopterin and, in the absence of this key cofactor, eNOS catalyzes the transfer of electrons from NADPH to molecular oxygen, generating superoxide (Milstien and Katusic 1999). This uncoupling also entails decreased efficiency of nitric oxide production. There is considerable evidence for a functional deficiency of tetrahydrobiopterin in the vascular endothelium and renal mesangium of diabetic or insulin-resistant rats, associated with deficient eNOS activity (Alp et al. 2003; Cai et al. 2005; Okumura et al. 2006; Pannirselvam et al. 2003; Pieper 1997; Prabhakar 2001; Satoh et al. 2005; Shinozaki et al. 1999, 2000); furthermore, intra-arterial infusion of tetrahydrobiopterin in type 2 diabetic patients rapidly improves endothelium-dependent vasodilation (Heitzer et al. 2000).

Increased production of superoxide by the endothelial mitochondria of diabetics presumably reflects increased substrate availability that leads to an excess of electrons in mitochondrial respiratory chains (Nishikawa et al. 2000); however, it is suspected that reduced efficiency of the distal portion of the respiratory chain, possibly secondary to oxidant stress, contributes to this phenomenon (Green et al. 2004).

Increased superoxide production may act in a number of interacting ways to promote vascular complications. Of key importance in this regard is the fact that superoxide antagonizes effective NO bioactivity—by directly quenching NO, by promoting uncoupling of eNOS, and by impairing the efficiency of the insulin-PI3K-Akt pathway that induces an activating phosphorylation of eNOS (Kim et al. 2005; Storz and Toker 2003). Moreover, the peroxynitrite generated when superoxide reacts with NO is a highly active oxidant that mediates much of the oxidant stress associated with diabetes and metabolic syndrome (Obrosova et al. 2005; Pacher and Szabo 2006). In particular, peroxynitrite, as well as the hydrogen peroxide derived from superoxide dismutation, can promote activation of NF-kappaB, inducing an "activated" phenotype conducive to macrovascular disease and other inflammatory complications (Cooke and Davidge 2002; Du et al. 1999; Pueyo et al. 2000; Sohn et al. 2003). These oxidants can also promote the synthesis and activation of TGF-beta, thereby acting as mediators of the profibrotic complications of diabetes such as glomerulosclerosis (McCarty 2006, 2007a, b; Xia et al. 2006). The utility of PARP inhibitors for mitigating endothelial dysfunction and related complications in diabetic rodents presumably reflects a key role for peroxynitrite-mediated DNA damage in the genesis of these complications (Szabo 2005). Pericyte apoptosis, an early event in the evolution of diabetic neuropathy, appears to be triggered by oxidant stress (Cacicedo et al. 2005).

Recent studies employing the NADPH oxidase inhibitor apocynin point to a role for this enzyme complex in the early stages of diabetic neuropathy in rats (Cotter and Cameron 2003) and demonstrate that the oxidative stress generated by chronic activation of NADPH oxidase in the hypertrophied adipocytes of obese mice renders these adipocytes insulin-resistant and "inflamed" such that they secrete the adipokines characteristic of insulin resistance syndrome (e.g., IL-6, TNF-alpha, leptin) (Furukawa et al. 2004). In the pancreas of diabetics, activation of this enzyme complex via local up regulation of angiotensin II activity promotes beta cell dysfunction and apoptosis (Chu et al. 2006; Inoguchi and Nawata 2005; Shao et al. 2006).

Brownlee and colleagues have demonstrated that oxidant stress works in several other ways to boost the activity of pathways thought to mediate diabetic vascular complications. In particular, peroxynitrite, by inducing DNA damage that activates PARP, leads to polyADP-ribosylation of GAPDH, diminishing the activity of this enzyme (Du et al. 2003). As a result, intermediates in the upper portion of the glycolytic chain build up—an effect that is of particular importance when hyperglycemia is promoting increased tissue glucose uptake. Thus, more substrate becomes available for de novo DAG synthesis, for aldose reductase, and for glutamine:fructose-6-phosphate amidotransferase, the first and rate-limiting enzyme for the hexosamine pathway that generates UDP-N-acetylglucosamine (Du et al. 2000, 2003; Nishikawa et al. 2000). Studies with diabetic rodents have demonstrated that PKC activation, the polyol pathway, and the hexosamine pathway all contribute to the induction of diabetic complications (Buse 2006; Suzen and Buyukbingol 2003; Way et al. 2001). Furthermore, oxidant stress can boost aldose reductase activity by blunting the tonic inhibitory impact of physiological levels of NO on this enzyme (Chandra et al. 1997, 2002; Ramana et al. 2003); as a result, accelerated utilization of NADPH reduces the availability of this cofactor to eNOS, compounding the NO deficiency, while also impairing NADPH-dependent antioxidant mechanisms.

There is recent evidence that vascular-bound myeloperoxidase (MPO) is dramatically increased in diabetic rats, and that its interaction with hydrogen peroxide diffusing from neighboring cells can give rise to hypochlorous acid (HOCl) and other chlorinating compounds that may be more detrimental to endothelial function than is hydrogen peroxide per se (Zhang et al. 2004). In vitro, the HOCl scavenger methionine alleviates the severe impairment of endothelium-dependent vasodilation noted in rat aorta exposed to high glucose and MPO—thus incriminating HOCl as the likely mediator of the endothelial toxicity associated with MPO (Zhang et al. 2004).

In aggregate, these considerations suggest that measures which suppress superoxide production by NADPH oxidase, eNOS, and—in diabetics—mitochondria may go a long way toward "getting" to the root of the macro- and microvascular complications associated with diabetes and metabolic syndrome. Practical nutraceutical strategies for blunting the production or downstream consequences of excessive superoxide would be of particular interest, as these might be more convenient, affordable, and safer than drug therapies. In this regard, a regimen composed of phycobilins, high-dose folate, lipoic acid, taurine, pyridoxamine, and benfotiamine may have great potential.

Phycobilins—Phytonutrient Inhibitors of NADPH Oxidase

The vascular protection associated with relatively high plasma levels of unconjugated bilirubin (Mayer 2000; Rigato et al. 2005) as well as that afforded by the antioxidant enzyme heme-oxygenase-1 (Exner et al. 2004; Immenschuh and Schroder 2006; Lanone et al. 2005; Morita 2005) most likely reflects the ability of bilirubin to serve as an endogenous inhibitor of NADPH oxidase (Lanone et al. 2005; McCarty 2006, 2007a, b); although bilirubin has usually been characterized as an oxidant scavenger, its intracellular concentrations are in the low nanomolar range (Sedlak and Snyder 2004) and, moreover, it is a relatively inefficient scavenger of superoxide (Stocker 2004). Thus, it is now suspected that the important antioxidant activity of bilirubin in tissues reflects, not scavenging activity, but rather direct inhibition of NAPDH oxidase (McCarty 2006, 2007a, b).

Bilirubin arises by reduction of biliverdin, a product of heme-oxygenase-mediated catabolism of heme; this reduction is mediated by the ubiquitously expressed enzyme biliverdin reductase (Baranano et al. 2002; Sedlak and Snyder 2004). Plants and blue–green algae generate chromophores known as "phycobilins" that are derivatives and close structural analogs of biliverdin; following conjugation to apoproteins, these chromophores function as light harvesters, much like chlorophyll (Beale 1994). These phycobilins are readily susceptible to reduction by biliverdin reductase, giving rise to "phycorubins" that are close analogs of bilirubin (Terry et al. 1993). Recently, Inoguchi has demonstrated that PCB, the chromophore responsible for the blue pigmentation of spirulina, is a potent inhibitor of NADPH oxidase activity in human endothelial, smooth muscle, and mesangial cell cultures; the effects are quite parallel to those of biliverdin in these cell lines, and are dose-dependent in the range of 1–20 μM (Toyoshi Inoguchi, personal communication). These findings may rationalize previous reports that oral administration of phycocyanin—the spirulina protein that includes PCB as a chromophore—exerts a wide range of anti-inflammatory effects in rodents (Romay et al. 2003). Thus, it has been suggested that PCB-enriched spirulina extracts may have potential as antioxidant nutraceuticals that, in appropriate oral doses, could partially inhibit endogenous NADPH oxidase activity (McCarty 2006, 2007a, b). The implications of such a strategy for controlling the excess oxidant stress associated with diabetes or metabolic syndrome are clear.

Recently, Inoguchi and colleagues have examined risk for vascular complications in diabetics who have Gilbert syndrome, a genetic variant in which free unconjugated bilirubin levels are 2–3-fold elevated owing to reduced hepatic expression of the isoform of UDP-glucuronosyl transferase (1A1) responsible for conjugation of bilirubin (Bosma et al. 1995). These researchers identified 96 diabetics with

Gilbert syndrome, and compared them with 425 diabetics without this syndrome; all diabetics included in the analysis had been diabetic for at least 5 years. Risks for retinopathy, macroalbuminuria, and coronary disease were significantly lower in the former group. After multiple regression analyses adjusting for a number of relevant factors, including parameters related to insulin resistance and diabetic control (control tended to be somewhat better in the Gilbert group, possibly reflecting a role for NADPH oxidase activation in insulin resistance and beta cell failure) (Furukawa et al. 2004; Inoguchi and Nawata 2005), relative risk for retinopathy, macroalbuminuria, and coronary disease in the Gilbert group was calculated to be 0.22, 0.20, and 0.21, respectively ($P < 0.05$) (Inoguchi et al. 2007). These findings suggest that the partial inhibition of NADPH oxidase activity associated with elevated unconjugated bilirubin in diabetics with Gilbert syndrome provides important protection from the vascular complications of diabetes. It is reasonable to anticipate that prompt administration of PCB, in a dosage schedule sufficient to replicate the degree of NADPH oxidase inhibition experienced by subjects with Gilbert syndrome, would likewise have an important effect on risk for major complications in diabetics.

While it is evident that excessive inhibition of NADPH oxidase could compromise immune defenses, it should be noted that subjects with Gilbert syndrome do not experience any evident increase in risk for infection or other disorders, and indeed appear to be at a notably decreased risk for vascular disorders (Kalousova et al. 2005; Vitek et al. 2002, 2006); thus, it may be feasible to achieve moderate inhibition of NADPH oxidase without compromising health. Furthermore, if a serious infection developed during the course of PCB therapy, PCB administration could be temporarily discontinued to quickly restore full NADPH oxidase activity.

Pending the availability of PCB in supplemental form (extracted from spirulina or synthesized), it may be feasible to use whole spirulina as a clinical source of PCB. A heaped tablespoon of spirulina (approximately 15 g) provides about 100 mg PCB, and can be rendered palatable by inclusion in "smoothies" made with ingredients such as soy milk, fruit juices, or whole fruit. It is pertinent to note that inclusion of whole spirulina in the diet of rodents has been reported to exert potent anti-inflammatory effects (Chamorro et al. 2006; Khan et al. 2005, 2006; Mohan et al. 2006; Rasool et al. 2006; Remirez et al. 2002).

Vitamin E Has Not Fulfilled Expectations

An alternative approach to suppressing activation of NADPH oxidase in diabetics was proposed some years ago by King and colleagues. These researchers demonstrated that, in diabetic rodents, high-dose injections of vitamin E (d-alpha-tocopherol, 40 mg/kg i.p., every other day) prevented activation of PKC (and hence its downstream target NADPH oxidase) by lowering the elevated concentrations of DAG in tissues giving rise to diabetic complications (retina, glomeruli, aorta); as might be expected, this effect was associated with an amelioration of the early impacts of diabetes on retinal and glomerular structure and function (Ishii et al. 1998; Koya et al. 1997, 1998; Kunisaki et al. 1994, 1995, 1996, 1998). In cell cultures exposed to high glucose, the impact of vitamin E on DAG levels was traced to an increase in DAG kinase activity, possibly indicating that oxidized lipids inhibit this enzyme.

Unfortunately, these benefits may have been contingent on exceptionally high tissue levels of vitamin E achieved through parenteral administration, as in clinical studies entailing oral administration of vitamin E, the benefits have been equivocal at best. In type 1 diabetics (average duration of diabetes 4 years), 1,800 IU d-alpha-tocopheryl acetate daily for 4 months was associated with a significant increase in retinal blood flow, reversing the trend toward decreased flow seen in unsupplemented diabetics (Bursell et al. 1999). Studies examining the impact of vitamin E supplementation on the depressed endothelium-dependent vasodilation of diabetics observed an improvement in type 1 subjects (1,000 IU daily for 3 months) but no change in type 2 subjects (1,600 IU daily for 8 weeks) (Gazis et al. 1999; Skyrme-Jones et al. 2000). More recently, the largest and longest study to evaluate high-dose vitamin E (1,800 IU daily for 1 year) in diabetics (of both types) failed to observe any improvement in endothelial function during vitamin E, and indeed trends in response were slightly but significantly better during placebo than vitamin E (Economides et al. 2005). Significant rises, relative to placebo, were seen in systolic blood pressure and plasma endothelin during vitamin E—although C-reactive protein fell slightly in the vitamin E group. The authors concluded that "because vitamin E-treated patients had a worsening in some vascular reactivity measurements when compared to control subjects, the use of high dosages of vitamin E cannot be recommended." These disappointing results parallel those seen in randomized prevention trials with vitamin E in patients at a high risk for coronary events; in particular, in the HOPE study, 400 IU vitamin E daily for an average of 4.5 years did not influence the subsequent incidence of coronary events or stroke in diabetic subjects, nor influence the onset of overt nephropathy (Lonn et al. 2002). While we cannot exclude the possibility that high-dose vitamin E might favorably influence some aspect of diabetic complications during some stage of the disease, the overall impression is disappointing compared to the clear-cut beneficial results seen in rats with recent-onset diabetes.

Recoupling eNOS with High-Dose Folate and Ascorbate

With respect to uncoupled eNOS, high-dose folate may offer a simple remedy. Several years ago, Rabelink and colleagues discovered that 5-methyltetrahydrofolate (the chief metabolite of folic acid circulating in plasma) can "pinch-hit" for the function of tetrahydrobiopterin in eNOS when concentrations of the latter are insufficient for coupled eNOS activity (Hyndman et al. 2002; Moat et al. 2006; Stroes et al. 2000). In other words, 5-methyltetrahydrofolate prevents uncoupled eNOS from generating superoxide, and restores its normal capacity to generate NO. Thus, acute infusions of 5-methyltetrahydrofolate have been shown to have a favorable impact on endothelium-dependent, NO-mediated vasodilation in various disorders associated with oxidant-mediated endothelial dysfunction (McCarty 2004a, b)—including diabetes (Van Etten et al. 2002). Of greater practical interest are studies demonstrating that relatively high daily oral intakes of folic acid—5–10 mg per day have usually been used—can improve endothelium-dependent vasodilation in both type 1 and type 2 diabetics (Mangoni et al. 2005; Pena et al. 2004). It seems likely, however, that higher doses would produce greater benefit. Recently, Tawakol et al. have shown that, in patients with ischemic heart disease, preadministration of 30 mg folic acid (two 15 mg doses, 12 h apart) produces a marked augmentation of adenosine-stimulated myocardial blood flow in ischemic regions of the heart (Tawakol et al. 2005); this phenomenon is thought to reflect a normalization of shear-induced, NO-mediated vasodilation, attributable to recoupling of eNOS. Remarkably, over 30 years ago, Oster reported that daily mega-doses of folate—40–80 mg per day—had a very favorable clinical impact on angina, intermittent claudication, and the healing of ischemic ulcers (Oster 1968; Oster et al. 1983); most likely, these benefits reflected the fact that Oster inadvertently had repaired eNOS function in his patients (McCarty 2006, 2007a, b). (This phenomenon most likely is unrelated to modulation of homocysteine levels; in any case, recent evidence from prospective trials suggests that a moderate elevation of homocysteine is a marker for, rather than a mediator of, vascular risk (Bonaa et al. 2006).)

Recent evidence suggests that the remarkable ability of 5-methyltetrahydrofolate to reconstitute the normal activity of eNOS in oxidatively-stressed endothelium reflects the fact that reduced forms of folate are efficient scavengers for the peroxynitrite-derived radicals (presumably nitrogen dioxide and carbonate radicals) that attack tetrahydrobiopterin (Rezk et al. 2003). This effect becomes functionally significant in the low micromolar concentrations of reduced folates that can be achieved in endothelial cells exposed to pharmacological concentrations of folate (Antoniades et al. 2006). It is intriguing to speculate that high-dose folate, in addition to protecting tetrahydrobiopterin from inactivating oxidation, may have the potential to ameliorate other adverse effects mediated by peroxynitrite-derived radicals in diabetes and other disorders—at least in tissues that are capable of concentrating folate to low micromolar concentrations (McCarty 2007c, d). In particular, high-dose folate may suppress oxidant-mediated PARP activation and its downstream consequences.

An alternative strategy for preserving tetrahydrobiopterin function in cells subjected to oxidant stress is to ensure optimal intracellular concentrations of ascorbic acid. Exposure of cultured human endothelial cells to ascorbic acid increases their eNOS activity; this effect has been traced to ascorbate's ability to enhance intracellular levels of tetrahydrobiopterin, while decreasing the levels of oxidized forms of this cofactor (Baker et al. 2001; Heller et al. 1999, 2001; Huang et al. 2000). Since ascorbate does not enhance tetrahydrobiopterin synthesis (Heller et al. 2001), it seems to be protecting this cofactor from oxidation by peroxynitrite or other oxidants; presumably, it readily donates electrons to oxidized forms of this cofactor, restoring its proper tetrahydro structure. This phenomenon has also been demonstrated in vivo, in the aortas of ApoE-deficient mice treated with high-dose dietary ascorbate (d'Uscio et al. 2003).

However, there may be a limited scope for the impact of ascorbate on eNOS activity in humans. Uptake of ascorbate by endothelial cells is saturated at a plasma concentration of 100 μM (Ek et al. 1995; Heller et al. 1999), which is readily achieved and maintained by ascorbate intakes of 500 mg daily (Gokce et al. 1999). This suggests that supplemental ascorbate would only benefit endothelial function in subjects with mediocre baseline ascorbate status. Possibly, this accounts for the inconsistent findings of studies which have examined the impact of supplemental ascorbate on vascular endothelial function in subjects at increased coronary risk. (The favorable effects of intravenous vitamin C infusion on endothelial function in various studies, including those involving diabetics (Timimi et al. 1998; Ting et al. 1996), are not likely to be germane, as the supraphysiological plasma concentrations of ascorbate achieved have superoxide scavenging activity that is insignificant at the lower physiological concentrations achievable through oral dosing (Sherman et al. 2000).) Several studies have found that supplemental ascorbate (at least 800 mg daily) does not benefit endothelium-dependent vasodilation or markers of endothelial inflammation in patients with diabetes (Chen et al. 2006; Darko et al. 2002; Tousoulis et al. 2006). One study reported a favorable effect of oral ascorbate on endothelial function in type 1 diabetics but not in type 2 (Beckman et al. 2003), another study found benefit only in diabetic patients concurrently afflicted with coronary disease (Antoniades et al. 2004). A study evaluating the acute negative impact of a fatty meal on endothelial function in type 2 diabetics found that prior supplementation with ascorbate alleviated the endothelial dysfunction (Anderson et al. 2006).

The possible impact of ascorbate status on risk for diabetic microvascular complications, either in animal models or clinically, remains largely unexplored. There are several reports that plasma ascorbate levels tend to be lower in diabetics with microangiopathy than those without (Ali and Chakraborty 1989; Chakraborty 1992; Sinclair et al. 1991); it is unclear whether ascorbate deficiency predisposes to microangiopathy, or whether these findings simply reflect the fact that high oxidative stress associated with a predilection to microangiopathy degrades ascorbate status.

In a recent analysis pooling nine large prospective cohort studies (involving a total of over 293,000 subjects), Harvard researchers concluded that use of vitamin C supplementation (700 mg or more daily) was associated with a significant 25% reduction in risk for coronary heart disease (Knekt et al. 2004). Multiple regression analysis was employed to account for lifestyle factors associated with supplementation, and vitamin E supplementation was found to be without benefit in this analysis. In light of the facts that coronary disease is the chief cause of mortality in diabetes, and that adequate ascorbate status helps to preserve effective eNOS function, it would seem prudent to include a moderate dose of ascorbate in any nutraceutical program intended to ameliorate diabetic complications.

Flavanol-rich cocoa has recently attracted considerable attention for its cardio-vascular protective potential. Epicatechin, richly supplied by unprocessed cocoa, is now known to provoke or up regulate release of NO from vascular endothelium (Fisher et al. 2003; Schroeter et al. 2006). Such an effect would presumably be more clinically beneficial in subjects with diabetes or insulin resistance syndrome if uncoupling of eNOS was concurrently corrected by administration of high-dose folate, in conjuction with other measures to limit endothelial superoxide production. Of particular interest is the fact that hypertension is essentially absent among the Kuna Indians of Panama, so long as they follow their traditional lifestyle that includes ingestion of 3–4 servings of unprocessed cocoa per day (Hollenberg et al. 1997). In clinical studies, cocoa administration has been shown to lower elevated blood pressure and to improve muscle insulin sensitivity (Grassi et al. 2005; Taubert et al. 2003). Its impact in clinical diabetes has not yet been studied.

Addressing Mitochondrial Superoxide—A Role for Lipoic Acid?

As noted, increased mitochondrial production of superoxide contributes to oxidant stress when endothelial cells are exposed to elevated glucose levels. The extent of this contribution remains in dispute. Brownlee and colleagues have reported that uncoupling agents and certain inhibitors of the mitochondrial respiratory chain can virtually normalize superoxide production in endothelial cells exposed to hyper-glycemia (Du et al. 2000, 2003; Nishikawa et al. 2000). Other researchers have presented data suggesting that NADPH oxidase has primacy as a superoxide source (Christ et al. 2002; Cosentino et al. 2003; Inoguchi et al. 2000; Weidig et al. 2004; Yano et al. 2004). More studies may be required to clarify this issue. One possible way to resolve this controversy would be to liken mitochondrial superoxide production to "kindling" that boosts DAG production, enabling NADPH oxidase to then provide the main "blaze"; once NADPH oxidase is active, the resulting superoxide production would be adequate to sustain inhibition of GAPDH and thus DAG production.

A strategy for blunting mitochondrial superoxide production in diabetic endothe-lium is suggested by the recent discovery that PPARgamma coactivator-1α (PGC-1α) boosts endothelial expression of a diverse group of antioxidant enzymes specific to mitochondria—including manganese-dependent superoxide dismutase, UCP2, several peroxiredoxins (which detoxify hydrogen peroxide and peroxyni-trite), thioredoxin, and a thioredoxin reductase—as well as catalase (Valle et al. 2005). This effect of PGC-1α reflects its participation in the transcriptional com-plexes binding to the promoter regions of the affected genes. Overexpression of PGC-1α in endothelial cells reduced their production of oxidants by about 50% under both basal and high-glucose conditions (Valle et al. 2005). This finding has practical significance in light of other recent work demonstrating that activators of AMP-activated kinase (AMPK)—including AICAR and the antidiabetic drug metformin—increase PGC-1α expression in endothelial cells (Kukidome et al. 2006). Not surprisingly, these agents were shown to reduce oxidant production in endothelial cells exposed to high glucose—an effect abrogated by dominant negative AMPK. Moreover, there is other recent evidence that metformin (likely acting through AMPK) can also suppress superoxide production by NADPH oxidase in endothelial cells exposed to angiotensin II or high glucose (Ouslimani et al. 2005). Furthermore, by boosting fatty acid oxidation (via suppression of malonyl-coA production), AMPK can diminish the availability of substrate for DAG synthesis— a less direct way to diminish NADPH oxidase activation (Cacicedo et al. 2004; McCarty 2005a, b; Ruderman et al. 2003). Thus, stimulation of endothelial AMPK may have great potential for control of oxidant stress in metabolic syndrome or diabetes. This may go a long way toward rationalizing the observation that metformin therapy in type 2 diabetics has a much more dramatic impact on macrovascular risk than do injectible insulin or sulphonylureas, despite comparable effects on glycemic control (UKPDS 1998a; UKPDS 1998b; Ruderman et al. 2003). In this regard, recent evidence that the readily available and inexpensive natural agent berberine can activate AMPK in rat adipocytes and myotubes, as well as in human hepatocytes, is of particular interest (Brusq et al. 2006; Lee et al. 2006). Berberine is used to treat diabetes in China, and is said to be without apparent side effects (unlike metformin) in doses up to 2 g daily (Ni 1988; Zeng et al. 2003).

The "metavitamin" lipoic acid, whether administered orally or parenterally, has recently been shown to activate AMPK in endothelial cells and skeletal muscle of rodents, whereas it inhibits this enzyme in the hypothalamus (Kim et al. 2004; Lee et al. 2005a, b). These effects are precisely parallel to those of the hormone leptin, but can be observed in the absence of leptin or its receptor—suggesting that lipoic acid may somehow be activating leptin's postreceptor signaling mechanism (Kim et al. 2004). Remarkably, the inhibitory effect of lipoic acid on hypothalamic AMPK results in diminished appetite and weight loss (or decreased weight gain) in rodents, although no analogous effect has yet been reported in humans; the dose of lipoic acid used in rodent studies (usually 0.5% of diet) is high relative to the doses so far tested in humans. Lipoic acid has the further advantage that it acts as a phase II inducer—boosting cellular levels of reduced glutathione and inducing HO-1 (Flier et al. 2002; Ogborne et al. 2005; Suh et al. 2004). Thus, the versatile antioxidant activity of lipoic acid may reflect AMPK-mediated reductions in oxidant production by mitochondria and NADPH oxidase, complemented by an increase in intracellular glutathione and bilirubin levels.

Not surprisingly, lipoic acid has long been used for control of diabetes complications. In particular, its clinical utility in diabetic neuropathy is well established. Doses of 600–1,800 mg lipoic acid daily have been shown to have a favorable impact on diabetic neuropathy in controlled clinical trials—albeit these trials included an initial phase in which lipoic acid was administered intravenously (Reljanovic et al. 1999; Ziegler 2004; Ziegler et al. 1999, 2004). Recent open trials suggest that oral lipoic acid per se (600 mg daily) may indeed have efficacy in this regard (Hahm et al. 2004; Negrisanu et al. 1999). Lipoic acid may also have utility for prevention of diabetic nephropathy; in an open, nonrandomized trial, therapy with lipoic acid (600 mg/day orally) was associated with a trend toward decreased urinary albumin over an 18-month follow-up, whereas urinary albumin increased significantly during this time in controls not receiving this agent (Morcos et al. 2001). Analogously, thrombomodulin—a serum protein often used as a marker for diabetic microangiopathy—rose in the control group, but fell in those receiving lipoic acid. Rodent studies likewise suggest that lipoic acid can have a favorable impact on diabetic nephropathy (Siu et al. 2006). And dietary lipoic acid was shown to prevent the formation of acellular capillaries and mitigate oxidant stress in the retinas of diabetic rats, leading the authors to suggest that "alpha-lipoic acid supplementation represents an achievable adjunct to help prevent vision loss in diabetic patients" (Kowluru and Odenbach 2004).

Of course, improving glycemic control is the most definitive strategy for controlling mitochondrial superoxide production in diabetes. In this regard, PCB may have potential for improving insulin function in obesity-related metabolic syndrome and diabetes, since there is now evidence that excessive NADPH activation in hypertrophied adipocytes may be largely responsible for systemic insulin resistance (Furukawa et al. 2004; Talior et al. 2005); indeed, the NADPH-inhibitory drug apocynin has been shown to improve insulin sensitivity and glycemic control in obese diabetic mice (Furukawa et al. 2004). Furthermore, there is evidence that lipoic acid can improve the insulin sensitivity of skeletal muscle in rats by activating AMPK (Lee et al. 2005a, b). There is an analogous report that oral lipoic acid can improve muscle insulin sensitivity in type 2 diabetics—albeit the fact that this effect did not show dose-dependency is puzzling, and lipoic acid has not had an evident impact on glycemic control in the clinical studies evaluating its efficacy in diabetic neuropathy.

In light of the fact that lipoic acid is known to be a phase II inducer, it is of interest to consider the possibility that other nutraceuticals or foods with such activity might have an impact in diabetes. Phase II induction is believed to mediate the anticarcinogenic activity of green tea flavonoids—most notably EGCG—in rodents (Khan et al. 1992). While green tea or supplemental EGCG has indeed shown some favorable effects on glycemic control and various other parameters in diabetic rodents (Babu et al. 2006a, b; Koyama et al. 2004; Tsuneki et al. 2004; Vinson and Zhang 2005; Wolfram et al. 2006; Wu et al. 2004; Yamabe et al. 2006; Yokozawa et al. 2005), heavy consumption of green tea by human diabetics has not so far shown any evident benefit (Fukino et al. 2005; Ryu et al. 2006); conceivably; this reflects a proportionately lower intake of EGCG relative to the rodent studies. Although one recent Japanese epidemiological study noted a reduced risk for diabetes in regular users of green tea, no dose-dependency was observed, suggesting that green tea use might be serving as a marker for other lifestyle factors (Iso et al. 2006).

Benfotiamine—Draining the Upper Glycolytic Pool

The build-up of glycolytic intermediates in the proximal portion of the glycolytic pathway quite clearly plays a role in the induction of diabetic complications. The enzyme transketolase has the ability to convert glyceraldehydes-3-phosphate and fructose-6-phosphate to pentose phosphates and other intermediates in the pentose phosphate cycle (a.k.a. hexose-monophosphate shunt). Furthermore, it is usually possible to enhance the activity of transketolase by boosting cellular levels of its cofactor thiamine diphosphate (Takeuchi et al. 1990). This increase in activity reflects not only greater cofactor binding, but also a feed-forward impact on transketolase expression (Pekovich et al. 1998). Thus, greater thiamine availability, by boosting transketolase activity, has the potential to draw substrate from the proximal glycolytic pathway into the pentose phosphate pathway—an effect which would be expected to diminish production of DAG (thus blunting activation of PKC and NADPH oxidase), polyols, and hexosamines (Berrone et al. 2006; Hammes et al. 2003).

Unfortunately, the utility of high-dose thiamine therapy is limited by the intestine's low capacity for thiamine absorption—a carrier-mediated uptake mechanism is effectively saturated at modest physiological intakes of this vitamin (Rindi and Laforenza 2000). For this reason, Japanese chemists over 40 years ago developed highly absorbable lipophilic analogs of thiamine that convert spontaneously to thiamine following absorption. The most effective of these is benfotiamine, which appears to have no more toxic risk than thiamine (indeed, its acute toxicity in rodents is less than that of thiamine), and is far better absorbed (Bitsch et al. 1991; Loew 1996; Schreeb et al. 1997). Benfotiamine has long been legally available as a nutritional supplement in Japan and the United States, and in Germany it is currently prescribable as a clinically validated treatment for diabetic neuropathy (Haupt et al. 2005; Stracke et al. 1996; Winkler et al. 1999).

Indeed, oral benfotiamine has shown good efficacy for alleviating a range of diabetic complications in rodents, including neuropathy, nephropathy, and retinopathy (Babaei-Jadidi et al. 2003; Beltramo et al. 2004; Gadau et al. 2006; Karachalias et al. 2005; Pomero et al. 2001; Stracke et al. 2001; Wu and Ren 2006). Its clinical utility as a treatment for diabetic neuropathy—usually in divided doses providing 150–300 mg daily—is well documented, and there appear to be no discernible side effects at these doses (Haupt et al. 2005; Stracke et al. 1996; Winkler et al. 1999). Further clinical studies will be required to assess the long-term impact of benfotiamine on progression of retinopathy or nephropathy in diabetics.

To date, benfotiamine has been tested in the context of diabetes—but might it be possible that this agent could also influence macrovascular risk in normoglycemic metabolic syndrome? In this circumstance, free fatty acid excess is primarily responsible for the enhanced endothelial generation of DAG that activates PKC (Inoguchi et al. 2000; Krebs and Roden 2005; Rask-Madsen and King 2005; Shankar and Steinberg 2005), but benfotiamine administration could be expected to decrease levels of the glycerol-3-phosphate to which fatty acids are conjugated—albeit the magnitude of this effect would not be as great as in the context of hyperglycemia. Thus, the impact of benfotiamine on the endothelial dysfunction associated with metabolic syndrome merits investigation.

Taurine—Physiological Antidote to Hypochlorous Acid

Taurine is a "metavitamin" that has antioxidant and osmoregulatory properties, and it also modulates transmembrane calcium flux. Taurine may have potential for controlling the oxidative damage attributable to MPO deposition in the vasculature. Recent evidence indicates that MPO, released by infiltrating phagocytes, accumulates in the arterial intima of rats with insulin resistance syndrome or diabetes (Zhang et al. 2004). This MPO can then make use of hydrogen peroxide released by inflamed endothelium or intimal macrophages to generate HOCl, a highly reactive oxidant that can compromise endothelium-dependent vasodilation (Zhang et al. 2004), presumably by modifying the plasma membrane in a way that compromises activation of eNOS (Jaimes et al. 2001). HOCl can also be generated and released by intimal foam cells. Taurine functions physiologically as a detoxicant of HOCl, reacting with it to generate taurine chloramine and innocuous hydroxyl ion. Taurine chloramine, while still an oxidant, is less promiscuously reactive than HOCl, and hence may be viewed as a detoxification product (McCarty 2004a, b; Schuller-Levis and Park 2003; Wright et al. 1985). Moreover, taurine chloramine formed in phagocytes, which maintain high intracellular levels of taurine, down regulates phagocyte activation by inhibiting activation of NF-kappaB (Barua et al. 2001; Kanayama et al. 2002). Taurine may thus have utility for alleviating the contribution of activated phagocytes to atherogenesis and ischemic syndromes. Indeed, oral taurine has shown efficacy in rodent models of atherogenesis (Kondo et al. 2000; Murakami et al. 2002; Petty et al. 1990) (albeit in some studies this reflects, in part, a marked hypolipidemic activity not reported in humans), and improves endothelium-dependent vasodilation in rat aorta and in young human smokers (Abebe and Mozaffari 2000; Fennessy et al. 2003). Forty years ago, many Italian clinicians reported that high daily doses of taurine could alleviate coronary angina and intermittent claudication—an effect which conceivably could reflect a decreased tendency of activated leukocytes to clog the microvasculature downstream from stenotic lesions (McCarty 1999). Taurine supplementation has a safe positive inotropic effect on cardiac function in clinical congestive heart failure (Azuma et al. 1985, 1992), has platelet-stabilizing activity complementary to that of aspirin (Franconi et al. 1992, 1995; Hayes et al. 1989), and has antihypertensive properties that are well documented in rodent models (Anuradha and Balakrishnan 1999; Harada et al. 2000; Ideishi et al. 1994; Nara et al. 1978; Sato et al. 1991) and supported by limited clinical evidence (Fujita et al. 1987; Kohashi et al. 1983; Yamori et al. 1996); these latter effects likely reflect an impact of taurine on calcium flux.

In rodent models of both type 1 and type 2 diabetes, taurine alleviates development of diabetic neuropathy—improving nerve conduction velocity, blunting the decline in endoneurial perfusion, and preventing hyperalgesia (Li et al. 2005, 2006; Obrosova et al. 2001a, b; Pop-Busui et al. 2001). No comparable clinical studies have been reported. Taurine also exerts antioxidant effects in the retina of diabetic rats, suggestive of a favorable impact on the course of diabetic retinopathy (Di Leo et al. 2002, 2003; Obrosova et al. 2001a, b). And taurine administration prolongs the survival of rats with streptozotocin-induced diabetes (Di Leo et al. 2004). However, 1-year-long clinical study found that supplemental taurine (3 g daily) failed to improve albuminuria in diabetics (Nakamura et al. 1999)—not surprising in light of the limited utility of taurine in rodent models of diabetic nephropathy (Franconi et al. 2006). Since taurine may have some utility in atherogenesis, hypertension, platelet hyperaggregability, congestive heart failure, and ischemic syndromes— common complications of insulin resistance syndrome and diabetes—and also may have potential for moderating the course of diabetic neuropathy and retinopathy, its inclusion in nutraceutical programs for alleviating the complications of insulin resistance and diabetes appears warranted. Taurine is highly soluble, flavorless, inexpensive, and apparently without a toxic risk even in high chronic doses—properties which make it an ideal ingredient for functional foods.

Protective Potential of Selenium

In light of taurine's potential to down regulate NF-kappaB activation in phagocytes, it is germane to cite a recent French clinical study demonstrating that high-dose supplemental selenium (960 mcg daily) normalizes such activation in the monocytes of type 2 diabetics (Faure et al. 2004). Since selenium is an essential component of key antioxidant enzymes—thioredoxin reductase, as well as several isoforms of glutathione peroxidase (Stadtman 2000)—and selenium status is often suboptimal in regions with low soil selenium (Hartikainen 2005), it stands to reason that insuring optimal selenium status through supplementation is appropriate in disorders such as diabetes in which oxidants play a prominent pathogenic role. An impact of selenium status on NF-kappaB activation conceivably could reflect the ability of glutathione peroxidase to degrade hydrogen peroxide as well as peroxynitrite (Sies et al. 1998), each of which can promote NF-kappaB activation; moreover, selenium-dependent thioredoxin reductase helps to restore protein-bound cysteines to a proper reduced configuration, thereby reversing the impact of peroxides on signaling pathways (Nordberg and Arner 2001).

An anecdotal clinical report that high-dose supplemental selenium appears to slow progression of diabetic retinopathy (Crary and McCarty 1984; McCarty 2005a, b) has not been followed up. A report that high-dose selenium suppresses expression of VEGF in rat mammary carcinomas (Jiang et al. 1999) may be relevant, since induction of VEGF expression in the hypoxic retina plays a key role in the induction of proliferative retinopathy. The effect of selenium on

VEGF expression in cancer was mediated by methylselenol rather than selenium-dependent enzymes (Jiang et al. 2000); this suggests that the effect could be dose-dependent beyond the modest intakes of selenium (i.e., 100 mcg daily) required to optimize the expression of these enzymes.

Melatonin—Inducer of Antioxidant Enzymes

Physiological (low nanomolar) levels of the pineal hormone melatonin have been reported to increase the expression of a range of antioxidant enzymes in various tissues; these enzymes include melatonin, superoxide dismutase (types 1 and 2), catalase, glutathione peroxidase, glutathione reductase, glucose-6-phosphate dehydrogenase, and gamma-glutamylcysteine synthetase (rate-limiting for glutathione synthesis); an increase in the ratio of reduced to oxidized glutathione, and a decrease in tissue markers of oxidation (e.g., malondialdehyde) are also often demonstrated in these studies (Antolin et al. 1996; Hardeland 2005; Mayo et al. 2002; Reiter et al. 1999; Rodriguez et al. 2004; Tomas-Zapico and Coto-Montes 2005). Melatonin regulates these enzymes at the transcriptional level, and membrane and/or nuclear receptors for melatonin appear likely to mediate this effect, although the precise mechanisms involved remain unclear. Melatonin's impact in this regard varies from tissue to tissue; in some studies, it does not influence baseline expression of these enzymes, but prevents pro-oxidant drugs from down-regulating them. Since melatonin is usually well tolerated when administered orally prior to bedtime (a schedule which presumably will not interfere with melatonin's chronotropic role), its use as an adjuvant in diabetes management can be contemplated.

Indeed, exogenous melatonin has been found to exert protective antioxidant effects in diabetic rodents, as well as in cultured cells exposed to hyperglycemia (albeit some of the in vitro studies must be viewed circumspectly in light of the micromolar concentrations of melatonin employed) (Aksoy et al. 2003; Anwar and Meki 2003; Derlacz et al. 2007; Ha et al. 1999, 2006; Montilla et al. 1998; Nishida 2005; Paskaloglu et al. 2004; Reyes-Toso et al. 2004; Vural et al. 2001; Winiarska et al. 2006). Most of the work in diabetic rodents has focused on renal structure and function; melatonin administration has been shown to help maintain normal kidney function, while also blunting the effects of diabetes on renal histology and protein expression. Other studies show that melatonin has a favorable impact on the endothelium-dependent vasodilation of diabetic rodents ex vivo (Paskaloglu et al. 2004; Reyes-Toso et al. 2004). In a recent clinical study with type 2 diabetics (from the University of Baghdad), nightly oral melatonin (10 mg, plus 50 mg zinc acetate) was found to decrease macroalbuminuria; favorable effects on glycemic control and serum lipid profile were also noted (Hussain et al. 2006; Kadhim et al. 2006). Other clinical research concludes that oral melatonin has a favorable effect on nocturnal blood pressure in type 1 diabetics (Cavallo et al. 2004a, b). These considerations suggest that melatonin deserves further evaluation as an adjuvant in diabetes management.

Inhibiting Formation of AGE/ALEs and Angiotensin II

Although accelerated de novo synthesis of DAG plays a crucial role in the over-activation of NADPH oxidase in diabetes, other factors also contribute in this regard. As noted above, advanced glycation and lipoxidation endproducts (AGEs and ALEs) trigger activation of NADPH oxidase via RAGE receptors expressed by many tissues, including key targets of diabetic damage such as vascular endothelium, retinal pericytes, and renal podocytes and mesangial cells (Hori et al. 1996; Okamoto et al. 2002; Tanji et al. 2000; Yamamoto et al. 2000). Furthermore, up-regulation of the local renin–angiotensin system (Cooper 2004; Engeli et al. 2003; Gilbert et al. 2003; Ko et al. 2006; Sodhi et al. 2003), reflecting at least in part a PKC-mediated increase in AT1 expression (Holzmeister et al. 1997), also promotes activation of NADPH oxidase via AT1 receptors (Mollnau et al. 2002; Nakayama et al. 2005; Seshiah et al. 2002). Fortunately, nutraceutical strategies are at hand for suppressing the production of AGEs and of angiotensin II.

The increased production of AGEs and ALEs noted in diabetic tissues presumably reflects the joint impact of hyperglycemia and oxidant stress. Major AGEs/ALEs, including carboxymethyl-lysine, carboxyethyl-lysine, and hydroimidazolones, act as potent agonists for RAGE receptor. The formation of AGEs/ALEs serves as an amplification mechanism for oxidant stress—oxidant stress catalyzes the production of these compounds, which in turn interact with RAGE to exacerbate the oxidant stress. Moreover, RAGE is oxidant inducible, reflecting the fact that NF-kappaB promotes transcription of the RAGE gene (Fujita 2006; Li and Schmidt 1997; Yamagishi et al. 2005).

Fortunately, certain compounds can block the formation of AGEs/ALEs by spontaneously forming adducts with highly reactive intermediates—such as alpha-dicarbonyls—that give rise to these toxins. In particular, the vitamer pyridoxamine has excellent efficacy in this regard, and is well tolerated. Studies with supplemental pyridoxamine in diabetic rodents confirm the efficacy of pyridoxamine as an antagonist of AGE/ALE formation in vivo, while providing further evidence that AGEs/ALEs play an important role in the induction of major diabetic complications such as nephropathy, retinopathy, neuropathy, cataracts, and atherosclerosis (Baynes and Thorpe 2000; Cameron et al. 2005; Degenhardt et al. 2002; Heidland et al. 2001; Jain et al. 2002; Metz et al. 2003; Padival and Nagaraj 2006; Stitt et al. 2002; Thomas et al. 2005; Williams 2004; Zheng et al. 2006). Pyridoxamine is effective in diabetic rats when administered in drinking water at 1 g per liter. Phase III studies with pyridoxamine in diabetic nephropathy are currently in progress. Although pyridoxamine is being developed as a drug, it is sold as a nutritional supplement in the United States, as it is a naturally occurring form of vitamin B6.

An agent which may have "sleeper" potential in this regard is the amino acid glycine. Glycine has the potential to inhibit glycation reactions, the earliest stage in the formation of AGEs. Diets highly enriched in glycine have been reported to lower glycated hemoglobin levels both in diabetic rats and diabetic humans, without influencing the underlying hyperglycemia; in the clinical study, glycine was administered in drinking water, 5 g four times daily (Carvajal et al. 1999a, b). These studies did not examine the impact of glycine on AGE production, but glycine supplementation in diabetic rats (1% in drinking water) has been reported to have favorable effects on kidney, lens, and retinal microvasculature (Alvarado-Vasquez et al. 2003, 2006). Glycine may have the potential to more directly inhibit NADPH oxidase activity in certain tissues, possibly because it exerts a hyperpolarizing effect inimical to this activity in cells that express glycine-activated chloride channels (Ikejima et al. 1997; McCarty 2007c, d; Wheeler and Thurman 1999); this effect may underlie the wide-ranging anti-inflammatory effects of high-glycine diets in rodents (Wheeler et al. 1999). Although high intakes of glycine presumably would be required for clinical activity, this amino acid is highly soluble, has a pleasant sweet taste, and is quite inexpensive—ideal qualities for use in a functional food.

Local upregulation of the renin–angiotensin system is believed to contribute to macrovascular, renal, and retinal complications of diabetes, and to promote the conversion of metabolic syndrome to overt diabetes; thus, even though the systemic RAS is typically not elevated in diabetics, drugs which suppress angiotensin II production or activity have been found to have a favorable impact on prognosis in diabetics, even in patients with "normal" blood pressure (Burnier and Zanchi 2006; Deinum and Chaturvedi 2002; Engeli et al. 2003; Gilbert et al. 2003; Hughes and Britton 2005; McFarlane et al. 2003; Strain and Chaturvedi 2002). The utility of these agents for control of diabetic nephropathy is especially well documented, and their effects in this regard are greater than would be expected from reduction of blood pressure per se (Burnier and Zanchi 2006). From the standpoint of developing nutraceuticals and functional foods for diabetics, it is intriguing to note that certain types of protein digests, derived from soy, fish, milk, and other foods, contain oligopeptides which can be absorbed intact and which exert moderate ACE-inhibitory activity in vivo (Cha and Park 2005; Fujita and Yoshikawa 1999; Kuba et al. 2003; Lo and Li-Chan 2005; Marczak et al. 2003; Mizuno et al. 2005; Murakami et al. 2004; Seppo et al. 2003; Yoshikawa et al. 2000).

Nicotinamide for PARP Inhibition

A key role for oxidant-mediated activation of PARP (poly(ADP-ribose) polymerase) in the genesis of diabetic complications has been noted above. In addition to inhibition of glyceraldehyde-3-phosphate dehydrogenase (Du et al. 2003), PARP activation may contribute to cellular dysfunction by inducing NAD+ deficiency; this can compromise ATP generation while also decreasing the availability of NAD(P)H for eNOS and glutathione reductase (albeit for NADPH oxidase as well) (Soriano et al. 2001). Furthermore, PARP serves as a promoter-specific coactivator for NF-kappaB in the transcription of certain proinflammatory genes, including iNOS and TNF-alpha; there is disagreement as to whether PARP activation is required for optimal coactivator activity (Hassa et al. 2005; Nakajima et al. 2004; Zheng et al. 2004). Potent inhibitors of PARP have been shown to have a favorable impact on a range of diabetic complications in rodents, including endothelial dysfunction, neuropathy, nephropathy, and retinopathy (Li et al. 2005; Obrosova et al. 2004a, b; Pacher and Szabo 2005; Piconi et al. 2004; Soriano et al. 2001; Szabo 2005; Szabo et al. 2006).

Unfortunately, potent pharmaceutical PARP inhibitors are not yet available for clinical use. However, nicotinamide is a relatively weak PARP inhibitor that has the twin merits of low toxicity and ready availability (Knip et al. 2000; Stevens et al. 2007). It also inhibits certain mono-ADP-ribosyltransferases, and presumably could aid reconstitution of the NAD+ pool when PARP is active. Very recently, high-dose nicotinamide administration (200–400 mg/kg/day, i.p.) has been reported to have a favorable dose-dependent impact on neuropathy in diabetic rats, improving neural perfusion while attenuating the decline in nerve conduction velocity and preventing hyperalgesia (Stevens et al. 2007). Whether these benefits were mediated by PARP inhibition is not clear. The authors concluded that "nicotinamide deserves consideration as an attractive, nontoxic therapy for diabetic peripheral neuropathy." This report is complemented by a much earlier study which found that niacinamide could slow the progression of nephropathy in diabetic rats (Wahlberg et al. 1985).

High-dose nicotinamide (usually 1.2 g/m (Du et al. 2003) daily, or 1 g tid) has already been evaluated clinically for prevention of beta cell damage in first-degree relatives of patients with type 1 diabetes, as well as for treatment of osteoarthritis; induction of iNOS plays a key role in each of these disorders. Despite initial positive reports, results in diabetes prevention have been inconsistent and largely disappointing (Cabrera-Rode et al. 2006; Gale 1996; Gale et al. 2004; Lampeter et al. 1998; Manna et al. 1992; Olmos et al. 2006), whereas limited data appear to confirm its utility in osteoarthritis (Jonas et al. 1996; McCarty and Russell 1999). It remains uncertain as to whether current clinical high-dose niacinamide therapy achieves tissue concentrations adequate for meaningful inhibition of PARP. The peak plasma concentrations seen with the 1.2 g/m^2 regimen is 100–120 μM; this appears meaningful relative to the IC50 for nicotinamide inhibition of PARP, which is around 30 μM with the purified enzyme (Pociot et al. 1993; Southan and Szabo 2003). Nonetheless, low millimolar concentrations of nicotinamide are required to protect cells from PARP-activating stressors in vitro, and some authorities doubt that feasible clinical intakes of nicotinamide can achieve useful inhibition of PARP (Elliott et al. 1993; Southan and Szabo 2003). Unfortunately, daily intakes in excess of 3 g are prone to induce nausea, and thus are not practical for long-term use (Southan and Szabo 2003). A convenient way to assess the likely utility and dose-dependency of feasible intakes of niacinamide in human diabetics would be to evaluate its impact on diabetic endothelial dysfunction; to date, no such studies have been published. A dose schedule which proved effective in this regard would quite likely have a favorable impact on diabetic complications.

The possibility that high-dose folate may suppress oxidant-mediated PARP activation, by scavenging the peroxynitrite-derived oxidants that attack DNA, has been cited above.

Summing up

These considerations suggest that a nutraceutical regimen composed of PCB, high-dose folate, lipoic acid, taurine, pyridoxamine, and benfotiamine, in appropriate doses, would likely have a substantial favorable impact on the vascular complications of diabetes; PCB, high-dose folate, and taurine may also have potential for slowing the progression of atherosclerosis in patients with normoglycemic metabolic syndrome. Supplementation with vitamin C, selenium, niacinamide, melatonin, and oligopeptide ACE inhibitors might also be of some benefit in these syndromes, whereas there is little present evidence that vitamin E could be helpful. Naturally, these agents could be used in concert with other nutraceuticals that aid glycemic control, or that help to ameliorate major risk factors—for example, hypertension and hyperlipidemia—that can greatly compound the risk associated with diabetes per se.

Why Nutraceuticals?

The particular merit of nutrients or phytonutrients (as opposed to drugs) is that it is feasible to combine a number of them together in a single nutraceutical product or functional food. Moreover, nutraceuticals tend to be cheaper than drugs, and do not entail the inconvenience and expense of requiring a physician's prescription. Finally, nutrients tend to be well tolerated—a consideration that is particularly trenchant when primary prevention is the aim. Thus, it would be highly appropriate to develop comprehensive nutraceutical regimens targeted to prevention of the macro- and microvascular complications of diabetes and metabolic syndrome.

References

Abebe W, Mozaffari MS. 2000. Effects of chronic taurine treatment on reactivity of the rat aorta. Amino Acids 19(3–4):615–623.

Aksoy N, Vural H, Sabuncu T, Aksoy S. 2003. Effects of melatonin on oxidative-antioxidative status of tissues in streptozotocin-induced diabetic rats. Cell Biochem Funct 21(2):121–125.

Ali SM, Chakraborty SK. 1989. Role of plasma ascorbate in diabetic microangiopathy. Bangladesh Med Res Counc Bull 15(2):47–59.

Alp NJ, Mussa S, Khoo J, Cai S, Guzik T, Jefferson A, Goh N, Rockett KA, Channon KM. 2003. Tetrahydrobiopterin-dependent preservation of nitric oxide-mediated endothelial function in diabetes by targeted transgenic GTP-cyclohydrolase I over-expression. J Clin Invest 112(5):725–735.

Alvarado-Vasquez N, Lascurain R, Ceron E, Vanda B, Carvajal-Sandoval G, Tapia A, Guevara J, Montano LF, Zenteno E. 2006. Oral glycine administration attenuates diabetic complications in streptozotocin-induced diabetic rats. Life Sci 79(3):225–232.

Alvarado-Vasquez N, Zamudio P, Ceron E, Vanda B, Zenteno E, Carvajal-Sandoval G. 2003. Effect of glycine in streptozotocin-induced diabetic rats. Comp Biochem Physiol C Toxicol Pharmacol 134(4):521–527.

Anderson RA, Evans LM, Ellis GR, Khan N, Morris K, Jackson SK, Rees A, Lewis MJ, Frenneaux MP. 2006. Prolonged deterioration of endothelial dysfunction in response to postprandial lipaemia is attenuated by vitamin C in Type 2 diabetes. Diabet Med 23(3):258–264.

Antolin I, Rodriguez C, Sainz RM, Mayo JC, Uria H, Kotler ML, Rodriguez-Colunga MJ, Tolivia D, Menendez-Pelaez A. 1996. Neurohormone melatonin prevents cell damage: Effect on gene expression for antioxidant enzymes. FASEB J 10(8):882–890.

Antoniades C, Shirodaria C, Warrick N, Cai S, de BJ, Lee J, Leeson P, Neubauer S, Ratnatunga C, Pillai R, Refsum H, Channon KM. 2006. 5-methyltetrahydrofolate rapidly improves endothelial function and decreases superoxide production in human vessels: Effects on vascular tetrahydrobiopterin availability and endothelial nitric oxide synthase coupling. Circulation 114(11):1193–1201.

Antoniades C, Tousoulis D, Tountas C, Tentolouris C, Toutouza M, Vasiliadou C, Tsioufis C, Toutouzas P, Stefanadis C. 2004. Vascular endothelium and inflammatory process, in patients with combined Type 2 diabetes mellitus and coronary atherosclerosis: The effects of vitamin C. Diabet Med 21(6):552–558.

Anuradha CV, Balakrishnan SD. 1999. Taurine attenuates hypertension and improves insulin sensitivity in the fructose-fed rat, an animal model of insulin resistance. Can J Physiol Pharmacol JID - 0372712 77(10):749–754.

Anwar MM, Meki AR. 2003. Oxidative stress in streptozotocin-induced diabetic rats: Effects of garlic oil and melatonin. Comp Biochem Physiol A Mol Integr Physiol 135(4):539–547.

Azuma J, Sawamura A, Awata N. 1992. Usefulness of taurine in chronic congestive heart failure and its prospective application. Jpn Circ J 56(1):95–99.

Azuma J, Sawamura A, Awata N, Ohta H, Hamaguchi T, Harada H, Takihara K, Hasegawa H, Yamagami T, Ishiyama T. 1985. Therapeutic effect of taurine in congestive heart failure: A double- blind crossover trial. Clin Cardiol 8(5):276–282.

Babaei-Jadidi R, Karachalias N, Ahmed N, Battah S, Thornalley PJ. 2003. Prevention of incipient diabetic nephropathy by high-dose thiamine and benfotiamine. Diabetes 52(8):2110–2120.

Babu PV, Sabitha KE, Shyamaladevi CS. 2006a. Therapeutic effect of green tea extract on advanced glycation and cross-linking of collagen in the aorta of streptozotocin diabetic rats. Clin Exp Pharmacol Physiol 33(4):351–357.

Babu PV, Sabitha KE, Shyamaladevi CS. 2006b. Therapeutic effect of green tea extract on oxidative stress in aorta and heart of streptozotocin diabetic rats. Chem Biol Interact 162(2):114–120.

Baker TA, Milstien S, Katusic ZS. 2001. Effect of vitamin C on the availability of tetrahydrobiopterin in human endothelial cells. J Cardiovasc Pharmacol 37(3):333–338.

Baranano DE, Rao M, Ferris CD, Snyder SH. 2002. Biliverdin reductase: A major physiologic cytoprotectant. Proc Natl Acad Sci USA 99(25):16093–16098.

Barua M, Liu Y, Quinn MR. 2001. Taurine chloramine inhibits inducible nitric oxide synthase and TNF-alpha gene expression in activated alveolar macrophages: Decreased NF-kappaB activation and IkappaB kinase activity. J Immunol 167(4):2275–2281.

Basta G, Lazzerini G, Del Turco S, Ratto GM, Schmidt AM, De Caterina R. July 2005. At least 2 distinct pathways generating reactive oxygen species mediate vascular cell adhesion molecule-1 induction by advanced glycation end products. Arterioscler Thromb Vasc Biol 25(7):1401–1407.

Baynes JW, Thorpe SR. 2000. Glycoxidation and lipoxidation in atherogenesis. Free Radic Biol Med 28(12):1708–1716.

Beale SI. 1994. Biosynthesis of open-chain tetrapyrroles in plants, algae, and cyanobacteria. Ciba Found Symp 180:156–168.

Beckman JA, Goldfine AB, Gordon MB, Garrett LA, Keaney JF, Jr., Creager MA. 2003. Oral antioxidant therapy improves endothelial function in Type 1 but not Type 2 diabetes mellitus. Am J Physiol Heart Circ Physiol 285(6):H2392–H2398.

Beltramo E, Berrone E, Buttiglieri S, Porta M. 2004. Thiamine and benfotiamine prevent increased apoptosis in endothelial cells and pericytes cultured in high glucose. Diabetes Metab Res Rev 20(4):330–336.

Berrone E, Beltramo E, Solimine C, Ubertalli AA, Porta M. 2006. Regulation of intracellular glucose and polyol pathway by thiamine and benfotiamine in vascular cells cultured in high glucose. J Biol Chem 281(4):9307–9313.

Bitsch R, Wolf M, Moller J, Heuzeroth L, Gruneklee D. 1991. Bioavailability assessment of the lipophilic benfotiamine as compared to a water-soluble thiamin derivative. Ann Nutr Metab 35(5):292–296.

Bonaa KH, Njolstad I, Ueland PM, Schirmer H, Tverdal A, Steigen T, Wang H, Nordrehaug JE, Arnesen E, Rasmussen K. 2006. Homocysteine Lowering and Cardiovascular Events after Acute Myocardial Infarction. N Engl J Med 354(15):1578–1588.

Bosma PJ, Chowdhury JR, Bakker C, Gantla S, de Boer A, Oostra BA, Lindhout D, Tytgat GN, Jansen PL, Oude Elferink RP. 1995. The genetic basis of the reduced expression of bilirubin UDP-glucuronosyltransferase 1 in Gilbert's syndrome. N Engl J Med 333(18):1171–1175.

Brusq JM, Ancellin N, Grondin P, Guillard R, Martin S, Saintillan Y, Issandou M. 2006. Inhibition of lipid synthesis through activation of AMP kinase: An additional mechanism for the hypolipidemic effects of berberine. J Lipid Res 47(6):1281–1288.

Burnier M, Zanchi A. January 2006. Blockade of the renin–angiotensin–aldosterone system: A key therapeutic strategy to reduce renal and cardiovascular events in patients with diabetes. J Hypertens 24(1):11–25.

Bursell SE, Clermont AC, Aiello LP, Aiello LM, Schlossman DK, Feener EP, Laffel L, King GL. 1999. High-dose vitamin E supplementation normalizes retinal blood flow and creatinine clearance in patients with type 1 diabetes. Diabetes Care 22(8):1245–1251.

Buse MG. 2006. Hexosamines, insulin resistance, and the complications of diabetes: Current status. Am J Physiol Endocrinol Metab 290(1):E1–E8.

Cabrera-Rode E, Molina G, Arranz C, Vera M, Gonzalez P, Suarez R, Prieto M, Padron S, Leon R, Tillan J, Garcia I, Tiberti C, Rodriguez OM, Gutierrez A, Fernandez T, Govea A, Hernandez J, Chiong D, Dominguez E, Di Mario U, Diaz-Diaz O, Diaz-Horta O. 2006. Effect of standard nicotinamide in the prevention of type 1 diabetes in first degree relatives of persons with type 1 diabetes. Autoimmunity 39(4):333–340.

Cacicedo JM, Benjachareowong S, Chou E, Ruderman NB, Ido Y. 2005. Palmitate-induced apoptosis in cultured bovine retinal pericytes: Roles of NAD(P)H oxidase, oxidant stress, and ceramide. Diabetes 54(6):1838–1845.

Cacicedo JM, Yagihashi N, Keaney JF, Jr, Ruderman NB, Ido Y. 2004. AMPK inhibits fatty acid-induced increases in NF-kappaB transactivation in cultured human umbilical vein endothelial cells. Biochem Biophys Res Commun 324(4):1204–1209.

Cai S, Khoo J, Mussa S, Alp NJ, Channon KM. 2005. Endothelial nitric oxide synthase dysfunction in diabetic mice: Importance of tetrahydrobiopterin in eNOS dimerisation. Diabetologia 48(9):1933–1940.

Cameron NE, Gibson TM, Nangle MR, Cotter MA. 2005. Inhibitors of advanced glycation end product formation and neurovascular dysfunction in experimental diabetes. Ann N Y Acad Sci 1043:784–792.

Carvajal SG, Juarez E, Ramos MG, Carvajal Juarez ME, Medina-Santillan R. 1999a. Inhibition of hemoglobin glycation with glycine in induced diabetes mellitus in rats. Proc West Pharmacol Soc 42:35–36.

Carvajal SG, Medina SR, Juarez E, RamosMartinez G, Carvajal Juarez ME. 1999b. Effect of glycine on hemoglobin glycation in diabetic patients. Proc West Pharmacol Soc 42:31–32.

Cavallo A, Daniels SR, Dolan LM, Bean JA, Khoury JC. 2004a. Blood pressure-lowering effect of melatonin in type 1 diabetes. J Pineal Res 36(4):262–266.

Cavallo A, Daniels SR, Dolan LM, Khoury JC, Bean JA. 2004b. Blood pressure response to melatonin in type 1 diabetes. Pediatr Diabetes 5(1):26–31.

Cha M, Park JR. 2005. Production and characterization of a soy protein-derived angiotensin I-converting enzyme inhibitory hydrolysate. J Med Food 8(3):305–310.

Chakraborty SK. 1992. Plasma ascorbate status in newly diagnosed diabetics exhibiting retinopathy—a finding that alarms. Bangladesh Med Res Counc Bull 18(1):30–35.

Chamorro G, Perez-Albiter M, Serrano-Garcia N, Mares-Samano JJ, Rojas P. 2006. Spirulina maxima pretreatment partially protects against 1-methyl-4-phenyl-1,2,3,6-tetrahydropyridine neurotoxicity. Nutr Neurosci 9(5–6):207–212.

Chandra D, Jackson EB, Ramana KV, Kelley R, Srivastava SK, Bhatnagar A. 2002. Nitric oxide prevents aldose reductase activation and sorbitol accumulation during diabetes. Diabetes 51(10):3095–3101.

Chandra A, Srivastava S, Petrash JM, Bhatnagar A, Srivastava SK. 1997. Active site modification of aldose reductase by nitric oxide donors. Biochim Biophys Acta 1341(2):217–222.

Chen H, Karne RJ, Hall G, Campia U, Panza JA, Cannon RO, III, Wang Y, Katz A, Levine M, Quon MJ. 2006. High-dose oral vitamin C partially replenishes vitamin C levels in patients with Type 2 diabetes and low vitamin C levels but does not improve endothelial dysfunction or insulin resistance. Am J Physiol Heart Circ Physiol 290(1):H137–H145.

Christ M, Bauersachs J, Liebetrau C, Heck M, Gunther A, Wehling M. 2002. Glucose increases endothelial-dependent superoxide formation in coronary arteries by NAD(P)H oxidase activation: Attenuation by the 3-hydroxy-3-methylglutaryl coenzyme A reductase inhibitor atorvastatin. Diabetes 51(8):2648–2652.

Chu KY, Lau T, Carlsson PO, Leung PS. 2006. Angiotensin II type 1 receptor blockade improves beta-cell function and glucose tolerance in a mouse model of type 2 diabetes. Diabetes 55(2):367–374.

Cooke CL, Davidge ST. 2002. Peroxynitrite increases iNOS through NF-kappaB and decreases prostacyclin synthase in endothelial cells. Am J Physiol Cell Physiol 282(2):C395–C402.

Cooper ME. 2004. The role of the renin–angiotensin–aldosterone system in diabetes and its vascular complications. Am J Hypertens 17(11 Pt 2):16S–20S.

Cosentino F, Eto M, De Paolis P, Van Der LB, Bachschmid M, Ullrich V, Kouroedov A, Delli GC, Joch H, Volpe M, Luscher TF. 2003. High glucose causes upregulation of cyclooxygenase-2 and alters prostanoid profile in human endothelial cells: Role of protein kinase C and reactive oxygen species. Circulation 107(7):1017–1023.

Cotter MA, Cameron NE. 2003. Effect of the NAD(P)H oxidase inhibitor, apocynin, on peripheral nerve perfusion and function in diabetic rats. Life Sci 73(14):1813–1824.

Crary EJ, McCarty MF. 1984. Potential clinical applications for high-dose nutritional antioxidants. Med Hypotheses 13(1):77–98.

d'Uscio LV, Milstien S, Richardson D, Smith L, Katusic ZS. 2003. Long-term vitamin C treatment increases vascular tetrahydrobiopterin levels and nitric oxide synthase activity. Circ Res 92(1):88–95.

Darko D, Dornhorst A, Kelly FJ, Ritter JM, Chowienczyk PJ. 2002. Lack of effect of oral vitamin C on blood pressure, oxidative stress and endothelial function in Type II diabetes. Clin Sci (Lond) 103(4):339–344.

Degenhardt TP, Alderson NL, Arrington DD, Beattie RJ, Basgen JM, Steffes MW, Thorpe SR, Baynes JW. 2002. Pyridoxamine inhibits early renal disease and dyslipidemia in the streptozotocin-diabetic rat. Kidney Int 61(3):939–950.

Deinum J, Chaturvedi N. 2002. The Renin–Angiotensin system and vascular disease in diabetes. Semin Vasc Med 2(2):149–156.

Derlacz RA, Sliwinska M, Piekutowska A, Winiarska K, Drozak J, Bryla J. 2007. Melatonin is more effective than taurine and 5-hydroxytryptophan against hyperglycemia-induced kidney-cortex tubules injury. J Pineal Res 42(2):203–209.

Di Leo MA, Ghirlanda G, Gentiloni SN, Giardina B, Franconi F, Santini SA. 2003. Potential therapeutic effect of antioxidants in experimental diabetic retina: A comparison between chronic taurine and vitamin E plus selenium supplementations. Free Radic Res 37(3):323–330.

Di Leo MA, Santini SA, Cercone S, Lepore D, Gentiloni SN, Caputo S, Greco AV, Giardina B, Franconi F, Ghirlanda G. 2002. Chronic taurine supplementation ameliorates oxidative stress and Na+ K+ ATPase impairment in the retina of diabetic rats. Amino Acids 23(4):401–406.

Di Leo MA, Santini SA, Silveri NG, Giardina B, Franconi F, Ghirlanda G. 2004. Long-term taurine supplementation reduces mortality rate in streptozotocin-induced diabetic rats. Amino Acids 27(2):187–191.

Du XL, Edelstein D, Rossetti L, Fantus IG, Goldberg H, Ziyadeh F, Wu J, Brownlee M. 2000. Hyperglycemia-induced mitochondrial superoxide overproduction activates the hexosamine pathway and induces plasminogen activator inhibitor-1 expression by increasing Sp1 glycosylation. Proc Natl Acad Sci U S A 97(22):12222–12226.

Du X, Matsumura T, Edelstein D, Rossetti L, Zsengeller Z, Szabo C, Brownlee M. 2003. Inhibition of GAPDH activity by poly(ADP-ribose) polymerase activates three major pathways of hyperglycemic damage in endothelial cells. J Clin Invest 112(7):1049–1057.

Du X, Stocklauser-Farber K, Rosen P. 1999. Generation of reactive oxygen intermediates, activation of NF-kappaB, and induction of apoptosis in human endothelial cells by glucose: Role of nitric oxide synthase? Free Radic Biol Med 27(7–8):752–763.

Economides PA, Khaodhiar L, Caselli A, Caballero AE, Keenan H, Bursell SE, King GL, Johnstone MT, Horton ES, Veves A. 2005. The effect of vitamin E on endothelial function of micro- and macrocirculation and left ventricular function in type 1 and type 2 diabetic patients. Diabetes 54(1):204–211.

Ek A, Strom K, Cotgreave IA. October 26, 1995. The uptake of ascorbic acid into human umbilical vein endothelial cells and its effect on oxidant insult. Biochem Pharmacol 50(9):1339–1346.

Elliott RB, Pilcher CC, Stewart A, Fergusson D, McGregor MA. November 30, 1993. The use of nicotinamide in the prevention of type 1 diabetes. Ann N Y Acad Sci 696:333–341.

Engeli S, Schling P, Gorzelniak K, Boschmann M, Janke J, Ailhaud G, Teboul M, Massiera F, Sharma AM. June 2003. The adipose-tissue renin–angiotensin–aldosterone system: Role in the metabolic syndrome? Int J Biochem Cell Biol 35(6):807–825.

Exner M, Minar E, Wagner O, Schillinger M. October 15, 2004. The role of heme oxygenase-1 promoter polymorphisms in human disease. Free Radic Biol Med 37(8):1097–1104.

Faure P, Ramon O, Favier A, Halimi S. July 2004. Selenium supplementation decreases nuclear factor-kappa B activity in peripheral blood mononuclear cells from type 2 diabetic patients. Eur J Clin Invest 34(7):475–481.

Fennessy FM, Moneley DS, Wang JH, Kelly CJ, Bouchier-Hayes DJ. January 28, 2003. Taurine and vitamin C modify monocyte and endothelial dysfunction in young smokers. Circulation 107(3):410–415.

Fisher ND, Hughes M, Gerhard-Herman M, Hollenberg NK. December 2003. Flavanol-rich cocoa induces nitric-oxide-dependent vasodilation in healthy humans. J Hypertens 21(12):2281–2286.

Flier J, Van Muiswinkel FL, Jongenelen CA, Drukarch B. June 2002. The neuroprotective antioxidant alpha-lipoic acid induces detoxication enzymes in cultured astroglial cells. Free Radic Res 36(6):695–699.

Fontayne A, Dang PM, Gougerot-Pocidalo MA, el Benna J. June 18, 2002. Phosphorylation of p47phox sites by PKC alpha, beta II, delta, and zeta: Effect on binding to p22phox and on NADPH oxidase activation. Biochemistry 41(24):7743–7750.

Franconi F, Bennardini F, Mattana A, Miceli M, Ciuti M, Mian M, Gironi A, Anichini R, Seghieri G. May 1995. Plasma and platelet taurine are reduced in subjects with insulin-dependent diabetes mellitus: Effects of taurine supplementation. Am J Clin Nutr JID - 0376027 61(5):1115–1119.

Franconi F, Loizzo A, Ghirlanda G, Seghieri G. January 2006. Taurine supplementation and diabetes mellitus. Curr Opin Clin Nutr Metab Care 9(1):32–36.

Franconi F, Miceli M, Bennardini F, Mattana A, Covarrubias J, Seghieri G. 1992. Taurine potentiates the antiaggregatory action of aspirin and indomethacin. Adv Exp Med Biol JID - 0121103 315:181–186.

Fujita T, Ando K, Noda H, Ito Y, Sato Y. March 1987. Effects of increased adrenomedullary activity and taurine in young patients with borderline hypertension. Circulation 75(3):525–532.

Fujita M, Okuda H, Tsukamoto O, Asano Y, Hirata YL, Kim J, Miyatsuka T, Takashima S, Minamino T, Tomoike H, Kitakaze M. October 2006. Blockade of angiotensin II receptors reduces the expression of receptors for advanced glycation end products in human endothelial cells. Arterioscler Thromb Vasc Biol 26(10):e138–e142.

Fujita H, Yoshikawa M. October 15, 1999. LKPNM: A prodrug-type ACE-inhibitory peptide derived from fish protein. Immunopharmacology 44(1–2):123–127.

Fukino Y, Shimbo M, Aoki N, Okubo T, Iso H. October 2005. Randomized controlled trial for an effect of green tea consumption on insulin resistance and inflammation markers. J Nutr Sci Vitaminol (Tokyo) 51(5):335–342.

Furukawa S, Fujita T, Shimabukuro M, Iwaki M, Yamada Y, Nakajima Y, Nakayama O, Makishima M, Matsuda M, Shimomura I. December 2004. Increased oxidative stress in obesity and its impact on metabolic syndrome. J Clin Invest 114(12):1752–1761.

Gadau S, Emanueli C, Van Linthout S, Graiani G, Todaro M, Meloni M, Campesi I, Invernici G, Spillmann F, Ward K, Madeddu P. February 2006. Benfotiamine accelerates the healing of ischaemic diabetic limbs in mice through protein kinase B/Akt-mediated potentiation of angiogenesis and inhibition of apoptosis. Diabetologia 49(2):405–420.

Gale EA. May 1996. Theory and practice of nicotinamide trials in pre-type 1 diabetes. J Pediatr Endocrinol Metab 9(3):375–379.

Gale EA, Bingley PJ, Emmett CL, Collier T. March 20, 2004. European Nicotinamide Diabetes Intervention Trial (ENDIT): A randomised controlled trial of intervention before the onset of type 1 diabetes. Lancet 363(9413):925–931.

Gazis A, White DJ, Page SR, Cockcroft JR. April 1999. Effect of oral vitamin E (alpha-tocopherol) supplementation on vascular endothelial function in Type 2 diabetes mellitus. Diabet Med 16(4):304–311.

Gilbert RE, Krum H, Wilkinson-Berka J, Kelly DJ. August 2003. The renin–angiotensin system and the long-term complications of diabetes: Pathophysiological and therapeutic considerations. Diabet Med 20(8):607–621.

Gokce N, Keaney JF, Jr, Frei B, Holbrook M, Olesiak M, Zachariah BJ, Leeuwenburgh C, Heinecke JW, Vita JA. June 29, 1999. Long-term ascorbic acid administration reverses endothelial vasomotor dysfunction in patients with coronary artery disease. Circulation 99(25):3234–3240.

Grassi D, Necozione S, Lippi C, Croce G, Valeri L, Pasqualetti P, Desideri G, Blumberg JB, Ferri C. August 2005. Cocoa reduces blood pressure and insulin resistance and improves endothelium-dependent vasodilation in hypertensives. Hypertension 46(2):398–405.

Green K, Brand MD, Murphy MP. February 2004. Prevention of mitochondrial oxidative damage as a therapeutic strategy in diabetes. Diabetes 53(Suppl 1):S110–S118.

Ha E, Yim SV, Chung JH, Yoon KS, Kang I, Cho YH, Baik HH. August 2006. Melatonin stimulates glucose transport via insulin receptor substrate-1/phosphatidylinositol 3-kinase pathway in C2C12 murine skeletal muscle cells. J Pineal Res 41(1):67–72.

Ha H, Yu MR, Kim KH. April 1999. Melatonin and taurine reduce early glomerulopathy in diabetic rats. Free Radic Biol Med 26(7–8):944–950.

Hahm JR, Kim BJ, Kim KW. March 2004. Clinical experience with thioctacid (thioctic acid) in the treatment of distal symmetric polyneuropathy in Korean diabetic patients. J Diabetes Complications 18(2):79–85.

Hammes HP, Du X, Edelstein D, Taguchi T, Matsumura T, Ju Q, Lin J, Bierhaus A, Nawroth P, Hannak D, Neumaier M, Bergfeld R, Giardino I, Brownlee M. March 2003. Benfotiamine blocks three major pathways of hyperglycemic damage and prevents experimental diabetic retinopathy. Nat Med 9(3):294–299.

Harada H, Kitazaki K, Tsujino T, Watari Y, Iwata S, Nonaka H, Hayashi T, Takeshita T, Morimoto K, Yokoyama M. May 2000. Oral taurine supplementation prevents the development of ethanol-induced hypertension in rats. Hypertens Res JID - 9307690 23(3):277–284.

Hardeland R. July 2005. Antioxidative protection by melatonin: Multiplicity of mechanisms from radical detoxification to radical avoidance. Endocrine 27(2):119–130.

Hartikainen H. 2005. Biogeochemistry of selenium and its impact on food chain quality and human health. J Trace Elem Med Biol 18(4):309–318.

Hassa PO, Haenni SS, Buerki C, Meier NI, Lane WS, Owen H, Gersbach M, Imhof R, Hottiger MO. December 9, 2005. Acetylation of poly(ADP-ribose) polymerase-1 by p300/CREB-binding protein regulates coactivation of NF-kappaB-dependent transcription. J Biol Chem 280(49):40450–40464.

Haupt E, Ledermann H, Kopcke W. February 2005. Benfotiamine in the treatment of diabetic polyneuropathy—a three-week randomized, controlled pilot study (BEDIP study). Int J Clin Pharmacol Ther 43(2):71–77.

Hayes KC, Pronczuk A, Addesa AE, Stephan ZF. June 1989. Taurine modulates platelet aggregation in cats and humans. Am J Clin Nutr JID - 0376027 49(6):1211–1216.

Heidland A, Sebekova K, Schinzel R. October 2001. Advanced glycation end products and the progressive course of renal disease. Am J Kidney Dis 38(4 Suppl 1):S100–S106.

Heitzer T, Krohn K, Albers S, Meinertz T. November 2000. Tetrahydrobiopterin improves endothelium-dependent vasodilation by increasing nitric oxide activity in patients with Type II diabetes mellitus. Diabetologia 43(11):1435–1438.

Heller R, Munscher-Paulig F, Grabner R, Till U. March 19, 1999. L-Ascorbic acid potentiates nitric oxide synthesis in endothelial cells. J Biol Chem 274(12):8254–8260.

Heller R, Unbehaun A, Schellenberg B, Mayer B, Werner-felmayer G, Werner ER. January 5, 2001. L-ascorbic acid potentiates endothelial nitric oxide synthesis via a chemical stabilization of tetrahydrobiopterin. J Biol Chem 276(1):40–47.

Hollenberg NK, Martinez G, McCullough M, Meinking T, Passan D, Preston M, Rivera A, Taplin D, Vicaria-Clement M. January 1997. Aging, acculturation, salt intake, and hypertension in the Kuna of Panama. Hypertension 29(1 Pt 2):171–176.

Holzmeister J, Graf K, Warnecke C, Fleck E, Regitz-Zagrosek V. August 1997. Protein kinase C-dependent regulation of the human AT1 promoter in vascular smooth muscle cells. Am J Physiol 273(2 Pt 2):H655–H664.

Hori O, Yan SD, Ogawa S, Kuwabara K, Matsumoto M, Stern D, Schmidt AM. 1996. The receptor for advanced glycation end-products has a central role in mediating the effects of advanced glycation end-products on the development of vascular disease in diabetes mellitus. Nephrol Dial Transplant 11(Suppl 5):13–16.

Huang A, Vita JA, Venema RC, Keaney JF, Jr. June 9, 2000. Ascorbic acid enhances endothelial nitric-oxide synthase activity by increasing intracellular tetrahydrobiopterin. J Biol Chem 275(23):17399–17406.

Hughes DB, Britton ML. November 2005. Angiotensin-converting enzyme inhibitors or angiotensin II receptor blockers for prevention and treatment of nephropathy associated with type 2 diabetes mellitus. Pharmacotherapy 25(11):1602–1620.

Hussain SA, Khadim HM, Khalaf BH, Ismail SH, Hussein KI, Sahib AS. October 2006. Effects of melatonin and zinc on glycemic control in type 2 diabetic patients poorly controlled with metformin. Saudi Med J 27(10):1483–1488.

Hyndman ME, Verma S, Rosenfeld RJ, Anderson TJ, Parsons HG. June 2002. Interaction of 5-methyltetrahydrofolate and tetrahydrobiopterin on endothelial function. Am J Physiol Heart Circ Physiol 282(6):H2167–H2172.

Ideishi M, Miura S, Sakai T, Sasaguri M, Misumi Y, Arakawa K. June 1994. Taurine amplifies renal kallikrein and prevents salt-induced hypertension in Dahl rats. J Hypertens JID - 8306882 12(6):653–661.

Ikejima K, Qu W, Stachlewitz RF, Thurman RG. June 1997. Kupffer cells contain a glycine-gated chloride channel. Am J Physiol 272(6 Pt 1):G1581–G1586.

Immenschuh S, Schroder H. June 2006. Heme oxygenase-1 and cardiovascular disease. Histol Histopathol 21(6):679–685.

Inoguchi T, Li P, Umeda F, Yu HY, Kakimoto M, Imamura M, Aoki T, Etoh T, Hashimoto T, Naruse M, Sano H, Utsumi H, Nawata H. November 2000. High glucose level and free fatty acid stimulate reactive oxygen species production through protein kinase

C—dependent activation of NAD(P)H oxidase in cultured vascular cells. Diabetes 49(11):1939–1945.

Inoguchi T, Nawata H. June 2005. NAD(P)H oxidase activation: A potential target mechanism for diabetic vascular complications, progressive beta-cell dysfunction and metabolic syndrome. Curr Drug Targets 6(4):495–501.

Inoguchi T, Sasaki S, Kobayashi K, Takayanagi R, Yamada T. September 26, 2007. Relationship between Gilbert syndrome and prevalence of vascular complications in patients with diabetes. JAMA 298(12):1398–1400.

Inoguchi T, Sonta T, Tsubouchi H, Etoh T, Kakimoto M, Sonoda N, Sato N, Sekiguchi N, Kobayashi K, Sumimoto H, Utsumi H, Nawata H. August 2003. Protein kinase C-dependent increase in reactive oxygen species (ROS) production in vascular tissues of diabetes: Role of vascular NAD(P)H oxidase. J Am Soc Nephrol 14(8 Suppl 3):S227–S232.

Ishii H, Koya D, King GL. January 1998. Protein kinase C activation and its role in the development of vascular complications in diabetes mellitus. J Mol Med 76(1):21–31.

Iso H, Date C, Wakai K, Fukui M, Tamakoshi A. April 18, 2006. The relationship between green tea and total caffeine intake and risk for self-reported type 2 diabetes among Japanese adults. Ann Intern Med 144(8):554–562.

Jaimes EA, Sweeney C, Raij L. October 2001. Effects of the reactive oxygen species hydrogen peroxide and hypochlorite on endothelial nitric oxide production. Hypertension 38(4):877–883.

Jain AK, Lim G, Langford M, Jain SK. December 15, 2002. Effect of high-glucose levels on protein oxidation in cultured lens cells, and in crystalline and albumin solution and its inhibition by vitamin B6 and N-acetylcysteine: Its possible relevance to cataract formation in diabetes. Free Radic Biol Med 33(12):1615–1621.

Jiang C, Ganther H, Lu J. December 2000. Monomethyl selenium—specific inhibition of MMP-2 and VEGF expression: Implications for angiogenic switch regulation. Mol Carcinog 29(4):236–250.

Jiang C, Jiang W, Ip C, Ganther H, Lu J. December 1999. Selenium-induced inhibition of angiogenesis in mammary cancer at chemopreventive levels of intake. Mol Carcinog 26(4):213–225.

Jonas WB, Rapoza CP, Blair WF. July 1996. The effect of niacinamide on osteoarthritis: A pilot study. Inflamm Res 45(7):330–334.

Kadhim HM, Ismail SH, Hussein KI, Bakir IH, Sahib AS, Khalaf BH, Hussain SA. September 2006. Effects of melatonin and zinc on lipid profile and renal function in type 2 diabetic patients poorly controlled with metformin. J Pineal Res 41(2):189–193.

Kalousova M, Novotny L, Zima T, Braun M, Vitek L. September 30, 2005. Decreased levels of advanced glycation end-products in patients with Gilbert syndrome. Cell Mol Biol 51(4):387–392.

Kanayama A, Inoue J, Sugita-Konishi Y, Shimizu M, Miyamoto Y. July 5, 2002. Oxidation of Ikappa Balpha at methionine 45 is one cause of taurine chloramine-induced inhibition of NF-kappa B activation. J Biol Chem 277(27):24049–24056.

Karachalias N, Babaei-Jadidi R, Kupich C, Ahmed N, Thornalley PJ. June 2005. High-dose thiamine therapy counters dyslipidemia and advanced glycation of plasma protein in streptozotocin-induced diabetic rats. Ann N Y Acad Sci 1043:777–783.

Khan SG, Katiyar SK, Agarwal R, Mukhtar H. July 15, 1992. Enhancement of antioxidant and phase II enzymes by oral feeding of green tea polyphenols in drinking water to SKH-1 hairless mice: Possible role in cancer chemoprevention. Cancer Res 52(14):4050–4052.

Khan M, Shobha JC, Mohan IK, Naidu MU, Sundaram C, Singh S, Kuppusamy P, Kutala VK. December 2005. Protective effect of Spirulina against doxorubicin-induced cardiotoxicity. Phytother Res 19(12):1030–1037.

Khan M, Shobha JC, Mohan IK, Rao Naidu MU, Prayag A, Kutala VK. September 2006. Spirulina attenuates cyclosporine-induced nephrotoxicity in rats. J Appl Toxicol 26(5):444–451.

Kim MS, Park JY, Namkoong C, Jang PG, Ryu JW, Song HS, Yun JY, Namgoong IS, Ha J, Park IS, Lee IK, Viollet B, Youn JH, Lee HK, Lee KU. July 2004. Anti-obesity effects of alpha-lipoic acid mediated by suppression of hypothalamic AMP-activated protein kinase. Nat Med 10(7):727–733.

Kim F, Tysseling KA, Rice J, Pham M, Haji L, Gallis BM, Baas AS, Paramsothy P, Giachelli CM, Corson MA, Raines EW. May 2005. Free fatty acid impairment of nitric oxide production in endothelial cells is mediated by IKKbeta. Arterioscler Thromb Vasc Biol 25(5):989–994.

Knekt P, Ritz J, Pereira MA, O'Reilly EJ, Augustsson K, Fraser GE, Goldbourt U, Heitmann BL, Hallmans G, Liu S, Pietinen P, Spiegelman D, Stevens J, Virtamo J, Willett WC, Rimm EB, Ascherio A. December 2004. Antioxidant vitamins and coronary heart disease risk: A pooled analysis of 9 cohorts. Am J Clin Nutr 80(6):1508–1520.

Knip M, Douek IF, Moore WP, Gillmor HA, McLean AE, Bingley PJ, Gale EA. November 2000. Safety of high-dose nicotinamide: A review. Diabetologia 43(11):1337–1345.

Ko SH, Hong OK, Kim JW, Ahn YB, Song KH, Cha BY, Son HY, Kim MJ, Jeong IK, Yoon KH. January 11, 2006. High glucose increases extracellular matrix production in pancreatic stellate cells by activating the renin–angiotensin system. J Cell Biochem 98(2):343–355.

Kohashi N, Okabayashi T, Hama J, Katori R. 1983. Decreased urinary taurine in essential hypertension. Prog Clin Biol Res 125:73–87.

Kondo Y, Murakami S, Oda H, Nagate T. 2000. Taurine reduces atherosclerotic lesion development in apolipoprotein E-deficient mice. Adv Exp Med Biol JID - 0121103 483:193–202.

Kowluru RA, Odenbach S. December 2004. Effect of long-term administration of alpha-lipoic acid on retinal capillary cell death and the development of retinopathy in diabetic rats. Diabetes 53(12):3233–3238.

Koya D, Haneda M, Kikkawa R, King GL. 1998. d-alpha-tocopherol treatment prevents glomerular dysfunctions in diabetic rats through inhibition of protein kinase C-diacylglycerol pathway. Biofactors 7(1–2):69–76.

Koya D, Lee IK, Ishii H, Kanoh H, King GL. March 1997. Prevention of glomerular dysfunction in diabetic rats by treatment with d-alpha-tocopherol. J Am Soc Nephrol 8(3):426–435.

Koyama Y, Abe K, Sano Y, Ishizaki Y, Njelekela M, Shoji Y, Hara Y, Isemura M. November 2004. Effects of green tea on gene expression of hepatic gluconeogenic enzymes in vivo. Planta Med 70(11):1100–1102.

Krebs M, Roden M. November 2005. Molecular mechanisms of lipid-induced insulin resistance in muscle, liver and vasculature. Diabetes Obes Metab 7(6):621–632.

Kuba M, Tanaka K, Tawata S, Takeda Y, Yasuda M. June 2003. Angiotensin I-converting enzyme inhibitory peptides isolated from tofuyo fermented soybean food. Biosci Biotechnol Biochem 67(6):1278–1283.

Kukidome D, Nishikawa T, Sonoda K, Imoto K, Fujisawa K, Yano M, Moto-shima H, Taguchi T, Matsumura T, Araki E. January 2006. Activation of AMP-activated protein kinase reduces hyperglycemia-induced mitochondrial reactive oxygen species production and promotes mitochondrial biogenesis in human umbilical vein endothelial cells. Diabetes 55(1):120–127.

Kunisaki M, Bursell SE, Clermont AC, Ishii H, Ballas LM, Jirousek MR, Umeda F, Nawata H, King GL. August 1995. Vitamin E prevents diabetes-induced abnormal retinal blood flow via the diacylglycerol-protein kinase C pathway. Am J Physiol 269(2 Pt 1):E239–E246.

Kunisaki M, Bursell SE, Umeda F, Nawata H, King GL. November 1994. Normalization of diacylglycerol-protein kinase C activation by vitamin E in aorta of diabetic rats and cultured rat smooth muscle cells exposed to elevated glucose levels. Diabetes 43(11):1372–1377.

Kunisaki M, Bursell SE, Umeda F, Nawata H, King GL. 1998. Prevention of diabetes-induced abnormal retinal blood flow by treatment with d-alpha-tocopherol. Biofactors 7(1–2):55–67.

Kunisaki M, Fumio U, Nawata H, King GL. July 1996. Vitamin E normalizes diacylglycerol-protein kinase C activation induced by hyperglycemia in rat vascular tissues. Diabetes 45(Suppl 3):S117–S119.

Lampeter EF, Klinghammer A, Scherbaum WA, Heinze E, Haastert B, Giani G, Kolb H. June 1998. The Deutsche Nicotinamide Intervention Study: An attempt to prevent type 1 diabetes. DENIS Group. Diabetes 47(6):980–984.

Lanone S, Bloc S, Foresti R, Almolki A, Taille C, Callebert J, Conti M, Goven D, Aubier M, Dureuil B, el Benna J, Motterlini R, Boczkowski J. November 2005. Bilirubin decreases nos2 expression via inhibition of NAD(P)H oxidase: Implications for protection against endotoxic shock in rats. FASEB J 19(13):1890–1892.

Lee YS, Kim WS, Kim KH, Yoon MJ, Cho HJ, Shen Y, Ye JM, Lee CH, Oh WK, Kim CT, Hohnen-Behrens C, Gosby A, Kraegen EW, James DE, Kim JB. August 2006. Berberine, a natural plant product, activates AMP-activated protein kinase with beneficial metabolic effects in diabetic and insulin-resistant states. Diabetes 55(8):2256–2264.

Lee WJ, Lee IK, Kim HS, Kim YM, Koh EH, Won JC, Han SM, Kim MS, Jo I, Oh GT, Park IS, Youn JH, Park SW, Lee KU, Park JY. December 2005a. Alpha-lipoic acid prevents endothelial dysfunction in obese rats via activation of AMP-activated protein kinase. Arterioscler Thromb Vasc Biol 25(12):2488–2494.

Lee WJ, Song KH, Koh EH, Won JC, Kim HS, Park HS, Kim MS, Kim SW, Lee KU, Park JY. July 8, 2005b. Alpha-lipoic acid increases insulin sensitivity by activating AMPK in skeletal muscle. Biochem Biophys Res Commun 332(3):885–891.

Li F, Abatan OI, Kim H, Burnett D, Larkin D, Obrosova IG, Stevens MJ. April 17, 2006. Taurine reverses neurological and neurovascular deficits in Zucker diabetic fatty rats. Neurobiol Dis 22(3):669–676.

Li F, Drel VR, Szabo C, Stevens MJ, Obrosova IG. May 2005. Low-dose poly(ADP-ribose) polymerase inhibitor-containing combination therapies reverse early peripheral diabetic neuropathy. Diabetes 54(5):1514–1522.

Li F, Obrosova IG, Abatan O, Tian D, Larkin D, Stuenkel EL, Stevens MJ. January 2005. Taurine replacement attenuates hyperalgesia and abnormal calcium signaling in sensory neurons of STZ-D rats. Am J Physiol Endocrinol Metab 288(1):E29–E36.

Li J, Schmidt AM. June 27, 1997. Characterization and functional analysis of the promoter of RAGE, the receptor for advanced glycation end products. J Biol Chem 272(26):16498–16506.

Lo WM, Li-Chan EC. May 4, 2005. Angiotensin I converting enzyme inhibitory peptides from in vitro pepsin-pancreatin digestion of soy protein. J Agric Food Chem 53(9):3369–3376.

Loew D. February 1996. Pharmacokinetics of thiamine derivatives especially of benfotiamine. Int J Clin Pharmacol Ther 34(2):47–50.

Lonn E, Yusuf S, Hoogwerf B, Pogue J, Yi Q, Zinman B, Bosch J, Dagenais G, Mann JF, Gerstein HC. November 2002. Effects of vitamin E on cardiovascular and microvascular outcomes in high-risk patients with diabetes: Results of the HOPE study and MICRO-HOPE substudy. Diabetes Care 25(11):1919–1927.

Mangoni AA, Sherwood RA, Asonganyi B, Swift CG, Thomas S, Jackson SH. February 2005. Short-term oral folic acid supplementation enhances endothelial function in patients with type 2 diabetes. Am J Hypertens 18(2 Pt 1):220–226.

Manna R, Migliore A, Martin LS, Ferrara E, Ponte E, Marietti G, Scuderi F, Cristiano G, Ghirlanda G, Gambassi G. 1992. Nicotinamide treatment in subjects at high risk of developing IDDM improves insulin secretion. Br J Clin Pract 46(3):177–179.

Marczak ED, Usui H, Fujita H, Yang Y, Yokoo M, Lipkowski AW, Yoshikawa M. June 2003. New antihypertensive peptides isolated from rapeseed. Peptides 24(6):791–798.

Mayer M. November 2000. Association of serum bilirubin concentration with risk of coronary artery disease. Clin Chem 46(11):1723–1727.

Mayo JC, Sainz RM, Antoli I, Herrera F, Martin V, Rodriguez C. October 2002. Melatonin regulation of antioxidant enzyme gene expression. Cell Mol Life Sci 59(10):1706–1713.

McCarty MF. October 1999. The reported clinical utility of taurine in ischemic disorders may reflect a down-regulation of neutrophil activation and adhesion. Med Hypotheses 53(4):290–299.

McCarty MF. 2004a. Coping with endothelial superoxide: Potential complementarity of arginine and high-dose folate. Med Hypotheses 63(4):709–718.

McCarty MF. 2004b. Supplementary taurine may stabilize atheromatous plaque by antagonizing the activation of metalloproteinases by hypochlorous acid. Med Hypotheses 63(3):414–418.

McCarty MF. 2005a. AMPK activation as a strategy for reversing the endothelial lipotoxicity underlying the increased vascular risk associated with insulin resistance syndrome. Med Hypotheses 64(6):1211–1215.

McCarty MF. 2005b. The putative therapeutic value of high-dose selenium in proliferative retinopathies may reflect down-regulation of VEGF production by the hypoxic retina. Med Hypotheses 64(1):159–161.

McCarty MF. 2006. Adjuvant strategies for prevention of glomerulosclerosis. Med Hypotheses 67(6):1277–1296.

McCarty MF. 2007a. "Iatrogenic Gilbert syndrome"—a strategy for reducing vascular and cancer risk by increasing plasma unconjugated bilirubin. Med Hypotheses 69(5):974–994.

McCarty MF. 2007b. Oster rediscovered—mega-dose folate for treatment of symptomatic atherosclerosis. Med Hypotheses 69(2):325–332.

McCarty MF. 2007c. High-dose folate may have clinical potential as a scavenger of peroxynitrite-derived radicals. Med Hypotheses, submitted for publication.

McCarty MF. 2007d. The hyperpolarizing impact of glycine on endothelial cells may be anti-atherogenic. Med Hypotheses, in press.

McCarty MF, Russell AL. October 1999. Niacinamide therapy for osteoarthritis—does it inhibit nitric oxide synthase induction by interleukin 1 in chondrocytes? Med Hypotheses 53(4):350–360.

McFarlane SI, Kumar A, Sowers JR. June 2003. Mechanisms by which angiotensin-converting enzyme inhibitors prevent diabetes and cardiovascular disease. Am J Cardiol 91(12A):30H–37H.

Metz TO, Alderson NL, Thorpe SR, Baynes JW. November 1, 2003. Pyridoxamine, an inhibitor of advanced glycation and lipoxidation reactions: A novel therapy for treatment of diabetic complications. Arch Biochem Biophys 419(1):41–49.

Milstien S, Katusic Z. October 5, 1999. Oxidation of tetrahydrobiopterin by peroxynitrite: Implications for vascular endothelial function. Biochem Biophys Res Commun 263(3):681–684.

Mizuno S, Matsuura K, Gotou T, Nishimura S, Kajimoto O, Yabune M, Kajimoto Y, Yamamoto N. July 2005. Antihypertensive effect of casein hydrolysate in a placebo-controlled study in subjects with high-normal blood pressure and mild hypertension. Br J Nutr 94(1):84–91.

Moat SJ, Clarke ZL, Madhavan AK, Lewis MJ, Lang D. January 20, 2006. Folic acid reverses endothelial dysfunction induced by inhibition of tetrahydrobiopterin biosynthesis. Eur J Pharmacol 530(3):250–258.

Mohan IK, Khan M, Shobha JC, Naidu MU, Prayag A, Kuppusamy P, Kutala VK. December 2006. Protection against cisplatin-induced nephrotoxicity by Spirulina in rats. Cancer Chemother Pharmacol 58(6):802–808.

Mollnau H, Wendt M, Szocs K, Lassegue B, Schulz E, Oelze M, Li H, Bodenschatz M, August M, Kleschyov AL, Tsilimingas N, Walter U, Forstermann U, Meinertz T, Griendling K, Munzel T. March 8, 2002. Effects of angiotensin II infusion on the expression and function of NAD(P)H oxidase and components of nitric oxide/cGMP signaling. Circ Res 90(4):E58–E65.

Montilla PL, Vargas JF, Tunez IF, Munoz de Agueda MC, Valdelvira ME, Cabrera ES. September 1998. Oxidative stress in diabetic rats induced by streptozotocin: Protective effects of melatonin. J Pineal Res 25(2):94–100.

Morcos M, Borcea V, Isermann B, Gehrke S, Ehret T, Henkels M, Schiekofer S, Hofmann M, Amiral J, Tritschler H, Ziegler R, Wahl P, Nawroth PP. June 2001. Effect of alpha-lipoic acid on the progression of endothelial cell damage and albuminuria in patients with diabetes mellitus: An exploratory study. Diabetes Res Clin Pract 52(3):175–183.

Morita T. September 2005. Heme oxygenase and atherosclerosis. Arterioscler Thromb Vasc Biol 25(9):1786–1795.

Murakami S, Kondo Y, Sakurai T, Kitajima H, Nagate T. July 2002. Taurine suppresses development of atherosclerosis in Watanabe heritable hyperlipidemic (WHHL) rabbits. Atherosclerosis JID - 0242543 163(1):79–87.

Murakami M, Tonouchi H, Takahashi R, Kitazawa H, Kawai Y, Negishi H, Saito T. July 2004. Structural analysis of a new anti-hypertensive peptide (beta-lactosin B) isolated from a commercial whey product. J Dairy Sci 87(7):1967–1974.

Nakajima H, Nagaso H, Kakui N, Ishikawa M, Hiranuma T, Hoshiko S. October 8, 2004. Critical role of the automodification of poly(ADP-ribose) polymerase-1 in nuclear factor-kappaB-dependent gene expression in primary cultured mouse glial cells. J Biol Chem 279(41):42774–42786.

Nakamura T, Ushiyama C, Suzuki S, Shimada N, Ohmuro H, Ebihara I, Koide H. 1999. Effects of taurine and vitamin E on microalbuminuria, plasma metalloproteinase-9, and serum type IV collagen concentrations in patients with diabetic nephropathy. Nephron 83(4):361–362.

Nakayama M, Inoguchi T, Sonta T, Maeda Y, Sasaki S, Sawada F, Tsubouchi H, Sonoda N, Kobayashi K, Sumimoto H, Nawata H. July 15, 2005. Increased expression of NAD(P)H oxidase in islets of animal models of Type 2 diabetes and its improvement by an AT1 receptor antagonist. Biochem Biophys Res Commun 332(4):927–933.

Nara Y, Yamori Y, Lovenberg W. 1978. Effect of dietary taurine on blood pressure in spontaneously hypertensive rats. Biochem Pharmacol JID - 0101032 27(23):2689–2692.

Negrisanu G, Rosu M, Bolte B, Lefter D, Dabelea D. July 1999. Effects of 3-month treatment with the antioxidant alpha-lipoic acid in diabetic peripheral neuropathy. Rom J Intern Med 37(3):297–306.

Ni YX. December 1988. Therapeutic effect of berberine on 60 patients with type II diabetes mellitus and experimental research. Zhong Xi Yi Jie He Za Zhi 8(12):711–713, 707.

Nishida S. July 2005. Metabolic effects of melatonin on oxidative stress and diabetes mellitus. Endocrine 27(2):131–136.

Nishikawa T, Edelstein D, Du XL, Yamagishi S, Matsumura T, Kaneda Y, Yorek MA, Beebe D, Oates PJ, Hammes HP, Giardino I, Brownlee M. April 13, 2000. Normalizing mitochondrial superoxide production blocks three pathways of hyperglycaemic damage. Nature 404(6779):787–790.

Nordberg J, Arner ES. December 1, 2001. Reactive oxygen species, antioxidants, and the mammalian thioredoxin system. Free Radic Biol Med 31(11):1287–1312.

Obrosova IG, Fathallah L, Stevens MJ. November 2001a. Taurine counteracts oxidative stress and nerve growth factor deficit in early experimental diabetic neuropathy. Exp Neurol 172(1):211–219.

Obrosova IG, Li F, Abatan OI, Forsell MA, Komjati K, Pacher P, Szabo C, Stevens MJ. March 2004a. Role of poly(ADP-ribose) polymerase activation in diabetic neuropathy. Diabetes 53(3):711–720.

Obrosova IG, Mabley JG, Zsengeller Z, Charniauskaya T, Abatan OI, Groves JT, Szabo C. March 2005. Role for nitrosative stress in diabetic neuropathy: Evidence from studies with a peroxynitrite decomposition catalyst. FASEB J 19(3):401–403.

Obrosova IG, Minchenko AG, Frank RN, Seigel GM, Zsengeller Z, Pacher P, Stevens MJ, Szabo C. July 2004b. Poly(ADP-ribose) polymerase inhibitors counteract diabetes- and hypoxia-induced retinal vascular endothelial growth factor overexpression. Int J Mol Med 14(1):55–64.

Obrosova IG, Minchenko AG, Marinescu V, Fathallah L, Kennedy A, Stockert CM, Frank RN, Stevens MJ. September 2001b. Antioxidants attenuate early up regulation of retinal vascular endothelial growth factor in streptozotocin-diabetic rats. Diabetologia 44(9):1102–1110.

Ogborne RM, Rushworth SA, O'Connell MA. October 2005. Alpha-lipoic acid-induced heme oxygenase-1 expression is mediated by nuclear factor erythroid 2-related factor 2 and p38 mitogen-activated protein kinase in human monocytic cells. Arterioscler Thromb Vasc Biol 25(10):2100–2105.

Okamoto T, Yamagishi S, Inagaki Y, Amano S, Koga K, Abe R, Takeuchi M, Ohno S, Yoshimura A, Makita Z. December 2002. Angiogenesis induced by advanced glycation end products and its prevention by cerivastatin. FASEB J 16(14):1928–1930.

Okumura M, Masada M, Yoshida Y, Shintaku H, Hosoi M, Okada N, Konishi Y, Morikawa T, Miura K, Imanishi M. August 2006. Decrease in tetrahydrobiopterin as a possible cause of nephropathy in type II diabetic rats. Kidney Int 70(3):471–476.

Olmos PR, Hodgson MI, Maiz A, Manrique M, De Valdes MD, Foncea R, Acosta AM, Emmerich MV, Velasco S, Muniz OP, Oyarzun CA, Claro JC, Bastias MJ, Toro LA. March 2006. Nicotinamide protected first-phase insulin response (FPIR) and prevented clinical disease in first-degree relatives of type-1 diabetics. Diabetes Res Clin Pract 71(3):320–333.

Oster KA. October 1968. Treatment of angina pectoris according to a new theory of its origin. Cardiol Dig 3:29–34. Oster KA, Ross DJ, Dawkins HHR. 1983. The XO Factor. New York: Park City Press.

Ouslimani N, Peynet J, Bonnefont-Rousselot D, Therond P, Legrand A, Beaudeux JL. June 2005. Metformin decreases intracellular production of reactive oxygen species in aortic endothelial cells. Metabolism 54(6):829–834.

Pacher P, Szabo C. November 2005. Role of poly(ADP-ribose) polymerase-1 activation in the pathogenesis of diabetic complications: Endothelial dysfunction, as a common underlying theme. Antioxid Redox Signal 7(11–12):1568–1580.

Pacher P, Szabo C. April 2006. Role of peroxynitrite in the pathogenesis of cardiovascular complications of diabetes. Curr Opin Pharmacol 6(2):136–141.

Padival S, Nagaraj RH. 2006. Pyridoxamine inhibits maillard reactions in diabetic rat lenses. Ophthalmic Res 38(5):294–302.

Pannirselvam M, Simon V, Verma S, Anderson T, Triggle CR. October 2003. Chronic oral supplementation with sepiapterin prevents endothelial dysfunction and oxidative stress in small mesenteric arteries from diabetic (db/db) mice. Br J Pharmacol 140(4):701–706.

Paskaloglu K, Sener G, Ayangolu-Dulger G. September 24, 2004. Melatonin treatment protects against diabetes-induced functional and biochemical changes in rat aorta and corpus cavernosum. Eur J Pharmacol 499(3):345–354.

Pekovich SR, Martin PR, Singleton CK. April 1998. Thiamine deficiency decreases steady-state transketolase and pyruvate dehydrogenase but not alpha-ketoglutarate dehydrogenase mRNA levels in three human cell types. J Nutr 128(4):683–687.

Pena AS, Wiltshire E, Gent R, Hirte C, Couper J. April 2004. Folic acid improves endothelial function in children and adolescents with type 1 diabetes. J Pediatr 144(4):500–504.

Petty MA, Kintz J, DiFrancesco GF. May 3, 1990. The effects of taurine on atherosclerosis development in cholesterol-fed rabbits. Eur J Pharmacol 180(1):119–127.

Piconi L, Quagliaro L, Da Ros R, Assaloni R, Giugliano D, Esposito K, Szabo C, Ceriello A. August 2004. Intermittent high glucose enhances ICAM-1, VCAM-1, E-selectin and interleukin-6 expression in human umbilical endothelial cells in culture: The role of poly(ADP-ribose) polymerase. J Thromb Haemost 2(8):1453–1459.

Pieper GM. January 1997. Acute amelioration of diabetic endothelial dysfunction with a derivative of the nitric oxide synthase cofactor, tetrahydrobiopterin. J Cardiovasc Pharmacol 29(1):8–15.

Pociot F, Reimers JI, Andersen HU. June 1993. Nicotinamide—biological actions and therapeutic potential in diabetes prevention. IDIG Workshop, Copenhagen, Denmark, December 4–5, 1992. Diabetologia 36(6):574–576.

Pomero F, Molinar MA, La Selva M, Allione A, Molinatti GM, Porta M. 2001. Benfotiamine is similar to thiamine in correcting endothelial cell defects induced by high glucose. Acta Diabetol 38(3):135–138.

Pop-Busui R, Sullivan KA, Van Huysen C, Bayer L, Cao X, Towns R, Stevens MJ. April 2001. Depletion of taurine in experimental diabetic neuropathy: Implications for nerve metabolic, vascular, and functional deficits. Exp Neurol 168(2):259–272.

Prabhakar SS. July 2001. Tetrahydrobiopterin reverses the inhibition of nitric oxide by high glucose in cultured murine mesangial cells. Am J Physiol Renal Physiol 281(1):F179–F188.

Pueyo ME, Gonzalez W, Nicoletti A, Savoie F, Arnal JF, Michel JB. March 2000. Angiotensin II stimulates endothelial vascular cell adhesion molecule-1 via nuclear factor-kappaB activation induced by intracellular oxidative stress. Arterioscler Thromb Vasc Biol 20(3):645–651.

Ramana KV, Chandra D, Srivastava S, Bhatnagar A, Srivastava SK. March 2003. Nitric oxide regulates the polyol pathway of glucose metabolism in vascular smooth muscle cells. FASEB J 17(3):417–425.

Rask-Madsen C, King GL. March 2005. Proatherosclerotic mechanisms involving protein kinase C in diabetes and insulin resistance. Arterioscler Thromb Vasc Biol 25(3):487–496.

Rasool M, Sabina EP, Lavanya B. December 2006. Anti-inflammatory effect of Spirulina fusiformis on adjuvant-induced arthritis in mice. Biol Pharm Bull 29(12):2483–2487.

Reiter RJ, Tan DX, Cabrera J, D'Arpa D, Sainz RM, Mayo JC, Ramos S. January 1999. The oxidant/antioxidant network: Role of melatonin. Biol Signals Recept 8(1–2):56–63.

Reljanovic M, Reichel G, Rett K, Lobisch M, Schuette K, Moller W, Tritschler HJ, Mehnert H. September 1999. Treatment of diabetic polyneuropathy with the antioxidant thioctic acid (alpha-lipoic acid): A two year multicenter randomized double-blind placebo-controlled trial (ALADIN II). Alpha Lipoic Acid in Diabetic Neuropathy. Free Radic Res 31(3):171–179.

Remirez D, Gonzalez R, Merino N, Rodriguez S, Ancheta O. April 2002. Inhibitory effects of Spirulina in zymosan-induced arthritis in mice. Mediators Inflamm 11(2):75–79.

Reyes-Toso CF, Linares LM, Ricci CR, Aran M, Pinto JE, Rodriguez RR, Cardinali DP. May 7, 2004. Effect of melatonin on vascular reactivity in pancreatectomized rats. Life Sci 74(25):3085–3092.

Rezk BM, Haenen GR, Van dV, Bast A. December 18, 2003. Tetrahydrofolate and 5-methyltetrahydrofolate are folates with high antioxidant activity. Identification of the antioxidant pharmacophore. FEBS Lett 555(3):601–605.

Rigato I, Ostrow JD, Tiribelli C. June 2005. Bilirubin and the risk of common non-hepatic diseases. Trends Mol Med 11(6):277–283.

Rindi G, Laforenza U. September 2000. Thiamine intestinal transport and related issues: Recent aspects. Proc Soc Exp Biol Med 224(4):246–255.

Rodriguez C, Mayo JC, Sainz RM, Antolin I, Herrera F, Martin V, Reiter RJ. January 2004. Regulation of antioxidant enzymes: A significant role for melatonin. J Pineal Res 36(1):1–9.

Romay C, Gonzalez R, Ledon N, Remirez D, Rimbau V. June 2003. C-phycocyanin: A biliprotein with antioxidant, anti-inflammatory and neuroprotective effects. Curr Protein Pept Sci 4(3):207–216.

Ruderman NB, Cacicedo JM, Itani S, Yagihashi N, Saha AK, Ye JM, Chen K, Zou M, Carling D, Boden G, Cohen RA, Keaney J, Kraegen EW, Ido Y. February 2003. Malonyl-CoA and AMP-activated protein kinase (AMPK): Possible links between insulin resistance in muscle and early endothelial cell damage in diabetes. Biochem Soc Trans 31(Pt 1):202–206.

Ryu OH, Lee J, Lee KW, Kim HY, Seo JA, Kim SG, Kim NH, Baik SH, Choi DS, Choi KM. March 2006. Effects of green tea consumption on inflammation, insulin resistance and pulse wave velocity in type 2 diabetes patients. Diabetes Res Clin Pract 71(3):356–358.

Sato Y, Ogata E, Fujita T. May 1991. Hypotensive action of taurine in DOCA-salt rats—involvement of sympathoadrenal inhibition and endogenous opiate. Jpn Circ J JID-7806868 55(5):500–508.

Satoh M, Fujimoto S, Haruna Y, Arakawa S, Horike H, Komai N, Sasaki T, Tsujioka K, Makino H, Kashihara N. June 2005. NAD(P)H oxidase and uncoupled nitric oxide synthase are major sources of glomerular superoxide in rats with experimental diabetic nephropathy. Am J Physiol Renal Physiol 288(6):F1144–F1152.

Schreeb KH, Freudenthaler S, Vormfelde SV, Gundert-Remy U, Gleiter CH. 1997. Comparative bioavailability of two vitamin B1 preparations: Benfotiamine and thiamine mononitrate. Eur J Clin Pharmacol 52(4):319–320.

Schroeter H, Heiss C, Balzer J, Kleinbongard P, Keen CL, Hollenberg NK, Sies H, Kwik-Uribe C, Schmitz HH, Kelm M. January 24, 2006. (-)-Epicatechin mediates beneficial effects of flavanol-rich cocoa on vascular function in humans. Proc Natl Acad Sci U S A 103(4):1024–1029.

Schuller-Levis GB, Park E. September 26, 2003. Taurine: New implications for an old amino acid. FEMS Microbiol Lett 226(2):195–202.

Sedlak TW, Snyder SH. June 2004. Bilirubin benefits: Cellular protection by a biliverdin reductase antioxidant cycle. Pediatrics 113(6):1776–1782.

Seppo L, Jauhiainen T, Poussa T, Korpela R. February 2003. A fermented milk high in bioactive peptides has a blood pressure-lowering effect in hypertensive subjects. Am J Clin Nutr 77(2):326–330.

Seshiah PN, Weber DS, Rocic P, Valppu L, Taniyama Y, Griendling KK. September 6, 2002. Angiotensin II stimulation of NAD(P)H oxidase activity: Upstream mediators. Circ Res 91(5):406–413.

Shankar SS, Steinberg HO. February 2005. FFAs: Do they play a role in vascular disease in the insulin resistance syndrome? Curr Diab Rep 5(1):30–35.

Shao J, Iwashita N, Ikeda F, Ogihara T, Uchida T, Shimizu T, Uchino H, Hirose T, Kawamori R, Watada H. June 16, 2006. Beneficial effects of candesartan, an angiotensin II type 1 receptor blocker, on beta-cell function and morphology in db/db mice. Biochem Biophys Res Commun 344(4):1224–1233.

Sherman DL, Keaney JF, Jr, Biegelsen ES, Duffy SJ, Coffman JD, Vita JA. April 2000. Pharmacological concentrations of ascorbic acid are required for the beneficial effect on endothelial vasomotor function in hypertension. Hypertension 35(4):936–941.

Shinozaki K, Kashiwagi A, Masada M, Okamura T. March 2004. Molecular mechanisms of impaired endothelial function associated with insulin resistance. Curr Drug Targets Cardiovasc Haematol Disord 4(1):1–11.

Shinozaki K, Kashiwagi A, Nishio Y, Okamura T, Yoshida Y, Masada M, Toda N, Kikkawa R. December 1999. Abnormal biopterin metabolism is a major cause of impaired endothelium-dependent relaxation through nitric oxide/O2- imbalance in insulin-resistant rat aorta. Diabetes 48(12):2437–2445.

Shinozaki K, Nishio Y, Okamura T, Yoshida Y, Maegawa H, Kojima H, Masada M, Toda N, Kikkawa R, Kashiwagi A. September 29, 2000. Oral administration of tetrahydrobiopterin prevents endothelial dysfunction and vascular oxidative stress in the aortas of insulin-resistant rats. Circ Res 87(7):566–573.

Shinozaki K, Nishio Y, Yoshida Y, Koya D, Ayajiki K, Masada M, Kashiwagi A, Okamura T. October 2005. Supplement of tetrahydrobiopterin by a gene transfer of GTP cyclohydrolase I cDNA improves vascular dysfunction in insulin-resistant rats. J Cardiovasc Pharmacol 46(4):505–512.

Sies H, Klotz LO, Sharov VS, Assmann A, Briviba K. March 1998. Protection against peroxynitrite by selenoproteins. Z Naturforsch [C] 53(3–4):228–232.

Sinclair AJ, Girling AJ, Gray L, Le Guen C, Lunec J, Barnett AH. March 1991. Disturbed handling of ascorbic acid in diabetic patients with and without microangiopathy during high dose ascorbate supplementation. Diabetologia 34(3):171–175.

Siu B, Saha J, Smoyer WE, Sullivan KA, Brosius FC, III. March 15, 2006. Reduction in podocyte density as a pathologic feature in early diabetic nephropathy in rodents: Prevention by lipoic acid treatment. BMC Nephrol 7:6.

Skyrme-Jones RA, O'Brien RC, Berry KL, Meredith IT. July 2000. Vitamin E supplementation improves endothelial function in type I diabetes mellitus: A randomized, placebo-controlled study. J Am Coll Cardiol 36(1):94–102.

Sodhi CP, Kanwar YS, Sahai A. March 2003. Hypoxia and high glucose upregulate AT1 receptor expression and potentiate ANG II-induced proliferation in VSM cells. Am J Physiol Heart Circ Physiol 284(3):H846–H852.

Sohn HY, Krotz F, Zahler S, Gloe T, Keller M, Theisen K, Schiele TM, Klauss V, Pohl U. March 2003. Crucial role of local peroxynitrite formation in neutrophil-induced endothelial cell activation. Cardiovasc Res 57(3):804–815.

Soriano FG, Pacher P, Mabley J, Liaudet L, Szabo C. October 12, 2001. Rapid reversal of the diabetic endothelial dysfunction by pharmacological inhibition of poly(ADP-ribose) polymerase. Circ Res 89(8):684–691.

Southan GJ, Szabo C. February 2003. Poly(ADP-ribose) polymerase inhibitors. Curr Med Chem 10(4):321–340.

Stadtman TC. 2000. Selenium biochemistry. Mammalian selenoenzymes. Ann N Y Acad Sci 899:399–402.

Stevens MJ, Li F, Drel VR, Abatan OI, Kim H, Burnett D, Larkin D, Obrosova IG. January 2007. Nicotinamide reverses neurological and neurovascular deficits in streptozotocin diabetic rats. J Pharmacol Exp Ther 320(1):458–464.

Stitt A, Gardiner TA, Alderson NL, Canning P, Frizzell N, Duffy N, Boyle C, Januszewski AS, Chachich M, Baynes JW, Thorpe SR. September 2002. The AGE inhibitor pyridoxamine inhibits development of retinopathy in experimental diabetes. Diabetes 51(9):2826–2832.

Stocker R. October 2004. Antioxidant activities of bile pigments. Antioxid Redox Signal 6(5):841–849.

Storz P, Toker A. January 2, 2003. Protein kinase D mediates a stress-induced NF-kappaB activation and survival pathway. EMBO J 22(1):109–120.

Stracke H, Hammes HP, Werkmann D, Mavrakis K, Bitsch I, Netzel M, Geyer J, Kopcke W, Sauerland C, Bretzel RG, Federlin KF. 2001. Efficacy of benfotiamine versus thiamine on function and glycation products of peripheral nerves in diabetic rats. Exp Clin Endocrinol Diabetes 109(6):330–336.

Stracke H, Lindemann A, Federlin K. 1996. A benfotiamine-vitamin B combination in treatment of diabetic polyneuropathy. Exp Clin Endocrinol Diabetes 104(4):311–316.

Strain WD, Chaturvedi N. December 2002. The renin–angiotensin–aldosterone system and the eye in diabetes. J Renin Angiotensin Aldosterone Syst 3(4):243–246.

Stroes ES, van Faassen EE, Yo M, Martasek P, Boer P, Govers R, Rabelink TJ. June 9, 2000. Folic acid reverts dysfunction of endothelial nitric oxide synthase. Circ Res 86(11):1129–1134.

Suh JH, Shenvi SV, Dixon BM, Liu H, Jaiswal AK, Liu RM, Hagen TM. March 9, 2004. Decline in transcriptional activity of Nrf2 causes age-related loss of glutathione synthesis, which is reversible with lipoic acid. Proc Natl Acad Sci U S A 101(10):3381–3386.

Suzen S, Buyukbingol E. August 2003. Recent studies of aldose reductase enzyme inhibition for diabetic complications. Curr Med Chem 10(15):1329–1352.

Szabo C. July 2005. Roles of poly(ADP-ribose) polymerase activation in the pathogenesis of diabetes mellitus and its complications. Pharmacol Res 52(1):60–71.

Szabo C, Biser A, Benko R, Bottinger E, Susztak K. November 2006. Poly(ADP-ribose) polymerase inhibitors ameliorate nephropathy of type 2 diabetic Leprdb/db mice. Diabetes 55(11):3004–3012.

Takeuchi T, Jung EH, Nishino K, Itokawa Y. 1990. The relationship between the thiamin pyrophosphate effect and the saturation status of the transketolase with its coenzyme in human erythrocytes. Int J Vitam Nutr Res 60(2):112–120.

Talior I, Tennenbaum T, Kuroki T, Eldar-Finkelman H. February 2005. PKC-delta-dependent activation of oxidative stress in adipocytes of obese and insulin-resistant mice: Role for NADPH oxidase. Am J Physiol Endocrinol Metab 288(2):E405–E411.

Tanji N, Markowitz GS, Fu C, Kislinger T, Taguchi A, Pischetsrieder M, Stern D, Schmidt AM, D'Agati VD. September 2000. Expression of advanced glycation end products and their cellular receptor RAGE in diabetic nephropathy and nondiabetic renal disease. J Am Soc Nephrol 11(9):1656–1666.

Taubert D, Berkels R, Roesen R, Klaus W. August 27, 2003. Chocolate and blood pressure in elderly individuals with isolated systolic hypertension. JAMA 290(8):1029–1030.

Tawakol A, Migrino RQ, Aziz KS, Waitkowska J, Holmvang G, Alpert NM, Muller JE, Fischman AJ, Gewirtz H. May 17, 2005. High-dose folic acid acutely improves coronary vasodilator function in patients with coronary artery disease. J Am Coll Cardiol 45(10):1580–1584.

Terry MJ, Maines MD, Lagarias JC. December 15, 1993. Inactivation of phytochrome- and phycobiliprotein-chromophore precursors by rat liver biliverdin reductase. J Biol Chem 268(35):26099–26106.

Thomas MC, Baynes JW, Thorpe SR, Cooper ME. June 2005. The role of AGEs and AGE inhibitors in diabetic cardiovascular disease. Curr Drug Targets 6(4):453–474.

Timimi FK, Ting HH, Haley EA, Roddy MA, Ganz P, Creager MA. March 1, 1998. Vitamin C improves endothelium-dependent vasodilation in patients with insulin-dependent diabetes mellitus. J Am Coll Cardiol 31(3):552–557.

Ting HH, Timimi FK, Boles KS, Creager SJ, Ganz P, Creager MA. January 1, 1996. Vitamin C improves endothelium-dependent vasodilation in patients with non-insulin-dependent diabetes mellitus. J Clin Invest 97(1):22–28.

Tomas-Zapico C, Coto-Montes A. September 2005. A proposed mechanism to explain the stimulatory effect of melatonin on antioxidative enzymes. J Pineal Res 39(2):99–104.

Tousoulis D, Antoniades C, Vasiliadou C, Kourtellaris P, Koniari K, Marinou K, Charakida M, Ntarladimas I, Siasos G, Stefanadis C. August 16, 2006. Effects of atorvastatin and vitamin C on forearm hyperaemic blood flow, asymmentrical dimethylarginine levels and inflammatory process, in patients with type 2 diabetes mellitus. Heart 93:244–246.

Tsuneki H, Ishizuka M, Terasawa M, Wu JB, Sasaoka T, Kimura I. August 26, 2004. Effect of green tea on blood glucose levels and serum proteomic patterns in diabetic (db/db) mice and on glucose metabolism in healthy humans. BMC Pharmacol 4:18.

UK Prospective Diabetes Study (UKPDS) Group. 1998a. Intensive blood-glucose control with sulphonylureas or insulin compared with conventional treatment and risk of complications in patients with type 2 diabetes (UKPDS 33). Lancet 352(9131):837–853.

UK Prospective Diabetes Study (UKPDS) Group. 1998b. Effect of intensive blood-glucose control with metformin on complications in overweight patients with type 2 diabetes (UKPDS 34) . Lancet 352(9131):854–865.

Valle I, Alvarez-Barrientos A, Arza E, Lamas S, Monsalve M. June 1, 2005. PGC-1alpha regulates the mitochondrial antioxidant defense system in vascular endothelial cells. Cardiovasc Res 66(3):562–573.

Van Etten RW, De Koning EJ, Verhaar MC, Gaillard CA, Rabelink TJ. June 2002. Impaired NO-dependent vasodilation in patients with Type II (non-insulin-dependent) diabetes mellitus is restored by acute administration of folate. Diabetologia JID - 0006777 45(7):1004–1010.

Vinson JA, Zhang J. May 4, 2005. Black and green teas equally inhibit diabetic cataracts in a streptozotocin-induced rat model of diabetes. J Agric Food Chem 53(9):3710–3713.

Vitek L, Jirsa M, Brodanova M, Kalab M, Marecek Z, Danzig V, Novotny L, Kotal P. February 2002. Gilbert syndrome and ischemic heart disease: A protective effect of elevated bilirubin levels. Atherosclerosis 160(2):449–456.

Vitek L, Novotny L, Sperl M, Holaj R, Spacil J. March 9, 2006. The inverse association of elevated serum bilirubin levels with subclinical carotid atherosclerosis. Cerebrovasc Dis 21(5–6):408–414.

Vural H, Sabuncu T, Arslan SO, Aksoy N. October 2001. Melatonin inhibits lipid peroxidation and stimulates the antioxidant status of diabetic rats. J Pineal Res 31(3):193–198.

Wahlberg G, Carlson LA, Wasserman J, Ljungqvist A. November 1985. Protective effect of nicotinamide against nephropathy in diabetic rats. Diabetes Res 2(6):307–312.

Wautier MP, Chappey O, Corda S, Stern DM, Schmidt AM, Wautier JL. May 2001. Activation of NADPH oxidase by AGE links oxidant stress to altered gene expression via RAGE. Am J Physiol Endocrinol Metab 280(5):E685–E694.

Wautier JL, Schmidt AM. August 6, 2004. Protein glycation: A firm link to endothelial cell dysfunction. Circ Res 95(3):233–238.

Way KJ, Katai N, King GL. December 2001. Protein kinase C and the development of diabetic vascular complications. Diabet Med 18(12):945–959.

Weidig P, McMaster D, Bayraktutan U. November 2004. High glucose mediates pro-oxidant and antioxidant enzyme activities in coronary endothelial cells. Diabetes Obes Metab 6(6):432–441.

Wheeler MD, Ikejema K, Enomoto N, Stacklewitz RF, Seabra V, Zhong Z, Yin M, Schemmer P, Rose ML, Rusyn I, Bradford B, Thurman RG. November 30, 1999. Glycine: A new anti-inflammatory immunonutrient. Cell Mol Life Sci 56(9–10):843–856.

Wheeler MD, Thurman RG. November 1999. Production of superoxide and TNF-alpha from alveolar macrophages is blunted by glycine. Am J Physiol 277(5 Pt 1):L952–L959.

Williams ME. December 2004. Clinical studies of advanced glycation end product inhibitors and diabetic kidney disease. Curr Diab Rep 4(6):441–446.

Winiarska K, Fraczyk T, Malinska D, Drozak J, Bryla J. March 2006. Melatonin attenuates diabetes-induced oxidative stress in rabbits. J Pineal Res 40(2):168–176.

Winkler G, Pal B, Nagybeganyi E, Ory I, Porochnavec M, Kempler P. March 1999. Effectiveness of different benfotiamine dosage regimens in the treatment of painful diabetic neuropathy. Arzneimittelforschung 49(3):220–224.

Wolfram S, Raederstorff D, Preller M, Wang Y, Teixeira SR, Riegger C, Weber P. October 2006. Epigallocatechin gallate supplementation alleviates diabetes in rodents. J Nutr 136(10):2512–2518.

Wright CE, Lin TT, Lin YY, Sturman JA, Gaull GE. 1985. Taurine scavenges oxidized chlorine in biological systems. Prog Clin Biol Res 179:137–147.

Wu LY, Juan CC, Hwang LS, Hsu YP, Ho PH, Ho LT. April 2004. Green tea supplementation ameliorates insulin resistance and increases glucose transporter IV content in a fructose-fed rat model. Eur J Nutr 43(2):116–124.

Wu S, Ren J. February 13, 2006. Benfotiamine alleviates diabetes-induced cerebral oxidative damage independent of advanced glycation end-product, tissue factor and TNF-alpha. Neurosci Lett 394(2):158–162.

Xia L, Wang H, Goldberg HJ, Munk S, Fantus IG, Whiteside CI. February 2006. Mesangial cell NADPH oxidase upregulation in high glucose is protein kinase C dependent and required for collagen IV expression. Am J Physiol Renal Physiol 290(2):F345–F356.

Yamabe N, Yokozawa T, Oya T, Kim M. October 2006. Therapeutic potential of (-)-epigallocatechin 3-O-gallate on renal damage in diabetic nephropathy model rats. J Pharmacol Exp Ther 319(1):228–236.

Yamagishi S, Takeuchi M, Matsui T, Nakamura K, Imaizumi T, Inoue H. August 15, 2005. Angiotensin II augments advanced glycation end product-induced pericyte apoptosis through RAGE overexpression. FEBS Lett 579(20):4265–4270.

Yamamoto Y, Yamagishi S, Yonekura H, Doi T, Tsuji H, Kato I, Takasawa S, Okamoto H, Abedin J, Tanaka N, Sakurai S, Migita H, Unoki H, Wang H, Zenda T, Wu PS, Segawa Y, Higashide T, Kawasaki K, Yamamoto H. May 2000. Roles of the AGE-RAGE system in vascular injury in diabetes. Ann N Y Acad Sci 902:163–170.

Yamori Y, Nara Y, Ikeda K, Mizushima S. 1996. Is taurine a preventive nutritional factor of cardiovascular diseases or just a biological marker of nutrition? Adv Exp Med Biol 403:623–629.

Yano M, Hasegawa G, Ishii M, Yamasaki M, Fukui M, Nakamura N, Yoshikawa T. 2004. Short-term exposure of high glucose concentration induces generation of reactive oxygen species in endothelial cells: Implication for the oxidative stress associated with postprandial hyperglycemia. Redox Rep 9(2):111–116.

Yokozawa T, Nakagawa T, Oya T, Okubo T, Juneja LR. June 2005. Green tea polyphenols and dietary fibre protect against kidney damage in rats with diabetic nephropathy. J Pharm Pharmacol 57(6):773–780.

Yoshikawa M, Fujita H, Matoba N, Takenaka Y, Yamamoto T, Yamauchi R, Tsuruki H, Takahata K. 2000. Bioactive peptides derived from food proteins preventing lifestyle-related diseases. Biofactors 12(1–4):143–146.

Zeng XH, Zeng XJ, Li YY. July 15, 2003. Efficacy and safety of berberine for congestive heart failure secondary to ischemic or idiopathic dilated cardiomyopathy. Am J Cardiol 92(2):173–176.

Zhang C, Yang J, Jennings LK. November 2004. Leukocyte-derived myeloperoxidase amplifies high-glucose—induced endothelial dysfunction through interaction with high-glucose—stimulated, vascular non—leukocyte-derived reactive oxygen species. Diabetes 53(11):2950–2959.

Zheng L, Szabo C, Kern TS. November 2004. Poly(ADP-ribose) polymerase is involved in the development of diabetic retinopathy via regulation of nuclear factor-kappaB. Diabetes 53(11):2960–2967.

Zheng F, Zeng YJ, Plati AR, Elliot SJ, Berho M, Potier M, Striker LJ, Striker GE. August 2006. Combined AGE inhibition and ACEi decreases the progression of established diabetic nephropathy in B6 db/db mice. Kidney Int 70(3):507–514.

Ziegler D. 2004. Thioctic acid for patients with symptomatic diabetic polyneuropathy: A critical review. Treat Endocrinol 3(3):173–189.

Ziegler D, Hanefeld M, Ruhnau KJ, Hasche H, Lobisch M, Schutte K, Kerum G, Malessa R. August 1999. Treatment of symptomatic diabetic polyneuropathy with the antioxidant alpha-lipoic acid: A 7-month multicenter randomized controlled trial (ALADIN III Study). ALADIN III Study Group. Alpha-Lipoic Acid in Diabetic Neuropathy. Diabetes Care 22(8):1296–1301.

Ziegler D, Nowak H, Kempler P, Vargha P, Low PA. February 2004. Treatment of symptomatic diabetic polyneuropathy with the antioxidant alpha-lipoic acid: A meta-analysis. Diabet Med 21(2):114–121.

Chapter 12

Ginseng in Type 2 Diabetes Mellitus: A Review of the Evidence in Humans

John L Sievenpiper, MD, PhD, Alexandra L Jenkins, RD, PhD,
Anamarie Dascalu, MD, MSc, P. Mark Stavro, PhD, and
Vladimir Vuksan, PhD

Introduction

There is a clear and present need for more effective medications to control type 2 diabetes. This is reflected in the five new classes of medications that have been approved for diabetes since metformin was last approved in the United States in 1995 (Nathan 2006). Despite this growing armamentarium of medications, diabetes management has become one of the most important unmet treatment challenges. Trends in the prevalence of obesity and diabetes suggest that the burden has reached epidemic proportions affecting nearly 10% of adults in North America with a rate of growth that has outpaced earlier projections in the United States (Mokdad et al. 2003) and Canada (Lipscombe and Hux 2007). Progression of the disease has been no better attenuated. Irrespective of medication or treatment policy, glycemic control has been shown to worsen over time with eventual beta-cell fatigue, (Turner et al. 1998). Monotherapy failure with medications has been shown to be more than 75% at 9 years (Turner et al. 1999). Medication side effects and complications have further compromised outcomes. A recent meta-analysis showed that rosiglitazone, one of the more durable treatments (Kahn et al. 2006), increased myocardial infarction and cardiovascular death (Nissen and Wolski 2007).

Complementary and alternative medicine (CAM) is flourishing in this treatment vacuum (Eisenberg et al. 1998). Estimates suggest that 75% of general medical patients in the United States are now using CAM therapies (Winslow and Shapiro 2002). It has been reported that adults who do not use medications to treat their diabetes, coronary artery disease (CAD), stroke, hypertension, or high cholesterol are more likely to use herbs and other supplements (Buettner et al. 2007). A strong independent determinant of this behavior is the use of CAM to treat diabetes. Over a third of North Americans with diabetes report using CAM (Yeh et al. 2002). Despite this consumer endorsement, there has been a paucity of clinical efficacy and safety evidence to support diabetes indications for herbs. The biguanide metformin remains the only example of an approved drug that was derived from an herb (French lilac (*Galega officinalis*)).

Following a concerted call in the late 1990s by medical journal editors and commentators for more rigorous controlled clinical investigations to evaluate health claims attributed to CAM treatments (Angell and Kassirer 1998), a database of randomized (RCTs) and nonrandomized clinical trials has begun to emerge to support the use of specific herbs in diabetes. A 2003 systematic review of 42 randomized and 16 nonrandomized clinical trials (Yeh et al. 2003) showed that the herbs with a minimum of supporting clinical data in diabetes included ginseng (*Panax* spp.), ivy gourd (*Coccinia Indica*), garlic (*Allium sativum and Allium cepa*), holy basil (*Ocimum sanctum*), fenugreek (*Trigonella foenum graecum*), prickly pear cactus (*Opuntia streptacantha*), milk thistle (*Silibum marianum*), fig leaf (*Ficus carica*), gurmar (*Gymnema sylvestre*), bitter melon (*Momordica charantia*), *Aloe vera*, *Gingko biloba*, and various herb combinations in Traditional Chinese, Native American, and Tibetan medicine. American ginseng (*Panax quinquefolius* L.) and ivy gourd were found to have the best evidence from adequately designed RCTs to support clinical efficacy in diabetes (Yeh et al. 2003).

Despite having some of the best evidence for efficacy among herbs, ginseng has remained largely untested in diabetes. Before the millennium, there was only a small group of flawed and poorly reported studies in humans to support the antidiabetic effects of ginseng reported anecdotally over the past 2,000 years. At the same time, the direction and magnitude of effects seen in in vitro and animal models were also quite variable for different types of ginseng (*Panax* and non-*Panax* spp.) and their active components, represented chiefly as ginsenosides. This state of the research raised several important questions. It was not clear whether glycemic lowering effects were reproducible across different ginseng sources and protocols and whether ginsenosides, for which the most compelling evidence exists for biological activity, were mediating the glycemic effects. We and other investigators have begun to address these issues over the last decade.

This chapter addresses the current state of the clinical research supporting the use of ginseng in the management of diabetes. It starts with an introduction to the variability in ginseng source and composition. It then proceeds to a review of the evidence for our *American and Korean Red Ginseng Clinical Testing Programs*, emphasizing the need for reproducible ginseng sources. This extends into an exploration of ginseng source factors that affect the regulation of glycemia and other features of metabolism in humans. Present and future research evaluating the effect of specific ginseng components on glycemia in humans is then considered. Finally, as a means of explaining outcomes, possible mechanism of action of ginseng and its components are discussed.

Variability in Ginseng Source and Composition

Whole Ginseng Sources

Ginseng is an herb derived from several species of the plant family Araliaceae and genus *Panax* indigenous both to Asia and North America. It is usually consumed as dried powder in capsules, but is also available as a gel, tea, or tincture. Thirteen distinct species of ginseng have been identified with numerous different cultivars (Figure 12.1). The two most popular species of ginseng are American and Asian ginsengs. Other commercial species include Korean red (steamed *Panax ginseng* C.A. Meyer), Japanese (*Panax japonicus* CA Meyer), Sanchi, Vietnamese (*Panax vietnamensis*), and the non-*Panax* species Siberian ginseng. American ginseng is indigenous to Ontario, Quebec, British Columbia, and Wisconsin, while the other species are indigenous to Asia. All are genetically distinct (Mihalov et al. 2000).

Saponins (Ginsenosides)

Various components of ginseng have been shown to have pharmacological activity. These include its saponin (Attele et al. 1999), polysaccharide (Ng and Yeung 1985), and peptide fractions (Kajiwara et al. 1996). Of these components, most of the pharmacological action of ginseng is attributed to its saponins, referred to as ginsenosides (Attele et al. 1999). These compounds belong principally to a family of steroids called dammarane-type triterpene glycosides. Other classes include the acidic oleanane-type triterpene glycosides and the octillilol-type glycosides. Techniques for their measurement include largely high performance liquid chromatography (HPLC) techniques. Various other techniques have also been described. These techniques and some of the most common dammarane-type triterpene glycoside type ginsenosides identified are shown in Figure 12.1.

Steroidal Ginsenosides

Five distinguishing features characterize the >30 steroidal or dammarane-type triterpene glycosidal ginsenosides. These include (1) a four *trans*-ring rigid steroid skeleton (the aglycone) which shares similarities with the classical steroid hormones; (2) hydroxyl (–OH) groups which give rise to the two major classes, the protopanaxadiol (PPD) ginsenosides with two –OH groups (C-3, C-20) and protopanaxatriol (PPT) ginsenosides with three –OH groups (C-3, C-6, C-20) (Attele et al. 1999); (3) sugar moieties bound to these –OH groups, the type and number of which give rise to the different individual ginsenosides in each class (Attele et al. 1999); (4) malonyl groups at the terminus of the sugar residue at C-3 in some PPDs, giving rise to the four malonic acid derivatives called malonyl ginsenosides (m): m-Rb$_1$, m-Rb$_2$, m-RC, and m-Rd (Li et al. 2000; Awang 2000); and (5) stereochemistry differences occurring at C-20 giving rise to *S* and *R* isomers, although most ginsenosides are naturally present as enantiomeric mixtures (Attele et al. 1999). Together these distinguishing characteristics allow for

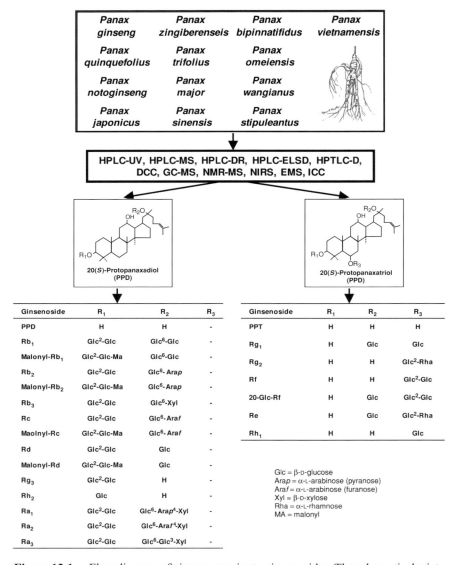

Figure 12.1. Flow diagram of ginseng species to ginsenosides. The schematic depicts the 13 *Panax* species of ginseng identified to date, the analytical techniques applied to these ginseng sources, the two main classes of dammarane-type triterpene glycosides derived from these analyses, and the most common individual ginsenosides belonging to each class. HPLC-UV denotes high performance liquid chromatography-ultraviolet; HPLC-MS, high performance liquid chromatography-mass spectroscopy; HPLC-DR, high performance liquid chromatography-differential refractometry; HPLC-ELSD, high performance liquid chromatography-electrospray light scattering detection;

248

considerable structural variation, affecting their potency, pharmacokinetic profile, and biological activity (Attele et al. 1999; Li et al. 2000).

Interspecies Differences in Ginsenosides

Several differences in the profile of these steroidal ginsenosides occur between species (Figure 12.1). First, the ginsenoside content has been shown to be different. For example, Sanchi and American ginseng have been shown to have higher total ginsenosides than that of Asian ginseng, while Siberian ginseng does not contain ginsenosides (Attele et al. 1999).

Second, some ginsenosides are absent in various species. American ginseng does not contain ginsenosides Rf (Awang 2000; Li et al. 2000) and Rg_2 (Awang 2000), Asian ginseng does not contain the ocotillol-type dammarane saponin $24(R)$-pseudoginsenoside F-11 (Awang 2000; Li et al. 2000), and red Asian ginseng does not contain malonyl ginsenosides (Li et al. 2000).

Third, the ratios of various ginsenoside classes and common individual ginsenosides will differ between species. A ratio smaller than one of the ginsenoside ratios Rg_1/Re and Rb_2/Rc has been shown to distinguish American ginseng from Asian ginseng (Wang et al. 1999). Other ratio differences are those between Rb_1 and Rg_1 and those between neutral and malonyl ginsenosides. In both Asian and Sanchi ginseng, the ratio of $Rb_1:Rg_1$ is usually from 1 to 3, while in American ginseng it is 3–10 (Awang 2000; Li et al. 1996; Ma et al. 1996). The proportion of malonyl ginsenosides is also smaller in Asian ginseng ($\sim$10% of total) than in American ginseng ($\sim$40% of total) (Awang 2000).

Finally, some ginsenosides are unique to a species. Vietnamese ginseng, for example, has had two novel acetylated ocotillol-type dammarane saponins and seven novel dammarane-type triterpene glycosides identified (Nguyen et al. 1994). Both American and Sanchi ginseng have also had numerous novel ginsenosides isolated.

Together these differences are used in the authentication of ginseng and may contribute to a high degree of variability in pharmacology between species.

Intraspecies Differences in Ginsenosides Are Very Important

The same differential applies within species. Wild-Asian ginseng has been shown to have >1.5-fold higher total ginsenosides than its cultivated counterpart

---

Figure 12.1. (*continued*) GC-MS, gas chromatography-mass spectroscopy; DCC, droplet counter-current chromatography; HPTLC-D, high performance thin layer chromatography desnsitometry; LC-EMS, liquid Chromatography-electrospray MS; NMR-MS, Nuclear magnetic resonance-MS; NIRS, near-infrared reflectance spectroscopy; and ICC, immunoaffinity column chromatography. (Adapted from Vuksan and Sievenpiper (2005).)

(Yamaguchi et al. 1988a). Cultivated older Asian ginseng has been shown to have higher total ginsenosides than younger Asian ginseng, with the largest increases (as much as 5-fold) coming between the 4th and 6th years of growth (Soldati and Tanaka 1984). Differences have also been shown between ginsengs harvested from different locations. The CV of the total and main PPD and PPT ginsenosides for American ginseng calculated for nine British Columbia farms ranged from $\sim$16 to 40% (Li et al. 1996). There are even differences seen within the same plant. For example, the leaves of American ginseng have been shown to have $\sim$1.4-fold higher total ginsenosides and >8-fold smaller $Rb_1:Rg_1$ ratio than the roots from the same plant (Li et al. 1996). Ginsenoside differences have also been observed among root parts. For example, the rootlets of Asian and American ginseng were both found to have >3-fold higher total ginsenosides than the main root body, with the rootlets of Asian ginseng also found to have >2-fold higher PPD:PPT ratio and >6-fold higher $Rb_1:Rg_1$ ratio). In a similar analysis, the fine rootlets of American ginseng were found to have >5-fold more total ginsenosides and >1.5-fold more PPD:PPT and $Rb_1:Rg_1$ ratios (Ren and Chen 1999). These differences may again contribute to a high degree of variability in pharmacology within species.

Ginsenoside Degradation Products

Biomodification of ginsenosides produces additional ginsenoside species and degradation products. In vitro, animal and human models have shown that ginsenosides are degraded in the proximal gut by gastric pH and then in the distal gut by microfloral enzymes. For example, PPT ginsenosides through hydrolytic cleavage of sugar moieties can be modified either to form ginsenosides Rh_1 and Rg_2 and their hydrated species by the mild acidic conditions of the stomach or to ginsenoside F_1 or Rg_1 in the case of Re by colonic bacteria (Tawab et al. 2003). On the other hand, PPD ginsenosides appear to undergo very little modification in the stomach. But these ginsenosides through a stepwise cleavage of their sugar moieties can be transformed to ginsenoside Rd, Rg_3, PPD, or compound K by bacteria in the colon (Hasegawa et al. 1996). The formation of these different compounds increases the already diverse pharmacokinetic and pharmacodynamic possibilities for ginsenosides measured in the plant.

Ginsenoside in Vivo Pharmacokinetics

An important consideration in mediating the effect of ginsenosides and their degradation products is their ability to reach systemic circulation and the rate with which this occurs. Differences in absorption, distribution, metabolism, and excretion of ginsenosides and their degradation products have been observed. Pharmacokinetic and pharmacodynamic studies show that ginsenosides in American ginseng reach systemic circulation with a half-life of less than 8 h in the rabbit (Chen et al. 1980) and a time-course of effects not lasting beyond 24 h in the rat (Yokozawa et al. 1984).

Others have investigated PPT and PPD ginsenosides in the plasma and urine of humans post consumption of various ginseng containing products (Tawab et al. 2003; Wang et al. 1999). Estimates from these studies show that the pharmacokinetic time course of ginsenosides is quite variable. The PPT ginsenosides Rg_1 and Re and PPD ginsenosides Rb_2, Rc, Rd, and Rg_3 appear to be absorbed, metabolized, distributed, and excreted quickly with appearance in urine of humans 0–3 h post consumption (Tawab et al. 2003; Wang et al. 1999). The t_{max} for Rg_3 was 0.66 ± 0.1 h with no detection beyond 10 h (Wang et al. 1999). Intermediate in its time course is Rh_1 with an appearance in plasma within $\sim$40–200 min and elimination in urine within 3–6 h. Ginseoside Rb_1 and colonic degradation products, such as compound K, and ginsenosides F have a longer time course, appearing within 6–12 h and are eliminated within 6–24 h (Tawab et al. 2003). Absorption of ginsenosides is estimated to be <5% (Tawab et al. 2003; Wang et al. 1999). The relative handling of these different ginsenosides and their degradation products may help to explain their pharmacodynamic time course of effects and biological variability.

Polysaccharides and Peptidoglycans

Polysaccharides (ginsenans) and peptidoglycans (panaxans/quinquefolans/eleutherans) have also received some study. Various polysaccharides have been isolated by NMR and GC/MS analyses. They contain neutral sugars (glucose, rhamnose, arabinose, galactose) and acidic sugars (galacturonic acid, glucuronic acid) in different molar ratios (Tomoda et al. 1993, 1994). These ratios define ginsenans S-IA, SII-A, and PA. Various homogenous peptidoglycans have also been isolated from ginseng. The peptidoglycan fraction of ginseng is referred to as quinquefolans for American ginseng, panaxans for the other *Panax* species of ginseng and eleutherans for *Eleutherococcus senticosus*. Although the structure of these molecules is not known, their compositions have been reported. Hydrolysis and TLC analyses have shown that these molecules are composed of 1–5% amino acids including histidine, leucine, alanine, tryptophan, glycine, aspartic acid, serine, and threonine in different molar ratios, while the rest is composed of polysaccharide, including various neutral sugars (glucose, rhamnose, arabinose, galactose, xylose) and acidic sugars (galacturonic acid, glucuronic acid) in different molar ratios (Ng and Yeung 1985). The different peptidoglycans defined by these ratios include panaxans A–U from Asian ginseng, eleutherans A–G from Siberian ginseng, and quinquefolans A–C from American ginseng.

The in vivo effects of these compounds have only been shown when administered intraperitoneally (Ng and Yeung 1985. Whether the main route of administration for commercial products, oral administration, will result in the same effects is uncertain. The main reason for the uncertainty is that proteolytic enzymes secreted by the gut and by colonic microflora would likely degrade these components to their base sugar and amino acid units, preventing their reaching the circulation intact.

Neither pharmacokinetic nor pharmacodynamic data from oral administration of these polysaccharide or peptidoglycan compounds are available to support or refute these possibilities. There are, nevertheless, data that suggest that fractional amounts of intact peptides can be absorbed (Yamamoto 2001), making their interaction in effects a possibility.

Standardization of Ginseng

Ginseng Standardization Remains Controversial

Ginsenosides, the principal markers of pharmacological activity and basis for ginseng standardization (Cui 1995; Fitzloff et al. 1998), suffer from high variability and a lack of consensus regarding adequate assay criteria. To quantify the extent of ginsenoside variability and the role of the assay, we undertook a meta-analysis. The coefficient-of-variation (CV) in ginsenosides was assessed across ginseng type (batch, preparation, variety, species), assay technique, and ginsenoside type (Sievenpiper et al. 2004a). Thirty-two articles met the inclusion criteria. Together these articles reported ginsenoside concentrations for 317 batches of ginseng, in which ginseng type comprised 10 levels of *Panax* species, their preparations, and their varieties; assay type comprised six levels of different assay techniques; and ginsenoside type comprised 21 levels of ginsenoside indices including individual ginsenosides and their classes, sums, and ratios. The meta-analysis demonstrated a high CV in ginsenosides across the three main factors: 26–103% for ginseng type, 31–81% for assay technique, and 36–112% for ginsenoside type with the differences in ginseng type dependent on the assay type. This analysis demonstrated that the ginsenoside composition is highly variable across different ginseng source parameters (Sievenpiper et al. 2004a). It was concluded that the implication of this high variability in ginsenosides is that its efficacy and safety in diabetes may be equally highly variable. This may even apply to ginsenoside standardized products, as it is not clear which ginsenosides should be standardized for different indications. In the absence of adequate assay criteria to assess ginsenosides and data that tie specific ginsenoside and nonginsenoside components (ginsenans, panaxans, quinquefolans, eleutherans, peptides, degradation products, etc.) to its effects, the attainment of reproducible and sustainable efficacy and safety for specific indications is uncertain.

These limitations prompted us to develop clinical screening models with sequential screening steps to select batches and treatment protocols (dosing, timing, and modes of administration) that achieve reproducible and sustainable efficacy from the acute to the long-term in type 2 diabetes. It also prompted us to explore ginseng source factors such as species, variety, batch, and preparation that affect the reproducibility of glycemic results and identify active ginsenoside and nonginsenoside components with the eventual goal of defining specific component profiles for diabetes indications.

American Ginseng Clinical Testing Program

To test the hypothesis that a source of ginseng selected and administered appropriately can produce reproducible and sustainable glycemic reductions, we initiated an *American Ginseng Clinical Testing Program* in 1997 (Vuksan and Sievenpiper 2005). The main aim of the program was to explore the "acute-to-chronic" effects of American ginseng in humans, as depicted in Figures 12.2 and 12.3. The program was conducted in two phases: an acute phase and a long-term phase. We started by conducting a stepwise series of 5, acute, single-bolus, single-blind, placebo-controlled, multiple-crossover, randomized controlled trials (RCTs) to evaluate the efficacy of a single batch of American ginseng root provided by *Chai-Na-Ta Corp.*, BC, Canada (source #1). The ginseng was given as single doses from 1 to 9 g with times of administration from 0 to -120 min preprandially in subjects with normoglycemia and subjects with type 2 diabetes using a 25 g oral glucose tolerance test (OGTT) protocol. These studies represented "Phase-I" type batch-, dose-, time-finding studies, in which each level of batch, dose, and time of administration were crossed in a randomized block design. The batch, treatment protocol, and ginsenoside profile information gained from these acute studies was then applied to a long-term, double-blind, RCT in people with type 2 diabetes.

The first two acute, "Phase I," RCTs (Vuksan et al. 2000a; Vuksan and Sievenpiper 2000) assessed the efficacy of American ginseng root (source #1) taken as a single 3 g dose given orally either 0 min or -40 min preprandially in lowering the glycemic response to a 25 g OGTT in 10 normoglycemic (Figure 12.2, panel 1) and nine type 2 diabetic subjects (Figure 12.3, panel 1). Our rationale for the dose and times was derived ethnopharmacologically. A starting dose of 3 g was chosen, as herbs are treated much like diluted drugs with usual daily doses of approximately 10 g and minimum daily doses of 3 g in traditional systems of Chinese medicine. Modes of administration that included preprandial dosing schedules were investigated, as ginseng is usually taken fasting or between meals. Reductions in the incremental glycemic response to the 25 g OGTT were seen when the 3 g of the selected batch of American ginseng was administered both -40 min and 0 min preprandially in the type 2 diabetic subjects but only when administered -40 min preprandially in the normoglycemic subjects. There was no significant reduction in postprandial incremental glycemia, when the 3 g of the selected batch of American ginseng was administered 0 min preprandially in the normoglycemic subjects. Reductions in area under the curve (AUC) were 19% and 22%, respectively, for the type 2 diabetic subjects and 18% for the normoglycemic subjects. It was concluded that single, 3 g doses of the selected batch of American ginseng root demonstrated postprandial glycemic lowering efficacy in the normoglycemic and type 2 diabetic subjects. This efficacy was time dependent in the case of the normoglycemic subjects.

The third (Vuksan et al. 2000b) and fourth (Vuksan et al. 2000c) acute, "Phase I," RCTs assessed the effects of dosing and timing escalation of single doses of

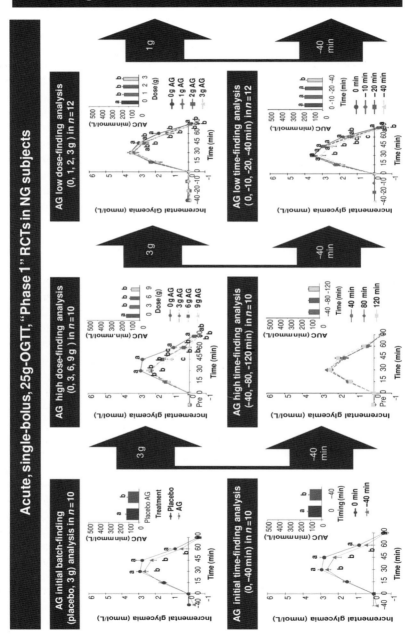

254

American ginseng root (source #1) at higher doses (0 g (placebo), 3, 6, 9 g) and more distal times of administration (0, −40, −80, −120 min in the type 2 diabetic subjects and −40, −80, −120 min in the normoglycemic subjects) in 10 normoglycemic subjects (Figure 12.2, panel 2) and 10 type 2 diabetic (Figure 12.3, panel 2) subjects. The 3, 6, and 9 g doses of the selected batch of American ginseng root significantly reduced the incremental glycemic response to the 25 g OGTT compared with placebo in both the normoglycemic subjects and type 2 diabetic. These reductions in incremental glycemia were reflected in significant reductions in the AUC from 15.3 to 19.7% in the type 2 diabetic subjects and from 26.6 to 38.5% in the normoglycemic subjects. There were no differences in glycemia between the doses or any of the times of administration. No interaction was observed between the usual oral hypoglycemic agents and the selected batch of American ginseng root in the type 2 diabetic subjects. The conclusion was that the selected batch of American ginseng root decreased the glycemic response to the 25 g OGTT irrespective of dose and time of administration. No more than 3 g of the selected batch of American ginseng was required preprandially from −40 to −120 min in normoglycemic subjects and from 0 to −120 min in type 2 diabetic subjects to achieve reductions.

The fifth and final acute, "Phase I," RCTs in the series (Vuksan et al. 2001b) assessed the escalation of single, lower doses (0 (placebo), 1, 2, 3 g) and more proximal times of administration (0, −10, −20, −40 min) of the selected batch of American ginseng root (source #1) in 12 normoglycemic subjects (Figure 12.2, panel 3). The 1, 2, and 3 g doses of the selected batch of American

←——————————————————————————————————————

Figure 12.2. Schematic of the *American Ginseng Clinical Testing Program* in normoglycemic (NG) subjects. The overall program consisted of a stepwise series of acute, single-bolus, "Phase I" randomized controlled trials (RCTs) in normoglycemic and DM2 subjects to evaluate the most efficacious dose and time of administration of a single-batch of American ginseng root, followed by a long-term, "Phase II," RCT, in which this information was applied to a representative American ginseng extract in DM2 subjects. The three acute, "Phase I," RCTs in normoglycemic subjects are depicted. Using a randomized, single-blind, placebo-controlled, multiple-crossover design, each RCT evaluated the effects of different single doses (1 to 9 g) and times of administration (0 to −120 min preprandially) on the glycemic response to a 25 g OGTT over 90 min, in which the levels of dose and time were crossed. Each RCT in the series informed the next RCT regarding the most efficacious dose and time of administration to advance. Panel 1 shows the initial batch plus time finding analyses. Panel 2 shows the high dose plus high time finding analyses. Panel 3 shows the low dose plus low time finding analyses. Points or bars with different letters are significantly different ($P < 0.05$). Doses and times of administration written in arrows indicate the dosing and timing schedules that advanced to the next study. These represented the most efficacious doses and times identified by the previous study or in cases where there was no difference, the lowest dose or time of administration closest to the meal. (Adapted from Vuksan et al. (2000a–c, 2001a, b) and Vuksan and Sievenpiper (2000).)

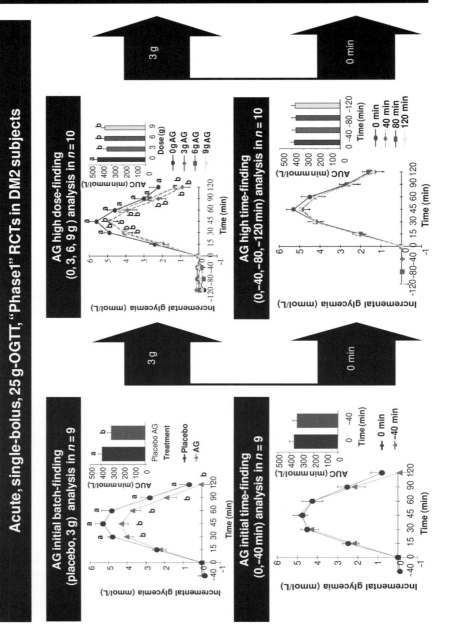

256

ginseng significantly reduced the incremental glycemic response to the 25 g OGTT compared with placebo. This was reflected in significant reductions in the AUC from 9.1 to 14.4%. There were no differences in glycemia between the doses. There was also a significant effect of time of administration. Incremental glycemia was lower, when ginseng was administered −40 min, compared with −20, −10, and 0 min preprandially. This was reflected in significant reductions in the AUC from 9.2 to 15%. The conclusion was that the selected batch of American ginseng root decreased the glycemic response to the 25 g OGTT in a time but not dose dependent manner: an effect was seen only when administration at −40 min preprandially and doses within the range of 1–3 g were equally efficacious.

Throughout the preceding series of acute studies, attempts were made to establish which aspects of the profile of the selected American ginseng root (source #1) gave rise to the acute antihyperglycemic effects. The ginsenoside composition of the selected batch of American ginseng was assessed (Table 12.1). It was noticed that it had a high proportion of PPD (Rb_1, Rb_2, Rc, and Rd) relative to PPT (Rg_1, Re, Rf) ginsenosides, a ratio smaller than 1 for the ginsenoside ratios Rg_1/Re and Rb_2/Rc, and Rf was absent. All these features of its composition indicated that the selected American ginseng root was of the genus and species selected (Awang 2000; Ma et al. 1996; Wang et al. 1999). Although it was tempting to suggest that these features might be responsible for its effects, other unmeasured components could have played an independent or interactive role. These include the >25 additional unmeasured ginsenosides, >10 peptidoglycans (quinquefolans for American ginseng), various ginsenans, numerous peptides and fatty acids, and countless other organic compounds. Without a basis for comparison, it was concluded that the ginsenoside profile was interpretable only for authentication (Vuksan et al. 2000a–c, 2001b; Vuksan and Sievenpiper 2000). To offer evidence for the efficacy of this ginsenoside profile, replication of the findings with an

---

Figure 12.3. Schematic of the *American Ginseng Clinical Testing Program* in type 2 diabetic (DM2) subjects. The 2 acute, "Phase I," RCTs in DM2 subjects are depicted. Using a randomized, single-blind, placebo-controlled, multiple-crossover design, each study evaluated the effects of different single doses (3 to 9 g) and times of administration (0 to −120 min preprandially) on the glycemic response to a 25 g OGTT over 120 min, in which the levels of dose and time were crossed. Each RCT in the series informed the next RCT regarding the most efficacious dose and time of administration to advance. Panel 1 shows the initial batch plus time finding analyses. Panel 2 shows the high dose plus high time-finding analyses. Panel 3 represents the long-term, "Phase II," RCT. Points or bars with different letters are significantly different ($P < 0.05$). Doses and times of administration written in arrows indicate the dosing and timing schedules that advanced to the next study. These represented the most efficacious doses and times identified by the previous study or in cases where there was no difference, the lowest dose or time of administration closest to the meal. (Adapted from Vuksan et al. (2000a–c, 2001a, b) and Vuksan and Sievenpiper (2000).)

Table 12.1. Ginsenoside profiles of the American ginseng batch (source #1) and the derived extract used in the American Ginsneg Clinical Testing Program.

Ginsenosides	Content (% wt/wt) and ratios	
	American (source #1)	American-extract (source #2)
Protopanaxadiols (PPD)		
Rb1	1.53	1.34
Rb2	0.06	0.08
Rc	0.24	0.46
Rd	0.44	0.62
PPD subtotal	2.27	2.50
Protopanaxatriols (PPT)		
Rg_1	0.1	0.13
Re	0.83	0.91
Rf	0	0
PPT subtotal	0.93	1.04
Total	3.21	3.54
Ratios		
PPD/PPT	2.44	2.40
Rb_1/Rg_1	15.30	10.31
Rb_2/Rc	0.25	0.17
Rg_1/Re	0.12	0.14

Ginsenoside content was determined by HPLC–UV analyses (Fitzloff et al. 1998). Subtotals, totals, and ratios were derived from the measured values for the individual ginsenoside concentrations. Some of the subtotals and totals may not equal the sum of the individual ginsenosides due to rounding to the nearest hundredth. Data taken from Vuksan et al. (2000a–c, 2001a–c) and Vuksan and Sievenpiper (2000, 2005).

American ginseng selected or designed to have a similar profile was considered to be the most viable approach. Specific ginsenoside markers to be targeted included total ginsenosides ≥3.21% and a PPD:PPT ginsenoside ratio ≥2.44.

Several conclusions were drawn from the stepwise series of acute, "Phase I," RCTs. A total of 32 normoglycemic and 29 type 2 diabetic subjects underwent 52 treatments of the selected batch of American ginseng root (source #1), in which the levels of dose (1 to 9 g) and time (0 to −120 min) were crossed across 5 acute clinical studies, using a randomized, single-blind, placebo-controlled, multiple-crossover design. Taken together, this robust design yielded five important conclusions:

(1) the selected batch of American ginseng given as a single, preprandial, oral agent reproducibly decreased the glycemic response to the 25 g OGTT across the dose range studied from 9 to 39% in the normoglycemic subjects and from 15 to 22% in the type 2 diabetic subjects; (2) doses from 1 to 9 g were equally efficacious; (3) times of administration from 0 to −120 min preprandially were equally efficacious in the type 2 diabetic subjects without interaction with their usual anti-hyperglycemic therapy; (4) only administration of the selected American ginseng batch ≥ −40 min preprandially reduced the glycemic response to the 25 g OGTT in normoglycemic subjects; and (5) the postprandial glycemic lowering efficacy of the selected batch of American ginseng coincided with a ginsenoside profile that included total ginsenosides of 3.2% and a PPD:PPT ginsenoside ratio of 2.44. It was generalized that the selected batch of American ginseng at a dose of 1 g given as an oral agent −40 min preprandially was sufficient to reduce acute post-prandial glycemia in both normoglycemic and type 2 diabetic subjects (Vuksan et al. 2000a–c, 2001b; Vuksan and Sievenpiper 2000). The implication was that an American ginseng extract selected to have a similar ginsenoside profile at a dose of 1 g given as an oral agent −40 min preprandially tid should have sustainable long-term effects in people with type 2 diabetic subjects.

The information gleaned from this acute phase of our *American Ginseng Clinical Testing Program* was applied to a long-term, "Phase II," RCT. An American ginseng extract (source #2;*Chai-Na-Ta Corp.,* Langley, BC) was designed to have a ginsenoside profile that matched or exceeded that of the American ginseng (source #1) used in the five acute studies (Vuksan et al. 2000a–c, 2001b; Vuksan and Sievenpiper 2000) (Table 12.1). An 8-week double-blind, placebo-controlled crossover trial was then undertaken, in which the designed American ginseng extract or placebo at a dose of 1 g was administered as an oral agent −40 min preprandially thrice daily (TID; 3 g/day) in 24 subjects with type 2 diabetes (Vuksan et al. 2001a). The primary outcomes were markers of glycemic control. Fasting glucose and HbA$_{1c}$ were decreased on the American ginseng extract compared with placebo after 8 weeks. There was also an observable but insignificant increase in insulin suggesting a possible improvement in β-cell function. These benefits occurred without increasing adverse events or altering hepatic, renal, haemostatic, or blood pressure function. Taken together, the data represented proof of two concepts. First, our acute clinical postprandial model used to select the most efficacious ginsenoside profile, dose, and time of administration successfully predicted long-term safety and efficacy of a source of American ginseng in type 2 diabetes. The glycemic lowering efficacy seen acutely (Vuksan et al. 2000a–c, 2001b; Vuksan and Sievenpiper 2000) was sustained in the long-term clinical investigation. Second, standardization of key features of the ginsenoside profile was related to achieving these sustainable effects, although it may have only been acting as a proxy for other >25 unmeasured ginsenosides or nonginsenoside components such as the peptidoglycan (quinquefolan) and polysaccharide (ginsenan) fractions.

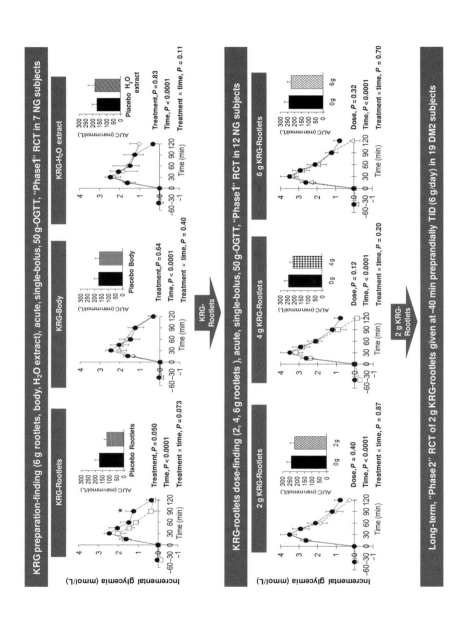

KRG preparation-finding (6 g rootlets, body, H₂O extract), acute, single-bolus, 50 g-OGTT, "Phase1" RCT in 7 NG subjects

KRG-Rootlets

KRG-Body

KRG-H₂O extract

Treatment, P = 0.050
Time, P < 0.0001
Treatment × time, P = 0.073

Treatment, P = 0.64
Time, P < 0.0001
Treatment × time, P = 0.40

Treatment, P = 0.83
Time, P < 0.0001
Treatment × time, P = 0.11

KRG-rootlets dose-finding (2, 4, 6g rootlets), acute, single-bolus, 50 g-OGTT, "Phase1" RCT in 12 NG subjects

2 g KRG-Rootlets

4 g KRG-Rootlets

6 g KRG-Rootlets

Dose, P = 0.40
Time, P < 0.0001
Treatment × time, P = 0.87

Dose, P = 0.12
Time, P < 0.0001
Treatment × time, P = 0.20

Dose, P = 0.32
Time, P < 0.0001
Treatment × time, P = 0.70

Long-term, "Phase 2" RCT of 2 g KRG-rootlets given at -40 min preprandially TID (6 g/day) in 19 DM2 subjects

260

Korean Red Ginseng Clinical Testing Program

To test whether the batch, preparation, dosing, and timing of another species of ginseng could be selected to have long-term efficacy using the same acute postprandial testing program, a similar approach was applied to Korean red ginseng (steam-treated *Panax ginseng* C.A. Meyer) in the *Korean Red Ginseng Clinical Testing Program* (Vuksan and Sievenpiper 2005). This species was considered to be a good candidate for efficacy screening, as it is the only other type that has been reported to decrease glycemia in humans (Tetsutani et al. 2000). It also shares a similar PPD:PPT ratio with the selected batch of American ginseng used in our *American Ginseng Clinical Testing Program* (Sievenpiper et al. 2004a, b). We hypothesized that a batch of Korean red ginseng selected to have a close PPD:PPT ginsenoside ratio would lower postprandial glycemia. Fractionation and extraction were applied to a single batch of Korean red ginseng (Korean Ginseng and Tobacco Research Institute, Daejeon, South Korea) to produce different preparations with a wide range in ginsenoside profiles. These included three preparations: Korean red ginseng rootlets (source #1), root body (source #2), and whole root H_2O extract (source #3). We used this starting material to initiate the *Korean Red Ginseng Clinical Testing Program* consisting of an acute and a long-term phase. A depiction is provided in Figure 12.4. Sequential, acute, single-bolus, "Phase I," type,

---

Figure 12.4. Schematic of the *Korean Red Ginseng Clinical Testing Program*. The program consisted of a series of acute, single-bolus, "Phase I," RCTs to evaluate the optimal preparation and dose derived from a single batch of Korean red ginseng (KRG) in normoglycemic (NG) subjects, followed by a long-term, "Phase II," RCT, in which this information was applied in type 2 diabeteic (DM2) subjects. Randomized, single-blind, placebo-controlled, multiple-crossover designs were used, in which the acute effects were assessed on the glycemic response to 50 g OGTT compared with placebo. Each study informed the next regarding the most efficacious preparation and dose to advance (written in arrows). Panel 1 shows the preparation-finding study, evaluating the effects of single 6 g doses of 3 KRG preparations (rootlets, root body, and whole root H_2O extract) given preprandially (-40 min). Panel 2 shows the dose-finding study, evaluating the effects of single 0 g (placebo), 2 g, 4 g, and 6 g doses of the most efficacious KRG preparation (rootlets). Although there was no effect of any of the three doses individually versus placebo, the mean of the three doses reduced the glycemic response to the 50 g OGTT significantly compared with placebo, implying that the three doses were equally efficacious. By this interpretation, the lowest dose advanced. Panel 3 represents the long-term, double "Phase II," RCT, in which the most efficacious KRG preparation (rootlets), dose (2 g), and mode of administration (oral agent given at -40 min preprandially) were applied TID (6 g/day) in 19 DM2 subjects. Asterisk indicates significant difference ($P < 0.05$). P-values are for independent and interactive effects assessed by two-way ANOVA. (Adapted from Sievenpiper et al. (2006) and Vuksan et al. (2006).)

double-blind, randomized, placebo-controlled, multiple-crossover, clinical studies were conducted to identify an efficacious KRG preparation, dose, mode, and timing of administration. The dose range (2–6 g), mode (periprandial oral agent), and timing (−40 min) were based on that used in our *American Ginseng Clinical Testing Program* described above. The studies consisted of a preparation-finding study (Korean red ginseng rootlets, root body, whole root H_2O extract), followed by a dose-finding study (0 (placebo), 2, 4, 6 g) of the most efficacious fraction. A 50 g OGTT protocol was used in which single boluses of the selected Korean red ginseng preparations, and doses were given as an oral agent −40 min preprandially in people with normoglycemia. The preparation and treatment protocol gained from these acute studies was then applied to a long-term, double-blind, "Phase II," RCT in subjects with type 2 diabetes.

In the first, acute, preparation-finding, "Phase I," RCT (Sievenpiper et al. 2006), we assessed the efficacy of single 6 g doses of the Korean red ginseng rootlets (source #1), root body (source #2), and whole root H_2O extract (source #3) given as an oral agent −40 min preprandially in lowering the glycemic response to a 50 g OGTT relative to placebo in seven normoglycemic subjects (Figure 12.4, panel 1). A wide variation in the ginsenoside profiles was achieved across the three root fractions (Table 12.2). This variation coincided with differential effects, although the effects did not appear to be related to differences in the PPD:PPT ginsenoside ratio. Korean red ginseng rootlets decreased the glycemic response to the 50 g OGTT significantly at 90 min compared with placebo, while neither the Korean red ginseng root body nor whole root H_2O extract affected glycemia significantly. This was reflected in a significant 29% reduction in AUC by the Korean red ginseng rootlets compared with placebo. It was concluded that Korean red ginseng rootlets represented the most efficacious preparation and would be advanced to the next step of the acute phase of the clinical testing program.

In the acute, dose-finding, "Phase I," RCT (Sievenpiper et al. 2006), 12 normoglycemic subjects received single doses of 0 g (placebo), 2 g, 4 g, and 6 g of the selected Korean red ginseng rootlets (source #1) following the same 50 g OGTT protocol described for the preparation-finding study (Figure 12.4, panel 2). A significant effect of Korean red ginseng rootlets treatment (mean of three doses) was found. The mean of all three doses decreased mean postprandial incremental glycemia by 17% compared with placebo. This was reflected in the AUC, in which there was a tendency for the mean of all three doses of the Korean red ginseng rootlets (source #1) to decrease the AUC by 14% compared with placebo. There was, however, no effect of individual doses compared with placebo. It was concluded that the selected Korean red ginseng rootlets decreased the glycemic response to the 50 g OGTT irrespective of dose. The selected Korean red ginseng rootlets were equally efficacious in lowering postprandial glycemia at doses from 2 to 6 g, when given as an oral agent −40 min preprandially.

Several conclusions were drawn from the sequential acute, "Phase I," RCTs for extrapolation to the long-term, "Phase II," RCT of the *Korean Red Ginseng Clinical Testing Program*. The studies were successful in identifying an efficacious

Table 12.2. Ginsenoside profiles of the three Korean red ginseng (KRG) fractions studied in the *Korean Red Ginseng Clinical Testing Program.*

Ginsenoside	Concentration (% wt/wt) and ratios		
	Rootlets (source #1)	Body (source #2)	H_2O extract (source #3)
Protopanaxadiols (PPD)			
Rb_1	0.479	0.188	0.71
Rb_2	0.249	0.071	0.436
Rc	0.289	0.063	0.484
Rd	0.097	0.014	0.321
Rg_3	0.055	0.03	0.445
PPD subtotal	1.168	0.367	2.396
Protopanaxatriols (PPT)			
Rg_1	0.225	0.175	0.146
Re	0.506	0.132	0.463
Rf	0.026	0.012	0.035
PPT subtotal	0.756	0.319	0.645
Total	1.924	0.564	3.041
Ratios			
PPD:PPT	1.545	1.150	3.721
Rb_1:Rg_1	2.129	1.074	4.863
Rb_2:Rc	0.862	1.127	0.901
Rg_1:Re	0.445	1.326	0.315

Ginsenoside content was determined by HPLC–UV analyses (Fitzloff et al. 1998). Subtotals, totals, and ratios were derived from the measured values for the individual ginsenoside concentrations. Some of the subtotals and totals may not equal the sum of the individual ginsenosides due to rounding to the nearest hundredth. Data are adapted from Sievenpiper et al. (2006), Vuksan et al. (2006), and Vuksan and Sievenpiper (2005).

preparation and dose of Korean red ginseng. It was concluded that the selected Korean red ginseng rootlets at a dose of 2 g, given as an oral agent −40 min preprandially, were sufficient to achieve reproducible reductions in acute postprandial glycemia. It was extrapolated that the selected Korean red ginseng rootlets at a dose of 2 g, given as an oral agent −40 min preprandially tid (6 g/day) would have sustainable long-term effects in people with type 2 diabetic subjects. This treatment protocol was indicated for long-term study.

The most efficacious preparation (Korean red ginseng rootlets (source #1)), dose (2 g), and mode of administration (oral agent given at −40 min preprandially) selected from the sequential acute preparation and dose-finding studies were applied to the long-term, "Phase II," RCT (Figure 12.5) (Vuksan et al. 2006).

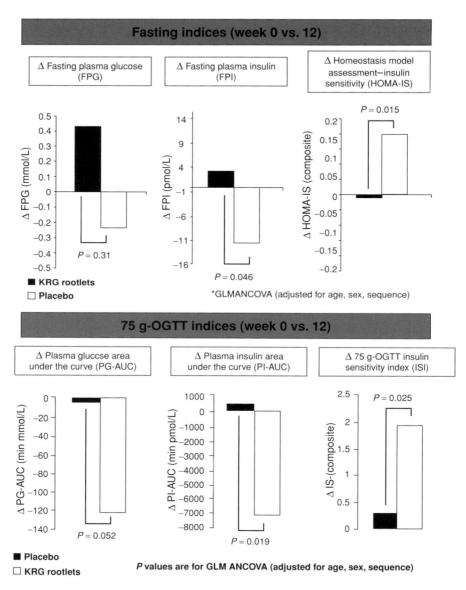

Figure 12.5. Main findings from the long-term phase of of the *Korean Red Ginseng Clinical Testing Program*. Using a double-blind, randomized, placebo-controlled, crossover design, the effect of 12 weeks supplementation with the most efficacious preparation (rootlets), dose (2 g), and mode of administration (oral agent given at −40 min preprandially) selected from the acute phase of testing and applied TID (6 g/day) was assessed on fasting and 75 g OGTT plasma glucose and insulin indices in 19 type 2 diabetic (DM2) subjects. (Adapted from Vuksan et al. (2006).)

A double-blind, randomized, placebo-controlled, crossover trial was conducted. Nineteen type 2 diabetic subjects received 2 g placebo or the selected Korean red ginseng rootlets as an oral agent −40 min preprandially tid (6 g/day) for 12 weeks, while maintained on their conventional diabetes treatment. Fasting plasma insulin and 75 g OGTT derived AUC plasma insulin were significantly decreased on the selected Korean red ginseng rootlets treatment compared with placebo. This occurred while fasting plasma glucose was unchanged and 75 g OGTT derived AUC plasma glucose was significantly decreased. The combination was reflected in an identical 33% increase in both the homeostasis model assessment (HOMA-IS) and the 75 g OGTT derived (75 g OGTT ISI) insulin sensitivity indices on the selected Korean red ginseng treatment compared with placebo. These benefits occurred without increasing adverse events or altering hepatic, renal, haemostatic, or blood pressure function. We concluded that our *Korean Red Ginseng Clinical Testing Program* successfully identified a Korean red ginseng preparation (rootlets (source #1)), dose (2 g), and mode of administration (oral agent given −40 min preprandially) that resulted in sustainable improvements in long-term glucose and insulin regulation safely beyond conventional treatment in type 2 diabetes.

Ginseng Source and Glucose Regulation

Variable pharmacological effects of ginseng appear secondary to differences in source and composition. Both the *American and Korean Red Ginseng Clinical Testing Programs* (Figures 12.2–12.5) emphasized the importance of a consistent source of ginseng in the reproducibility and sustainability of postprandial glycemic reductions. We conducted another series of acute, clinical studies, to assess the effects of increasing ginsenoside variability across ginseng source parameters of progressively greater ginsenoside variability (batch, preparation, variety, and species) on the postprandial regulation of glycemia and insulinemia using a 75 g oral glucose tolerance test (75 g OGTT) protocol, in which the mode (preprandial oral agent) and time (−40 min) of administration were based on our previous studies (Table 12.3). Other investigators have also evaluated the glycemic effects of various extracts in subjects with and without type 2 diabetes. A discussion follows of the evidence in humans for the effect of different source factors of ginseng on the variability in glycemic outcomes.

Differential Effects of American Ginseng Batches

A single analysis of two studies (Sievenpiper et al. 2003a) was used to compare the results for two distinct batches of American ginseng root from the same supplier (*Chai-Na-Ta Corp*, BC, Canada): source #1 (Table 12.1) and source #3 (Table 12.4). In the first study (Table 12.3), the effect of a single 6 g dose given −40 min preprandially of the efficacious batch of American ginseng root from the *American*

Table 12.3. Differential effects of ginseng species, variety, preparation, batch, farm, and dose compared with control on the plasma glucose and insulin responses to a 75 g oral glucose tolerance test (75 g OGTT) in normoglycemic subjects across 6 acute studies.

Study	Sample (n)	Ginseng type	Ginseng dose	Δ incremental plasma glucose	Δ incremental plasma insulin
Sievenpiper et al. (2003a)	8 NGT	American (source #1)	6 g	↓ 45, 60, 90 min, Peak, AUC	↔
	12 NGT	American (source #3)	6 g	↔	↔
Sievenpiper et al. 2003b	11 NGT	Asian (source #1)	1 g	↔	↔
		Asian (source #1)	2 g	↔	↔
		Asian (source #1)	3 g	↔	↔
Sievenpiper et al. (2003b)	11 NGT	Asian (source #1)	3 g	↔	↔
		Asian (source #1)	6 g	↔	↔
		Asian (source #1)	9 g		↔
		Mean	1 g-9g	↑ 2 h (absolute)	↔
Sievenpiper et al. (2004b)	12 NGT	American (source #4)	3 g	↓ 90 min	↔
		Asian (source #1)	3 g	↑ Peak, AUC	↔
		Sanchi	3 g		↑ 0 min
		Vietnamese-wild	3 g	↓ 90 min	↓ 90 min*, AUC*
		Japanese-rhizome	3 g		↔
		Siberian	3 g	↑ 90, 120 min, AUC	↔
		Korean red (source #4)	3 g		↔
		American-wild	3 g	↑ 120 min	↔
Dascalu et al. (2007)	12 NGT	American Farm A (source #5)	9 g	↓ 30, 45, 60 min, peak, AUC	↓ 60 min
		American Farm B (source #6)	9 g	↔	↔
		American Farm C (source #7)	9 g	↓ 15, 30, 45, 60 min, peak, AUC	↓ 60 min, AUC
		American Farm D (source #8)	9 g	↔	↔
		American Farm E (source #9)	9 g	↓ 45, 60 min, peak, AUC	↔
		Mean	9 g	↓ 15, 30, 45, 60 min, peak, AUC	↓ 45, 60 min, AUC

Arrows indicate direction of significant differences ($P < 0.05$) compared with control by one-way ANOVA adjusted for multiple comparisons using the Tukey-Kramer procedure or nonorthogonal contrasts. Asterisk indicates significant difference in a subset of overweight subjects.

Table 12.4. Ginsenoside profiles across ginseng source parameters of species, variety, preparation, and batch of ginsneg treatments reported in more than 10 clinical studies.

Ginsenosides	Content (% wt/wt) and ratios													
	American (source #3)	American (source #4)	American (source #5)	American (source #6)	American (source #7)	American (source #8)	American (source #9)	Asian (source #1)	Sanchi	Siberian	Korean red (source #4)	Japanese rhizome	American-wild	Vietnamese-wild
Protopanaxadiols (PPD)														
Rb_1	0.65	1.60	4.09	4.15	4.19	3.25	3.29	0.20	1.90	0	0.35	0.12	4.04	2.30
Rb_2	0.02	0.13	0.06	0.04	0.04	0.04	0.05	0.09	0	0	0.14	0	0.25	0.01
Rc	0.11	0.78	0.33	0.22	0.25	0.24	0.29	0.07	0	0	0.15	0	0.05	0.02
Rd	0.12	0.50	0.25	0.27	0.36	0.24	0.32	0.03	0.42	0	0.04	0	0.18	0.95
PPD subtotal	0.90	3.01	4.73	4.68	4.84	3.77	3.95	0.38	2.33	0	0.68	0.12	4.52	3.27
Protopanaxatriols (PPT)														
Rg1	0.08	0.13	0.17	0.08	0.11	0.11	0.13	0.16	1.74	0	0.17	0.05	0.64	1.43
Re	0.67	1.29	1.67	1.52	1.39	1.29	1.25	0.17	0.32	0	0.17	0	1.50	0.36
Rf	0	0	0	0	0	0	0	0.09	0	0	0.06	0	0	0
PPT subtotal	0.75	1.42	1.84	1.60	1.50	1.40	1.38	0.41	2.06	0	0.40	0.05	2.14	1.79
Total	1.65	4.43	6.57	6.28	6.34	5.17	5.33	0.79	4.39	0	1.09	0.17	6.66	5.06
Ratios														
PPD:PPT	1.20	2.13	2.57	2.93	3.23	2.69	2.86	0.94	1.13	0	1.70	2.33	2.11	1.82
Rb1:Rg1	8.13	12.76	24.06	51.88	38.09	29.55	25.31	1.29	1.09	0	2.09	2.33	6.27	1.60
Rb2:Rc	0.18	0.16	0.18	0.18	0.16	0.17	0.17	1.21	0	0	0.92	0	5.23	0.59
Rg1:Re	0.12	0.10	0.10	0.05	0.08	0.09	0.10	0.94	5.44	0	0.97	0	0.43	3.98

Ginsenoside content was determined by HPLC–UV analyses (Fitzoff et al. 1998). Subtotals, totals, and ratios were derived from the measured values for the individual ginsenoside concentrations. Some of the subtotals and total may not equal the sum of the individual ginsenosides due to rounding to the nearest. hundredth. Data for Asian ginseng represent the mean of separate analyses for the same batch of Asian ginseng reported in two studies [Sievenpiper et al. 2003b, 2004b]. Data taken from Sievenpiper et al. (2003a, b, 2004b), Jenkins et al. (2002a, b, 2003a, b), Stavro et al. (2005, 2006), and dascalu (2007).

267

Ginseng Clinical Testing Program (source #1) was assessed on plasma glucose and insulin indices during a 75 g OGTT relative to an historical control (75 g OGTT done alone previously) using a nonrandomized, crossover design, in eight normoglycemic subjects. The results of this first study were then qualitatively compared with the results of the second study (Table 12.3), in which the effect of a third batch of American ginseng root (source #3) with a lower ginsenoside profile (lower total ginsenosides and specific individual ginsenosides with altered ginsenoside ratios including a low PPD:PPT ratio) was assessed as a single 6 g dose given as an oral agent −40 min preprandially relative to placebo on indices of plasma glucose and insulin during a 75 g OGTT, using a single-blind, randomized, crossover design in a separate sample of 12 normoglycemic subjects. A direct comparison between the batches in the same subjects was precluded by insufficient quantities of our original efficacious batch of American ginseng at the time of the second study.

The two batches showed differential effects. Pairwise comparisons showed that incremental plasma glucose at 45, 60, and 90 min were significantly decreased by our original efficacious batch of American ginseng relative to control. This was reflected in a significant 43% decrease in the AUC. Neither plasma insulin AUC nor the whole body 75 g OGTT insulin sensitivity index (75 g OGTT ISI) nor first release insulin secretion index (Δ plasma insulin$_{30-0}$/Δ plasma glucose$_{30-0}$) was affected. There was no effect of the second American ginseng batch on any of the plasma glucose or insulin parameters. It was concluded that while our original efficacious batch of American ginseng root (source #1) from the *American Ginseng Clinical Testing Program* again demonstrated acute postprandial glycemic lowering efficacy, this new batch (source #3) with a lower ginsenoside profile including a low PPD:PPT ratio was ineffective (Sievenpiper et al. 2003a).

Differential Effects of Asian Ginseng Species

Having noticed differential effects of batch within the same species, we moved to an investigation of the effect of species, a ginseng source parameter that gives an even wider variation in ginsenosides (Sievenpiper et al. 2003b). Asian ginseng root was chosen, as it contains a ginsenoside not present in American ginseng, Rf, and various inverted ginsenoside ratios: Rb1:RDg1 <3 and Rg1:Re and Rb2:Rc >1 (Sievenpiper et al. 2004a). Two separate studies of the same batch of Asian ginseng root (source #1) given as an oral agent −40 min preprandially were consolidated in a single analysis that looked at the effect of escalating single doses over a 9-fold dose range on indices of plasma glucose and insulin during a 75 g OGTT, using a single-blind, placebo-controlled, multiple-crossover design (Table 12.3). Two dose ranges were assessed: low doses (0 (placebo), 1, 2, 3 g) in 11 normoglycemic subjects in the first study and high doses (0 (placebo), 3, 6, 9 g) in a separate sample of 11 normoglycemic subjects in the second study. The aggregate of all treatments was then assessed in a pooled analysis across all subjects.

Equivocal effects of Asian ginseng were seen. Ginsenoside analyses showed that the selected Asian ginseng root (source #1) had lower total ginsenosides with up to 96% lower and 7-fold higher specific individual ginsenosides and altered ginsenoside ratios including the PPD:PPT ratio <1 (Table 12.4). This profile coincided with no effect of either the doses or the pooled-treatment on incremental plasma glucose and insulin concentrations, AUC, 75 g OGTT ISI, or first release insulin secretion index (Δ plasma insulin$_{30-0}$/Δ plasma glucose$_{30-0}$). But the diagnostically and therapeutically relevant 2 h plasma glucose (2 h PG) value was almost 10% higher for pooled Asian ginseng treatment than placebo. It was concluded that the selected batch of Asian ginseng root (source #1) provided species-driven differences in ginsenosides, including a lower and inverted ginsenoside profile, over a 9-fold dose range that coincided with null and increasing effects on indices of acute postprandial plasma glucose and insulin in a combined data set of 22 normoglycemic subjects.

Differential Effects of Ginseng Species, Variety, Preparation, and Batch

A limitation of the preceding two analyses (Sievenpiper et al. 2003a, b) was that the effects of batch and species on postprandial plasma glucose and insulin were inferred indirectly. We therefore compared species, variety, preparation, and batch relative to placebo using a head-to-head design allowing for direct comparisons across a wider ginseng source-driven range in ginsenoside profiles (Table 12.3) (Sievenpiper et al. 2004b). Eight of the most popular types of *Panax* and non-*Panax* species of ginseng were selected to provide a cross-section of species, variety, preparation, and batch and, with it, approximate the spectrum of possible ginsenoside profiles to which consumers are exposed. A randomized, double-blind, double-placebo-controlled design was used, in which 12 normoglycemic subjects received 10 single 3 g doses treatments consisting of a fourth batch of American ginseng root (source #4), the same batch of Asian ginseng root (source #1) from the previous two combined, dose escalation studies, a fourth batch of Korean red ginseng root (source #4), American-wild ginseng root, Vietnamese-wild ginseng (*Panax vietnamensis*) root, the non-*Panax* species Siberian ginseng (*Eleutherococcus senticosus*), Japanese ginseng (*Panax japonicus* C.A. Meyer) rhizome, Sanchi ginseng (*Panax notoginseng*) root, and the same placebo done on two separate occasions. Each treatment was consumed as an oral agent −40 min preprandially relative to a 75 g OGTT. Results were expressed relative to the mean of the placebo done on two separate occasions.

Variable effects were seen across the eight different ginseng types. The study achieved a wide range in ginsenoside content and ratios across the ginseng source parameters of species, variety, preparations, and batches of the eight different types of ginseng (Table 12.4). This coincided with an equally wide range in effects on 75 g OGTT plasma glucose and insulin indices. The fourth batch of American ginseng root (source #4), with a similar ginsenoside profile to our original efficacious batch (source #1), lowered incremental plasma glucose during the 75 g OGTT, while the fourth batch of Korean red ginseng (source #4) along with batches of Japanese-rhizome, Sanchi, and Vietnamese-wild root had null effects. The same batch of Asian ginseng root (source #1) from the previous two, combined, dose escalation studies (Sievenpiper et al. 2004b) again significantly raised acute plasma glucose during the 75 g OGTT (37% increase in AUC compared with placebo), as did batches of the non-*Panax* species, Siberian ginseng root (35% increase in AUC compared with placebo), and American-wild ginseng root. Insulin effects were seen only for Asian ginseng, which decreased incremental plasma insulin at 0 min before the start of the 75 g OGTT.

These results were found to be dependent upon by the weight status of the subjects. Subgroup analyses showed that Vietnamese-wild ginseng root significantly lowered 90 min and AUC plasma insulin during the 75 g OGTT compared with placebo in a subset of overweight (BMI > 25 kg/m^2, $n = 6$) but not normal weight (BMI < 25 kg/m^2, $n = 6$) subjects. The overweight subjects had a higher degree of insulin resistance compared with their normal weight counterparts, as indicated by a 27% lower OGTT ISI. This suggested that there might be a phenotypic difference in the response to ginseng, such that overweight, insulin resistant subjects might derive more benefit from efficacious ginseng sources. This was interpreted as being consistent with our previous observation that the effect of our original efficacious batch of American ginseng root (source #1) was more robust in the subjects with type 2 diabetes than normoglycemic subjects, lowering glycemia during the 25 g OGTT irrespective of time of administration from 0 to −120 min preprandially.

Several conclusions were drawn based on these observations. First, that head-to-head comparison of ginseng source parameters across species, variety, preparation, and batch resulted in a wide range in ginsenoside profiles. Second, decreasing, null, and increasing effects were observed on 75 g OGTT derived plasma glucose and insulin indices in normoglycemic subjects that appeared secondary to this variability in ginseng source parameters and ginsenoside composition. Finally, this relationship between ginseng source parameters, ginsenosides, and indices of acute postprandial glycemic control may be modified by the overweight and insulin resistance status of subjects.

Differential Effects of American Ginseng from Different Farms

Source parameters of ginseng that predict reproducible efficacy were not clear from the preceding series of studies. There was a need to identify the active component(s) involved in mediating the effects on plasma glucose and insulin regulation so that specific component profiles could be defined for antihyperglycemic indications. We initiated the *Systematic Clinical Testing Program of Ginseng Fraction* that was modeled on our *American and Korean Red Ginseng Clinical Testing Programs* (Figures 12.2–12.5), following their success at identifying sources of ginseng with reproducible and sustainable clinical efficacy. Starting from an efficacious whole ginseng source, progressively smaller fractions were to be assessed in a stepwise series of acute, "Phase I," type studies, using a 75 g OGTT protocol in normoglycemic subjects, until specific fractions or individual components were identified that yielded reproducible efficacy. American ginseng was chosen as the whole ginseng source, as it had proven the most consistent species: three (sources #1, #2, #4) out of the four batches of American ginseng we tested had demonstrated acute, postprandial, glycemia lowering efficacy.

To identify an efficacious batch of American ginseng for systematic fractionation, we evaluated the effect of batches of American ginseng root from five Ontario farms on postprandial plasma glucose and insulin indices during a 75 g OGTT in 12 normoglycemic subjects. The batches were selected by the Ontario Ginseng Growers Association to represent the spectrum of American ginseng on the market and a range of compositional profiles, according to quality, determined by age, price, physical appearance such as body shape (short, long, thin, etc.), the ratio of root body to fiber (rootlets), and growing conditions (location, soil type, age at harvest, etc.). The expectation was that this selection procedure would produce a CV for the main PPD and PPT ginsenosides similar to the ~16–40% that was calculated for American ginseng analyzed from nine different BC farms (Li et al. 1996). A double-blind, randomized, multiple-crossover design was used, in which subjects consumed single, 9 g doses of the American ginseng root from the five Ontario farms (A–E: source #5–9) taken as oral agents −40 min preprandially in random order followed by an unblinded water control done on the last visit (Dascalu et al. 2007).

Differential effects were seen across the batches of American ginseng root from the five Ontario farms. The ginseng sources achieved a representative range in quality as assessed by age (3–4 years) and price ($33–55/kg) and ginsenoside composition with an interbatch CV in ginsenosides from 6 to 34% (Table 12.3). Batches from the three of the five farms had a significant effect on indices of plasma glucose and insulin during the 75 g OGTT compared with placebo: farm A decreased peak plasma glucose and incremental plasma glucose AUC by 35.2%; farm C decreased peak plasma glucose, incremental plasma glucose AUC by 32.6%, and incremental plasma insulin AUC by 28%; and farm E decreased incremental peak plasma glucose. There were no differences detected between the five farms. The mean effect of the five farms on the 75 g OGTT response decreased peak plasma glucose, plasma glucose AUC by 28%, and incremental plasma insulin AUC by 24% compared with control.

It was concluded that the batches of American ginseng root from five Ontario farms (A–E: source #5–9) achieved sufficiently variable profiles to achieve differential effects on 75 g OGTT derived plasma glucose and insulin indices in normoglycemic subjects. Irrespective of their ginsenoside composition, two of the five batches did not affect plasma glucose with the anticipated magnitude. American ginseng root from farm A (source #5) was identified as the most efficacious batch for systematic fractionation and evaluation in the proposed *Systematic Clinical Testing Program of Ginseng Fractions*.

Consistent Effects of American Ginseng in Combination With Viscous Soluble Dietary Fiber

Sufficient quantities of this batch of American ginseng from farm A (source #5) were also budgeted for inclusion as an efficacious American ginseng source to be used in a series of studies assessing the combination of American ginseng and a viscous soluble fiber blend containing konjac mannan. A combination of the two treatments was conceived as a corollary to the increasing trend to polyphar-macy with oral hypoglycemic agents to meet stricter treatment targets for glycemic control (American Diabetes Association 2007). American ginseng (Vuksan et al. 2000a–c, 2001a, b; Vuksan and Sievenpiper 2000) and a viscous soluble fiber blend containing konjac mannan (Vuksan et al. 1999, 2001a) were chosen as both have resulted in sustainable long-term improvements in glycemic control by apparently complementary mechanisms. Two studies were conducted to assess the acute and long-term effect of the combination therapy in type 2 diabetes.

The first study (Jenkins et al. 2002a, b) investigated the acute "first" and "second meal" effects of each intervention alone or in combination on the postprandial plasma glucose and insulin responses. Using a single-blind, randomized, two-by-two factorial design, 10 subjects with type 2 diabetes underwent four treatment sets: (1) Control (7 g of wheat bran); (2) American ginseng (3 g American ginseng from farm A (source#5) + 7 g of wheat bran); (3) viscous soluble fiber blend (4 g viscous soluble fiber blend containing Konjac mannan); and (4) their combination. Treatments were administered as preprandial agents at −30 min relative to a mixed first meal (breakfast) containing 50 g available carbohydrate from Ensure™. This was followed at 240 min by a second meal (lunch) given without treatment that consisted of soup, a cheese sandwich, and fruit. A significant "first meal" effect was observed. American ginseng from farm A alone decreased the postprandial incremental plasma glucose at 15 min and in combination with the viscous sol-uble fiber blend decreased both the postprandial incremental plasma glucose and insulin at 30 min, compared with control. This was followed by a significant "sec-ond meal" effect. The combination of the two treatments significantly decreased the postprandial incremental plasma glucose AUC response to the second meal, compared with control. It was concluded that 3 g of American ginseng from farm A alone and in combination with 4 g of the viscous soluble fiber blend containing konjac mannan showed reproducible acute glycemic effects in using a different clinical model. The most robust effects were seen for the combination of the two treatments with reductions in the acute postprandial plasma glucose and insulin responses to the first and second meals in subjects with type 2 diabetes.

This combination therapy with a modified treatment protocol was indicated for long-term study (Jenkins et al. 2003a, b). Using a randomized, double-blind, placebo-controlled, crossover design, 30 subjects with well-controlled type 2 diabetes were randomized to 12 weeks of treatment with either combination therapy (3 g of the selected batch of American ginseng from farm A + 7 g viscous soluble fiber blend containing Konjac mannan) or placebo. The subjects were then washed out for 8 weeks and crossed to the alternate treatment for another 12 weeks. Treatments were administered as preprandial oral agents at -30 min tid and added adjunctively to their usual diabetes therapy of diet and/or medications. The combination significantly reduced HbA_{1c} compared with placebo for an absolute end-difference reduction in HbA_{1c} of 0.3% at 12 weeks. It was concluded that the batch of American ginseng from farm A (source #5) in combination with viscous soluble fiber containing konjac mannan improved long-term glycemic control safely beyond usual diabetes treatment with an efficacy comparable to that seen with the alpha-glucosidae inhibitor, acarbose, in subjects with well-controlled type 2 diabetes. Although the conclusions were for the combination therapy, these studies again demonstrated that our model of selecting an American ginseng source and treatment protocol using an acute clinical screening program is able to achieve both reproducible and sustainable glycemic effects.

Consistent Effects of Various Ginseng Extracts

Other studies of mixed quality have described improvements in indices of long-term glycemic control with various extracts of ginseng in subjects with type 2 diabetes. Sotaniemi and coworkers (1995) reported that 8 weeks of treatment with 100 and 200 mg/day of an unspecified ginseng extract improved fasting glycemia and HbA_{1c}, respectively, using a randomized, placebo-controlled parallel design in 36 type 2 diabetic subjects. But this study was confounded by significant weight loss in the ginseng group, poorly described statistics, and lack of characterization of the ginsenoside profile of the unspecified ginseng source. Similarly, Tetsutani and coworkers (2000) reported that four months of treatment with a Korean red ginseng extract at doses from 3 to 4.5 g decreased HbA_{1c} in 34 people with type 2 diabetes compared with controls (Tetsutani et al. 2000). But limitations of the trial included poor descriptions of subject selection, allocation to treatment, application of statistics, follow-up, and characterization of the ginsenoside profile of the Korean red ginseng. Various secondary sources have also reported glycemic benefits of Korean red ginseng in type 2 diabetes (Anonymous 2005), but information is not available to verify these effects.

Several studies from a single investigator group have shown differential effects of an Asian ginseng extract in normoglycemic subjects. The Newcastle-upon-Tyne group showed that the well-characterized (Cui 1995) extract G115 (Pharmaton SA, Lugano, Switzerland) in single doses from 200 to 400 mg consistently decreased acute fasting plasma glucose at 60 min and 120 min postdose compared with placebo in three double-blind, placebo-controlled, crossover studies which included a total of 114 normoglycemic subjects (Reay et al. 2005, 2006a, b). On the other hand, there was no effect of 200 mg on the plasma glucose response to a 25 g OGTT compared with the 25 g OGTT dose alone (Reay et al. 2005, 2006a, b).

Despite the cited limitations, these studies viewed as a whole were remarkably consistent in showing that Korean red ginseng extracts improved long-term markers of glycemic control across different study designs in subjects with type 2 diabetes and that the extract G115 decreased acute fasting plasma glucose in normoglycemic subjects.

Interpretation

Overall, the compositional (ginsenoside) variability achieved experimentally across the ginseng source parameters of progressively greater variability (batch, farm, preparation, variety, species) in the preceding studies was equal to the actual variability in ginsenosides reported across similar source parameters in the meta-analysis of all published ginsenoside analyses (Sievenpiper et al. 2004). This high and representative compositional variability in ginsenosides coincided with highly variable effects on glycemia in terms of the significance, magnitude, and direction of the effects. Some sources of ginseng, however, performed more reliably than others. Interpreted together with the data from our *American and Korean Red Ginseng Clinical Testing Programs*, the most robust evidence from these studies for the reproducibility and sustainability of glycemic lowering efficacy across the different source parameters, doses, study designs, protocols, lengths of follow-up, and patient cohorts was for American and Korean red ginseng sources. Five distinct sources of American ginseng and more than three distinct sources of Korean red ginseng improved markers of glycemic regulation. Reductions in acute postprandial glycemia as assessed by AUC ranged from 9.1 to 43% for American ginseng and from 14 to 29% for Korean red ginseng compared with control. This translated into similar long-term improvements in markers of long-term glycemic regulation.

Ginseng Components and Glucose Regulation

Although total ginsenosides and the ratio of PPD:PPT ginsenosides were identified as possible markers of efficacy in explaining the batch differences in American ginseng and species differences between American and Asian ginseng, the role of ginsenosides remains uncertain. The relationship between ginsenoside composition and glycemic effects was poorly defined by the categorical analyses in the preceding studies. To assess the independent ginsenoside predictors of plasma glucose and insulin regulation, continuous analyses were attempted in three of the studies.

Models were constructed first for the acute, double-blind, randomized, placebo-controlled, multiple-crossover study, in which we compared head-to-head eight of the most popular types of ginseng in 12 normoglycemic subjects (Sievenpiper et al. 2004b) (Table 12.3). This study provided a cross-section of species, variety,

Table 12.5. Stepwise-multiple regression analyses of the independent ginsenoside predictors of 75 g -OGTT plasma glucose (PG) and insulin (PI) indices in 12 normoglycemic subjects.

Index	Model	β	*P*	Partial[2]
90min-PG	Rd	−0.64	0.080	0.033
	PPD/PPT	−0.43	0.088	0.072
Peak-PG	—	—	—	—
Mean-PG	PPD/PPT	−0.22	0.028	0.050
AUC-PG	PPD/PPT	−0.25	0.016	0.060
90min-PI	PPD/PPT	−0.26	0.012	0.065
Mean-PI	—	—	—	—
AUC-PI	PPD/PPT	−0.20	0.025	0.040

AUC denotes area under the curve. 90min-PG and -PI and AUC-PG data are expressed as incremental values. AUC-PI data are expressed as incremental values and were logarithmically transformed for statistical analyses to normalize their distribution. Peak-PG and Peak-PI data are expressed as absolute values. Each stepwise regression model for the PG and PI indices contained all ginsenoside parameters analyzed and derived from the eight types of ginseng: the four individual protopanaxadiol (PPD) ginsenosides (Rb_1, Rb_2, Rc, Rd), the three individual protopanaxatriol (PPT) ginsenosides (Rg_1, Re, Rf), and the derived values for the PPD and PPT ginsenoside subtotals, total ginsenosides, and the four ginsenoside ratios (PPD/PPT, Rb_1/Rg_1, Rb_2/Rc, Rg_1/Re). The ginsenosides shown for each model are those selected by the stepwise regression and entered into the multiple regression procedure. Table adapted from Sievenpiper et al. (2004b).

preparation, batch, and ginsenoside profiles. The wide range in ginsenoside content expressed as individual ginsenosides, their subtotals, total, and ratios among the eight different types of ginseng were correlated against the seven indices of plasma glucose (90 min, peak, AUC, and mean plasma glucose) and insulin (90 min, AUC, and mean plasma insulin) found to be significant in categorical analyses. Pearson correlations showed that all the plasma glucose and insulin indices, with the exception of peak plasma glucose and mean plasma insulin, were significantly correlated with ginsenoside indices. The ratio of PPD:PPT ginsenosides was the strongest correlate in all five cases. Other correlates included Re, Rc, Rb_2, Rd, PPD, total, and the ratio of Rb_1:Rg_1 ginsenosides for 90 min plasma glucose and Rd and the ratio of Rg_1:Rb_1 ginsenosides for AUC plasma glucose. Stepwise multiple regression models were applied to these data. The models selected for each of the indices of plasma glucose and insulin are shown in Table 12.5. Four of the seven models were significant. The ratio of PPD:PPT ginsenosides was assessed as the sole independent predictor of the effects on AUC and mean plasma glucose and 90 min and AUC plasma insulin. The direction of the relationship was negative in each case, such that the higher the ratio of PPD:PPT ginsenosides the lower the plasma glucose and insulin indices. The variance in these indices explained by the differences in the ratio of PPD:PPT ginsenosides was from only 4 to 7%. The suggestion was that other unmeasured ginsenoside or nonginsenoside components were likely playing independent or interactive role in the effects.

This possibility was explored further in the acute phase of the *Korean Red Ginseng Clinical Testing Program* (Figures 12.3 and 12.4). Ginsenoside compositions (Table 12.2) were modeled against glycemic effects in the acute, single-blind, randomized, controlled, multiple-crossover, preparation-finding study (Figure 12.3), in which the effects of single, 6 g doses of three Korean red ginseng preparations (rootlets (source #1), root body (source #2), and H_2O extract (source #3)) given as oral agents −40 min preprandially were assessed on the glycemic response to a 50 g OGTT compared with placebo in seven normoglycemic subjects (Sievenpiper et al. 2006). Stepwise multiple regression models were constructed that included all the measured individual ginsenoside concentrations and their derived subtotals, totals, and ratios. The PPD:PPT ginsenoside ratio was not related to postprandial glycemic effects. Instead the ginsenoside Rg1 was identified as the sole independent predictor of these effects. It explained 23% of the change in AUC blood glucose and 3% of the change in mean blood glucose. Although the former value represented a >3-fold improvement over the predictive value of the PPD:PPT ginsenoside ratio reported earlier (Sievenpiper et al. 2004b), >75% of the variance in postprandial glycemia remained to be explained. Again, the suggestion was that other unmeasured ginsenoside or nonginsenoside components were involved.

The first study from the *Systematic Clinical Testing Program of Ginseng Fractions* offered no further clarification. Ginsenoside compositions were modeled against indices of plasma glucose and insulin in the double-blind, randomized, controlled, head-to-head, multiple-crossover comparison of the effect of batches of American ginseng root from five Ontario farms (A–E: sources #5–9) at doses of 9 g given as single, oral doses −40 min preprandially on postprandial plasma glucose and insulin indices during a 75 g OGTT in 12 normoglycemic subjects. Pearson correlations and stepwise multiple regressions identified no significant correlates or multivariate models of independent predictors of plasma glucose and insulin indices (Dascalu et al. 2007).

Single and multivariate regression models implicating the PPD:PPT ginsenoside ratio and ginsenoside Rg1 are clearly not enough to explain the full glycemic effects. There is a need to disentangle the relative contributions the different ginsenoside and nonginsenoside components (ginsenans, panaxans, quinquefolans, eleutherans, peptides, degradation products, etc.) have on glycemic effects through direct clinical investigations. One option would be to perform direct testing on the most promising components identified from the present set of studies, namely, on isolated ginsenoside Rg1 and pure extracts of PPD and PPT ginsenosides with a PPD:PPT ratio ≥2.1. Another option is to undertake the *Systematic Clinical Testing Program of Ginseng Fractions* in humans that was briefly described above. This program includes five acute "Phase I" type studies in healthy subjects that follow the same placebo-controlled acute 75 g OGTT protocol used in the present set of clinical studies. Each of the studies proceeds in a stepwise fashion deriving from the last. As described above, the first step of selecting an efficacious batch of American ginseng for fractionation has already been completed (Dascalu et al. 2007) with selection of American ginseng root from farm A (source#5). With this starting material, the exciting next steps are now under way. The proposed second study will compare five-dose levels of this batch of American ginseng root from farm A (source#5). The most efficacious dose will then advance to the third study, in which isolated *macrocomponents* (total saponins fraction, total polysaccharide fraction, and their combination) will be compared at a dose that is bioequivalent to the most efficacious dose of whole ginseng determined by the second study. *Medial-components* (classes of ginsenosides and polysaccharides) isolated from the macrocomponents will then advance to the fourth acute study for comparison. Using regression analyses, the strongest candidate *microcomponents* (ginseno-sides, panaxans, ginsenans) will be assessed. These will then advance to the final acute study for comparison. The results of the *Systematic Clinical Testing Program of Ginseng Fractions* will provide invaluable information in elucidating the active components in ginseng that relate to glucose regulation in humans.

Ginseng Source and Metabolic Regulation

Additional effects of various ginseng sources have been observed in humans on related features of metabolic syndrome: dyslipidemia, hypertension, obesity, and impaired fibrinolysis. Again the results appear dependent on the source of ginseng studied. A discussion follows.

Differential Lipid Effects of Ginseng across Sources and when Combined with Dietary Fiber

Cholesterol and lipid lowering effects have been observed in three trials across different ginseng species, batches, and combinations with viscous soluble dietary fiber in assorted patient populations. Korean red ginseng at a dose of 4.5 g/day (1.5 g before each meal tid) improved dyslipidemia, decreasing triglycerides and increasing HDL cholesterol after 7 days in an uncontrolled pilot study of five normal and six hyperlipidemic men (Yamamoto et al. 1983). Conversely, we noticed that 12 weeks of supplementation with the selected Korean red ginseng rootlets (source #1) from our *Korean Red Ginseng Clinical Testing Program* (Figures 12.3 and 12.4) given at a dose of 2 g as an oral agent −40 min preprandially tid (6 g/day) showed no significant effects on cholesterol, lipid, or apolipoprotein markers in 19 type 2 diabetic subjects. There was only a nonsignificant tendency to a reduction in LDL-cholesterol values from week 0 to week 12. More consistent results were noticed for American ginseng sources. In our long-term, randomized, double-blind, crossover study from the *American Ginseng Clinical Testing Program* (Figure 12.2), the selected American ginseng extract (source #2) given at a dose of

1 g as an oral agent −40 min preprandially TID (3 g/day), decreased serum total cholesterol and LDL-cholesterol values, compared with placebo, with an insignificant increase in HDL-cholesterol values in subjects with type 2 diabetes (Vuksan et al. 2001c). We reported similar results in our study investigating the efficacy and safety of the combination of 3 g of American ginseng plus 7 g of a viscous soluble fiber blend, but these effects may relate more to the konjac mannan than to the ginseng (Jenkins et al. 2003b).

Differential Antihypertensive Effects of Ginseng Batch, Preparation, and Species

Effects on indices of blood pressure function have been observed across different ginseng species, batches, and preparations in assorted patient populations. Some of the strongest evidence for blood pressure lowering effects exists for Korean red ginseng. Han and coworkers (1998) showed that Korean red ginseng at a dose of 4.5 g/day (1.5 g before each meal TID) decreased 24 h mean systolic blood pressure compared with placebo after eight weeks in 26 subjects with essential hypertension in a nonrandomized, unblinded, crossover study with a shortened placebo phase (4 vs. 8 weeks). The same authors in a subsequent acute, nonrandomized study showed that 610 mg of a Korean red ginseng water extract significantly decreased blood pressure at 45, 60, and 75 min post intake in normotensive adults (Han et al. 2005). In contrast, the selected Korean red ginseng rootlets (source #1) from our *Korean Red Ginseng Clinical Testing Program* (Figures 12.3 and 12.4) given as a preprandial oral agent at a dose of 2 g −40 min preprandially tid (6 g/day) had no effect on office or 24 h ambulatory blood pressure in a subset of the type 2 diabetes patients. This lack of effect, however, may not apply to the others preparations studied in our *Korean Red Ginseng Clinical Testing Program*. We showed that of the three preparations studied (Table 12.2), the Korean red ginseng root body (source #2) at a dose of 3 g decreased ambulatory blood pressure over 160 min post intake compared with placebo in an acute, randomized, double-blind study in hypertensive subjects (Stavro et al. 2004). The consistency of these observations showing a blood pressure lowering benefit of korean red ginseng is supported by mechanistic studies in humans. Han and coworkers (2005) in their nonrandomized study with 610 mg of a Korean red ginseng water extract showed that the blood pressure effects were concomitant with an increase in breath nitric oxide. Sung and coworkers (2000) further showed that another source of Korean red ginseng root significantly increased forearm blood flow responses, consistent with a nitric oxide mechanism, in seven hypertensive test subjects compared with 10 untreated hypertensive control subjects.

Blood pressure lowering effects have also been reported for Asian ginseng sources in humans. The *Panax ginseng* extract *Ginsana G115* (Pharmaton, Ridgefield, CT) significantly decreased acute blood pressure 2 h after ingestion compared with baseline (Caron et al. 2002). American ginseng has shown less consistent effects on blood pressure. In our long-term study from the *American Ginseng Clinical Testing Program* (Figure 12.2), 8 weeks of supplementation with the selected American ginseng extract (source #2) given at a dose of 1 g as an oral agent −40 min preprandially tid (6 g/day) significantly reduced systolic and diastolic blood pressures compared with placebo in 24 subjects with type 2 diabetes (Stavro et al. 2000). In contrast, two follow-up studies from our clinic showed null effects of six batches of American ginseng selected for an *American Ginseng Clinical Testing Program for Hypertension* (Stavro et al. 2005, 2006). The program used the same systematic "acute-to-chronic" clinical screening model applied successfully in our other clinical testing programs (Figures 12.2–12.4). In the first acute, batch-finding study, the effect of single 3 g doses of batches of American ginseng root from six Ontario farms (A–F) was assessed on ambulatory systolic and diastolic blood pressure measured every 5 min for 160 min, using a double-blind, placebo-controlled design in 16 hypertensive subjects. Five of the six batches of American ginseng were the same as those tested in our *Systematic Clinical Testing Program of Ginseng Fractions*. Neither the individual batches nor the mean of all six batches affected blood pressure compared with the mean of placebo done on two separate occasions. Despite the lack of effect, the batch with the greatest tendency to decrease blood pressure was evaluated in a long-term study. Supplementation for 12 weeks with American ginseng from farm B (source #6) administered as an oral dose of 1.5 g mané and 1.5 g nocté (3 g/day) did not affect 24 h ambulatory blood pressure indices or kidney function compared with placebo in 52 hypertensive subjects. It was concluded that the selected batch of American ginseng had neutral effects in hypertensive subjects.

Differential Antiobesity Effects across Ginseng Species and Batch

Weight-controlling effects of ginseng have been reported in one clinical trial of an unspecified ginseng extract. As described above, Sotaniemi and coworkers (1995) reported that Supplementation with 100 and 200 mg/day of the unspecified ginseng extract resulted in a significant decrease in weight after 8 weeks of treatment compared with baseline in 36 type 2 diabetic subjects. These results, however, could not be reproduced in any of our long-term randomized, placebo-controlled, crossover trials across different sources of American and Korean red ginseng. No changes in weight were reported with longterm supplementation with American ginseng extract given as an oral agent at a dose of 1 g −40 min preprandially tid (3 g/day) in 24 type 2 diabetic subjects (Vuksan et al. 2001a, b); 3 g of American ginseng given in combination with 7 g of the viscous soluble fiber blend in 30 type 2 diabetic subejcts (Jenkins et al. 2003a, b); American ginseng as an oral dose of 1.5 g in the morning and 1.5 g in the evening (3 g/day) in 52 hypertensive subjects (Stavro et al. 2006); or Korean red ginseng rootlets given as a preprandial oral agent at a dose of 2 g −40 min preprandially tid (6 g/day) in 19 type diabetic subjects (Vuksan et al. 2006).

Possible Effects of An American Ginseng Extract on Impaired Fibrinolysis

Improvements in haemostatic parameters have been observed with an American ginseng extract in subjects with type 2 diabetes. In the long-term study with American ginseng extract was given as an oral agent at a dose of 1 g −40 min preprandially tid (3 g/day), a significant reduction in plasminogen activator inhibitor-1 (PAI-1) was observed from week 0 to week 8 in 24 type 2 diabetic subjects. But the comparison with placebo was only approaching significance (Vuksan et al. 2001d).

Mechanisms of Action of Different Ginseng Sources and Components

The mechanisms through which ginseng sources and their various fractions have an effect on glucose and metabolic regulation are not clear. A growing database of in vitro and animal data support four possibilities: modulation of (A) glucose absorption, (B) insulin secretion and binding, (C) glucose transport, and/or (D) glucose disposal (Figure 12.6). A discussion of each follows.

Modulation of Glucose Absorption

There is indirect in vitro and animal evidence to suggest that different sources of ginseng may affect the rate of digestion. An inhibition of neuronal discharge frequency from the gastric compartment of the brain stem in rats by American ginseng root has been observed (Yuan et al. 1998). Inhibition of gastric secretion by Asian ginseng root has also been observed in rats (Suzuki et al. 1991). The result of both may be to slow the digestion of food, decreasing the rate of carbohydrate absorption into portal hepatic circulation. This suggestion is supported by more direct evidence. It was shown that the addition of Asian ginseng inhibited both glucose and maltose stimulated duodenal muscle movement, as assessed by short circuit current in isolated rat and human duodenal mucosa (Onomura et al. 1999).

Modulation of Insulin Secretion and Binding

Both in vitro and animal data indicate that various ginseng extracts and ginsenosides increase insulin secretion and binding. DPG-3-2 stimulated insulin biosynthesis in mice islets and rat pancreas at concentrations from 0.5 to 1.0 mg/mL (Hitonobu et al. 1982; Kimura et al. 1981a, b). EPG-3-2 increased glucose-stimulated insulin secretion in alloxan diabetic mice at doses from 10 to 50 mg/kg after intraperitoneal injection (Waki et al. 1982). Total ginsenosides from Asian ginseng increased glucose and nonglucose stimulated insulin release from rat islets in a dose-dependent manner at doses from 0.10 to 0.25 mg/mL (Li and Lu 1987).

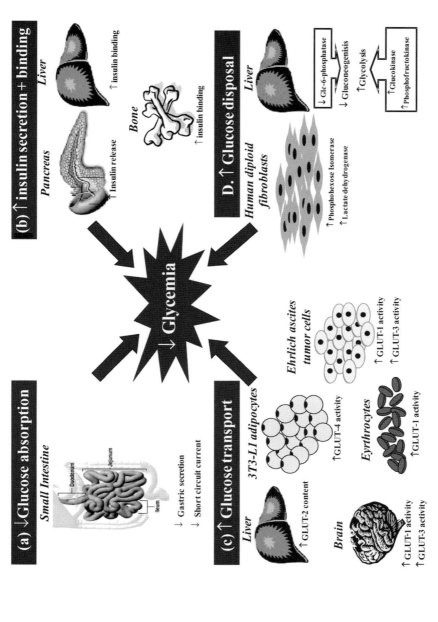

Figure 12.6. Proposed antidiabetic mechanisms of ginseng and its components: (a) modulation of glucose absorption, (b) modulation of insulin secretion and binding, (c) modulation of glucose transport, and (d) modulation of glucose disposal.

Rg_1 increased insulin binding with a $\sim$2-fold increase in the total number of binding sites in rat liver and brain 3 days and 5 days after 3 consecutive days of intraperitoneal injections at 10 mg/kg/day (Tchilian et al. 1991). An Asian ginseng extract with 40% total ginsenosides/saponins increased insulin binding by $\sim$4-fold in the bone marrow of older rats at an intragastric dose of 100 mg/kg/day for 12 days (Yushu and Yuzhen 1988). Finally, panaxan B increased glucose stimulated insulin secretion in a dose-dependent manner at doses from 0 to 10 mg/kg with a maximal increase that was 6-fold higher than control and increased insulin binding 5 h after intraperitoneal injection in normal mice (Suzuki and Hikino 1989).

Modulation of Glucose Transport

In vitro studies suggest that various ginseng extracts and components may increase glucose transport in various cell lines. A water extract of Asian ginseng increased glucose transporter (GLUT) GLUT-2 protein in the livers of normal and hyper-glycemic mice (Ohnishi et al. 1996). PPT, (20R)-PPD, Rg_1, Rc, Rd, Re, Rf, Rg_2, Rh_1, Rb_1, and Rb_2 increased 2-deoxy-D-(2-^{3}H) glucose uptake in isolated sheep erythrocytes by GLUT-1 in a dose dependent manner at doses from 0.01 to 10 μM (Hasegawa et al. 1994). The standardized Asian ginseng extract G115$^{\circledR}$, which has a high ratio of PPD:PPT (Cui 1995), increased 2-deoxy-D-(2-^{3}H) glucose uptake in a dose dependent manner in rabbit brain at doses from 23 to 46 μg/mL (Samira et al. 1985), and Ehrlich ascites tumor cells at doses from 0.5 to 10 μg/mL (Yamasaki et al. 1993). A water extract of Asian ginseng increased basal (nonin-sulin stimulated) 2-deoxy-D-(2-^{3}H) glucose uptake in 3T3-L1 adipocytes at a dose of 100 μg/ml, presumably through GLUT-4 (Hong et al. 2000). Finally, panaxan B increased glucose disappearance during an insulin suppression test at an intraperi-toneal dose of 10 mg/kg after 5 h in normal mice (Yushu and Yuzhen 1988).

These direct effects on glucose transport are further supported by primary effects of ginseng on related intracellular signaling systems. An effect on the peroxisome proliferator activated receptor-γ (PPAR-γ), the target of the thiazolidinediones, has been observed. The aqueous extract of Asian ginseng rootlets increased PPAR-γ protein comparably with rosiglitazone in KKAy mice (Chung et al. 2001). Gin-senosides Rh1 and Rh2 also decreased PKC activity at concentrations from 10 to 100 μM, via a concomitant decrease in its allosteric activator diaclyglycerol (DAG), in NIH 3T3 fibroblasts (Byun et al. 1997). Other effects included decreas-ing effects of ginsenosides Rb1, Rb2, and Rc on tumor necrosis factor-α (TNF-α) in human macrophages at concentrations from 10 to 100 μM (Cho et al. 2001).

Modulation of Glucose Disposal

Data from various animal models of diabetes suggest that different ginseng extracts and ginsenosides may increase glucose disposal. The saponin fraction from G115$^{\circledR}$ at a dose of 200 μg/mL in human diploid fibroblasts increased the glycolytic enzyme, phosphohexose isomerase, and several isozymes of the pyruvate–lactate shunt enzyme, lactate dehydrogenase, although no changes in intracellular or extracellular glucose concentrations or glucose uptake accompanied these changes (Shia et al. 1982). Oral administration of aqueous extracts of Asian ginseng root body and rootlets at a dose of 500 mg/kg for 28 days decreased the activity of the rate limiting gluconeogenic enzyme glucose-6-phophatase (G6Pase) in liver preparations of KKAy diabetic mice by 46% and 20% (Chung et al. 2001). Asian ginseng berry extract at 150 mg/kg and American ginseng leaf at 50 and 150 mg/kg significantly increased glucose disappearance in ob/ob mice after 12 days (Attele et al. 2002; Xie et al. 2004). Finally, Rb_2 increased the activity of the rate-limiting glycolytic enzymes phosphofructokinase and pyruvate kinase while decreasing the activity of the rate-limiting gluconeogenic enzyme G6Pase in liver from normal rats after a single 10 mg intraperitoneal injection (Yokozawa et al. 1984) and increased the activity of glucokinase while decreasing the activity of G6Pase in liver from streptozotocin diabetic rats after intraperitoneal injection at 10 mg/day for 6 days (Yokozawa et al. 1985). These mechanisms are also supported by the same primary cellular signaling mechanisms described above for improvements in glucose transport.

Interpretation

Taken together, these mechanistic investigations using in vitro and animal models of diabetes (Figure 12.6) provide biological plausibility for the various species, extracts, and preparations of ginseng that have been shown to improve glycemic and metabolic regulation in clinical models. These investigations also provide direct evidence for the role of ginsenosides and panaxans in mediating these improvements. A large spectrum of possible active components has been generated. Some of the most promising candidates for which there are consistent data across different models, species, doses, modes of administration, and investigator groups include Re (Attele et al. 2002; Hasegawa et al. 1994), Rb_2 (Hasegawa et al. 1994; Yokozawa et al. 1984, 1985), and panaxan B (Konno et al. 1984; Suzuki and Hikino 1989). The data are not so robust for other components, as most have not yet had their effects reproduced, while in other cases the data are conflicting, with hypoglycemic and hyperglycemic effects having been shown in different studies with the same component. Examples of components that fall into this latter category include Rg_1 (Joo and Kim 1992; Hasegawa et al. 1994; Martinez and Staba 1984; Tchilian et al. 1991) and PPT (Han et al. 1999; Hasegawa et al. 1994).

Conclusions

Although the traditional and clinical evidence for generic ginseng is largely aligned in supporting a diabetes indication, the safety and efficacy of specific ginseng sources in improving glycemic and metabolic outcomes in humans remain questionable. We have described three robust systematic, "acute-to-chronic," clinical screening models that identified ginseng sources with reproducible and sustainable safety and efficacy: *The American Ginseng Clinical Testing Program*, *Korean Red Ginseng Clinical Testing Program*, and *Systematic Clinical Testing Program of Ginseng Fractions* (as applied to the study of the combination therapy of American ginseng with a viscous soluble fiber blend). All three programs provide evidence that a stepwise series of well-controlled, blinded, acute, "Phase I," RCTs in humans can identify reproducible efficacious and safe ginseng sources, doses, and modes of administration that when applied to long-term, "Phase II," RCTs achieve sustainable efficacy and safety in subjects with type 2 diabetes. Despite the success of these models, their results are not necessarily representative. Not all ginseng sources exhibit the same magnitude or direction of glycemic effects in subjects with and without diabetes. The high variability in ginsenosides reported across different sources of ginseng, assay techniques, and types of ginsenosides may explain these discrepancies. Highly variable acute glycemic effects have been observed secondary to the ginsenoside profile, as it varies across ginseng source parameters of increasing variability (batch, preparation, farm, variety, and species). Variability in other unmeasured ginsenoside and nonginsenoside components (ginsenans, panaxans, quinquefolans, eleutheranns, peptides, etc.) also cannot be excluded. The clinical consequences of this compositional variability may be that the antidiabetic efficacy and safety of different batches will be equally highly variable and, as such, may not be generalizable to other over-the-counter sources of ginseng. In the absence of generalizable effects, neither the consumer nor the healthcare professional can be assured of the safety and efficacy from one ginseng product to the next and the call for more RCTs of generic ginseng products is rendered moot.

This problem necessitates a basis for ginseng standardization. In the present review of the literature, the best evidence for reproducibility and sustainability of efficacy on glycemic and metabolic regulation was for American and Korean red ginseng across the different source parameters (batches, preparations, extracts, plant parts, farms, varieties, and species), doses (100 mg – 9 g), modes of administration (prandial or nonprandial), study designs (crossover vs. parallel randomized vs. nonrandomized, and controlled vs. uncontrolled), protocols (acute (fasting, postprandial, or second meal) vs. chronic dosing), length of long-term follow-up (7 days–6 months), and patient cohorts (normal, diabetic, hyperlipidemic, and hypertensive subjects). Seven distinct sources of American ginseng and six distinct sources of Korean red ginseng showed improvement on at least one glycemic or metabolic marker. This was without any of the other sources of either type of ginseng showing an adverse effect on a single glycemic or metabolic marker. Reproducible and sustainable improvements were on the order of 9.1–43% for American ginseng and from 14 to 29% for Korean red ginseng compared with control. Their effects were also weakly but independently predicted by the PPD:PPT ginsenoside ratio and the content of the ginsenoside Rg1, explaining from 4 to 23% of the variation in postprandial glycemia. These studies are further supported by in vitro and animal models of diabetes that have provided biological plausibility for the role of these types of ginseng and components, which also include Re, Rb2, and panaxan B. Nevertheless, these observations do not provide a sufficient means for standardization. In the absence of confirmatory clinical efficacy testing, it remains unclear which sources of ginseng have antihyperglycemic efficacy and which ginsenoside or nonginsenoside components confer this efficacy. Direct systematic clinical testing of ginseng fractions in humans to identify the active components is the only viable means to support standardization claims for reproducible and sustainable efficacy in type 2 diabetes. To this end, our laboratory is presently undertaking a stepwise assessment of the effect of progressively smaller ginseng fractions isolated from a single efficacious batch of American ginseng on markers of acute postprandial glycemia in humans as part of the *Systematic Clinical Testing Program of Ginseng Fractions*. The results of this program are eagerly awaited.

Acknowledgments

This work was supported by grants from the Canadian Diabetes Association, Canadian Institutes of Health Research (CIHR), Korean Ministry of Agriculture and Forestry, Ontario ginseng growers Association, and Ontario ministry of Agriculture and rural affairs.

Duality of Interest

VV has received research funding from the Ontario Ginseng Growers Association, Simcoe, ON; Ginseng Growers of Canada, Simcoe, ON; Ontario Ministry of Agriculture and rural affairs, Toronto, ON, the Korean Ministry of Agriculture and Forestry, Seoul, South Korea.

References

American Diabetes Association. 2007. Diagnosis and classification of diabetes mellitus. Diabetes Care 29:s43–s48.

Angell M, Kassirer JP. 1998. Alternative Medicine—the risks of untested and unregulated remedies. N Engl J Med 339:839–841.

Anonymous. Efficacy of ginseng. 2005. Controls diabetes mellitus. The medicinal effects of Korean red ginseng. Available at http://www.kgc.or.kr/eng/04r&d/r&d_efficacy01.html. Accessed on February 1, 2005.

Attele AS, Wu JA, Yuan CS. 1999. Ginseng pharmacology: Multiple constituents and multiple actions. Biochem Pharmacol 58:1685–1693.

Attele AS, Zhou YP, Xie JT, Wu JA, Zhang L, Dey L, Pugh W, Rue PA, Polonsky KS, Yuan CS. 2002. Antidiabetic effects of *Panax ginseng* berry extract and the identification of an effective component. Diabetes 51:1851–1858.

Awang DVC. 2000. The neglected ginsenosides of North American ginseng (*Panax quinquefolius* L.). J Herbs Spices Med Plants 7:103–109.

Buettner C, Phillips RS, Davis RB, Gardiner P, Mittleman MA. 2007. Use of dietary supplements among United States adults with coronary artery disease and atherosclerotic risks. Am J Cardiol 99:661–666.

Byun BH, Shin I, Yoon YS, Kim SI, Joe CO. 1997. Modulation of protein kinase C activity in NIH 3T3 cells by plant glycosides from *Panax ginseng*. Planta Med 63:389–392.

Caron MF, Hotsko AL, Robertson S, Mandybur L, Kluger J, White CM. 2002. Electrocardiographic and hemodynamic effects of *Panax ginseng*. Ann Pharmacother 36:758–763.

Chen SE, Sawchuk RJ, Staba EJ. 1980. American ginseng. III. Pharmacokinetics of ginsenosides in the rabbit. Eur J Drug Metab Pharmacokinet 5:161–168.

Cho JY, Yoo ES, Baik KU, Park MH, Han BH. 2001. In vitro inhibitory effect of protopanaxadiol ginsenosides on tumor necrosis factor (TNF)-alpha production and its modulation by known TNF-alpha antagonists. Planta Med 67:213–218.

Chung SH, Choi CG, Park SH. 2001. Comparisons between white ginseng radix and rootlet for antidiabetic activity and mechanism in KKAy mice. Arch Pharm Res 24:214–218.

Cui JF. 1995. Identification and quantification of ginsenosides in various commercial ginseng preparations. Eur J Pharm Sci 3:77–85.

Dascalu A, Sievenpiepr JL, Arnason JT, Leiter LA, Vuksan V. 2007. Five Batches Representative of Ontario-grown American Ginseng Root (*Panax quinquefolius* L.) Produce Comparable Reductions of Postprandial Glycemia in Healthy Individuals. Can J Physiol Pharmacol 85:856–864

Eisenberg DM, Davis RB, Ettner SL, Appel S, Wilkey S, Van Rompay M, Kessler RC. 1998. Trends in Alternative Medicine use in the United States, 1900–1997: Results of a Follow-up National Survey. JAMA 280:1569–1575.

Fitzloff JF, Yat P, Lu ZZ, Awang DVC, Arnason JT, van Breeman RB, Hall T, Blumethal M, Fong HHS. 1998. Perspectives on the quality control assurance of ginseng products in North America. In Advances in Ginseng Research: Proceedings of the 7th International Symposium on Ginseng, edited by Huh H, Choi KJ, Kim YC, pp. 138–145. Seoul, Korea: Korean Society of Ginseng.

Han KH, Choe SC, Kim HS, Sohn DW, Nam KY, Oh BH, Lee MM, Park YB, Choi YS, Seo JD, Lee YW. 1998. Effect of red ginseng on blood pressure in patients with essential hypertension and white coat hypertension. Am J Chin Med 26:199–209.

Han HJ, Park SH, Koh HJ, Nah SY, Shin DH, Choi HS. 1999. Protopanaxatriol ginsenosides inhibit glucose uptake in primary cultured rabbit renal proximal tubular cells by arachidonic acid release. Kidney Blood Press Res 22:114–120.

Han KH, Shin IC, Choi KJ, Yun YP, Hoing JT, Oh KW. 2005. Korean red ginseng water extract increased nitric oxide concentrations in exhaled breath. Nitric Oxide 12:159–162.

Hasegawa H, Matsumiya S, Murakami C, Kurokawa T, Kasai R, Ishibashi S, Yamasaki K. 1994. Interactions of ginseng extract, ginseng separated fractions, and some triterpenoid saponins with glucose transporters in sheep erythrocytes. Planta Med 60:153–157.

Hasegawa H, Sung JH, Matsumiya S, Uchiyama M. 1996. Main ginseng saponin metabolites formed by intestinal bacteria. Planta Med 62:453–457.

Hitonobu IW, Masatoshi Y, Kimura M. 1982. Effects of hypoglycmic component of ginseng radix on insulin biosynthesis in normal and diabetic animals. J Pharm Dyn 5:547–554.

Hong SJ, Fong JC, Hwang JH. 2000. Effects of crude drugs on glucose uptake in 3T3-L1 adipocytes. Kaohsiung J Med Sci 16:445–451.

Jenkins AL, Morgan L, Bishop J, Zdravkovc U, Di Buono M, Vuksan V. 2002a. Insulin sparing effect of Konjac mannan fiber when used in combination with American ginseng in type 2 diabetes. FASEB J 16:A647.

Jenkins AL, Morgan L, Bishop J, Zdravkovc U, Vuksan V. 2002b. Reduced glycemia of second meal with Konjac mannan fiber with or without American ginseng in type 2 diabetes. Diabetes 51(Suppl 2):A601.

Jenkins AL, Morgan L, Bishop J, Zdravkovc U, Vuksan V. 2003a. Reduction in HbA1c after long-term administration of American ginseng and Konjac mannan fiber in type diabetes. Diabetes 52(Suppl 1):A386.

Jenkins AL, Morgan L, Zdravkovic U, Naeem, A, Sievenpiper JL, Vuksan V. 2003b. Improved metabolic control after long-term treatment with American Ginseng and Konjac mannan fiber in type 2 diabetes. Can J Diabetes 27:357.

Joo C-N, Kim JH. 1992. Study on the hypoglycemic action of ginseng saponin on strepozotocin induced diabetic rats. Korean J Ginseng Sci 16:190–197.

Kahn SE, Haffner SM, Heise MA, Herman WH, Holman RR, Jones NP, Kravitz BG, Lachin JM, O'Neill MC, Zinman B, Viberti G; ADOPT Study Group. 2006. Glycemic durability of rosiglitazone, metformin, or glyburide monotherapy. N Engl J Med 355:2427–2443.

Kajiwara H, Hemmings AM, Hirano H. 1996. Evidence of metal binding activities of pentadecapeptide from *Panax ginseng*. J Chromatogr B 687:443–448.

Kimura M, Isami W, Tanaka O, Nagi Y, Shibita S. 1981a. Pharmacological sequential trials for the fraction of components with hypoglycemic activity in alloxan diabetic mice from ginseng radix. J Pharm Dyn 4:402–409.

Kimura M, Waki I, Chujo T, Kikuchi T, Hiyama C, Yamazaki K, Tanaka O. 1981b. Effects of hypoglycemic components in ginseng radix on blood insulin level in alloxan diabetic mice and one insulin release from perfused rat pancreas. J Pharm Dyn 4:410–417.

Konno C, Sugimyama K, Kano M, Hikino H. 1984. Isolation and hypoglycemic activity of panaxans A, B, C, D, and E glycans of *Panax ginseng* roots. Planta Med 50:434–436.

Li G, Lu Z. 1987. Effect of ginseng saponins on insulin release from isolated pancreatic islets of rats. Zhongguo Zhong Xi Yi Jie He Za Zhi 7:357–359.

Li TSC, Mazza G, Cottrell AC, Gao L. 1996. Ginsenosides in roots and leaves of American ginseng. J Agric Food 44:717–720.

Li W, Gu C, Zhang H, Awang DV, Fitzloff JF, Fong HH, van Breemen RB. 2000. Use of high-performance liquid chromatography-tandem mass spectrometry to distinguish *Panax ginseng* C. A. Meyer (Asian ginseng) and *Panax quinquefolius* L. (North American ginseng). Anal Chem 72:5417–5422.

Lipscombe LL, Hux JE. 2007. Trends in diabetes prevalence, incidence, and mortality in Ontario, Canada 1995–2005: A population-based study. Lancet 369:750–756.

Ma Y, Mai L, Malley L, Doucet M. 1996. Distribution and proportion of major ginsenosides and quality control of ginseng products. Chin J Med Chem 6:11–21.

Martinez B, Staba EJ. 1984. The physiological effects of Aralia, Panax and Eleutherococcus on exercised rats. Jpn J Pharmacol 35:79–85.

Mihalov JJ, Marsderosian AD, Pierce JC. 2000. DNA identification of commercial ginseng samples. J Agric Food Chem 48:3744–3752.

Mokdad AH, Ford ES, Bowman BA, Dietz WH, Vinicor F, Bales VS, Marks JS. 2003. Prevalence of obesity, diabetes, and obesity-related health risk factors, 2001. JAMA 289:76–79.

Nathan DM. 2006. Thiazolidinediones for initial treatment of type 2 diabetes? N Engl J Med 355:2477–2479.

Ng TB, Yeung HW. 1985. Hypoglycemic constituents of *Panax ginseng*. Gen Pharmacol 16:549–552.

Nguyen MD, Kasai R, Ohtani K, Ito A, Nguyen TN, Yamasaki K, Tanaka O. 1994. Saponins from Vietnamese Ginseng, *Panax* vietnamensis HA et Grushv. Collected in central Vietnam. II. Chem Pharm Bull 42:115–122.

Nissen SE, Wolski K. 2007. Effect of rosiglitazone on the risk of myocardial infarction and death from cardiovascular causes. N Engl J Med 356:2457–2471. Erratum in: N Engl J Med 2007;357:100.

Ohnishi Y, Takagi S, Miura T, Usami M, Kako M, Ishihara E, Yano H, Tanigawa K, Seino Y. 1996. Effect of ginseng radix on GLUT2 protein content in mouse liver in normal and epinephrine-induced hyperglycemic mice. Biol Pharm Bull 19:1238–1240.

Onomura M, Tsukada H, Fukuda K, Hosokawa M, Nakamura H, Kodama M, Ohya M, Seino Y. 1999. Effects of ginseng radix on sugar absorption in the small intestine. Am J Chin Med 27:347–354.

Reay JL, Kennedy DO, Scholey AB. 2005. doses of *Panax ginseng* (G115) reduce blood glucose levels and improve cognitive performance during sustained mental activity. J Psychopharmacol. 19(4):357–365.

Reay JL, Kennedy DO, Scholey AB. 2006a. Effects of *Panax ginseng*, consumed with and without glucose, on blood glucose levels and cognitive performance during sustained "mentally demanding" tasks. J Psychopharmacol 20:771–781.

Reay JL, Kennedy DO, Scholey AB. 2006b. The glycaemic effects of single doses of *Panax ginseng* in young healthy volunteers. Br J Nutr 96:639–642.

Ren G, Chen F. 1999. Simultaneous quantification of ginsenosides in American ginseng (*Panax quinquefolium*) root powder by visible/near-infrared reflectance spectroscopy. J Agric Food Chem 47:2771–2775.

Samira MM, Attia MA, Allam M, Elwan O. 1985. Effect of the standardized Ginseng Extract G115 on the metabolism and electrical activity of the rabbit's brain. J Int Med Res 13:342–348.

Shia GT, Ali S, Bittles AH. 1982. The effects of ginseng saponins on the growth and metabolism of human diploid fibroblasts. Gerontology 28:121–124.

Sievenpiper JL, Arnason JT, Leiter LA, Vuksan V. 2003a. Differential effects of American ginseng: A batch of American ginseng (*Panax quinquefolius* L.) with a depressed ginsenoside profile does not affect postprandial glycemia. Eur J Clin Nutr 57:243–248.

Sievenpiper JL, Arnsoson JT, Leiter LA, Vuksan V. 2003b. Possible opposing effects of Asian ginseng (*Panax ginseng* C.A. Meyer) on glycemia: Results of two acute dose escalation studies. J Am Coll Nutr 22:524–532.

Sievenpiper JL, Arnasosn JT, Leiter LA, Vuksan V. 2004a. A quantitative systematic analysis of the literature of the high variability in ginseng (*Panax* spp.): Should it be trusted? Diabetes Care 27:839–840.

Sievenpiper JL, Arnsoson JT, Leiter LA, Vuksan V. 2004b. Decreasing, null, and increasing effects of eight popular types of ginseng on acute postprandial glycemic indices: The role of ginsenosides. J Am Coll Nutr 50:23:248–258.

Sievenpiper JL, Sung MK, Di Buono M, Lee KS, Nam KY, Arnason JT, Leiter LA, Vuksan V. 2006. Korean red ginseng rootlets decrease acute postprandial glycemia: results from sequential preparation- and dose-finding studies. J Am Coll Nutr 25:100–107.

Soldati F, Tanaka O. 1984. *Panax ginseng*: Relation between age of plant and content of ginsenosides. Planta Med 50:351–354.

Sotaniemi EA, Haapakoski E, Rautio A. 1995. Ginseng therapy in non-insulin-dependent diabetic patients. Diabetes Care 18:1373–1375.

Stavro PM, Woo M, Heim T, Leiter LA, Vuksan V. 2005. North American ginseng exerts a neutral effect on blood pressure in individuals with hypertension. Hypertension 46:406–411.

Stavro PM, Woo M, Heim T, Leiter LA, Vuksan V. 2006. Long-term intake of North American ginseng has no effect on 24-hour blood pressure and renal function. Hypertension 47:791–796.

Stavro PM, Woo M, Vuksan V. 2004. Korean red ginseng lowers blood pressure in individuals with hypertension (abstract). Am J Hypertension 17:S33.

Stavro PM, Xu Z, Beljan-Zdravkovic U, Jenkins AL, Sievenpiper JL, Vuksan V. 2000. American ginseng improves blood pressure in type 2 diabetes (abstract). Circulation 102:II-417.

Sung J, Han KH, Zo JH, Park HJ, Kim CH, Oh BH. 2000. Effects of red ginseng upon vascular endothelial function in patients with essential hypertension. Am J Chin Med 28:205–216.

Suzuki Y, Hikino H. 1989. Mechanisms of hypoglycaemic activity of panaxans A and B, glycans of *Panax ginseng* roots: Effects on plasma level, secretion, sensitivity and binding of insulin in mice. Phytother Res 3:20–24.

Suzuki Y, Ito Y, Konno C, Furuya T. 1991. Effects of tissue cultured ginseng on gastric secretion and pepsin activity (Japanese). Yakugaku Zasshi 111:770–774.

Tawab MA, Bahr U, Karas M, Wurglics M, Schubert-Zsilavecz M. 2003. Degradation of ginsenosides in humans after oral administration. Drug Metab Dispos 31:1065–1071.

Tchilian EZ, Zhelezarov IE, Hadjiivanova CI. 1991. Effect of ginsenoside Rg1 on insulin binding in mice liver and brain membranes. Phytother Res 5:46–48.

Tetsutani T, Yamamura M, Yamaguchi T, Onoyama O, Kono M. 2000. Can red ginseng control blood glucose in diabetic patients. Ginseng Rev 28:44–47.

Tomoda M, Hirabayashi K, Shimizu N, Gonda R, Ohara N. 1994. The core structure of ginsenan PA, a phagocytosis-activating polysaccharide from the root of *Panax ginseng*. Biol Pharm Bull 17:1287–1291.

Tomoda M, Hirabayashi K, Shimizu N, Gonda R, Ohara N, Takada K. 1993. Characterization of two novel polysaccharides having immunological activities from the root of *Panax ginseng*. Biol Pharm Bull 16:1087–1090.

Turner RC, Cull CA, Frighi V, Holman RR; UK Prospective Diabetes Study (UKPDS) Group. 1999. Glycemic control with diet, sulfonylurea, metformin, or insulin in patients with type 2 diabetes mellitus: Progressive requirement for multiple therapies (UKPDS 49). JAMA 281:2005–2012.

Turner RC, Holman R, Stratton I. 1998. Correspondence. Lancet 352:1934.

Vuksan V, Jenkins DJA, Spadafora P, Sievenpiper JL, Swilley J, Brighenti F, Josse RG, Leiter L. 1999. Bruce-Thompson C Konjac-mannan (glucomannan) improves glycemia and other associated risk factors for coronary heart disease in type 2 diabetes: A randomized controlled metabolic trial. Diabetes Care 22:913–919.

Vuksan V, Sievenpiper JL. 2000. The variable effects of whole-leaf digitalis is a paradigm of the glycemic effects of ginseng-Reply. Arch Intern Med 160:3330–3331.

Vuksan V, Sievenpiper JL. 2005. Herbal remedies in the management of diabetes: Lessons learned from the study of ginseng. Nutr Metab Cardiovasc Dis 15:149–160.

Vuksan V, Sievenpiper JL, Beljan-Zdravkovic U, Di Buono M, Stavro PM. 2001c. American ginseng improves lipid control in patients with type 2 diabetes (abstract). FASEB J 15:A632.

Vuksan V, Sieveniper JL, Koo VYY, Francis T, Beljan-Zdravkovic U, Xu Z, Vidgen E. 2000a. American ginseng reduces postprandial glycemia in nondiabetic and diabetic individuals. Arch Intern Med 160:1009–1013.

Vuksan V, Stavro MP, Sievenpiper JL, Beljan-Zdravkovic U, Leiter LA, Josse RG, Zheng Xu. 2000c. Similar postprandial glycemic reductions with escalation of dose and administration time of American ginseng in type 2 diabetes. Diabetes Care 23:1221–1226.

Vuksan V, Sievenpiper JL, Wong J, Xu Z, Beljan-Zdravkovic U, Arnason JT, Assinewe V, Stavro MP, Jenkins AL, Leiter LA, Francis T. 2001b. American ginseng (*Panax quinquefolius* L.) attenuates postprandial glycemia in a time, but not dose, dependent manner in healthy individuals. Am J Clin Nutr 73:753–758.

Vuksan V, Sievenpiper JL, Xu Z, Jenkins AL, Beljan-Zdravkovic U, Leiter LA, Josse RG, Stavro, MP. 2001a. Konjac-mannan and American ginseng: Emerging alternative therapies for type 2 diabetes. J Am Coll Nutr 20:370s–380s.

Vuksan V, Stavro MP, Sievenpiper JL, Koo VY, Wong E, Beljan-Zdravkovic U, Francis T, Jenkins AL, Leiter LA, Josse RG, Xu Z. 2000b. American ginseng improves glycemia in individuals with normal glucose tolerance: Effect of dose and time escalation. J Am Coll Nutr 19:738–744.

Vuksan V, Sung MK, Sievenpiper JL, Stavro PM, Jenkins AL, Di Buono M, Lee KS, Leiter LA, Nam KY, Arnason JT, Choi M, Naeem A. 2008. Korean red ginseng (Panax ginseng) improves glucose and insulin regulation in well-controlled, type 2 diabetes: results of a randomized, double-blind, placebo-controlled study of efficacy and safety. Nutr. Metab. Cardiovasc. Dis. 18:46–56.

Vuksan V, Xu Z, Sievenpiper JL, Stavro MP, Beljan-Zdravkovic U, Di Buono M, Jenkins AL. 2001d. American ginseng improves plasminogen activator inhibitor-1 concentrations in type 2 diabetes (abstract). Diabetes 50:A368.

Waki I, Kyo H, Yasuda M, Kimura M. 1982. Effects of a hypoglycemic component of ginseng radix on insulin biosynthesis in normal and diabetic animals. J Pharmacobiodyn 5:547–554.

Wang X, Sakuma T, Asafu-Adjaye E, Shiu GK. 1999. Determination of ginsenosides in plant extracts from *Panax ginseng* and *Panax quinquefolius* L. by LC/MS/MS. Anal Chem 71:1579–1584.

Winslow LC, Shapiro H. 2002. Physicians want education about complementary and alternative medicine to enhance communication with their patients. Arch Intern Med 162:1176–1181.

Xie JT, Mehendale SR, Wang A, Hana AH, Wua JA, Osinski J, Yuan CS. 2004. American ginseng leaf: Ginsenoside analysis and hypoglycemic activity. Pharmacol Res 49:113–117.

Yamaguchi H, Kasai R, Matsuura H, Tanaka O, Fuwa T. 1988a. High-performance liquid chromatographic analysis of acidic saponins of ginseng and related plants. Chem Pharm Bull 36:3468–3473.

Yamamoto A. 2001. Improvement of transmucosal absorption of biologically active peptide drugs. Yakugaku Zasshi 121:929–948.

Yamamoto M, Uemura T, Nakama S, Uemiya M, Kumagai A. 1983. Serum HDL-cholesterol-increasing and fatty liver-improving actions of *Panax ginseng* in high cholesterol diet-fed rats with clinical effect on hyperlipidemia in man. Am J Chin Med 11:96–101.

Yamasaki K, Murakami C, Ohtani K, Kasai R, Kurokawa T, Ishibashi S. 1993. Effects of standardized *Panax ginseng* extract G115 on the D-glucose transport by Ehrlich ascites tumour cells. Phytother Res 7:200–202.

Yeh GY, Eisenberg DM, Davis RB, Phillips RS. October 2002. Use of complementary and alternative medicine among persons with diabetes mellitus: Results of a national survey. Am J Public Health 92(10):1648–1652.

Yeh GY, Eisenberg DM, Kaptchuk TJ, Phillips RS. 2003. Systematic review of herbs and dietary supplements for glycemic control in diabetes. Diabetes Care 26:1277–1294.

Yokozawa T, Kobayashi T, Kawai A, Oura H, Kawashima Y. 1984. Stimulation of lipid and sugar metabolism in ginsenoside-Rb2 treated rats. Chem Pharm Bull 32:2766–2772.

Yokozawa T, Kobayashi T, Oura H, Kawashima Y. 1985. Studies on the mechanism of the hypoglycemic activity of ginsenoside-Rb2 in streptozotocin-diabetic rats. Chem Pharm Bull 33:869–872.

Yuan CS, Wu JA, Lowell T, Gu M. 1998. Gut and brain effects of American ginseng root on brainstem neuronal activities in rats. Am J Chin Med 26:47–55.

Yushu H, Yuzhen H. 1988. The effect of *Panax ginseng* extract (GS) on insulin and corticosteroid receptors. J Trad Chin Med 8:292–295.

Chapter 13

Traditional Chinese Medicine in the Management and Treatment of the Symptoms of Diabetes

Azadeh Lankarani-Fard, MD, and Zhaoping Li, MD, PhD

Introduction

Modern Chinese medical practice continues to incorporate elements of Traditional Chinese Medicine (TCM) in all aspects of care. The management and treatment of the symptom of diabetes in China involves the use of traditional Chinese herbs alongside medications commonly used in Western medicine. The evidence for the use of such herbal treatments is rooted in thousands of years of medical history and is undergoing more rigorous scientific investigation today. TCM is a range of traditional medical practices used in China that developed over several thousand years rooted in the philosophy of Taoism. TCM practices include herbal medicine (*Zhong-yao*), acupuncture (*Zhen-ju*), orthopedics surgery (*Zheng Gu, Tui-na*), and energy practice (*Qi-gong*). In the theory of TCM, the processes of the human body are interrelated and constantly interact with the environment. The practice of TCM focuses on the signs of disharmony in the external and internal environment of a person in order to understand, treat, and prevent illness.

One of the oldest medical texts, *Huangdi Neijing* (*The Yellow Emperor's Classic of Medicine*), compiled over 2000 years ago, describes a syndrome of *xiao ke* or "wasting–thirst" which occurred mostly in the wealthy who indulged in sweet and fatty foods (Ho and Lisowski 1997). In some cases, the symptoms reportedly resolved with restriction on dietary intake. The syndrome seems similar to the later stages of type 2 diabetes, the diabetes associated with insulin resistance and obesity.

A more definitive reference to diabetes is found in the eighth century. Wang Tao described a condition of thirst where the person excreted copious, sweet urine. The syndrome was treated with the consumption of pork pancreas and progress could be monitored with interval testing of the urine. The practice of treating diseases with the organ the patient was presumed deficient had started in the seventh century (Ho and Lisowski 1997). This approach is similar to the Western utilization of insulin for the treatment of diabetes and also serves as a basis for current trials on oral tolerance for certain autoimmune diseases such as type 1 diabetes (Achenbach and Fuchtenbusch 2004).

In TCM, diabetes is viewed as a deficiency of Yin, the cold, dark, and inward elements of the body with resulting excess heat or Yang. The Yin and Yang forces interact to form a person's Qi, or vital energy. The progression of diabetes is characterized by the progression of excess heat. It is divided into three stages, with the first stage being the consumption of Yin, the second stage with the loss of Qi, and finally the third stage with the destruction of Yang itself (Li et al. 2004a, b). As the dry heat progresses, blood stasis occurs with alterations in circulation leading to vascular complications such as atherosclerosis, neuropathy, and retinopathy. Different herbal preparations are used at each of these stages to nourish Yin, restore Qi, and remove Yang. Herbs known for their blood revitalizing properties are added if deficiencies in blood flow are noted.

The excess heat is also classified as involving the upper, middle, or lower burner. The upper burner involves excess heat within the lungs resulting in excessive thirst. The middle burner is characterized by excessive hunger or heat within the stomach. The lower burner involves the kidney and is associated with excess urination (Covington 2001). A person could experience symptoms from any and all burners at the same time.

Another disease paradigm involves the theory of five phases. The five basic elements are wood, fire, earth, metal, and water. Each element is in harmony with the next and represents an organ. Wood is associated with the liver; fire with the heart; earth with the spleen, pancreas, and stomach. Metal is associated with the lungs and water with the kidney. Each element is also connected to a color, season, or emotion. The treatment is based on an assessment of each individual's emotional state as well as their personal preferences for color and season.

Such individualized practices make TCM difficult to standardize and assess using modern scientific methods. However, as diabetes and obesity are increasing globally, there will be an increased demand for treatment beyond the current modalities available in Western medicine. As Chinese herbs have been observed to have therapeutic potential for the treatment of diabetes, a more rigorous examination of the formulations is warranted. This chapter will discuss the available evidence behind the use of specific herbs and herbal preparations for the treatment of diabetes in China.

Individual Herbs in Traditional Chinese Medicine

Fructus Balsampear (ku gua): Fruit of Momordica charantia, *Family Curcurbitaceae*

Bitter melon (*Momordica charantia*) is an alternative therapy that has primarily been used for lowering blood glucose levels in patients with diabetes mellitus. Components of bitter melon extract appear to have structural similarities to animal insulin. Antiviral and antineoplastic activities have also been reported in vitro.

In a study to assess the toxicity of the juice and alcohol extract from bitter melon, both extracts induced a significant decrease in serum glucose levels in normal and diabetic rats (Abd El Sattar et al. 2006). The extracts did not show any significant effect in liver and kidney parameters in normal rats. However, in diabetic rats, the two extracts caused a significant decrease in serum urea, creatinine, transaminases, alkaline phosphatase, cholesterol, and triglycerides. These results suggest that bitter melon extracts possess antidiabetic, hepato-renal protective, and hypolipidemic activity. The fruit may also potentiate the effect of other glucose-lowering agents (Tongia et al. 2004).

Some clinical trials have shown bitter melon to have moderate hypoglycemic activity (Ahmad et al. 1999; Khanna et al. 1981; Welihinda et al. 1986). However, these studies were preliminary as they were either small, not double-blinded, or not randomized. Therefore the fruit cannot be routinely recommended as yet.

Fructus Corni (a li shan wu wei zi): Fruit of Cornus officinalis, *Family Cornaceae*

Commonly known as the Dogwood fruit or the Asiatic Cornelian Cherry fruit, the flesh of *Cornus officinalis* has been used traditionally to treat diabetes and prevent diabetic complications. The active components of the extract, ursolic acid and oleanolic acid, have been shown to lower blood sugar in streptozotocin-induced diabetic rats (Liou et al. 2004; Yamahara et al. 1981). Its ability to lower blood sugar is potentially due to improved insulin sensitivity through the up-regulation of glucose transporter 4 (GLUT4) mRNA expression and its protein (Qian et al. 2001). Yamabe et al. (2007) demonstrated that *Fructus corni* played an important role in reducing glucose toxicities, up-regulating renal function, and consequently ameliorating glycation-associated renal damage in streptozotocin-treated diabetic rats. Recent studies further showed that *Fructus corni* attenuates oxidative stress and enhances antioxidant defense in endothelial cells (Peng et al. 1998a, b).

Fructus Ligustri Lucidi (nu zhen zi): Fruits of Ligustrum lucidum, *Family Oleaceae*

Commonly known as the glossy privet fruit, *Ligustrum lucidum* is being increasingly recognized for its antioxidant, hepatoprotective, and hypoglycemic effects.

Isolates from the fruit have been shown to provide protection against carbon tetrachloride hepatotoxicity by up-regulating glutathione regeneration. Oleanolic acid was identified as the active agent involved in preventing liver injury (Yim et al. 2001). Glucosides from the fruit have shown antioxidant activity against red blood cell lysis (He et al. 2001). The oleanolic acid fraction has shown some weak antioxidant activity as well (Lin et al. 2006). Evidence for the glucose lowering ability of the fruit is more sparse. Preparations of privet fruit have been shown to lower blood sugar in alloxan-treated diabetic mice at high doses such as 30 g/kg (Hao et al. 1992).

Fructus Lycii and Cortex Lycii Radicis (gou qi zi): Fruits and Root of Lycium barbarum, *Family Solanaceae*

Commonly known as goji or wolfberry, *Lycium barbarum* is being increasingly studied as an antioxidant, a hypoglycemic agent, and a lipid-lowering supplement. The berry is typically dried, ground into a powder, and reconstituted in boiling water. This hot-water preparation has been shown to lower the blood sugar of alloxan-induced diabetic rabbits. Further purification has identified the *Lycium barbarum* polysaccharide fraction as the active hypoglycemic agent (Luo et al. 2004). The same study also revealed the fruit to have significant lipid-lowering and antioxidant effects. This glucose lowering effect has been shown in animal models to be due to increased insulin sensitization as a result of up-regulation of GLUT4 in skeletal muscle (Zhao et al. 2005). GLUT4 is responsible for glucose uptake and utilization in skeletal muscle.

Fructus Mori, Folium Mori, and Cortex Mori Radicis (sang ye): Fruits, Leaves, and Bark of Morus alba, *Family Moraceae*

Alcohol extraction and further fractionation from the root and bark of *Morus alba*, or the white mulberry, yield a glycoprotein, moran A. This glycoprotein has been shown to have hypoglycemic properties in normal and chemically induced diabetic mice (Chen et al. 1995; Hikino et al. 1985). More recent studies have shown that extracts from the leaves of *M. alba* inhibit intestinal disaccharidases in humans and rats (Oku et al. 2006). It is this mechanism which may serve to lower blood sugar.

Fructus Schisandrae (wu wei zi): Fruits of Schisandra chinensis *and* Schisandra sphenanthera, *Family Mognoliaceae*

Chinese magnolia vine fruit is commonly used to invigorate the kidney, activate Yang, and increase blood flow especially to sexual organs. The fruit is used as a supplement in the treatment of diabetic complications such as excessive thirst or urination, or for the treatment of sexual dysfunction. Its ability to enhance circulation may be due in part to the ability to promote nitric oxide release and alter calcium influx in vascular smooth muscle (Lee et al. 2004; Rhyu et al. 2006). In addition, the fruit and its derivatives have been shown in rats to be protective against liver injury and are commonly used in China as treatment for chemical and viral-induced hepatitis (Ko et al. 1995; Mak and Ko 1997).

Lignans, such as schisanhenol, schisandrin B and C, have been identified as active components. The antioxidant effect of lignans may have a role in its hep-atoprotective properties (Ko et al. 1995; Lu and Liu 1991). Schisandrin B protects against carbon tetrachloride-induced liver damage primarily through the regener-ation of glutathione (Chiu and Ko 2006; Ip et al. 1996, 2000). There is evidence that this antioxidant effect also protects other tissues such as the brain and the heart (Kim et al. 2004; Ko and Yiu 2001; Xue et al. 1992).

Schisandrin B has also been shown to elevate triglycerides when given to mice at doses of 0.05–0.8 g/kg (Pan et al. 2006). This elevation may be attributed to the increased conversion of free fatty acids to triglycerides in the presence of an antioxidant.

Gynostemmae Herba (jiao gu lan): Stems and Leaves of Gynostemma pentaphyllum, *Family Cucurbitaceae*

Gynostemma pentaphyllum is sometimes called "southern ginseng" in China, because of the similarity of its action to ginseng. The active component of the drug in the treatment of diabetes is felt to be the saponins. These saponins are felt to be identical to components found in ginseng (Cui et al. 1999).

Several types of saponins have been isolated from *G. pentaphyllum*. These sub-stances are also known as gypenosides. Purified gypenosides from the herb have been found to reduce blood sugars in obese but not lean rats suggesting the extract has an insulin sensitizing effect (Megalli et al. 2006). There is also a dose-dependent reduction in triglycerides, cholesterol, and low-density lipoproteins. One gypeno-side, called phanoside, has been recently found to be an initiator and potentiator of insulin secretion in a rat model (Norberg et al. 2004). The hypoglycemic effect was shown to be similar or more potent than a sulfonylurea when compared in vitro. In addition to its blood sugar-lowering and lipid-lowering effects, extracts from *G. pentophyllum* have been shown to have hepatoprotective effects (Chen et al. 2000; Lin et al. 2000).

Hirudo (shui zhi): Whole Body of Whitmania pigra *and Several Species of the Same Family Hirudinidae*

Dried leeches are pulverized into powder form for medicinal use in conditions of blood stasis. Water and alcohol extracts have been shown to have fibrinolytic activity (Ding et al. 1994; Wang et al. 1989). In a rat model of blood stasis, hirudo has been shown to reduce vascular endothelial cell damage (Li et al. 2001). Leeches have long been known for their anticoagulant ability. The drug lep-irudin is commonly used in Western medicine as an anticoagulant in patients with heparin-induced thrombocytopenia. This drug is the recombinant form of the polypeptide hirudin which is derived from the saliva of the fresh leech, *Hirudo medicinalis* (Cheng-Lai 1999; Greinacher et al. 1999). Studies have shown that the dried leeches commonly used in TCM retain their fibrinolytic properties as well. The compound whitmanin, extracted from dried *Whitmania pigra,* has been shown to have potent anticoagulant activity as measured by plasma recalcification time (Zhong et al. 2007).

Poria (fu ling): **Poria cocos, Family Polyporaceae**

Poria is a fungus commonly found on pine trees and is used in TCM for conditions of fluid congestion. The fungus is often used as part of a formula such as Rehamania 8 or Hoelen. Studies of obese male *db/db* mice, a model for type 2 diabetes, suggest that poria has a hypoglycemic and insulin-sensitizing effect. The active isolate from the fungus was identified as a terpinoid, dehydrotrametenolic acid, which acts as a ligand for peroxisome proliferator-activated receptor gamma (PPAR gamma). This receptor is involved in insulin sensitization and is a target for diabetic drugs such as thiazolidinediones (Sato et al. 2002). *Poria cocos* has also been found to improve circulation and have antioxidant activity. Extracts from the fungus have been shown to decrease superoxide anion formation (Wu et al. 2004) and act as an anti-inflammatory by inhibiting leukotriene B4 release (Prieto et al. 2003). Mixtures containing *P. cocos* are associated with improvement in cerebral blood flow and decreased endothelial cell damage (Jingyi et al. 1997; Kim et al. 2003).

Radix Astragali (huang qi): Dried Roots of **Astragalus membranaceus, *Family Leguminosae***

Radix Astragali is the dried root of the perennial herbs, *Astragalus membranaceus* and *A. mongholicus*. This herb appears in almost every antidiabetic compound recipe. Radix Astragali is used in Chinese medicine to invigorate Qi and strengthen the body. The active constituents of Radix Astragali consist of primarily polysaccharides and flavonoids. Astragalus polysaccharides have been shown to significantly lower blood glucose (Tsutsumi et al. 2003; Wu et al. 2005), slow the progression of diabetic nephropathy (Yin et al. 2006; Zhang et al. 2006), protect against cardiomyopathy (Li et al. 2004a, b), and decrease lipid peroxidation (Toda et al. 2000) in diabetic rats and mice.

In clinical trials, *A. membranaceus* was associated with decreased urinary albumin excretion and improved endothelial function in type 2 diabetics with microalbuminuria (Lu et al. 2005).

Radix Ginseng (ren shen): Roots and Rhizomes of **Panax ginseng, *Family Araliaceae***

Several different plant species are often referred to as ginseng. These include Chinese or Korean ginseng (*Panax ginseng*), Siberian ginseng (*Eleutherococcus senticosus*), American ginseng (*P. quiquefolius*), and Japanese ginseng (*P. japonicus*).

Panax species (from the root panacea) are often touted for their "cure-all" adaptogenic properties, immune-stimulant effects, and their ability to increase stamina, concentration, longevity, and overall well being. Most preparations use the herb's root, and some sources report greater efficacy with roots that are greater than 3-year old. The principal components are believed to be the triterpenoid saponin glycosides known as ginsenosides or panaxosides.

Hypoglycemic effects have been shown in streptozotocin rat models (Shapiro and Gong 2002). Cho and colleagues (2006) demonstrated that ginsenoside could lower blood glucose and lipid levels, and protect against oxidative damage in the eye and kidney of diabetic rats. This finding was confirmed by another independent study (Kang et al. 2006). The proposed mechanisms of action include decreased rate of carbohydrate absorption into the portal hepatic circulation, increased glucose transport and uptake mediated by nitric oxide, increased glycogen storage, and modulation of insulin secretion.

Vuksan and colleagues (2000a, b) studied American ginseng and examined the short-term effects on patients with type 2 diabetes after a standard oral glucose tolerance test. Two longer-term trials involving a sample size of 24 and 36 patients, administered American ginseng for 8 weeks. Both reported decreases in fasting blood glucose and HbA$_{1c}$ (Sotaniemi et al. 1995; Vuksan et al. 2001b). Three other short-term metabolic trials in healthy volunteers also found decreases in postprandial glucose (Vuksan et al. 2000b, 2001a, 2008). The available evidence for American ginseng in diabetes suggests a possible hypoglycemic effect; however, the trials are small and longer-term studies are needed.

The mechanism of action of ginseng is to decrease intestinal glucose absorption (Jia et al. 2003b), promote insulin secretion (Lee et al. 2006a, b; Luo and Luo 2006), and enhance insulin sensitivity (Wang et al. 2003). Ginseng was also reported to protect against nephropathy (Kang et al. 2006).

Radix Ophiopogonis (Mai Dong): Root Tubers of Ophiopogon japonicus, *Family Liliaceae*

Ophiopogon japonicus, commonly known as the Dwarf lilyturf tuber, is typically combined with other herbs, especially ginseng, when used for the treatment of diabetes. A commonly used preparation, Xiao-ke-an, has *O. japonicus* as well as *Astragalus mongholicus*, *Pueraria lobata*, and *Hirudo nipponia* (Jia et al. 2003a). The tuber is felt to nourish Yin, moisten dryness, and treat diabetes due to internal heat. The alcohol extract from the herb has shown moderate hypoglycemic activity in reducing the blood sugar of streptozotocin-induced diabetic rates. Average blood sugar reduction was about 50 mg/dL (Kako et al. 1995). The mechanism for blood sugar reduction is unclear. The herb has been shown to have antithrombotic and anti-inflammatory properties (Kou et al. 2005a, b). Several homoisoflavones have been isolated from *O. japonicus* (Chang et al. 2002; Hoang Anh et al. 2003; Ye et al. 2005). Although the ability of these homoisoflavones to lower blood sugar is not known, other members of the Lily family have been shown to have homoisoflavones with insulin-sensitizing activity (Choi et al. 2004).

Radix Puerariae (ge gen): Roots of **Pueraria lobata,** *Family Leguminosae*

Radix Pueraiae is the root of *Pueraria lobata*. It has been reported to have comprehensive pharmacological action in treatment of diabetes and cardiovascular diseases. The active hypoglycemic components of the root are flavonoids. Puerarin is an isoflavone extracted from the roots of *P. lobata*. It has been shown to lower plasma glucose in streptozotocin-induced diabetic rats (Chen et al. 2004; Hsu et al. 2003). The potential mechanism of action includes decreasing insulin resistance by increasing cell-surface level of GLUT4 (Song and Bi 2004), preventing islet cells from the toxic action of reactive oxygen species (Xiong et al. 2006), potentiating insulin-induced preadipocyte differentiation, and promoting glucose uptake of adipocytes (Xu et al. 2005). There are also reports that puerarin decreases oxidative stress (Kang et al. 2005) and the production of atherogenic lipoproteins (Lee et al. 2002).

Radix Rehmanniae and Radix Rehmanniae Praeparata (sheng di huang and shu di huang): Roots of **Rehmannia glutinosa,** *Family Scrophulariaceae*

Commonly known as the Chinese foxglove root, Radix Rehmanniae is present in many herbal preparations for the treatment of diabetes and is derived from the roots of *Rehmannia glutinosa*. Radix Rehmanniae Praeparata is also extracted from the roots of *R. glutinosa* and is prepared by steaming it with wine and drying it repeatedly.

Preparations containing the herb have been shown to have hypoglycemic activity in normal and diabetic mice (Kimura and Suzuki 1981; Miura et al. 1997). The active components are most likely the polysaccharide extract which has been shown to increase insulin secretion. Although certain extractions of the polysaccharide have been shown to reduce glycogen stores (Kiho et al. 1992), other extractions have shown an increase in hepatic glucose storage (Zhang et al. 2004a). Therefore the mechanism of action is unclear. The stachycose extract from the herb has demonstrated hypoglycemic properties and there is evidence that the herb reduces oxidative stress as well (Kim et al. 2005; Zhang et al. 2004b). In addition to its glucose lowering effects, the root and its extract have shown some protection against diabetic complications such as nephropathy (Yokozawa et al. 2004).

Radix Salviae Miltiorrhizae (dan shen): Roots and Rhizomes of **Salvia miltiorrhiza,** *Family Labiatae*

Danshen is commonly used in Chinese medicine to promote blood circulation and reduce stasis. In this regard it is used as part of herbal preparations to treat the vascular complications of diabetes. Magnesium lithospermate B (MLB), an active constituent of Radix Salviae Miltiorrhizae, suppressed the progression of renal injury in streptozotocin-induced diabetic rats (Lee et al. 2003). Studies have shown that MLB possess antioxidant activity (Zhang et al. 2004c) as evidenced by its ability to scavenge free radicals, inhibit lipid peroxidation (Wu et al. 2000), and enhance nitric oxide release (Luo et al. 2002; Yokozawa and Chen 2000).

Radix Trichosanthis (tian hua fen, gua lou gen): The Roots of Trichosanthes kirilowii *Maxim., Family Cucurbitaceae*

The roots of *Trichosanthes kirilowii* have shown a hypoglycemic effect in animal studies. Bioactivity-guided fractionation determined by the active pharmacological agents were five glycans identified as trichosans A, B, C, D, and E. These fractions showed hypoglycemic activity in normal mice but only one glycan, trichosan A, exhibited activity in alloxan-induced hyperglycemic mice (Hikino et al. 1989).

Rhizoma Coptidis (huang lian): Rhizomes of Coptis chinensis *and* Coptis deltoidea, *Family Ranunculaceae*

The major hypoglycemic effect of *Coptis chinensis* is due to the compound berberine which lowered serum glucose levels in a clinical trial involving 60 patients with type 2 diabetes (Ni 1988). In animal models, berberine significantly inhibited the progression of diabetes, lowered cholesterol, and attenuated blood sugar elevations due to glucose or adrenaline administration. The compound also inhibited platelet aggregation in vitro (Chen and Xie 1986; Tang et al. 2006).

The antidiabetic effects of berberine may be related to its ability to stimulate insulin secretion, modulate lipids, decrease glucose absorption, and increase hepatic glucose uptake (Pan et al. 2003; Yin et al. 2002). Berberine has also been shown to reduce body weight and significantly improve glucose tolerance without altering food intake in *db/db* mice. Similarly, berberine reduced body weight and plasma triglycerides and improved insulin action in high-fat-fed Wistar rats (Lee et al. 2006a, b).

Rhizoma Polygonati (yu zhu): Rhizomes of Polygonatum odoratum, Polygonatum sibiricum, Polygonatum cyrtonema, *Family Liliaceae*

The rhizomes from Rhizoma Polygonati have shown a blood sugar lowering effect in animal models of hyperglycemic mice (Kato and Miura 1994; Miura et al. 1995). Although the species *Polygonatum officinale* acts as an insulin sensitizer, *P. sibricum* does not have this activity (Miura and Kato 1995). Studies suggest that intraperitoneal administration of the methanol extract from Rhizoma Polygonati decreases GLUT 2 expression in the liver (Miura et al. 1995). The active components have been identified as a spirostanol glycoside PO-1 and PO-2. PO-2 showed remarkable hypoglycemic activity when given as an intraperitoneal injection into normal mice. PO-1 only slightly lowered the blood glucose in the normal mice. However, both PO-1 and PO-2 exhibited significant hypoglycemic effects in streptozotocin-induced diabetic mice (Kato et al. 1995). Another species, *P. odoratum* has antihyperglycemic effects as well. Three steroidal glycosides (SG-100, SG-280, and SG-460) have been isolated from *P. odoratum*. SG-100 promotes peripheral insulin sensitivity without changing insulin secretion (Choi and Park 2002).

Rhizoma Anemarrhenae (zhi mu): Rhizomes of Anemarrhena asphodeloides, *Family Liliaceae*

In TCM, the rhizome from the *Anemarrhena asphodeloides* is felt to nourish Yin. The herb is found in the formula Byakko-ka-ninjin, which also contains ginseng, licorice, gypsum, and rice. Water extraction has identified two active components in anemarrhena: mangiferin, a xanthone, and its glucoside, mangiferin-7-0-beta *glucoside* (Ichiki et al. 1998). The compounds were shown to lower glucose in a dose-dependent fashion in a mouse model of type 2 diabetes. Normal mice were not as susceptible to the extract's hypoglycemic effect. Moreover, insulin levels in the mouse model were reduced when the extract was administered suggesting that its hypoglycemic effect was due to insulin sensitization (Miura et al. 2001). In contrast to the insulin sensitization of the water extract, the ethanol isolate from the rhizome, TH2, may act as an insulin secretagogue and stimulate insulin secretion in normal and diabetic rats (Hoa et al. 2004).

Rhizoma Atractylodis (cang zhu): Rhizomes of Atractylodes lancea *or* Atractylodes chinensis *or* Atractylodes japonica, *Family Compositae*

The rhizome is used for its ability to invigorate Qi, or the life force, by enhancing fluid movement throughout the body. It is commonly used as part of a formula designed to enhance the immune system, promote muscle growth, treat edema, and repair tissue. Extracts from *Atractylodes japonica* have been shown to have hypoglycemic activity in alloxan-induced hyperglycemic mice. The active pharmacologic agents from the extract were identified as atractans A, B, C (Konno et al. 1985). Formula 1, which contains Rhizoma Atractylodis, did not lower blood sugar in streptozotocin-induced diabetic rats model but did influence regional glucose metabolism in vitro (Chan et al. 2007). The authors felt that this local tissue effect may support the role of Formula 1 for enhancing blood flow and preventing tissue damage especially in diabetics. The isolate beta-eudesmol from *A. lancea* has been shown to desensitize the hypersensitive neuromuscular junctions in diabetic mice. Its ability to reduce neuromuscular overactivity may be the basis for its role in the treatment of diabetic complications such as neuropathy and tissue damage (Kimura 2006; Kimura et al. 1995). There is some evidence that if the rhizome is not processed properly, the resulting compound may have increased cytotoxic activity as well as an unacceptable level of heavy metals (Wang et al. 2007).

Summary

Although herbal preparations have been an integral part of TCM for thousands of years, the rigorous scientific studies examining their efficacy are relatively recent. Most studies focus on proximal end-points and use animal models. High-quality human studies should be investigated in the future. As there is regional variation of herbs and differences in preparation, further studies need to investigate the standardization of essential components. Some studies have shown that current available preparations may be adulterated with heavy metals or known Western medications (Ernst 2004). In addition, long-term safety studies evaluating side effects and drug–herb interactions need to be conducted.

References

Abd El Sattar EB, El-Gengaihi SE, El Shabrawy OA. 2006. Some toxicological studies of *Momordica charantia* L. on albino rats in normal and alloxan diabetic rats. J Ethnopharmacol 108:236–242.

Achenbach P, Fuchtenbusch M. 2004. Modulating the autoimmune response in Type 1 diabetes: A report on the 64th scientific sessions of the ADA, Orlando, FL, USA. Rev Diabet Stud 1:137–140.

Ahmad N, Hassan MR, Halder H, Bennoor KS. 1999. Effect of *Momordica charantia* (Karolla) extracts on fasting and postprandial serum glucose levels in NIDDM patients. Bangladesh Med Res Counc Bull 25:11–13.

Chan CM, Chan YW, Lau CH, Lau TW, Lau KM, , Lam FC, Che CT, Leung PC, Fung KP, Lau CB, Ho YY. 2007. Influence of an anti-diabetic foot ulcer formula and its component herbs on tissue and systemic glucose homeostasis. J Ethnopharmacol 109:10–20.

Chang JM, Shen CC, Huang YL, Chien MY, Ou JC, Shieh BJ, Chen CC. 2002. Five new homoisoflavonoids from the tuber of *Ophiopogon japonicus*. J Nat Prod 65:1731–1733.

Chen F, Nakashima N, Kimura I, Kimura M. 1995. Hypoglycemic activity and mechanisms of extracts from mulberry leaves (folium mori) and cortex mori radicis in streptozotocin-induced diabetic mice. Yakugaku Zasshi 115:476–482.

Chen JC, Tsai CC, Chen LD, Chen HH, Wang WC. 2000. Therapeutic effect of gypenoside on chronic liver injury and fibrosis induced by CCl4 in rats. Am J Chin Med 28:175–185.

Chen QM, Xie MZ. 1986. Studies on the hypoglycemic effect of *Coptis chinensis* and berberine. Yao Xue Xue Bao 21:401–406.

Chen WC, Hayakawa S, Yamamoto T, Su HC, Liu IM, Cheng JT. 2004. Mediation of beta-endorphin by the isoflavone puerarin to lower plasma glucose in streptozotocin-induced diabetic rats. Planta Med 70:113–116.

Cheng-Lai A. 1999. Cardiovascular drug highlight: Hirudin. Heart Dis 1:41–49.

Chiu PY, Ko KM. 2006. Schisandrin B-induced increase in cellular glutathione level and protection against oxidant injury are mediated by the enhancement of glutathione synthesis and regeneration in AML12 and H9c2 cells. Biofactors 26:221–230.

Cho WC, Chung WS, Lee SK, Leung AW, Cheng CH, Yue KK. 2006. Ginsenoside Re of *Panax ginseng* possesses significant antioxidant and antihyperlipidemic efficacies in streptozotocin-induced diabetic rats. Eur J Pharmacol 550:173–179.

Choi SB, Park S. 2002. A steroidal glycoside from *Polygonatum odoratum* (Mill.) Druce. improves insulin resistance but does not alter insulin secretion in 90% pancreatectomized rats. Biosci Biotechnol Biochem 66:2036–2043.

Choi SB, Wha JD, Park S. 2004. The insulin sensitizing effect of homoisoflavone-enriched fraction in *Liriope platyphylla* Wang et Tang via PI3-kinase pathway. Life Sci 75:2653–2664.

Covington MB. 2001. Traditional chinese medicine in the treatment of diabetes. Diabetes Spectr 14:154–159.

Cui J, Eneroth P, Bruhn JG. 1999. *Gynostemma pentaphyllum*: Identification of major sapogenins and differentiation from *Panax* species. Eur J Pharm Sci 8:187–191.

Ding JX, Ou XC, Zhang QH. 1994. Comparison among 5 different extracting methods of *Whitmania pigra* whitman. Zhongguo Zhong Xi Yi Jie He Za Zhi 14:134,165–166.

Ernst E. 2004. Risks of herbal medicinal products. Pharmacoepidemiol Drug Saf 13:767–771.

Greinacher A, Volpel H, Janssens U, Hach-Wunderle V, Kemkes-Matthes B, Eichler P, Mueller-Velten HG, Potzsch B. 1999. Recombinant hirudin (lepirudin) provides safe and effective anticoagulation in patients with heparin-induced thrombocytopenia: A prospective study. Circulation 99:73–80.

Hao Z, Hang B, Wang Y. 1992. Hypoglycemic effect of fructus Ligustri Lucidi. Zhongguo Zhong Yao Za Zhi 17:429–431, 447.

He ZD, But PPH, Chan TW, Dong H, Xu HX, Lau CP, Sun HD. 2001. Antioxidative glucosides from the fruits of *Ligustrum lucidum*. Chem Pharm Bull (Tokyo) 49:780–784.

Hikino H, Mizuno T, Oshima Y, Konno C. 1985. Isolation and hypoglycemic activity of moran A, a glycoprotein of *Morus alba* root barks. Planta Med 51:159–160.

Hikino H, Yoshizawa M, Suzuki Y, Oshima Y, Konno C. 1989. Isolation and hypoglycemic activity of trichosans A, B, C, D, and E: Glycans of *Trichosanthes kirilowii* roots. Planta Med 55:349–350.

Ho PY, Lisowski FP. 1997. A Brief History of Chinese Medicine, 2nd ed. Singapore: World Scientific.

Hoa NK, Phan DV, Thuan ND, Ostenson CG. 2004. Insulin secretion is stimulated by ethanol extract of *Anemarrhena asphodeloides* in isolated islet of healthy Wistar and diabetic Goto-Kakizaki Rats. Exp Clin Endocrinol Diabetes 112:520–525.

Hoang Anh NT, Van ST, Porzel A, Franke K, Wessjohann LA. 2003. Homoisoflavonoids from *Ophiopogon japonicus* Ker-Gawler. Phytochemistry 62:1153–1158.

Hsu FL, Liu IM, Kuo DH, Chen WC, Su HC, Cheng JT. 2003. Antihyperglycemic effect of puerarin in streptozotocin-induced diabetic rats. J Nat Prod 66:788–792.

Ichiki H, Miura T, Kubo M, Ishihara E, Komatsu Y, Tanigawa K, Okada M. 1998. New antidiabetic compounds, mangiferin and its glucoside. Biol Pharm Bull 21:1389–1390.

Ip SP, Poon MK, Che CT, Ng KH, Kong YC, Ko KM. 1996. Schisandrin B protects against carbon tetrachloride toxicity by enhancing the mitochondrial glutathione redox status in mouse liver. Free Radic Biol Med 21:709–712.

Ip SP, Yiu HY, Ko KM. 2000. Differential effect of schisandrin B and dimethyl diphenyl bicarboxylate (DDB) on hepatic mitochondrial glutathione redox status in carbon tetrachloride intoxicated mice. Mol Cell Biochem 205:111–114.

Jia W, Gao W, Tang L. 2003a. Antidiabetic herbal drugs officially approved in China. Phytother Res 17:1127–1134.

Jia W, Gao WY, Xiao PG. 2003b. Antidiabetic drugs of plant origin used in China: Compositions, pharmacology, and hypoglycemic mechanisms. Zhongguo Zhong Yao Za Zhi 28:108–113.

Jingyi W, Yasuhiro M, Naoya H, Seok RC, Yoshiharu Y, Nagara T, Fumiko T, Shigeru M, Junji K. 1997. Observation on the effects of Chinese medicine zhenxuanyin for improving cerebral blood flow in rats with cerebral ischemia. J Tradit Chin Med 17:299–303.

Kako M, Miura T, Usami M, Kato A, Kadowaki S. 1995. Hypoglycemic effect of the rhizomes of ophiopogonis tuber in normal and diabetic mice. Biol Pharm Bull 18:785–787.

Kang KA, Chae S, Koh YS, Kim JS, Lee JH, You HJ, Hyun JW. 2005. Protective effect of puerariae radix on oxidative stress induced by hydrogen peroxide and streptozotocin. Biol Pharm Bull 28:1154–1160.

Kang KS, Kim HY, Yamabe N, Nagai R, Yokozawa T. 2006. Protective effect of sun ginseng against diabetic renal damage. Biol Pharm Bull 29:1678–1684.

Kato A, Miura T. 1994. Hypoglycemic action of the rhizomes of *Polygonatum officinale* in normal and diabetic mice. Planta Med 60:201–203.

Kato A, Miura T, Fukunaga T. 1995. Effects of steroidal glycosides on blood glucose in normal and diabetic mice. Biol Pharm Bull 18:167–168.

Khanna P, Jain SC, Panagariya A, Dixit VP. 1981. Hypoglycemic activity of polypeptide-p from a plant source. J Nat Prod 44:648–655.

Kiho T, Watanabe T, Nagai K, Ukai S. 1992. Hypoglycemic activity of polysaccharide fraction from rhizome of *Rehmannia glutinosa* Libosch. f. hueichingensis Hsiao and the effect on carbohydrate metabolism in normal mouse liver. Yakugaku Zasshi 112:393–400.

Kim BJ, Kim YK, Park WH, Ko JH, Lee YC, Kim CH. 2003. A water-extract of the Korean traditional formulation Geiji-Bokryung-Hwan reduces atherosclerosis and hypercholesteremia in cholesterol-fed rabbits. Int Immunopharmacol 3:723–734.

Kim SR, Lee MK, Koo KA, Kim SH, Sung SH, Lee NG, Markelonis GJ, Oh TH, Yang J, Kim YC. 2004. Dibenzocyclooctadiene lignans from *Schisandra chinensis* protect primary cultures of rat cortical cells from glutamate-induced toxicity. J Neurosci Res 76:397–405.

Kim SS, Son YO, Chun JC, Kim SE, Chung GH, Hwang KJ, Lee JC. 2005. Antioxidant property of an active component purified from the leaves of paraquat-tolerant *Rehmannia glutinosa*. Redox Rep 10:311–318.

Kimura I. 2006. Medical benefits of using natural compounds and their derivatives having multiple pharmacological actions. Yakugaku Zasshi 126:133–143.

Kimura M, Diwan PV, Yanagi S, Kon-no Y, Nojima H, Kimura I. 1995. Potentiating effects of beta-eudesmol-related cyclohexylidene derivatives on succinylcholine-induced neuromuscular block in isolated phrenic nerve-diaphragm muscles of normal and alloxan-diabetic mice. Biol Pharm Bull 18:407–410.

Kimura M, Suzuki J. 1981. The pattern of action of blended Chinese traditional medicines to glucose tolerance curves in genetically diabetic KK-CAy mice. J Pharmacobiodyn 4:907–915.

Ko KM, Ip SP, Poon MK, Wu SS, Che CT, Ng KH, Kong YC. 1995. Effect of a lignan-enriched fructus schisandrae extract on hepatic glutathione status in rats: Protection against carbon tetrachloride toxicity. Planta Med 61:134–137.

Ko KM, Yiu HY. 2001. Schisandrin B modulates the ischemia-reperfusion induced changes in non-enzymatic antioxidant levels in isolated-perfused rat hearts. Mol Cell Biochem 220:141–147.

Konno C, Suzuki Y, Oishi K, Munakata E, Hikino H. 1985. Isolation and hypoglycemic activity of atractans A, B and C, glycans of *Atractylodes japonica* rhizomes. Planta Med (2):102–103

Kou J, Sun Y, Lin Y, Cheng Z, Zheng W, Yu B, Xu Q. 2005b. Anti-inflammatory activities of aqueous extract from Radix *Ophiopogon japonicus* and its two constituents. Biol Pharm Bull 28:1234–1238.

Kou J, Yu B, Xu Q. 2005a. Inhibitory effects of ethanol extract from Radix *Ophiopogon japonicus* on venous thrombosis linked with its endothelium-protective and anti-adhesive activities. Vasc Pharmacol 43:157–163.

Lee GT, Ha H, Jung M, Li H, Hong SW, Cha BS, Lee HC, Cho YD. 2003. Delayed treatment with lithospermate B attenuates experimental diabetic renal injury. J Am Soc Nephrol 14:709–720.

Lee YJ, Cho JY, Kim JH, Park WK, Kim DK, Rhyu MR. 2004. Extracts from *Schizandra chinensis* fruit activate estrogen receptors: A possible clue to its effects on nitric oxide-mediated vasorelaxation. Biol Pharm Bull 27:1066–1069.

Lee WK, Kao ST, Liu IM, Cheng JT. 2006a. Increase of insulin secretion by ginsenoside Rh2 to lower plasma glucose in Wistar rats. Clin Exp Pharmacol Physiol 33:27–32.

Lee JS, Mamo J, Ho N, Pal S. 2002. The effect of Puerariae radix on lipoprotein metabolism in liver and intestinal cells. BMC Complement Altern Med 2:12–19.

Lee YS, Kim WS, Kim KH, Yoon MJ, Cho HJ, Shen Y, Ye JM, Lee CH, Oh W, Kim CT, Hohnen-Behrens C, Gosby A, Kraegen EW, James DE, Kim JB. 2006b. Berberine, a natural plant product, activates AMP-activated protein kinase with beneficial metabolic effects in diabetic and insulin-resistant states. Diabetes 55:2256–2264.

Li C, Cao L, Zeng Q. 2004a. Astragalus prevents diabetic rats from developing cardiomyopathy by downregulating angiotensin II type2 receptors' expression. J Huazhong Univ Sci Technolog Med Sci 24:379–384.

Li FW, Zhang LS, Liu H, Cai YC, Pan JH, Jia XY, Ding JX. 2001. Study of protective effect of leech, radix Salviae miltiorrhizae and its composite recipe on vascular endothelial cells in rats with blood stasis syndrome. Zhongguo Zhong Yao Za Zhi 26:703–706.

Li WL, Zheng HC, Bukuru J, De KN. 2004b. Natural medicines used in the traditional Chinese medical system for therapy of diabetes mellitus. J Ethnopharmacol 92:1–21.

Lin CC, Huang PC, Lin JM. 2000. Antioxidant and hepatoprotective effects of *Anoectochilus formosanus* and *Gynostemma pentaphyllum*. Am J Chin Med 28:87–96.

Lin HM, Yen FL, Ng LT, Lin CC. April 20, 2006. Protective effects of *Ligustrum lucidum* fruit extract on acute butylated hydroxytoluene-induced oxidative stress in rats. J Ethnopharmacol 111(1), 129–136.

Liou SS, Liu IM, Hsu SF, Cheng JT. 2004. Corni fructus as the major herb of Die-Huang-Wan for lowering plasma glucose in Wistar rats. J Pharm Pharmacol 56:1443–1447.

Lu H, Liu GT. 1991. Effect of dibenzo[*a*, *c*]cyclooctene lignans isolated from Fructus schizandrae on lipid peroxidation and anti-oxidative enzyme activity. Chem Biol Interact 78:77–84.

Lu ZM, Yu YR, Tang H, Zhang XX. 2005. The protective effects of Radix Astragali and Rhizoma Ligustici chuanxiong on endothelial dysfunction in type 2 diabetic patients with microalbuminuria. Sichuan Da Xue Xue Bao Yi Xue Ban 36:529–532.

Luo Q, Cai Y, Yan J, Sun M, Corke H. 2004. Hypoglycemic and hypolipidemic effects and antioxidant activity of fruit extracts from Lycium barbarum. Life Sci 76:137–149.

Luo WB, Dong L, Wang YP. 2002. Effect of magnesium lithospermate B on calcium and nitric oxide in endothelial cells upon hypoxia/reoxygenation. Acta Pharmacol Sin 23:930–936.

Luo JZ, Luo L. 2006. American ginseng stimulates insulin production and prevents apoptosis through regulation of uncoupling protein-2 in cultured beta cells. Evid Based Complement Altern Med 3:365–372.

Mak DH, Ko KM. 1997. Alterations in susceptibility to carbon tetrachloride toxicity and hepatic antioxidant/detoxification system in streptozotocin-induced short-term diabetic rats: Effects of insulin and Schisandrin B treatment. Mol Cell Biochem 175:225–232.

Megalli S, Davies NM, Roufogalis BD. 2006. Anti-hyperlipidemic and hypoglycemic effects of *Gynostemma pentaphyllum* in the Zucker fatty rat. J Pharm Pharm Sci 9:281–291.

Miura T, Kako M, Ishihara E, Usami M, Yano H, Tanigawa K, Sudo K, Seino Y. 1997. Antidiabetic effect of seishin-kanro-to in KK-Ay mice. Planta Med 63:320–322.

Miura T, Kato A. 1995. The difference in hypoglycemic action between polygonati rhizoma and polygonati officinalis rhizome. Biol Pharm Bull 18:1605–1606.

Miura T, Kato A, Usami M, Kadowaki S, Seino Y. 1995. Effect of polygonati rhizoma on blood glucose and facilitative glucose transporter isoform 2 (GLUT2) mRNA expression in Wistar fatty rats. Biol Pharm Bull 18:624–625.

Miura T, Ichiki H, Iwamoto N, Kato M, Kubo M, Sasaki H, Okada M, Ishida T, Seino Y, Tanigawa K. 2001. Antidiabetic activity of the rhizoma of *Anemarrhena asphodeloides* and active components, mangiferin and its glucoside. Biol Pharm Bull 24:1009–1011.

Ni YX. 1988. Therapeutic effect of berberine on 60 patients with type II diabetes mellitus and experimental research. Zhong Xi Yi Jie He Za Zhi 8:711–713, 707.

Norberg A, Hoa NK, Liepinsh E, Van PD, Thuan ND, Jörnvall H, Sillard R, Ostenson CG. 2004. A novel insulin-releasing substance, phanoside, from the plant *Gynostemma pentaphyllum*. J Biol Chem 279:41361–41367.

Oku T, Yamada M, Nakamura M, Sadamori N, Nakamura S. 2006. Inhibitory effects of extractives from leaves of *Morus alba* on human and rat small intestinal disaccharidase activity. Br J Nutr 95:933–938.

Pan SY, Dong H, Han YF, Li WY, Zhao XY, Ko KM. 2006. A novel experimental model of acute hypertriglyceridemia induced by schisandrin B. Eur J Pharmacol 537:200–204.

Pan GY, Wang GJ, Sun JG, Huang ZJ, Zhao XC, Gu Y, Liu XD. 2003. Inhibitory action of berberine on glucose absorption. Yao Xue Xue Bao 38:911–914.

Peng Q, Wei Z, Lau BH. 1998a. *Fructus corni* attenuates oxidative stress in macrophages and endothelial cells. Am J Chin Med 26:291–300.

Peng Q, Wei Z, Lau BH. 1998b. *Fructus corni* enhances endothelial cell antioxidant defenses. Gen Pharmacol 31:221–225.

Prieto JM, Recio MC, Giner RM, Manez S, Giner-Larza EM, Rios JL. 2003. Influence of traditional Chinese anti-inflammatory medicinal plants on leukocyte and platelet functions. J Pharm Pharmacol 55:1275–1282.

Qian DS, Zhu YF, Zhu Q. 2001. Effect of alcohol extract of *Cornus officinalis* Sieb. et Zucc on GLUT4 expression in skeletal muscle in type 2 (non-insulin-dependent) diabetic mellitus rats. Zhongguo Zhong Yao Za Zhi 26:859–862.

Rhyu MR, Kim EY, Yoon BK, Lee YJ, Chen SN. 2006. Aqueous extract of *Schizandra chinensis* fruit causes endothelium-dependent and -independent relaxation of isolated rat thoracic aorta. Phytomedicine 13:651–657.

Sato M, Tai T, Nunoura Y, Yajima Y, Kawashima S, Tanaka K. 2002. Dehydrotrametenolic acid induces preadipocyte differentiation and sensitizes animal models of noninsulin-dependent diabetes mellitus to insulin. Biol Pharm Bull 25:81–86.

Shapiro K, Gong WC. 2002. Natural products used for diabetes. J Am Pharm Assoc (Wash) 42:217–226.

Song CY, Bi HM. 2004. Effects of puerarin on plasma membrane GLUT4 content in skeletal muscle from insulin-resistant Sprague-Dawley rats under insulin stimulation. Zhongguo Zhong Yao Za Zhi 29:172–175.

Sotaniemi EA, Haapakoski E, Rautio A. 1995. Ginseng therapy in non-insulin-dependent diabetic patients. Diabetes Care 18:1373–1375.

Tang LQ, Wei W, Chen LM, Liu S. 2006. Effects of berberine on diabetes induced by alloxan and a high-fat/high-cholesterol diet in rats. J Ethnopharmacol 108:109–115.

Toda S, Yase Y, Shirataki Y. 2000. Inhibitory effects of astragali radix, crude drug in Oriental medicines on lipid peroxidation and protein oxidative modification of mouse brain homogenate by copper. Phytother Res 14:294–296.

Tongia A, Tongia SK, Dave M. 2004. Phytochemical determination and extraction of *Momordica charantia* fruit and its hypoglycemic potentiation of oral hypoglycemic drugs in diabetes mellitus (NIDDM) Indian J Physiol Pharmacol 48:241–244.

Tsutsumi T, Kobayashi S, Liu YY, Kontani H. 2003. Anti-hyperglycemic effect of fangchinoline isolated from Stephania tetrandra Radix in streptozotocin-diabetic mice. Biol Pharm Bull 26:313–317.

Vuksan V, Sievenpiper JL, Koo VY, Francis T, Beljan-Zdravkovic U, Xu Z, Vidgen E. 2000b. American ginseng (*Panax quinquefolius* L) reduces postprandial glycemia in nondiabetic subjects and subjects with type 2 diabetes mellitus. Arch Intern Med 160:1009–1013.

Vuksan V, Sievenpiper JL, Wong J, Xu Z, Beljan-Zdravkovic U, Arnason JT, Assinewe V, Stavro MP, Jenkins AL, Leiter LA, Francis T. 2001a. American ginseng (*Panax quinquefolius* L.) attenuates postprandial glycemia in a time-dependent but not dose-dependent manner in healthy individuals. Am J Clin Nutr 73:753–758.

Vuksan V, Sievenpiper JL, Xu Z, Wong EY, Jenkins AL, Beljan-Zdravkovic U, Leiter LA, Josse RG, Stavro MP. 2001b. Konjac-Mannan and American ginseng: Emerging alternative therapies for type 2 diabetes mellitus. J Am Coll Nutr 20:370S–380S.

Vuksan V, Stavro MP, Sievenpiper JL, Beljan-Zdravkovic U, Leiter LA, Josse RG, Xu Z. 2000a. Similar postprandial glycemic reductions with escalation of dose and administration time of American ginseng in type 2 diabetes. Diabetes Care 23:1221–1226.

Vuksan V, Sung MK, Sievenpiper JL, Stavro PM, Jenkins AL, Di Buono M, Lee KS, Leiter LA, Nam KY, Arnason JT, Choi M, Naeem A. 2008. Korean red ginseng (*Panax ginseng*) improves glucose and insulin regulation in well-controlled, type 2 diabetes: Results of a randomized, double-blind, placebo-controlled study of efficacy and safety. Nutr Metab Cardiovasc Dis 18(1):46–56.

Wang KT, Chen LG, Yang LL, Ke WM, Chang HC, Wang CC. 2007. Analysis of the sesquiterpenoids in processed Atractylodis Rhizoma. Chem Pharm Bull (Tokyo) 55:50–56.

Wang L, Higashiura K, Ura N, Miura T, Shimamoto K. 2003. Chinese medicine, Jiang-Tang-Ke-Li, improves insulin resistance by modulating muscle fiber composition and muscle tumor necrosis factor-alpha in fructose-fed rats. Hypertens Res 26:527–532.

Wang JD, Narui T, Kurata H, Takeuchi K, Hashimoto T, Okuyama T. 1989. Hemato-logical studies on naturally occurring substances. II. Effects of animal crude drugs on blood coagulation and fibrinolysis systems. Chem Pharm Bull (Tokyo) 37:2236–2238.

Welihinda J, Karunanayake EH, Sheriff MH, Jayasinghe KS. 1986. Effect of *Momordica charantia* on the glucose tolerance in maturity onset diabetes. J Ethnopharmacol 17:277–282.

Wu SJ, Ng LT, Lin CC. 2004. Antioxidant activities of some common ingredients of traditional chinese medicine, *Angelica sinensis*, *Lycium barbarum* and *Poria cocos*. Phytother Res 18:1008–1012.

Wu Y, Ou-Yang JP, Wu K, Wang Y, Zhou YF, Wen CY. 2005. Hypoglycemic effect of Astragalus polysaccharide and its effect on PTP1B. Acta Pharmacol Sin 26:345–352.

Wu XJ, Wang YP, Wang W, Sun WK, Xu YM, Xuan LJ. 2000. Free radical scavenging and inhibition of lipid peroxidation by magnesium lithospermate B. Acta Pharmacol Sin 21:855–858.

Xiong FL, Sun XH, Gan L, Yang XL, Xu HB. 2006. Puerarin protects rat pancreatic islets from damage by hydrogen peroxide. Eur J Pharmacol 529:1–7.

Xu ME, Xiao SZ, Sun YH, Zheng XX, Ou-Yang Y, Guan C. 2005. The study of anti-metabolic syndrome effect of puerarin in vitro. Life Sci 77:3183–3196.

Xue JY, Liu GT, Wei HL, Pan Y. 1992. Antioxidant activity of two dibenzocyclooctene lignans on the aged and ischemic brain in rats. Free Radic Biol Med 12:127–135.

Yamabe N, Kang KS, Goto E, Tanaka T, Yokozawa T. 2007. Beneficial effect of Corni fructus, a constituent of Hachimi-jio-gan, on advanced glycation end-product-mediated renal injury in streptozotocin-treated diabetic rats. Biol Pharm Bull 30:520–526.

Yamahara J, Mibu H, Sawada T, Fujimura H, Takino S, Yoshikawa M, Kitagawa I. 1981. Biologically active principles of crude drugs. Antidiabetic principles of corni fructus in experimental diabetes induced by streptozotocin (author's transl). Yakugaku Zasshi 101:86–90.

Ye M, Guo D, Ye G, Huang C. 2005. Analysis of homoisoflavonoids in *Ophiopogon japonicus* by HPLC-DAD-ESI-MSn. J Am Soc Mass Spectrom 16:234–243.

Yim TK, Wu WK, Pak WF, Ko KM. 2001. Hepatoprotective action of an oleanolic acid-enriched extract of *Ligustrum lucidum* fruits is mediated through an enhancement on hepatic glutathione regeneration capacity in mice. Phytother Res 15:589–592.

Yin J, Hu R, Chen M, Tang J, Li F, Yang Y, Chen J. 2002. Effects of berberine on glucose metabolism in vitro. Metabolism 51:1439–1443.

Yin X, Zhang Y, Yu J, Zhang P, Shen J, Qiu J, Wu H, Zhu X. 2006. The antioxidative effects of astragalus saponin I protect against development of early diabetic nephropathy. J Pharmacol Sci 101:166–173.

Yokozawa T, Chen CP. 2000. Role of Salviae Miltiorrhizae Radix extract and its compounds in enhancing nitric oxide expression. Phytomedicine 7:55–61.

Yokozawa T, Kim HY, Yamabe N. 2004. Amelioration of diabetic nephropathy by dried Rehmanniae Radix (Di Huang) extract. Am J Chin Med 32:829–839.

Zhang Y, Akao T, Nakamura N, Hattori M, Yang XW, Duan CL, Liu JX. 2004c. Magnesium lithospermate B is excreted rapidly into rat bile mostly as methylated metabolites, which are potent antioxidants. Drug Metab Dispos 32:752–757.

Zhang RX, Jia ZP, Kong LY, Ma HP, Ren J, Li MX, Ge X. 2004b. Stachyose extract from *Rehmannia glutinosa* Libosch. to lower plasma glucose in normal and diabetic rats by oral administration. Pharmazie 59:552–556.

Zhang R, Zhou J, Jia Z, Zhang Y, Gu G. 2004a. Hypoglycemic effect of *Rehmannia glutinosa* oligosaccharide in hyperglycemic and alloxan-induced diabetic rats and its mechanism. J Ethnopharmacol 90:39–43.

Zhang YW, Xie D, Chen YX, Zhang HY, Xia ZX. 2006. Protective effect of Gui Qi mixture on the progression of diabetic nephropathy in rats. Exp Clin Endocrinol Diabetes 114:563–568.

Zhao R, Li Q, Xiao B. 2005. Effect of *Lycium barbarum* polysaccharide on the improvement of insulin resistance in NIDDM rats. Yakugaku Zasshi 125:981–988.

Zhong S, Cui Z, Sakura N, Wang D, Li J, Zhai Y. 2007. A rapid method for isolation and purification of an anticoagulant from *Whitmania pigra*. Biomed Chromatogr 21(5):439–445.

Chapter 14

Fenugreek and Traditional Antidiabetic Herbs

of Indian Origin

Krishnapura Srinivasan, PhD

Overview

Several herbs and tree plants of Indian origin have found potential use as antidiabetic in the folk medicine and in the Indian systems of medicine. The efficacious attributes of these have been experimentally validated only in the past few decades. The antidiabetic efficacy of several such plants, their crude extracts, and active constituents that have shown experimental or clinical antidiabetic activity has been reviewed periodically in recent years. Indian herbs and trees which are effective in diabetes and their complications as evidenced in different diabetic animal models and in clinical trials are: *Aegle marmelos, Allium cepa, Allium sativum, Aloe vera, Azadirachta indica, Boerhaavia diffusa, Brassica juncea, Coccinia indica, Caesalpinia bonducella, Cassia auriculata, Cinnamomum zeylanicum, Citrullus colocynthis, Cuminum cyminum, Curcuma longa, Syzigium cumini (Eugenia jambolana), Ficus bengalenesis,Gymnema sylvestre, Momordica charantia, Murraya koeingii, Musa sapientum, Nelumbo nucifera, Nigella sativa, Ocimum sanctum, Phylanthus amarus, Piper betle, Pterocarpus marsupium, Scoparia dulcis, Swertia chirayita, Tinospora cordifolia,* and *Trigonella foenum-graecum.* In this treatise, the available information on the effectiveness of these various antidiabetic plants in the glycemic control and in the management of diabetic complications is discussed.

Introduction

Although plants and herbal preparations are being used in traditional medicine for various ailments since ancient times, the efficacious attributes of these have been experimentally validated only in recent decades. Several herbs and trees have found potential use as antidiabetic in the folk medicine and in Indian systems of medicine. Many Indian plants that are mentioned in the ancient literature or used

311

traditionally for diabetes have been investigated for their beneficial use in diabetes, and the antidiabetic claims are more or less validated. The antidiabetic efficacy of several such plants, their extracts, and active constituents have been reviewed periodically in recent years (Grover et al. 2002a, b; Gupta et al. 2002; Krawinkel and Keding 2006; Mukherjee et al. 2006; Platel and Srinivasan 1997; Srinivasan 2005, 2006). In this treatise, the available information on the effectiveness of various Indian herbs and trees in the glycemic control and in the management of diabetic complications as evidenced in diabetic animal models and in clinical trials is discussed in the following sections roughly in the order of the exhaustiveness of the studies made.

Fenugreek (*Trigonella foenum-graecum*)

The seeds of fenugreek (*Trigonella foenum-graecum*) are employed as antidiabetic agent in traditional systems of medicine. The antidiabetic attribute of fenugreek has been experimentally evidenced in a large number of animal studies employing diabetic rats, mice, rabbits, and dogs wherein the oral intake of seeds or their extracts has consistently shown hypoglycemic effect (Srinivasan 2005). Besides animal studies, several human trials have unequivocally demonstrated the beneficial hypoglycemic potential and improvement in glucose tolerance in both type 1 and type 2 diabetes. Fenugreek seed is an excellent source of dietary fiber, which may be up to 51.7% comprising 19.2% mucilaginous fiber and 32.5% neutral fiber, and hence it is advantageous in the context of diabetes. Mechanism of fenugreek's antidiabetic action as understood by animal studies involves the ability of its dietary fiber to delay gastric emptying, suppress release of gastric inhibitory peptides and insulinotropic hormones (Srinivasan 2005).

Animal Studies

Defatted fenugreek seeds (which contains fiber and saponins) fed to normal or diabetic dogs for 8 days lowered blood glucose (Ribes et al. 1984) and also decreased cholesterolemia in diabetic dogs (Valette et al. 1984). Isolated fiber and saponin fraction from fenugreek seeds given for 3 weeks to alloxan-diabetic dogs showed significant antihyperglycemic and antiglycosuric effect along with reduction in plasma glucagon and somatostatin (Ribes et al. 1986). Dietary defatted fenugreek seeds checked the rise in fasting blood glucose in rats during 2 weeks following the administration of streptozotocin (Mondal et al. 2004). Hypoglycemic effect of alcoholic extract of fenugreek seeds has been evidenced in normal and alloxan-diabetic rats given daily for 3 weeks, the effect being more pronounced in diabetic animals (Vats et al. 2002). The soluble dietary fiber fraction of fenugreek seeds has been shown to reduce postprandial elevation in blood glucose level in diabetic rats and lower the serum fructosamine level with no effect on insulin level when administered for 4 weeks (twice daily; 0.5 g/kg) (Hannan et al. 2003). A decrease in atherogenic lipids and a tendency to lower platelet aggregation were also found to decrease in fenugreek fiber-fed rats.

Oral administration of fenugreek (100 mg/kg) for 15 days improved glucose tolerance and increased serum insulin in alloxan-diabetic rabbits (Murthy et al. 1990). Stimulation of peripheral utilization of glucose is inferred in addition to an effect at the pancreatic level. Beneficial decreases in serum total and LDL cholesterol and triglycerides were also evidenced. Oral administration of an aqueous extract of fenugreek seeds to the subdiabetic and mild diabetic rabbits (50 mg/kg for 15 days) resulted in an improved glucose tolerance and glucose induced insulin response, suggesting that the hypoglycemic effect is mediated through stimulated insulin synthesis and/or secretion from the pancreatic β-cells (Puri et al. 2002). Prolonged administration of the seed extract for 30 days to the severely diabetic rabbits lowered fasting blood glucose significantly, but could not elevate the serum insulin level, which suggests an extra-pancreatic mode of action, such as increasing the sensitivity of tissues to available insulin.

Therapeutic effectiveness of dietary fenugreek seed powder (5% for 3 weeks) in type 1 diabetes, as exemplified by studies on alloxan-diabetic rats has been attributed to the beneficial countering of the changes in the activities of enzymes of glucose and lipid metabolism in liver and kidney, thus stabilizing glucose homeostasis (Gupta et al. 1999; Raju et al. 2001). The decreased activities of glycolytic and lipogenic enzymes and elevated activities of gluconeogenic enzymes in diabetic animals were favorably restored by dietary fenugreek. Orally administered fenugreek seed (5% for 3 weeks) restored the increased activities of mitochondrial enzymes of tricarboxylic acid cycle, ketone body metabolism, urea cycle in liver, and kidney and brain tissues in alloxan-diabetic rats (Thakran et al. 2003). Countering of enzyme activities was even more effective when administered in combination with sodium orthovanadate (Genet et al. 1999).

Administration of fenugreek seed extract (2 g/kg per day) to alloxan-diabetic rats exerted a favorable effect on body weight and blood glucose and anticataract effect as evident from decreased opacity index (Vats et al. 2004a, b). Several single oral dose studies have also reported beneficial influence of fenugreek seeds. Administering whole fenugreek powder/methanol extract/water extract/soluble fiber produced blood glucose lowering in normal and STZ-induced diabetic rats (Nahar et al. 1992). Pretreatment of fenugreek suspension 1 h prior to glucose load in rats caused significant decrease in blood glucose at 30, 60, and 90 min indicating reduced rate of gastric emptying (Madar 1984). Oral administration of fenugreek at 2 and 8 g/kg to normal and alloxan-diabetic rats produced a dose-related fall in blood glucose (Khosla et al. 1995). Hypoglycemic and antihyperglycemic effects have also been observed with fenugreek leaves (0.06–1.0 g/kg, i.p. and 1, 2, 8 g/kg p.o.) in normal and alloxan-diabetic rats (Abdel-barry et al. 1997). A significant beneficial effect of fenugreek leaves on hyperglycemia, hypoinsulinemia, and glycosylated haemoglobin has been evidenced during supplementation of fenugreek leaves in STZ-diabetic rats (Devi et al. 2003).

Figure 14.1.   Fenugreek (*Trigonella foenum-graecum*) seeds.

The hypoglycemic effect of fenugreek was initially attributed to its major alkaloid—trigonelline (Mishkinsky et al. 1967) and the alkaloid-rich fraction (Jain et al. 1987). Now there is much evidence to believe that the hypoglycemic effect of fenugreek is attributable to the fiber which constitutes as much as 52% of fenugreek seeds. Dietary fiber of fenugreek delays gastric emptying by direct interference with glucose absorption. In addition, gel-forming dietary fiber reduces the release of insulinotropic hormones and gastric inhibitory polypeptides. 4-Hydroxy isoleucine isolated from fenugreek seeds has increased glucose-induced insulin release from the isolated islets of Langerhans in both rats and humans (Sauvaire et al. 1998) (Figure 14.1 and Tables 14.1 and 14.2).

Human Studies

Madar and Arad (1989) have observed significant decrease in blood glucose with no change in insulin levels in type 2 diabetic patients given 15 g fenugreek per day. Improvement in plasma glucose response was observed, while insulin levels were reduced when fenugreek seeds (25 g) were administered to diabetics for 21 days (Sharma 1986). Urinary glucose and serum cholesterol levels were reduced; polyuria, polydypsia, and polyphagia were controlled. In type 1 diabetic subjects, daily administration of 25 g fenugreek seed improved both plasma glucose profile and glycosuria, and reduced their insulin requirement after 8 weeks of treatment. In a long-term trial, 100 g defatted fenugreek seeds fed to either type 1 diabetics

Table 14.1. Effectiveness of fenugreek (*Trigonella foenum-graecum*) in diabetes mellitus in animal models.

Animal model	Effect demonstrated	Investigators
Rat		
Diabetic	Hypoglycaemic action of trigonelline	Mishkinsky et al. (1967)
Diabetic and normal	Mild hypoglycaemic effect of seeds and alkaloids	Shani et al. (1974)
Diabetic	Improved glucose tolerance by seeds	Madar (1984)
Diabetic and normal	Prevention of diabetes induction by seeds	Amin Riyad et al. (1988)
Diabetic and normal	Hypoglycaemic effect of seeds	Nahar et al. (1992)
Diabetic and normal	Hypoglycemic effect of seeds	Khosla et al. (1995)
Diabetic and normal	Hypoglycemic effect of seeds	Ali et al. (1995)
Diabetic and normal	Prevented increase in glucose during GTT	Ahmad et al. (1995)
Diabetic and normal	Antihyperglycemic effect of fenugreek leaves	Abdel-Barry et al. (1997)
Diabetic	Modulation of activities of gluconeogenic enzymes by dietary seeds	Gupta et al. (1999)
Diabetic	Modulation of activities of enzymes of carbohydrate metabolism and urea cycle by a combination of orthovanadate and fenugreek seeds	Genet et al. (1999)
Diabetic	Modulation of activities of gluconeogenic, glycolytic, and lipogenic enzymes by dietary seeds	Raju et al. (2001)
Diabetic	Hypoglycemic effect of alcoholic extract	Vats et al. (2002)
Diabetic	Fenugreek leaves countered hyperblycemia, hypoinsulinemia, and glycated hemoglobin	Devi et al. (2003)

(continued)

315

Table 14.1. (*continued*)

Animal model	Effect demonstrated	Investigators
Diabetic	Soluble fiber fraction lowered serum fructosamine, triglycerides, cholesterol	Hannan et al. (2003)
Diabetic	Restoration of the activities of enzymes of carbohydrate metabolism by fenugreek seeds	Thakran et al. (2003)
Diabetic	Prevention of rise in fasting blood glucose by defatted seeds	Mondal et al. (2004)
Diabetic	Alcoholic extract of seeds produced hypoglycemic and anticataract effect	Vats et al. (2004a, b)
Mice		
Diabetic and normal	Hypoglycaemic effect of decoction and ethanolic extract of seeds	Ajabnoor and Tilmisany (1988)
Normal	Hypoglycemic effect of water/methanolic extract of seeds	Zia et al. (2001)
Rabbits		
Normal	Hypoglycaemic effect of fiber and saponin fraction	Jain et al. (1987)
Diabetic	Improved glucose tolerance, Increased insulin level and hypolipidemic effect by fenugreek fraction	Murthy et al. (1990)
Diabetic	Lowered fasting blood glucose and higher insulin secretion by a fenugreek principle	Puri et al. (2002)
Dogs		
Diabetic and normal	Hypoglycemic effect of defatted fraction	Ribes et al. (1984)
Diabetic and normal	Hypoglycaemic effect of defatted fraction	Valette et al. (1984)
Diabetic	Antidiabetic effects of fiber and saponin fractions from fenugreek seeds	Ribes et al. (1986)
In vitro study	Insulinotropic action of 4-hydroxy isoleucine	Sauvaire et al. (1998)

Table 14.2. Antidiabetic efficacy of fenugreek (*Trigonella foenum-graecum*) in humans.

Human study	Effect demonstrated	Investigators
Normal subjects	Hypoglycemic effect of fenugreek seeds	Pahwa (1990)
Normal subjects	Hypoglycemic effect of fenugreek seeds/fiber	Nahar et al. (1992)
Type 2 diabetic and normal subjects	Hypoglycemic effect of fenugreek seeds	Sharma (1986)
Type 2 diabetic	Hypoglycemic effect of fenugreek seeds	Madar et al. (1988)
Type 2 diabetic	Hypoglycemic effect with no change in insulin	Madar and Arad (1989)
Type 1 diabetic	Hypoglycemic action, Improved glucose tolerance, reduced glucose excretion, Hypolipidemic effect of defatted seeds	Sharma et al. (1990)
Type 2 diabetic	Hypoglycemic action, improved glucose tole-rance; hypolipidemic effect of defatted seeds	Sharma and Raghuram (1990)
Type 2 diabetic	Fenugreek seeds improved peripheral glucose utilization	Raghuram et al. (1994)
Type 2 diabetic	Improved glucose tolerance, reduction in blood and urinary glucose, and glycated hemoglobin	Sharma et al. (1996)
Type 2 diabetic and normal subjects	Hypoglycemic effect of germinated seeds	Neeraja and Rajyalakshmi (1996)
Type 2 diabetic	Hypoglycemic and hypolipidemic effect of seed powder	Kuppu et al. (1998)
Type 2 diabetic	Favorable effect on glycaemic control and hypertriglyceridemia	Gupta et al. (2001)

(Sharma et al. 1990) or type 2 diabetics (Sharma and Raghuram 1990) for 10 days reduced fasting blood glucose levels and urinary glucose, and improved glucose tolerance; serum cholesterol levels were also significantly reduced. Sharma et al. (1996) have envisaged hypoglycemic effect in a human trial involving 60 type 2 diabetic patients who were given fenugreek seed at 25 g per day through diet over a period of 24 weeks, as evident from lowered fasting blood glucose, improved glucose tolerance, reduced sugar excretion, and diminished glycated hemoglobin.

In a placebo-controlled study, where type 2 diabetic patients received alcoholic extract of fenugreek seeds (1 g/day) for 2-months, oral glucose tolerance test, lipid levels, fasting C-peptide, glycosylated hemoglobin, and insulin resistance were similar to control initially (Gupta et al. 2001). There was also no difference with respect to fasting blood glucose levels or the 2 h postglucose tolerance test at the end of the 2-month treatment. However, a decrease in insulin secretion as well as improved sensitivity to insulin was observed after the fenugreek treatment. Decreased serum triglycerides and increased HDL cholesterol was also observed. It was thus concluded that fenugreek seeds improve glycemic control and decrease insulin resistance in mild type 2 diabetic patients, with attendant favorable effect on hypertriglyceridemia.

In order to understand if the improvement in glucose tolerance in diabetic patients is due to the effect of fenugreek on the absorption or metabolism of glucose, a study was carried out on type 2 diabetics given 25 g fenugreek seeds for 15 days, in a crossover design (Raghuram et al. 1994). An i.v. glucose tolerance test at the end of study indicated that fenugreek reduced the area under the plasma glucose curve and increased the metabolic clearance rate and erythrocyte insulin receptors, suggesting that fenugreek improves peripheral glucose utilization contributing to an improved glucose tolerance. Thus, fenugreek may exert its hypoglycemic effect by acting at the insulin receptor as well as at the gastrointestinal level. Blood glucose lowering effect was also observed in healthy humans given whole seed powder or the soluble fiber fraction as a single dose (Nahar et al. 1992). Fenugreek consumption (10 g) 3 h prior to glucose load (1 g/kg) by fasting human volunteers resulted in significant hypoglycemic effect in diabetic individuals, while it had no effect in normal individuals (Pahwa 1990). The hypoglycemic effectiveness of debitterized fenugreek in the form of germinated seeds has also been evidenced in type 2 diabetic subjects (Neeraja and Rajyalakshmi 1996).

In summary, fenugreek seed is established to diminish hyperglycemia in individuals with diabetes. Fasting blood glucose, urinary sugar excretion, and serum lipid levels in diabetics are significantly reduced, while clinical symptoms are improved. Antidiabetic properties of subfractions of fenugreek seeds are also studied. The antidiabetic effect was found to be highest with whole seeds, followed by the gum isolate, extracted seeds, and cooked seeds. The benefical effects of fenugreek seeds are due to the fiber present in them. Fenugreek seeds included in the daily diet in amounts of 25–50 g can be an effective supportive therapy in the management of diabetes.

Figure 14.2. Bitter melon (*Momordica charantia*) fruits.

Bitter Melon (*Momordica charantia*)

Bitter melon (*Momordica charantia*) of the family *Cucurbitacea* is commonly consumed as a vegetable in India. The fruit, leaves, seeds, and roots of *M. charanatia* all of which taste bitter are used in the Indian system of medicine and in the folk medicine as antidiabetic and for other ailments. (Grover and Yadav 2004). *M. charantia* has been widely claimed to be antidiabetic and over 100 studies have authenticated its use in diabetes and its complications—nephropathy and cataract. Various parts of *M. charantia*, namely, seed, fruit and, whole plant have been evaluated in these studies, most of them using a single dose administration (Figure 14.2 and Table 14.3).

Human Studies

Several human studies have demonstrated significant blood sugar control and overall blood sugar lowering effects by bitter melon. The first clinical report on the antidiabetic potential of bitter melon is of Lakholia (1956), who observed complete disappearance of all symptoms of diabetes (polyuria, polydypsia, and polyphagia) in 2 months as a result of consuming fresh juice (up to 1 oz/day) with attendant reduction in urinary sugar after 15 days in spite of discontinuing insulin injections. Fresh bitter melon juice significantly reduced blood glucose and improved the glucose tolerance in type 2 diabetic patients who underwent glucose tolerance test (Welihinda et al. 1986). Fried bitter melon fruits consumed as a daily

Table 14.3. Antidiabetic influence of *M. charantia* evidenced in various experimental models.

Plant part	Experimental model	Investigators
Fruit	Human type 2 diabetics	Lakholia (1956); Leatherdale et al. (1981); Srivastava et al. (1993); William et al. (1993); Welihinda et al. (1986); Ahmad et al. (1999)
	Diabetic rabbits	Sharma et al. (1960); Akhtar et al. (1981); Tiangda et al. (1987)
	Normal rabbits	Lotlikar and Rao (1966)
	Diabetic rats	Sharma et al. (1960); Gupta (1963); Singh et al. (1989) Upadhyaya and Pant (1986); Srivastava et al. (1987); Dubey et al. (1987); Karunanayke et al. (1990); Higashino et al. (1992); Srivastava et al. (1993); Shibib et al. (1993); Ali et al. (1993); Rathi et al. (2002); Virdi et al. (2003); Chaturvedi et al. (2004); Sekar et al. (2005)
	Normal rats	Karunanayake et al. (1984); Upadhyaya and Pant (1986); Dubey et al. (1987); Jeevathayaparan et al. (1991); Ali et al. (1993)
	Diabetic mice	Day et al. (1990); Bailey et al. (1985); Rathi et al. (2002)
	Normal mice	Bailey et al. (1985); Day et al. (1990); Cakici et al. (1994)
Seed	Diabetic rats	Kedar and Chakrabarti (1982); Jeevathayaparan et al. (1991)
Whole plant	Diabetic rats	Chandrasekhar et al. (1989)

supplement to the diet (250 g) significantly improved glucose tolerance in diabetic subjects without any increase in serum insulin (Leatherdale et al. 1981). Aqueous extract of bitter melon was hypoglycemic in diabetic patients at the end of a 3 week trial (Srivastava et al. 1993). Adaptogenic properties were indicated by the delay in appearance of cataract and relief in neurological symptoms even before the hypoglycemia occurred. Type 2 diabetic patients, who consumed cooked/fried bitter melon, had reduced plasma glucose at the end of 1 h (William et al. 1993).

Homogenized pulp of *M. charantia* fruit given to 100 moderate type 2 diabetic patients caused a significant reduction in postprandial blood glucose (Ahmad et al. 1999). Contradictory to these findings, absence of any beneficial influence of bitter melon (given as fresh juice or dry powder for 6–12 weeks) on blood glucose levels or glucose tolerance in diabetic subjects is also reported (Patel et al. (1968). This contradiction is attributable to the differences in methodology (different forms of test materials) and study design (with regard to number of subjects, their age group, and duration of disease) employed by the investigators.

Animal Studies

A large number of studies have validated the beneficial influence of bitter melon in diabetic animal models. Extracts of fruit pulp, seed, and whole plant of *M. charantia* have been shown to be hypoglycemic in various animal models. Bitter melon juice lowered the rise in blood sugar after an oral glucose load when administered (6–12 mL/kg) to alloxan-diabetic rats and rabbits, and also lowered blood sugar levels in a long-term study (6 mL/kg) for 23 days (Gupta 1963; Sharma et al. 1960). A decrease in blood glucose levels was observed when the dried fruit of *M. charantia* was administered to alloxan-diabetic rabbits (1.0 and 1.5 g/kg) with a maximum effect at 10 h (Akhtar et al. 1981). Improvement in glucose tolerance in rats was best with fresh juice, while this potential decreased upon storage (Karunanayake et al. 1984). A reduction in blood sugar and delay in cataract development is reported in alloxan-diabetic rats administered aqueous extract of the fruit (daily 2 mL for 2 months) (Srivastava et al. 1987).

Beneficial effects of aqueous and organic solvent extracts of bitter melon given as a single oral dose have also been reported in diabetic rabbits (Tiangda et al. 1987), rats (Ali et al. 1993; Higashino et al. 1992; Shibib et al. 1993; Singh et al. 1989; Srivastava et al. 1993; Virdi et al. 2003), and mice (Day et al. 1990). A significant decrease in triglyceride and LDL cholesterol and a significant improvement in glucose tolerance were noted in a long-term study in diabetic rats given methanolic extract of *M. charantia* fruit for 30 days (Chaturvedi et al. 2004). In a long-term study, alcohol and aqueous extracts of *M. charantia* (daily 50–200 mg/kg) were evidenced to have antihyperglycemic effect in alloxan-diabetic rats (120 days) and STZ-diabetic mice (60 days) (Rathi et al. 2002). The alteration in hepatic and skeletal muscle glycogen content and hepatic glucokinase, hexokinase, and phosphofructokinase activities in diabetic mice were partially restored by *M. charantia*.

Hypoglycemic influence of *M. charantia* has been investigated even in nondiabetic animals. Absence of any beneficial influence of the juice of bitter melon has been reported in normal rabbits (Kulkarni and Gaitande 1962) and in rats maintained on high-carbohydrate diet (Gupta and Seth 1962). Aqueous extract of *M. charantia* fruit improved glucose tolerance after 8 h in normal mice and chronic oral administration of the extract to normal mice for 2 weeks improved glucose tolerance without any effect on plasma insulin levels (Bailey et al. 1985). Aqueous and ether extracts (Upadhyaya and Pant 1986) or aqueous and methanolic extracts (Dubey et al. 1987) of bitter melon fed to normal rats for 1–3 weeks produced significant hypoglycemic effect. Fresh as well as freeze-dried bitter melon juice was found to improve glucose tolerance significantly in normal rats (Ali et al. 1993; Jeevathayaparan et al. 1991).

A few workers have found that the seed of *M. charantia* also possesses a hypoglycemic potential (Jeevathayaparan et al. 1991; Kedar and Chakrabarti 1982; Sekar et al. 2005). The seed powder fed (3 g/kg for 3 weeks) to STZ-diabetic rabbits produced an increase in liver glycogen and hypoglycemic effect (Kedar and Chakrabarti 1982). The hypoglycemic effect accompanied by hypolipemia was also produced by the seed treatment. The aqueous extract of seeds treated for 30 days to STZ-diabetic rats produced significant reduction in blood glucose, glycosylated hemoglobin, activity of gluconeogenic enzymes, and a concomitant increase in the levels of glycogen and activities of hexokinase and glycogen synthase (Sekar et al. 2005).

A few studies on diabetic animals have also evidenced absence of any beneficial hypoglycemic effect. Absence of either acute or cumulative beneficial effect has been reported for ethanolic extract of *M. charantia* fruit in diabetic rats (Chandrasekhar et al. 1989) and for bitter melon juice (10 mL/kg daily for 30 days) in STZ-diabetic rats (Karunanayke et al. 1990). This has led to the conclusion that viable β-cells capable of secreting insulin upon stimulation are probably required for bitter melon to exhibit its hypoglycemic potential. Dietary (0.5%) intake of freeze-dried bitter melon for 6 weeks did not have any beneficial influence in STZ-diabetic rats (Platel and Srinivasan 1995).

Thus, although contradictory results have been recorded as to the beneficial influence of bitter melon, significant hypoglycemic effect of this vegetable has been observed by a large number of researchers. In most of the cases, where such an effect is not seen, normoglycemic animals were experimented upon. The hypoglycemic effect of this fruit seems to be more acute and transient rather than cumulative. Fresh aqueous extract of the whole fruit appears to be more effective than the dried powder, while the seed portion is also found to be hypoglycemic in a limited number of studies. Some attempts have been made to isolate the active hypoglycemic principle from bitter melon. A non-nitrogenous neutral substance "charantin" from *M. charantia* fruits, presumed to be a phytosterol, has been isolated (Lotlikar and Rao 1966). Administration of this principle (15 mg/kg) to fasting rabbits transiently decreased blood sugar 4 h after i.v. injection and produced a marked fall in blood sugar of anaesthetized cats injected i.v. at 5 mg/kg. The cumulative hypoglycemic potency curve was linear as the dose was increased. The lower effect of charantin in depancreatized cats suggested that the hypoglycemic effect could be pancreatic as well as extra-pancreatic. Polypeptide-p isolated from the fruit and seeds of *M. charantia* showed potent hypoglycemic effect when administered s.c. to gerbils, langurs, and humans (Khanna et al. 1981). Mier and Yaniv (1985) found that *M. charantia* fruit contained two compounds—one which inhibited hexokinase activity, and the other which inhibited glucose uptake by rat intestine fragments.

These compounds were extractable in hot water, alcohol, and acetone. Galactose binding lectin possessing insulin-like activity was isolated from the seeds of *M. charantia* (Ng et al. 1986). This lectin demonstrated antilipolytic activity which was susceptible to destruction by heat, trypsin, chymotrypsin, glutathione and galactose, and lipogenic activity in isolated rat adipocytes (Mukherjee et al. 1972). The glycoalkaloid from *M. charantia* seeds which possess hypoglycemic activity in fasting rats has been identified as 2,6-diamino pyrimidinol-5-β-D-glucopyranoside (Handa et al. 1990).

Oral administration of *M. charantia* extracts showed a varying pattern of anti-hyperglycemic effect without altering the insulin response suggesting its action which is independent of intestinal glucose absorption and probably involves an extra-pancreatic effect (Day et al. 1984). The extra-pancreatic effects investigated in rats suggested that the fruit juice causes an increased glucose uptake by tissues in vitro without concomitant increase of tissue respiration (Welihinda and Karunanayake 1986). Oral treatment of juice prior to a glucose load increased the glycogen content of liver and muscle while it had no effect on the triglyceride of adipose tissue. An aqueous extract of unripe fruits was found to be a potent stimulator of insulin release from β-cell-rich pancreatic islets isolated from obese–hyperglycemic mice (Welihinda et al. 1982). It is also suggested that its insulin-releasing action is the result of perturbations of membrane functions similar to that of saponin. The fruit juice increased β-cells in diabetic rats (Ahmed et al. 1998). Bitter melon fruit juice exerted antihyperglycemic and antioxidant effect in the pancreas of STZ-diabetic mice (Sitasawad et al. 2000). Bitter melon juice markedly reduced the STZ-induced lipid peroxidation in RIN cells and isolated islets in vitro (Sitasawad et al. 2000). Further, it also reduced the STZ-induced apoptosis in RIN cells indicating the mode of protection of the juice on RIN cells, islets, and pancreatic β-cells.

The mechanism of hypoglycemic action of bitter melon involves an extra-pancreatic effect—enhancement of glucose utilization by the liver and peripheral tissues (Day et al. 1990; Leatherdale et al. 1981; Lotlikar and Rao 1966; Shibib et al. 1993; Welihinda and Karunanayake 1986). Bitter melon may directly influence hepatic or peripheral glucose disposal by increasing glucose oxidation and depressing glucose synthesis through activation or inhibition of key enzymes. The reduction in glycosylated hemoglobin in patients receiving fried bitter melon also suggested an extra-pancreatic effect. At the same time, it does not appear to be related to an insulin effect, since insulin levels are not increased (Sarkar et al. 1996; Shibib et al. 1993). Bitter melon may also reduce insulin resistance in type 2 diabetes through its ability to increase the content of glucose transporter proteins present in the plasma membrane of the muscles (Miura et al. 2001). The insulin secretagogue effect had also been suggested by a few investigators by indirect evidence (Gupta 1963; Karunanayke et al. 1990; Kedar and Chakrabarti 1982). Viable β-cells capable of secreting insulin upon stimulation may be necessary for hypoglycemic action of bitter melon (Karunanayke et al. 1990). In support of this, Kedar and Chakrabarti (1982) found that bitter melon seed activated pancreatic β-cells in mild STZ-diabetic animals in which some β-cells were viable. Absence of any hypoglycemic effect of bitter melon in animal models representing type 1 diabetes, as found by some workers (Ali et al. 1993; Chandrasekhar et al. 1989), could also be probably attributed to insulin secretagogue effect of bitter melon. In summary, the beneficial antidiabetic property of bitter melon fruits is demonstrated in animal and human studies and in cell culture. The mechanism of action which is still under debate involves regulation of insulin release, altered glucose metabolism, and its insulin-like effect. Several substances with antidiabetic properties present in bitter melon fruits are identified as charantin, vicine, and polypeptide-p. Bitter melon, a vegetable with proven antidiabetic potential could serve as a dietary component in the management of human diabetes.

Garlic (*Allium sativum*) and Onion (*Allium cepa*)

Garlic (*Allium sativa*) and onion (*Allium cepa*) are extensively cultivated in India for culinary uses. Bulbs of these two herbs are used in medicine for centuries, and the ancient literature indicates their use in treating diabetes. Reports that garlic and onion contain hypoglycemic agents are available for the past 75 years. It is only in recent 3–4 decades that such claims have been subjected to scientific scrutiny (Table 14.4).

Oral administration (250 mg/kg) of ethanol, petroleum ether, ethyl ether extracts of garlic caused a reduction in blood sugar in alloxan-diabetic rabbits which corresponded to 60–90% of the effectiveness of a standard dose of tolbutamide (Jain and Vyas 1975). Oral administration of di-(2-propenyl) thiosulfinate or allicin (250 mg/kg) (from garlic) produced hypoglycemia comparable to tolbutamide in mildly diabetic rabbits (blood sugar levels 180–300 mg/dL), while it was ineffective in severely diabetic animals (blood sugar levels >350 mg/dL), thus suggesting the dependency of the hypoglycemic effect of allicin on endogenous insulin (Mathew and Augusti 1973). Treatment with allicin also increased serum insulin and liver glycogen. While the hypoglycemic effect of this garlic constituent might be due to the direct or indirect stimulation of insulin secretion by the pancreas, the authors suggested that it may also have the effect of sparing insulin from sulfhydryl inactivation by reacting with endogenous thiol-containing molecules—cysteine, glutathione, and serum albumin. The hypoglycemic influence of garlic ether extract is associated with an increase in insulin level, which facilitates conversion of blood glucose to glycogen (Chang and Johnson 1980).

The blood sugar-lowering effect of allicin is shown to be dose related (Augusti 1975a, b). In young rats fed allicin (100 mg/kg/day) for 15 days, although no differences in fasting blood sugar levels were noted, certain liver enzymes were affected (Augusti and Mathew 1975). Further studies of the hypoglycemic effect of garlic in rats have shown it to be age-related, the effect being greater and long-lasting in suckling rats than in adults (Pushpendran et al. 1982). Aqueous homogenate of garlic (10 mL/kg/day) administered orally to sucrose-fed rabbits

Table 14.4. Effectiveness of garlic (*Allium sativum*) in diabetes mellitus.

Animal model	Effect demonstrated	Investigators
Normal rabbit	Hypoglycemic effect of petroleum ether extract	Brahmachari and Augusti (1962b)
	Hypoglycemic effect of juice	Jain et al. (1973)
	Hypoglycemic effect of garlic in sucrose-fed animals	Zacharias et al. (1980)
	Lowered glucose concentration in GTT	Roman-Ramos et al. (1995)
Diabetic rabbit	Hypoglycemic effect of petroleum ether extract	Brahmachari and Augusti (1962b)
	Hypoglycemic effect of allicin; improved glucose tolerance	Mathew and Augusti (1973)
	Dose related hypoglycemic effect of allicin	Augusti (1975)
	Hypoglycemic effect of extracts of garlic	Jain and Vyas (1975)
Normal rats	Lower dose of allicin decreased liver G-6-pase	Augusti and Mathew (1975)
	Hypoglycemic effect of garlic extract; increased insulin	Chang and Johnson (1980)
	Hypoglycemic effect of di-2-propenyl disulfide	Pushpendran et al. (1982)
	Lowered polyol-induced cataract by garlic	Srivastava and Afaq (1989)
	Dietary garlic decreased blood glucose	Ahmed and Sharma (1997)
Diabetic rats	Hypoglycemic and hypolipidemic effect of garlic	Farva et al. (1986)
	Hypoglycemic effect of SACS; lowered gluco-neogenic enzymes	Sheela and Augusti (1992)
	Amelioration of diabetes by SACS	Sheela et al. (1995)
	Antidiabetic effect; antiperoxide effect of SACS	Augusti and Sheela (1996)

(continued)

Table 14.4.   (*continued*)

Animal model	Effect demonstrated	Investigators
	Antiatherosclerotic effects of garlic extract	Patumraj et al. (2000)
	Countering of impaired antioxidant status by oil	Anwar and Meki (2003)
	Antioxidant activity and antihyperglycemic action	El-Demerdash et al. (2005)
Normal mice	Treatment with aged garlic extract countered hyperglycemia in stress-induced hyperglycemia	Kasuga et al. (1999)

(10 g/kg/day for 2 months) decreased fasting blood sugar, increased hepatic glycogen, and free amino acid contents, decreased triglyceride levels in serum, liver, and aorta (Zacharias et al. 1980). In healthy rabbits, the maximum glucose concentration following a subcutaneous glucose tolerance test was lowered after oral treatment with garlic (Roman-Ramos et al. 1995).

Studies in experimental diabetes have shown beneficial effect of S-allyl cysteine sulfoxide (SACS) which is the precursor of allicin. Significant blood glucose lowering activity of SACS isolated from garlic (200 mg/kg) observed in alloxandiabetic rats (Sheela and Augusti 1992) was accompanied by lowering of serum lipids and hepatic gluconeogenic enzymes and increased hexokinase activity. In another study, oral administration of SACS to alloxan-diabetic rats for 1 month ameliorated glucose intolerance, weight loss, depletion of liver glycogen in comparison to glibenclamide and insulin (Sheela et al. 1995). SACS isolated from garlic is reported to have ameliorated diabetic conditions in alloxan-diabetic rats to the same extent as glibenclamide and insulin and also controlled lipid peroxidation better than glibenclamide and insulin (Augusti and Sheela 1996). It also stimulated insulin secretion in isolated pancreatic β-cells. Thus, beneficial effects of SACS are attributable to its antioxidant and its insulin secretagogue actions.

Normal adult rats fed 2% garlic for 1 month showed a significant decrease in blood glucose (Ahmed and Sharma 1997). Pretreatment with aged garlic extract (5 and 10 mg/kg p.o.) in stress-induced hyperglycemia of mice significantly prevented adrenal hypertrophy, hyperglycemia, and elevation of cortisone without altering serum insulin levels (Kasuga et al. 1999). The efficacy of aged garlic extract was the same as that of diazepam (5 mg/kg). Daily oral feeding of garlic extract (100 mg/kg) increased cardiovascular functions in STZ-diabetic rats, prevented abnormality in lipid profile, and increased fibrinolytic activities with decreased platelet aggregation. Plasma insulin level increased with concomitant decrease in plasma glucose level. In addition, daily oral feeding of the same dose for 16 weeks showed antiatherosclerotic effects in STZ-diabetic rats. Thus, garlic may prevent diabetic cardiovascular complications (Patumraj et al. 2000). Significant increase of lipid peroxides and a concomitant decrease in the levels of antioxidants were found in the plasma and tissues of STZ-diabetic rats, while treatment of these rats with garlic oil (10 mg/kg) for 15 days increased plasma total thiols and ceruloplasmin and effectively normalized the impaired antioxidant status (Anwar and Meki 2003).

After the first experimental evidence on the hypoglycaemic effect of petroleum ether extract of onion in alloxan-diabetic rabbits (Brahmachari and Augusti 1962a), Augusti and his coworkers have exhaustively studied the hypoglycemic effects of onion or onion extracts in rats, rabbits, and humans (Augusti 1973, 1974, 1975a, b, 1976; Augusti et al. 1974; Augusti and Benaim 1975; Augusti and Mathew 1974). Various ether soluble fractions of onion given as a single oral dose (0.25 mg/kg) showed significant hypoglycemic effect in normal fasted rabbits, ethyl ether extract being the most potent among them (Augusti 1973). Petroleum ether insoluble fraction of ether extract of dried onion powder (100 mg/kg) given orally for 1 week to alloxan-diabetic rabbits caused a significant antihyperglycemic effect (Mathew and Augusti 1975). Oral administration of 250 mg/kg of ethanol extract of powder of dried onion showed maximal reduction of 19% in fasting blood glucose of alloxan-diabetic rabbit (Jain and Vyas 1974). Administration of onion oil to normal volunteers, after a 12 h fast, caused a significant decrease in blood glucose levels and a rise in serum insulin levels (Table 14.5) during the subsequent 4 h (Augusti and Benaim 1975).

Among different fractions from onion bulb, only petroleum ether and chloroform extracts significantly lowered blood sugar in oral glucose tolerance test in rabbits (Gupta et al. 1977). The hypoglycemic activity was tested by simultaneous oral administration along with oral glucose loading (2 g/kg) in diabetic rabbits. A significant reduction in blood sugar occurred, which was approximately 80% of the effect produced by tolbutamide. Oral doses of aqueous onion extract (25, 50, 100, and 200 g) to overnight fasted volunteers 30 min before, after, or simultaneously with oral glucose, resulted in a dose-dependent increase in glucose tolerance, and the effect was comparable to tolbutamide (Sharma et al. 1977). In addition, adrenaline-induced hyperglycemia was also inhibited in these patients. There was no difference in the antihyperglycemic effect of raw and boiled onion extract in these volunteers (Sharma et al. 1977). Beneficial effects of dietary onion (3 × 20 g) has also been shown in a crossover designed clinical study, wherein blood sugar levels were either decreased or maintained (Tjokroprawiro et al. 1983). Beneficial effects of dietary raw onion in diabetic individuals included significant reduction in the dose of antidiabetic drug (Bushan et al. 1984).

Dietary onion produced significant hypolipidemic effect besides hypoglycemic influence in STZ-diabetic rats (Babu and Srinivasan 1997a, b). The same authors have also evidenced the amelioration of diabetic nephropathy in STZ-diabetic rats

Table 14.5. Effectiveness of onion (*Allium cepa*) in diabetes mellitus.

Animal/human model	Effect demonstrated	Investigators
Healthy humans	Hypoglycemic effect; increased serum insulin	Augusti and Benaim (1975)
	Reduced response to oral glucose by onion	Sharma et al. (1977)
NIDDM patients	Hypoglycemic effect of fresh onion	Tjokroprawiro et al. (1983)
	Improved glucose tolerance; reduced drug need	Bushan et al. (1984)
Normal rabbits	Hypoglycemic effect of petroleum ether extracts	Brahmachari and Augusti (1961)
	Hypoglycemic effect of juice	Jain et al. (1973)
Diabetic rabbits	Hypoglycemic effect of petroleum ether extracts	Brahmachari and Augusti (1962a)
	Potentiation of glucose uptake by extracts	Augusti (1973)
	Hypoglycemic effect of ethanolic extract	Jain and Vyas (1974)
	Hypoglycemic effect of allyl propyl disulfide	Augusti (1974)
	Hypoglycaemic effect of chloroform extract	Gupta et al. (1977)
Diabetic rats	Hypoglycemic effect of allyl propyl disulfide	Augusti et al. (1974)
	Hypoglycemic effect of factors from onion	Mathew and Augusti (1975)
	Improved diabetic condition by SMCS	Sheela et al. (1995)
	Enzyme activity changes by SMCS	Kumari et al. (1995)
	Hypoglycemic effect; hypolipidemic effect	Babu and Srinivasan (1997)
	Amelioration of diabetic renal lesions	Babu and Srinivasan (1999)
	Antidiabetic activity of callus cultures of onion	Kelkar et al. (2001)

Table 14.5. (*continued*)

Animal/human model	Effect demonstrated	Investigators
	Amelioration of diabetic condition by SMCS; significant antioxidant effect	Kumari and Augusti (2002)
	Antioxidant activity of onion along with hypo-glycemic and hypolipidemic action	Campos et al. (2003)
	Antioxidant activity of onion associated with antihyperglycemic action	El-Demerdash et al. (2005)
Normal rats	Hypolipidemic effect of allicin	Augusti and Mathew (1974)
Normal mice	Hypoglycemic potency of onion principles	Augusti (1976)

by dietary onion as a result of hypolipidemic and antioxidant effects (Babu and Srinivasan 1999). Callus cultures fed to diabetic rats showed a much higher antidiabetic activity as compared to bulbs of onion (Kelkar et al. 2001). Administration of S-methyl cysteine sulfoxide (SMCS) isolated from onion (200 mg/kg for 45 days) to alloxan-diabetic rats significantly controlled blood glucose and lipids in serum and tissues and improved the activities of hepatic hexokinase and glucose-6-phosphate dehydrogenase, the effect being comparable to that of glibenclamide and insulin (Kumari et al. 1995). Similar beneficial effect of SMCS in alloxan-diabetic rats on glucose intolerance, weight loss, and liver glycogen has also been observed (Sheela et al. 1995).

The hypoglycemic and hypolipidemic actions of onion (0.4 g/rat/day) in STZ-diabetic rats were associated with antioxidant activity, onion treatment suppressing the increase in lipid hydroperoxide and lipoperoxide concentrations (Campos et al. 2003). Thiobarbituric acid reactive substances and the activity of glutathione-S-transferase in plasma and other organs, which were increased in alloxan-diabetic rats, were restored to normal levels by oral administration of onion juice (0.4 g/100 g/day for 4 weeks), thus suggesting that the antioxidant and antihyperglycemic effects may consequently alleviate liver and renal damage in diabetes (El-Demerdash et al. 2005). Besides ameliorating the diabetic condition in alloxan-diabetic rats, SMCS (treated for 2 months) also lowered the levels of malondialdehyde, hydroperoxide, and conjugated dienes in tissues exhibiting antioxidant effect on lipid peroxidation in experimental diabetes (Kumari and Augusti 2002). This is achieved by their stimulating effects on glucose utilization and the antioxidant enzymes through a stimulation of insulin secretion.

In summary, both garlic and onion are documented to be hypoglycemic in different diabetic animal models and in limited human trials. The hypoglycemic potency of garlic and onion has been attributed to the sulfur compounds present in them and are identified as S-allyl cysteine sulfoxide in garlic and SMCS in onion. Animal studies indicate that these isolated compounds possess as much as 60–90% hypoglycemic potency of tolbutamide. The mechanism of hypoglycemic action probably involves direct or indirect stimulation of secretion of insulin by the pancreas and, in addition, these compounds have the potential of sparing insulin from sulfhydryl inactivation by reacting with endogenous thiol-containing molecules.

Gymnema sylvestre

Gymnema (*Gymnema sylvestre*) is an herb widely used in folk medicine for the control of diabetes mellitus in India. Gymnema leaves have a profound action on the modulation of taste, particularly suppressing sweet taste sensations, and hence *G. sylvestre* is also known as gur-mar, or "sugar destroyer". In a clinical study, aqueous extract of *G. sylvestre* leaves (3 × 2 g daily for 15 days) given to type 2 diabetic patients significantly reduced the fasting blood glucose and improved the tolerance to an oral glucose load (Khare et al. 1983). The effectiveness of *G. sylvestre* leaf extract in controlling hyperglycemia has also been evidenced in type 2 diabetic patients by administering the extract (400 mg/day for 18–20 months) as a supplement to the conventional oral hypoglycemic drugs (Baskaran et al. 1990). The patients showed significant reduction in blood glucose, glycosylated hemoglobin, and glycosylated plasma proteins. The conventional drug dosage could also be reduced. These data suggest that the β-cells may be regenerated in type 2 diabetic patients on Gymnema supplementation, which is suggested by the appearance of raised insulin levels in the serum of patients. Antidiabetic effect of a water-soluble extract (GS4) of the leaves of *G. sylvestre* has been documented in type 1 diabetic patients on insulin therapy, where its treatment (400 mg/day) resulted in significant reduction in fasting blood glucose, glycosylated hemoglobin, glycosylated plasma proteins, and serum lipids and lowering of the insulin requirement (Shanmugasundaram et al. 1990a). GS4 therapy enhanced endogenous insulin, possibly by regeneration of the residual pancreatic β-cells in insulin-dependent diabetes (Figure 14.3 and Table 14.6).

Animal Studies

Gymnema sylvestre appears to correct the metabolic derangements in diabetic liver, kidney, and muscle. Administration of the dried leaf powder to alloxan-diabetic rabbits not only produced blood glucose homeostasis but also increased the activities of enzymes participating in the utilization of glucose, while decreasing gluconeogenic enzymes (Shanmugasundaram et al. 1983). Pathological changes in liver were reversed by controlling hyperglycemia. Feeding of

Figure 14.3. *Gymnema sylvestre.*

powdered leaves of *G. sylvestre* (0.5 g daily) for 10 days significantly prevented beryllium nitrate induced hyperglycemia in rats (Prakash et al. 1986). Administration of water soluble extracts of the leaves of *G. sylvestre* (20 mg/day) to STZ-diabetic rats resulted in blood glucose homeostasis (Shanmugasundaram et al. 1990b).*Gymnema sylvestre* therapy led to a rise in serum insulin levels which was due to an increase in the number of β-cells in the diabetic rat pancreas. Thus, *G. sylvestre* therapy appears to bring about blood glucose homeostasis through increased serum insulin levels provided by regeneration of the endocrine pancreas. Alcoholic extract of *G. sylvestre* (GS4) stimulated insulin secretion from rat islets of Langerhans and several pancreatic β-cell lines (Persaud et al. 1999). Single administration of *G. sylvestre* (1 g/kg) to fasted nondiabetic rats or STZ-induced mildly diabetic rats significantly attenuated the serum glucose response to oral glucose load (Okabayashi et al. 1990). In an-other study, chronic administration of *G. sylvestre* (fed for 35 days) to mildly diabetic rats resulted in reduced serum glucose and improved the glucose tolerance (Okabayashi et al. 1990). Administration of varying doses (50–500 mg/kg) of aqueous extract of *G. sylvestre* decreased blood sugar level in STZ-diabetic rats in a dose-dependent manner (Chattopadhyay 1999).

The active constituents of *G. sylvestre* have been isolated from the saponin fraction and identified as gymnemosides and gymnemic acids (Murakami et al. 1996; Yoshikawa et al. 1997). Gymnemic acid fractions are evidenced to inhibit increase in blood glucose level by interfering with the intestinal glucose absorption

Table 14.6. Antidiabetic influence of *Gymnema sylvestre* leaves.

Model	Effect demonstrated	Investigators
Type 2 human diabetics	Reduced blood glucose level and improved tolerance to glucose	Khare et al. (1983)
Type 2 human diabetics	Reduction in fasting blood glucose and glycosylated hemoglobin; decreased drug requirement	Baskaran et al. (1990)
Type 1 human diabetics	Reduction in fasting blood glucose and glycosylated hemoglobin; lowered requirement for insulin	Shanmugasundaram et al. (1990a)
Alloxan-diabetic rabbits	Produced blood glucose homeostasis; increased the activities of enzymes involved in utilization of glucose and decreased gluconeogenic enzymes	Shanmugasundaram et al. (1983)
STZ-diabetic rats	Produced blood glucose homeostasis; increased serum insulin level; enabled islet cells regeneration	Shanmugasundaram et al. (1990b)
STZ-diabetic rats	Improved glucose tolerance; reduced blood glucose concentration by chronic dietary treatment	Okabayashi et al. (1990)
STZ-diabetic rats	Dose-dependent decrease in blood glucose level by 50–500 mg/kg *G. sylvestre* extract	Chattopadhyay (1999)
Normal rats	Prevented beryllium nitrate induced hyperglycemia	Prakash et al. (1986)
Normal rats	Improved glucose tolerance by singe dose	Okabayashi et al. (1990)
STZ-diabetic mice	Saponin fraction reduced blood glucose levels	Sugihara et al. (2000)
Normal mice	Decreased corticosteroid-induced hyperglycaemia	Gholap and Kar (2003)
In vitro Rat islet cells	Alcoholic extract stimulated insulin secretion	Persaud et al. (1999)

(Shimizu et al. 1997). Among the five triterpene glycosides (gymnemic acids I–IV and gymnemasaponin V), a crude saponin fraction, the saponin fraction (60 mg/kg) reduced blood glucose levels 2–4 h after the i.p. administration in STZ-diabetic mice (Sugihara et al. 2000). Gymnemic acid IV at doses of 3.4–13.4 mg/kg reduced the blood glucose levels by 13.5–60% 6 h after the administration comparable to the potency of glibenclamide, and did not change the blood glucose levels of normal mice. The results indicate that insulin-releasing action of gymnemic acid IV may contribute to the antihyperglycemic effect by the leaves of *G. sylvestre*. The efficacy of *G. sylvestre* leaf extract has been evidenced in the amelioration of corticosteroid (dexamethasone) induced hyperglycemia in mice (Gholap and Kar 2003). As no marked changes in thyroid hormone concentrations were observed by the administration of *G. sylvestre* in dexamethasone treated animals, it is further suggested that it may not be effective in thyroid hormone mediated type 2 diabetes.

In summary, the leaves of *G. sylvestre* exert beneficial antidiabetic influence in two ways, namely, (1) they attenuate hyperglycemia by promoting glucose utilization and (2) they are insulinotropic and promote insulin secretion. By this two-pronged approach, *G. sylvestre* proves to be a valuable aid in diabetes control. The active ingredient of *G. sylvestre* happens to be the triterpene gymnemic acids.

Ivy Gourd (*Coccinia indica*)

Coccinia indica (Syn. *Coccinia cordifolia*) is a perennial climber plant (*Cucurbitacea*) used in the Indian and Unani systems of medicine for the treatment of diabetes mellitus. Its fruits (ivy gourd) are consumed as vegetable. Studies conducted both in human subjects and experimental animals have employed leaves, roots, fruits, and whole plant (Table 14.7). Diabetic patients who consumed tablets made from freeze-dried leaves of *C. indica* (3 × 3 tablets daily, for 6 weeks) showed significant improvement in glucose tolerance following an oral glucose load (Khan et al. 1980). Powdered stem of *C. indica* was highly effective hypoglycemic agent in human type 2 cases (Kuppurajan et al. 1986). Oral administration of dried extract of *C. indica* (500 mg/kg) for 6 weeks significantly restored the raised activity of lipoprotein lipase, glucose-6-phosphatase, and lactate dehydrogenase which are elevated in diabetes (Kamble et al. 1998) (Figure 14.4).

Animal Studies

The hypoglycemic potency of orally given ethanolic and aqueous extracts (1.25 g/kg) of the sun-dried roots of *C. indica* has been documented in rabbits (Brahmachari and Augusti 1963). Among the ethanolic extract of whole plant and the root fed to normal fed, fasted, glucose-loaded, and STZ-diabetic rats, the whole plant extract exerted a hypoglycemic effect in the fasted, glucose-loaded, and the diabetic rats while, the root was effective only in the glucose-loaded model

Table 14.7. Antidiabetic influence of *C. indica* evidenced in various experimental models.

Part of the plant (oral)	Experimental model	Effect demonstrated	Investigators
Whole plant	Diabetic rats	Hypoglycemic action	Mukherjee et al. (1988)
	Normal rats	Hypoglycemic action	Mukherjee et al. (1988)
		Hypoglycemic action	Chandrasekhar et al. (1989)
Leaf	Human type 2 diabetics	Improved glucose tolerance	Khan et al. (1980)
		Restoration of raised activity of gluconeogenic enzymes	Kamble et al. (1998)
	Diabetic dogs	Hypoglycemic action Improved glucose tolerance	Singh et al. (1985)
	Normal dogs	Hypoglycemic action	
		Improved glucose tolerance	Singh et al. (1985)
	Diabetic rats	Hypoglycemic action Increased plasma insulin	Pari and Venkaeswaran (2003)
		Hypoglycemic action Decreased activity of gluconeogenic enzymes	Shibib et al. (1993)
		Antioxidant effect in blood and other tissues	Venkateswaran and Pari (2003a, b)
	Normal rats	Hypoglycemic action Decreased activity of gluco-neogenic enzymes	Hossain et al. (1992); Shibib et al. (1993)
	Normal guinea pigs	Hypoglycemic action	Mukherjee et al. (1972)

Table 14.7. (*continued*)

Part of the plant (oral)	Experimental model	Effect demonstrated	Investigators
Stem	Human type 2 diabetics	Hypoglycemic action	Kuppurajan et al. (1986)
Root	Normal rabbits	Hypoglycemic action	Brahmachari and Augusti (1963)
	Diabetic rats	Not effective	Mukherjee et al. (1988)
	Glucose loaded rats	Hypoglycemic action	Mukherjee et al. (1988)
	Normal rats	Improved glucose tolerance	Chandrasekhar et al. (1989)
Fruit	Diabetic rats	Not effective	Chandrasekhar et al. (1989)
		Hypoglycemic action Increased hepatic glycogen	Kameswara Rao et al. (1999)
	Normal rats	Not effective	Chandrasekhar et al. (1989)

Figure 14.4. *Coccinia indica.*

(Mukherjee et al. 1988). Alcoholic extract of *C. indica* fruit (200 mg/kg i.m.) produced an insignificant reduction in the hypoglycemic effect of diabetogenic hormones—somatotropin and corticotrophin in rats (Gupta and Variyar 1964). Oral administration of an ethanolic extract of *C. indica* fruit (250 mg/kg) did not have any significant effect on the blood glucose levels or on the response to an oral glucose load either in normal or in STZ-diabetic rats (Chandrasekhar et al. 1989). However, the ethanolic extract of the plant lowered the blood sugar levels in normal rats fasted for 18 h and the root extract brought about a significant response to an oral glucose load (Chandrasekhar et al. 1989). *Coccinia indica* fruit powder given for 15 days to alloxan-diabetic rats produced a significant reduction in fasting blood glucose levels and improvement in hepatic glycogen (Kameswara Rao et al. 1999). Lowering of serum cholesterol and triglycerides was also observed in the treated animals, thus suggesting that *C. indica* fruit powder possesses antidiabetic and hypolipidemic effects in alloxan-diabetic rats.

Aqueous and alcoholic extracts of *C. indica* leaves given as a single oral dose (5 g and 100 mg/kg resply.) to normal fasting guinea pigs along with glucose (1 g/kg) brought about a transient reduction in blood glucose at second hour after administration (Mukherjee et al. 1972). Oral administration of pectin isolated from *C. indica* fruit showed a hypoglycemic action in normal rats due to stimulation of glycogen synthetase and reduction of phosphorylase activity (Kumar et al. 1993). A water soluble alkaloid isolated also had a definite positive effect on the response to an oral glucose load. Administration of dried leaves of *C. indica* in the form of suspension in milk (500 mg/kg) orally to normal and alloxan-diabetic dogs brought about a decrease in blood glucose levels 1–5 h after administration (Singh et al. 1985).

Ethanolic extract of *C. indica* leaf, when fed to normal, fed, and 48-h-starved rats, lowered blood glucose in both the groups, the hypoglycemic activity being attributed partly to the repression of key gluconeogenic enzymes (Hossain et al. 1992). Orally administered ethanolic extract of the fresh, freeze-dried leaves (200 mg/kg) significantly lowered blood glucose levels in both normal and STZ-diabetic rats 90 min after administration (Shibib et al. 1993). The levels of the key gluconeogenic enzymes, glucose-6-phosphatase, fructose-1,2-bisphosphatase, and glucose-6-phosphate dehydrogenase in the liver were also lowered. On the other hand, the shunt enzyme glucose-6-phosphate dehydrogenase in the red cell was enhanced. It was postulated that the possible mechanism by which *C. indica* leaves exert their hypoglycemic effect is through depression of gluconeogenic enzymes on one hand, and elevation of red cell glucose-6-phosphate dehydrogenase on the other, thus facilitating glucose oxidation through the shunt pathway. *Coccinia indica* leaves were evidenced to exhibit hypoglycemic and hypolipidemic effects in STZ-diabetic rats (Pari and Venkateswaran 2003). STZ-diabetic rats administered the ethanolic extract of *Coccinia indica* leaves (200 mg/kg) for 45 days showed decrease in blood glucose, lipids and fatty acids, namely, palmitic, stearic, and oleic acid, whereas linolenic and arachidonic acid and plasma insulin were elevated. *Coccinia indica* leaves also prevented the fatty acid changes produced during diabetes. In STZ-diabetic rats treated with *C. indica* leaf extract (200 mg/kg) for 45 days, there was reduced accumulation and cross-linking of collagen (Venkateswaran et al. 2003).

Oral administration of *C. indica* leaf extract (200 mg/kg) for 45 days to STZ-diabetic rats resulted in a significant reduction in plasma thiobarbituric acid reactive substances, hydroperoxides, vitamin E, and ceruloplasmin (Venkateswaran and Pari 2003a).. The extract also caused a significant increase in plasma vitamin C and reduced glutathione, which clearly shows the antioxidant property of *C. indica* leaves. The same extract also caused a significant increase in reduced glutathione, superoxide dismutase, catalase, glutathione peroxidase, and glutathione-S-transferase in liver and kidney of STZ-diabetic rats, indicating the antioxidant property of *C. indica* leaves (Venkateswaran and Pari 2003b). Thus, while the fruit of *C. indica* which is used as a vegetable does not have a significant hypoglycemic effect, the leaves, root, and whole plant of *C. indica* seem to be potential hypoglycemic agents, the mechanism of their action being mostly involving alterations in the activities of enzymes involved in glucose metabolism.

Ocimum sanctum

Ocimum sanctum, the Indian holy basil, an herb found throughout India has been mentioned in Indian system of traditional medicine to be of value in the treatment of diabetes mellitus. In traditional systems of medicine, different parts (leaves, stem, flower, root, seeds, and even whole plant) of *O. sanctum* have been recommended for its therapeutic use. Eugenol (1-hydroxy-2-methoxy-4-allylbenzene), the active constituent present in *O. sanctum*, has been found to be largely responsible for its therapeutic potentials. *Ocimum sanctum* leaves have been reported to reduce blood glucose when administered to experimental rats and humans with diabetes (Figure 14.5 and Table 14.8).

Human Studies

A decrease in fasting and postprandial blood glucose has been observed on treatment with *O. sanctum* leaves in type 2 diabetic subjects in a randomized, placebo-controlled, crossover single blind trial (Agrawal et al. 1996). Urine glucose and serum total cholesterol showed a similar decrease during this treatment period. The study thus suggested that basil leaves may be prescribed as adjunct to dietary therapy and drug treatment in mild to moderate type 2 diabetes.

Animal Studies

Experimental evidence on diabetic rats indicates that leaf of *O. sanctum* has significant beneficial effect. The hypoglycemic effectiveness of the leaves or their alcoholic extracts has been evidenced in normal, glucose-fed hyperglycemic,

Figure 14.5.   *Ocimum sanctum.*

STZ-diabetic, and in alloxan-diabetic rats (Chattopadhyay 1993, 1999; Dhar et al. 1968; Kar et al. 2003; Vats et al. 2002). Hypoglycemic and hypolipidemic effect of *O. sanctum* leaf powder (1% in diet for 1 month) has been documented in diabetic rats, wherein a significant reduction in fasting blood sugar, cholesterol, triglyceride, and phospholipids as well as hepatic lipids was observed (Rai et al. 1997). Tissue lipid content in kidney and heart was also significantly reduced.

Administration of *O. sanctum* extract (200 mg/kg for 30 days) led to a decrease in plasma glucose levels on 15th and 30th day in STZ-diabetic rats (Vats et al. 2004a, b). While glycogen content was not affected by *O. sanctum*, diabetes-induced alteration in the activity of enzymes of carbohydrate metabolism—glucokinase, phosphofructokinase, and hexokinase—was partially corrected. Oral administration of the ethanolic extract of *O. sanctum* resulted in a significant decrease in blood glucose and glycosylated hemoglobin, with a concomitant increase in hepatic glycogen content in STZ-diabetic rats (Narendhirakannan et al. 2006). It also resulted in an increase in insulin levels and glucose tolerance. Decrease in activities of carbohydrate metabolizing enzymes, glucose-6-phosphate dehydrogenase, glycogen synthase, hexokinase, and the increase in activities of gluconeogenic enzymes—lactate dehydrogenase, fructose-1,6-diphosphatase, glucose-6-phosphatase, and glycogen phosphorylase in diabetic rats—were significantly countered in rats treated with extracts of *O. sanctum*. The ethanol extract of *O. sanctum* leaf stimulated insulin secretion from perfused rat pancreas, isolated rat islets, and a clonal

Table 14.8. Antidiabetic influence of *Ocimum sanctum* leaves.

Experimental model	Effect demonstrated	Investigators
Type 2 human diabetics	Reduced fasting and postprandial glucose level; Reduced urinary suga	Agrawal et al. (1996)
STZ-diabetic rats	Hypoglycemic effect of 50% ethanolic extract	Dhar et al. (1968)
Normal rats	Lowering of blood sugar level by alcoholic extract in glucose-induced hyperglycemia; potentiated the action of exogenous insulin	Chattopadhyay (1993)
STZ-diabetic rats	Lowering of blood sugar level by alcoholic extract	Chattopadhyay (1993)
Alloxan-diabetic rats	Hypoglycemic and hypolipidemic effect of dietary supplementation of leaf powder	Rai et al. (1997)
STZ-diabetic rats	Lowering of blood sugar level by the leaves	Chattopadhyay (1999)
Alloxan-diabetic rats	Hypoglycemic effect of ethanolic extract of leaves	Vats et al. (2002)
Normal rats	Hypoglycemic effect of ethanolic extract of leaves	Vats et al. (2002)
	Lowering of blood sugar in glucose-induced hyper glycemia	
Alloxan-diabetic rats	Hypoglycemic effect of freeze-dried alcoholic extract given for 2 weeks	Kar et al. (2003)
STZ-diabetic rats	Decreased plasma glucose; Enhanced the activity of enzymes involved in glucose utilization	Vats et al. (2004a, b)
STZ-diabetic rats	Decreased blood glucose, glycated hemoglobin; increased liver glycogen, plasma insulin; Enhanced activity of glucose utilization enzymes; suppression of gluconeogenic enzymes	Narendhirakannan et al. (2006)

rat β-cell line in a concentration-dependent manner (Hannan et al. 2006). The stimulatory effect of this was potentiated by glucose, isobutyl-methylxanthine, tolbutamide, and KCl. Inhibition of the secretory effect was observed with diazoxide, verapamil, and Ca^{2+} removal. This indicated that constituents of *O. sanctum* leaf extract have stimulatory effects on physiological pathways of insulin secretion which may underlie its antidiabetic action. The extract of *O. sanctum* decreased the serum concentration of both cortisol and glucose in male mice, suggesting that the hypoglycemic effect of *O. sanctum* is mediated through its cortisol inhibiting potency (Gholap and Kar 2003). *Ocimum sanctum* exhibited antiperoxidative, hypoglycemic, and cortisol lowering activities, suggesting that it may potentially regulate corticosteroid induced diabetes mellitus. *Ocimum sanctum* seed oil evaluated in alloxan-diabetic rabbits treated for 2 weeks (0.8 g/kg/day) showed no significant hypoglycemic effect (Gupta et al. 2006).

Other Antidiabetic Herbs

Among a host of other herbs that are experimentally documented to potentially lower blood glucose and beneficially interfere in the management of diabetes, the important ones are *Murraya koenigii*, *Scoparia dulcis*, *Aloe vera*, *Nigella sativa*, *Curcuma longa*, *Cuminum cyminum*, *Caesalpinia bonducella*, *Swertia chirata*, *Cassia auriculata*, *Phylanthus amarus*, *Tinospora cordifolia*, *Boerhaavia diffusa*, *Musa sapientum*, *Piper betle*, *Nelumbo nucifera*, etc.

Murraya koenigii (Curry Leaf)

The aromatic leaves of *Murraya koenigii* (*Rutaceae*) are extensively used for food flavoring in curries and chutneys in India and are traditionally consumed by diabetics in India. The only clinical trial with *M. koenigii* has evidenced reduction in fasting and postprandial blood sugar in type 2 diabetics given the leaf powder supplementation (12 g) for 1 month (Iyer and Mani 1990).

Dietary *M. koenigii* leaves (10% for 2 months) to normal rats showed hypoglycemic effect associated with increased hepatic glycogen content due to increased glycogenesis and decreased glycogenolysis and gluconeogenesis (Khan et al. 1995). Feeding of *M. koenigii* leaves (5, 10, and 15% in the diet) to normal rats for 7 days as well as mild alloxan-diabetic and moderate STZ-diabetic rats for 5 weeks showed varying hypoglycemic and antihyperglycemic effect (Yadav et al. 2002). In normal and moderate diabetic rats, reduction in blood glucose was insignificant. In mild diabetic rats, feeding of 10 and 15% *M. koenigii* caused a maximal reduction in blood sugar. The effect of daily oral administration of aqueous/methanolic extract of *M. koenigii* leaves (600 and 200 mg/kg, respectively) for a period of 8 weeks to alloxan-diabetic rats resulted in reduction in blood glucose, while plasma insulin showed an increase (Vinuthan et al. 2004). This suggests that the hypoglycemic effect may be mediated through stimulating insulin synthesis and/or secretion from the β-cells of pancreatic islets of Langerhans. Daily oral administration of ethanolic extract of *M. koenigii* (200 mg/kg) for 30 days to STZ-diabetic rats significantly decreased the levels of blood glucose, glycosylated hemoglobin, urea, uric acid, and creatinine in diabetic *M. koenigii* treated animals, while plasma insulin levels increased (Arulselvan et al. 2006). Oral administration of the ethanolic extract of *M. koenigii* to STZ-diabetic rats decreased blood glucose, glycosylated hemoglobin, and urea, with a concomitant increase in glycogen, hemoglobin, and protein in diabetic rats (Narendhirakannan et al. 2006). Treatment with *M. koenigii* extract also increased insulin and C-peptide levels and glucose tolerance. The decreased activities of carbohydrate metabolizing enzymes—glycogen synthase, hexokinase, and glucose-6-phosphate dehydrogenase in diabetic rats—were reversed in rats treated with extracts of *M. koenigii*, while the increased activities of glucose-6-phosphatase, fructose-1,6-bisphosphatase, lactate dehydrogenase, and glycogen phosphorylase were markedly reduced following *M. koenigii* extract treatment in STZ-diabetic rats.

Oxidative stress in diabetes coexists with a reduction in the antioxidant status, which can further increase the deleterious effects of free radicals. The protective effect of *M. koenigii* leaf extract against β-cell damage and antioxidant defense systems of plasma and pancreas were evidenced in STZ-diabetic rats (Arulselvan and Subramanian 2007). The alterations in the levels of glucose, glycosylated hemoglobin, insulin, thiobarbituric acid reactive substances, antioxidant enzymes, and nonenzymatic antioxidants found in diabetic rats were considerably reverted after the treatment of *M. koenigii* leaf extract. The aqueous extract of *M. koenigii* leaves evaluated for the hypoglycemic activity in normal and alloxan-diabetic rabbits (200–400 mg/kg) revealed lowering of blood glucose in normal and diabetic animals (Kesari et al. 2005). The maximum fall was observed after 4 h of oral administration which also showed an improvement in glucose tolerance in subdiabetic and in mild diabetic rabbits after 2 h (Table 14.9).

Scoparia dulcis *(Sweet Broom Weed)*

Scoparia dulcis (*Scrophulariaceae*; Sweet Broom Weed), an indigenous plant has been documented as a traditional treatment for diabetes in India. Administration of aqueous extract of *S. dulcis* (200 mg/kg) for 15 days decreased the blood glucose with an increase in plasma insulin in STZ-diabetic rats (Latha et al. 2004). Insulin secretagogue action of *S. dulcis* plant extract has been evidenced in isolated pancreatic islets from mice (10 μg/mL evoked sixfold stimulation of insulin secretion). The insulin-receptor-binding effect of *S. dulcis* has been evidenced in STZ-diabetic rats in which *S. dulcis* treatment (200 mg/kg p.o.) improved total erythrocyte membrane insulin binding sites with a concomitant increase in plasma insulin (Pari et al. 2004).

Aqueous extracts of *S. dulcis* showed a modulatory effect on oxidative damage during diabetes by attenuating the lipid peroxidation in STZ-diabetes (Latha

Table 14.9. Antidiabetic influence of *Murraya koenigii Scoparia dulcis* and *Aloe vera*.

Animal/human model	Effect demonstrated	Investigators
Curry leaves (*Murraya koenigii*)		
Type 2 diabetic patients	Reduction in fasting and postprandial blood sugar	Iyer and Mani (1990)
Normal rats	Hypoglycemic effect; beneficial changes in enzymes of carbohydrate metabolism by dietary leaves	Khan et al. (1995)
Diabetic rats	Hypoglycemic and antihyperglycemic effect of dietary leaves	Yadav et al. (2002)
	Extract of *M. koenigii* leaves decreased blood sugar and increased insulin level	Vinuthan et al. (2004)
	Antihyperglycemic efficacy of *M. koenigii* extract	Arulselvan et al. (2006)
	Decreased blood glucose and glcosylated hemoglobin; increased glycogen content	Narendhirakannan et al. (2006)
	Hypoglycemic effect; improved antioxidant status by *M. koenigii* extract	Arulselvan and Subramanian (2007)
	Hypoglycemic effect; improved glucose tolerance	Kesari et al. (2005)
Sweet broom weed (*Scoparia dulcis*)		
Diabetic rats	Aqueous extract decreased the blood glucose; increased plasma insulin level	Latha et al. (2004)
	Increased insulin binding sites; increased plasma insulin	Pari et al. (2004)
	Reduced lipid peroxidation and increased antioxidant status by aqueous extract	Latha and Pari (2003)
	Increased plasma insulin and plasma antioxidants	Pari and Latha (2004b)
	Beneficial effect on oxidative stress in the brain	Pari and Latha (2004a)

Table 14.9. (*continued*)

Animal/human model	Effect demonstrated	Investigators
	Reduced blood glucose and glycosylated hemoglobin; decreased influx of glucose to polyol pathway	Latha et al. (2004)
	Reduction in blood glucose and increase in plasma insulin decreased free radical formation in liver and kidney	Pari and Latha (2005)
	Antihyperlipidemic effect in addition to antidiabetic effect	Pari and Latha (2006)
Aloe vera		
Normal humans	Hypoglycemic effect of continuous intake of dry sap	Ghannam et al. (1986)
Diabetic mice	Hypoglycemic effect of exudates and bitter principle	Ajabnoor (1990)
	Dose-dependent improvement in wound healing	Chitra et al. 1998; Davis and Maro (1989)
Diabetic rats	Gum extracts increased glucose tolerance	Al-Awadi and Gumaa (1987)
	Hypoglycemic activity of leaf pulp extract	Okyar et al. (2001)
	Pulp or gel extracts diminished degenerative changes in kidney tissue	Bolkent et al. (2004)
	Hypoglycemic effect and increased insulin content by ethanolic extract antihyperlipidemic activity	Rajasekaran et al. (2006)

and Pari 2003a). Significant increases in insulin, reduced glutathione, vitamin C, and vitamin E and activities of antioxidant enzymes were observed in tissues on treatment with *S. dulcis* extracts. The treatment also decreased blood glucose, thiobarbituric acid reactive substances, and hydroperoxide formation in tissues, suggesting its protective role against lipid peroxidation-induced membrane damage. Oral administration of the aqueous extract of *S. dulcis* (200 mg/kg) for 6 weeks to STZ-diabetic rats increased the plasma insulin and antioxidants (vitamin C, vitamin E, reduced glutathione, and ceruloplasmin) and decreased lipid peroxidation (Pari and Latha 2004b). The beneficial effect of aqueous extract of *S. dulcis* (200 mg/kg) for 6 weeks on oxidative stress in the brain of STZ-diabetic rats has also been reported (Pari and Latha 2004a). Oral administration of aqueous extract of *S. dulcis* (200 mg/kg) for 3 weeks also reduced blood glucose and increased plasma insulin in STZ-diabetic rats, while free radicals in liver and kidney were decreased, with an increase in the activities of antioxidant enzymes (Pari and Latha 2005).

Scoparia dulcis given orally for 6 weeks (200 mg/kg) to STZ-diabetic rats significantly reduced blood glucose, sorbitol dehydrogenase, glycosylated hemoglobin, thiobarbituric acid reactive substances, and hydroperoxides, and significantly increased plasma insulin, glutathione, glutathione peroxidase, and glutathione-S-transferase in liver (Latha et al. 2004). The effect of the extract may have been due to the decreased influx of glucose into the polyol pathway leading to increased activities of antioxidant enzymes and plasma insulin and decreased activity of sorbitol dehydrogenase. The antihyperlipidemic action in experimental diabetic rats in addition to its antidiabetic effect upon oral administration of the aqueous extract of *S. dulcis* (200 mg/kg) for 6 weeks has been evidenced, whereas a significant reduction has been found in serum and tissue lipids, HMG-CoA reductase activity, and VLDL and LDL cholesterol levels (Pari and Latha 2006). These studies suggest that the glucose-lowering effect of *S. dulcis* is associated with potentiation of insulin release from pancreatic islets, and the antidiabetic effect is accompanied by a significant countering of oxidative stress in blood and other tissues.

Aloe vera

Aloe vera (*Liliaceae*) which grows wildly as hedgerows in the arid regions all over India has long been used for various medicinal properties. It is used in the folk medicine of Arabian Peninsula for the management of diabetes. There have been controversial reports on the hypoglycemic activity of *Aloe* species, probably due to differences in the parts of the plant used or to the model of diabetes studied. The dried sap of the plant has shown significant hypoglycemic effect in clinical trials (Ghannam et al. 1986).

Extracts of Aloe gum is reported to effectively increase glucose tolerance in both normal and diabetic rats (Al-Awadi and Gumaa 1987). *Aloe vera* leaf pulp extract showed hypoglycemic activity in type 1 and type 2 diabetic rats, the effectiveness being higher for type 2 (Okyar et al. 2001). The degenerative changes observed in the kidney tissue of neonatal STZ-induced type 2 diabetic rats were diminished in animals given *Aloe* leaf gel and pulp extracts (Bolkent et al. 2004). Daily oral administration of *Aloe vera* gel extract (300 mg/kg) to STZ-diabetic rats for 21 days resulted in a significant reduction in fasting blood glucose and improvement in plasma insulin, which was accompanied by reduction in plasma and tissue lipids (Rajasekaran et al. 2006). The acute and chronic effects of the exudate of *Aloe barbadensis* leaves (500 g/kg p.o.) and its bitter principle (5 mg/kg i.p.)

were studied on plasma glucose levels of alloxan-diabetic mice (Ajabnoor 1990). The hypoglycemic effect of a single oral dose of bitter principle was significant over a period of 24 h with maximum hypoglycemia observed at 8 h. Chronic administration of aloes or the bitter principle resulted in significant reduction in plasma glucose level on the 5th day. The hypoglycemic effect of aloes may be mediated through stimulated synthesis and/or release of insulin from the β-cells. Both *Aloe vera* and *Aloe gibberellin* (2–100 mg/kg) inhibited inflammation in a dose-dependent manner and improve wound healing in STZ-diabetic mice (Chitra et al. 1998; Davis and Maro 1989).

Nigella sativa *(Black Cumin)*

Nigella sativa (*Ranunculaceae*) commonly called "black cumin" is used in Moroccan folk medicine for the treatment of various ailments including diabetes mellitus. The oil of *N. sativa* significantly lowered blood glucose concentrations in STZ-diabetic rats after 2, 4, and 6 weeks (El-Dakhakhny et al. 2002). A study of the effect of *N. sativa* oil on insulin secretion from isolated rat pancreatic islets in the presence of glucose indicated that hypoglycemic effect of *N. sativa* oil may be mediated by extra-pancreatic actions rather than by stimulated insulin release.

Defatted extract of *N. sativa* seed increased glucose induced insulin release from isolated rat pancreatic islets in vitro (Rchid et al. 2004). Thus the antidiabetic properties of *N. sativa* seeds may be partly mediated by stimulated insulin release. The possible insulinotropic property of *N. sativa* oil has also been studied in STZ plus nicotinamide induced diabetes mellitus in hamsters. A decrease in blood glucose and an increase in serum insulin were observed after *N. sativa* oil treatment for 4 weeks (Fararh et al. 2002). Immunohistochemical staining revealed areas with positive immuno-reactivity for the presence of insulin in the pancreas from *N.sativa* oil-treated group, suggesting that the hypoglycemic effect of *N. sativa* oil results, at least partly, from a stimulatory effect on β-cell function. The hypoglycemic effect of *N. sativa* oil (400 mg/kg by p.o.) is due to, at least in part, a decrease in hepatic gluconeogenesis, and the immuno-potentiating effect of *N. sativa* oil is mediated through stimulation of macrophage phagocytic activity (Fararh et al. 2004). *Nigella sativa* extract given orally for 2 months decreased lipid peroxidation and increased antioxidant defense system and also prevented the lipid peroxidation-induced liver damage in diabetic rabbits (Meral et al. 2001). Daily oral administration of ethanol extract of *N. sativa* seeds (300 mg/kg) to STZ-diabetic rats for 30 days reduced the elevated levels of blood glucose, lipids, plasma insulin, and improved altered levels of lipid peroxidation products and antioxidant enzymes in liver and kidney (Kaleem et al. 2006). This suggested that in addition to antidiabetic activity, *N. sativa* seeds may control diabetic complications through antioxidant effects. Treatment of *N. sativa* oil (0.2 mL/kg oil; i.p.) for 30 days decreased the elevation in serum glucose and restored lowered serum insulin with partial regeneration/proliferation of pancreatic β-cells in STZ-diabetic rats (Kanter et al. 2003). The possible protective effects of *N. sativa* (0.2 mL/kg i.p.) against β-cell damage from STZ-diabetes in rats have been evidenced by the observed decrease in lipid peroxidation and serum nitric oxide, and increase in the activities of antioxidant enzymes in pancreas (Kanter et al. 2004). Increased intensity of staining for insulin and preservation of β-cell numbers were apparent in *N. sativa* treated diabetic rats. This suggests that *N. sativa* treatment exerts a protective effect in diabetes by decreasing oxidative stress and preserving pancreatic β-cell integrity (Table 14.10).

Curcuma longa (Turmeric)

Rhizomes of turmeric (*Curcuma longa*), an important spice widely cultivated and consumed in India, are claimed to possess beneficial antidiabetic influence in a limited number of studies (Table 14.10). The daily intake of curcumin (coloring constituent of turmeric) not only reduced the fasting blood sugar level, but also lowered the dosage of insulin needed (Srinivasan 1972). The rhizome extract is reported to lower blood glucose in alloxan-diabetic rats (Tank et al. 1990). An extract of turmeric rhizome in combination with Emblica fruit (*Emblica officinalis*) is reported to produce good reduction in blood sugar and a satisfactory response to glucose tolerance test in fasting alloxan-diabetic rats (Singh et al. 1991). Dietary curcumin was evidenced to ameliorate kidney lesions attendant with diabetes in STZ-diabetic rats (Babu and Srinivasan, 1998). Hypocholesterolemic influence of curcumin and its ability to lower lipid peroxidation in diabetic condition (Babu and Srinivasan 1997a, b) are implicated in this amelioration of renal lesions by curcumin. Administration of turmeric/curcumin in alloxan-diabetic rats reduced the oxidative stress encountered in diabetes as evidenced by lowering of thio-barbituric acid reactive substances and elevated activity of antioxidant enzyme, glutathione peroxidase, and also decreased the influx of glucose into the polyol pathway (Arun and Nalini 2002). A close association between increased oxidative stress and hyperglycemia is believed to contribute significantly to the accelerated accumulation of advanced glycation end (AGE) products and cross-linking of col-lagen in diabetes mellitus. Accumulation of AGE products and cross-linking of collagen in diabetic animals were prevented by curcumin (200 mg/kg) for 8 weeks by reducing the oxidative stress (Sajithlal et al. 1998). Tissue repair and wound healing are complex processes that involve inflammation, granulation, and tissue remodeling. Curcumin enhances wound repair in diabetes impaired healing (Sidhu et al. 1999).

Cuminum cyminum *(Cumin)*

Seeds of cumin (*Cuminum cyminum*; *apiaceae*) are widely used as spice for their distinctive aroma. The beneficial effect of cumin seeds has been reported in human diabetics (Karnick 1991). Cumin seeds or their water/methanol extracts have been observed to be hypoglycemic in alloxan-diabetic rabbits (Akhtar and Ali 1985). Roman-Ramos (1995) and Ahmad et al. (2000) also report the antihyperglycemic influence of *Cuminum nigrum*, the latter attributing the effect to the flavonoid

Table 14.10. Antidiabetic influence of *Nigella sativa*, *Curcuma longa*, and *Cuminum cyminum*.

Animal/human model	Effect demonstrated	Investigators
Black cumin (*Nigella sativa*)		
Diabetic rats	*N. sativa* oil lowered blood glucose concentrations	El-Dakhakhny et al. (2002)
Diabetic hamsters	*N. sativa* oil decreased in blood glucose level; increased in serum insulin level	Fararh et al. (2002)
Diabetic hamsters	*N. sativa* oil decreased hepatic gluconeogenesis	Fararh et al. (2004)
Diabetic rabbits	Hypoglycemic effect of *N. sativa* extract; decreased lipid peroxidation	Meral et al. (2001)
Pancreatic cells	Enhanced insulin secretion by defatted black cumin	Rchid et al. (2004)
In vitro		
Diabetic rats	Antidiabetic activity of ethanolic extract of *N. sativa* seed; improved antioxidant status	Kaleem et al. (2006)
Diabetic rats	Hypoglycemic effect of *N. sativa* oil; partial regene-ration of islets of pancreas	Kanter et al. (2003)
Diabetic rats	Decreased lipid peroxidation and preservation of the integrity of pancreatic cells	Kanter et al. (2004)
Turmeric (*Curcuma longa*)		
NIDDM patient	Hypoglycemic effect of curcumin; reduced insulin requirement	Srinivasan (1972)
Diabetic rats	Hypoglycemic effect of ethanolic extract	Tank et al. (1990)
	Hypoglycemic effect of turmeric; improved glucose tolerance	Singh et al. (1991)

(continued)

Table 14.10. *(continued)*

Animal/human model	Effect demonstrated	Investigators
	Hypolipidemic effect of dietary curcumin	Babu and Srinivasan (1997a, b)
	Ameliorating influence of dietary curcumin on renal lesions	Babu and Srinivasan (1998)
	Oral/topical curcumin application enhanced wound healing	Sidhu et al. (1999)
	Turmeric reduced blood sugar and glycosylated hemoglobin; lowered oxidative stress; decreased polyol pathway	Arun and Nalini (2002)
Cumin (*Cuminum cyminum*)		
Type 2 diabetic patients	Hypoglycemic effect as component of drug	Karnick (1991)
Normal rabbits	Antihyperglycemic effect of cumin seeds	Roman-Ramos et al. (1995)
Diabetic rabbits	Whole seed/water extract: Hypoglycemic	Akhtar and Ali (1985)
	Hypoglycemic action of flavonoids of *C. nigurm*	Ahmad et al. (2000)
Diabetic rats	Whole seeds: hypoglycaemic	Willatgamuwa et al. (1998)
	Hypolipidemic effect in blood and tissues; reduction in blood sugar	Dandapani et al. (2002)

compounds present in cumin. Dietary cumin seeds were also observed to alleviate diabetes-related metabolic abnormalities in STZ-diabetic rats (Willatgamuwa et al. 1998). Hyperlipidemia associated with diabetes mellitus was also effectively countered by dietary cumin in alloxan-diabetic rats (Dandapani et al. 2002).

Caesalpinia bonducella

Caesalpinia bonducella (Leguminosae) is a shrub widely distributed throughout coastal India, and its seeds are ethnically used by the tribal people of Andaman and Nicobar Islands as a remedy for diabetes mellitus. The aqueous and alcoholic extract of *C. bonducella* seeds are reported to exhibit hypoglycemic and antihyperglycemic activities in normal and STZ-diabetic rats (Simon et al. 1987) and in rabbits (Rao et al. 1994). Hypoglycemic, antihyperglycemic, and hypolipidemic activities of the aqueous/ethanolic extracts of *C. bonducella* seeds have been evidenced in normal and STZ-diabetic rats (Sharma et al. 1997). The effect of aqueous and ethanolic extracts of the seeds of this plant was studied in both type 1 and 2 diabetes mellitus in Long Evans rats (Chakrabarti et al. 2003). Significant blood sugar lowering effect of *C. bonducella* was observed in type 2 diabetic model. A detailed investigation confirmed the hypoglycemic activity of *C. bonducella* seeds on type 1 and type 2 diabetic animal models by Chakrabarti et al. (2005), who used different extracts from *C. bonducella* and also evaluated the insulin secretagogue activity of five fractions isolated from the seed kernel. Both the aqueous and ethanolic extracts showed potent hypoglycemic activity in type 2 diabetic model. Two fractions could increase secretion of insulin from isolated islets.

Swertia chirayita *(Indian Gentian)*

Swertia chirayita (*Gentianaceae*) is an herb mainly found in temperate Himalayan region of India. Extracts of its leaf which has two bitter principles, ophelic acid and chiratin, are used as bitter tonic for general debility in *Ayurveda* and *Siddha* systems of medicine in India. Various crude extracts have shown hypoglycemic activity in different animal models. Oral administration of ethanolic extract and hexane fraction of *S. chirayita* (10, 50, and 100 mg/kg) to normal, glucose-fed, and STZ-diabetic rats lowered blood glucose (Sekar et al. 1987). Single oral dose of the *S. chirayita* hexane fraction (250 mg/kg) to normal rats reduced blood sugar and increased plasma insulin without influencing hepatic glycogen, while its administration for 4 weeks increased hepatic glycogen content in conjunction with other effects probably by releasing insulin (Chandrasekhar et al. 1990). Swerchirin (a xanthone isolated from hexane extract of this plant) lowered blood sugar in fasted, glucose-fed, and tolbutamide pretreated rats. Oral ED_{50} of swerchirin for 40% reduction in blood sugar in rats was found to be 23 mg/kg (Bajpai et al. 1991). Swerchirin fed at 35 and 65 mg/kg to STZ-diabetic rats showed significant antihyperglycemic effect at 1, 3, and 7 h after the dose (Saxena et al. 1991). Single oral administration of swerchirin (50 mg/kg) to rats maximally decreased blood sugar at 7 h post-treatment. In vitro, glucose uptake and glycogen synthesis by muscle were enhanced in swerchirin fed rat. Swerchirin stimulated insulin release from isolated pancreatic islets at 1, 10, and 100 mM concentration in the medium (Saxena et al. 1993).

Cassia auriculata *(Avaram)*

Cassia auriculata (*Cesalpinaceae*), commonly known as "Avaram," is an evergreen wildly growing Indian shrub with vivid yellow flowers. It is also known as tanner's senna, since its bark is used in tanning. It is used in traditional medicine as a cure for diabetes. The effect of *C. auriculata* flowers on blood glucose and lipid levels has been demonstrated in experimental diabetic rats (Pari and Latha 2002). Aqueous extract of *C. auriculata* flowers administered orally at doses of 0.15, 0.30, and 0.45 g/kg for 1 month to STZ-diabetic rats suppressed the elevated blood glucose, and the effect of 0.45 g/kg was found to be comparable to glibenclamide. The study also indicated that the *C. auriculata* flowers possess antihyperlipidemic effect in addition to antidiabetic activity. In experimental diabetes, enzymes of glucose and fatty acid metabolism are markedly altered. Administration of *C. auriculata* flower extract (0.45 g/kg) for 1 month decreased blood glucose, glycosylated hemoglobin, and gluconeogenic enzymes and increased plasma insulin, hemoglobin, and hexokinase activity (Latha and Pari 2003b). The antihyperglycemic effect of aqueous extract of *C. auriculata* was comparable to glibenclamide and the enhanced gluconeogenesis during diabetes was countered, and the extract enhances the utilization of glucose through increased glycolysis. The aqueous extract of the flowers of *C. auriculata* increased the activities of antioxidant enzymes and glutathione and decreased thiobarbituric reactive substances in the brain in STZ-diabetic rats suggesting antiperoxidative role of this medicinal herb (Latha and Pari 2003c).

Phyllanthus amarus

Phylanthus amarus (*Euphorbiaceae*; syn. *Phylanthus niruri*) is traditionally used in the management of various ailments and has been mentioned in the indigenous system of medicine in India as a potential hypoglycemic drug. In a clinical observation, oral administration of *P. amarus* (5 g/day) for 10 days reduced blood glucose both in diabetic and nondiabetic subjects along with significant reduction in systolic blood pressure (Srividya and Periwal 1995). Methanolic extract of *P. amarus* (1 g/kg) reduced the blood sugar maximally at 4th hour in alloxan-diabetic rats (Raphael et al. 2002). Continued administration of the extract for 15 days produced significant reduction in blood sugar. The antidiabetic and antilipidemic potentials of the aqueous leaf and seed extracts of *P. amarus* have been evidenced and it is suggested that the extract could be enhancing peripheral utilization of glucose (Adeneyea et al. 2006).

Tinospora cordifolia

Widely found creeaper *Tinospora cordifolia* (*Menispermaeae*; "Amarta") is extensively used in the Indian system of medicine as a remedy for diabetes mellitus. Studies for validating its antidiabetic potential are, however, limited. Oral intake of 400 mg/kg of aqueous extract of *T. cordifolia* root for 15 weeks showed hypoglycemic effect in mild and moderate diabetic rats, while it did not have benefit in severe diabetes (Grover et al. 2000). Thus, hypoglycemic effect of *T. cordifolia* depends upon the functional status of the pancreatic β-cells. Oral administration of aqueous extract of *T. cordifolia* root (2.5, 5, and 7.5 mg/kg) reduced blood glucose, liver glucose-6-phosphatase, and increased hepatic hexokinase in alloxan-diabetic rats (Stanley et al. 2000). The hypoglycemic and hypolipidemic effects of alcohol extract of *T. cordifolia* roots were established in alloxan-diabetic rats in which its oral administration for 6 weeks reduced blood and urine glucose and lipids in serum and tissues (Prince et al. 2003).

Boerhaavia diffusa *(Punarnava)*

Boerhaavia diffusa (*Nyctaginaceae*) commonly known as "Punarnava" is an herbaceous plant. The whole plant or its leaves/stem/roots are known to have medicinal properties and are in use in indigenous system of medicine and by tribal folk in India. Limited animal studies suggest that *B. diffusa* leaf extract has remarkable antidiabetic activity and can improve antioxidant status. Daily oral administration of aqueous extract of *B. diffusa* leaf (200 mg/kg) for 4 weeks to normal and alloxan-diabetic rats decreased blood glucose and increased plasma insulin levels (Pari and Satheesh 2004). Treatment with *B. diffusa* leaf extract also reduced glycosylated hemoglobin level. The activities of hepatic hexokinase was increased, while those of glucose-6-phosphatase and fructose-1,6-diphosphatase were decreased by the administration of *B. diffusa* leaf extract in normal and diabetic rats, and there was a significant improvement in glucose tolerance. Administration of *B. diffusa* leaf extract (200 mg/kg) for 4 weeks significantly reduced thiobarbutric acid reactive substances and hydroperoxides, attendant with an increase in reduced glutathione, and antioxidant enzymes in liver and kidney of alloxan-diabetic rats (Satheesh and Pari 2004).

Musa sapientum *(Banana)*

Musa sapientum commonly known as "Banana" is cultivated throughout India, and its flowers are used in Indian folk medicine for the treatment of diabetes mellitus. Significant improvement in glucose tolerance in hyperglycemic rabbits intragastrically treated with *M. sapientum* flower decoction (4 mL/kg) has been reported (Alarcon-Aguilara et al. 1998). Oral administration of chloroform extract of *M. sapientum* flowers (150, 200, and 250 mg/kg) for 30 days reduced blood glucose and glycosylated hemoglobin while increasing total hemoglobin in alloxan-diabetic rats (Pari and Umamaheswari 1999). Oral glucose tolerance test indicated a significant improvement in glucose tolerance in animals treated with banana flower extract. Banana flower extract treatment also resulted in a decrease in free radical formation in the tissues in addition to an antihyperglycemic action (Pari and Umamaheswari 2000). The decrease in thiobarbituric acid reactive substances and the increase in reduced glutathione and antioxidant enzymes clearly showed the antioxidant property of banana flower extract.

Piper betle

Leaves of *Piper betle* (*Piperaceae*) which are widely consumed as betel-quid after meals possess several bioactivities and are used in traditional medicinal systems. The antidiabetic activity of *Piper betle* leaves has been recently investigated in normoglycemic and STZ-diabetic rats (Arambewelaa et al. 2005). In normoglycemic rats, oral administration of hot water extract and cold ethanolic extract lowered the blood glucose level in a dose-dependent manner. In glucose tolerance test, both extracts markedly reduced the external glucose load, thus suggesting that the extract of *P. betle* leaves possess strong antidiabetic activity. Oral administration of the leaf of *P. betle* (75 and 150 mg/kg) for 30 days to STZ-diabetic rats resulted in significant reduction in blood glucose and glycosylated hemoglobin and decreased the activities of liver glucose-6-phosphatase and fructose-1,6-diphosphatase, while liver hexokinase increased (Santhakumari et al. 2006).

Nelumbo nucifera *(Lotus)*

Nelumbo nucifera (*Nymphaeaceae*) is an aquatic herb found throughout India. Oral intake of the extract of *N. nucifera* rhizome (300 and 600 mg/kg) caused a reduction of blood glucose levels in STZ-diabetic rats at the end of 12 h thus indicating that *N. nucifera* possesses hypoglycemic activity in hyperglycemic animals (Mukherjee et al. 1995). Orally given ethanolic extract of *N. nucifera* rhizomes markedly reduced the blood sugar of normal, glucose-fed hyperglycemic, and STZ-diabetic rats (Mukherjee et al. 1997). The extract improved glucose tolerance and potentiated the action of exogenously injected insulin in normal rats.

Citrullus colocynthis *(Bitter Apple)*

Citrullus colocynthis (*Cucurbitaceae*; "Indrayan") is found in wild throughout India in the warm areas. Infusions of *C. colocynthis* fruits are traditionally used as antidiabetic medication in Mediterranean countries. Hypoglycemic effects of aqueous, glycosidic, alkaloidal, and saponin extracts of the rind of *C. colocynthis* on the plasma glucose levels in normal rabbits, and of saponin extract on the fasting plasma glucose levels in alloxan-diabetic rabbits have been evidenced (Abdel-Hassan et al. 2000). Thus, the hypoglycemic effect of aqueous extract of the rind of *C. colocynthis* could be attributed more to the presence of saponin and glycosidic components. Extracts of *C. colocynthis* seed have been evidenced to possess insulinotropic effects in vitro in the isolated rat pancreas and isolated rat islets in the presence of glucose suggesting that the insulinotropic effect of *C. colocynthis* seed could at least partially account for the antidiabetic activities of these fruits (Nmila et al. 2000).

Brassia juncea *(Mustard)*

Mustard (*Brassica juncea*) is extensively cultivated and used as spice in Indian cusine. Mustard is reported to be hypoglycemic in rats when fed for 2 months at

10% level in the diet (Khan et al. 1995). This effect was attributed to an increase in hepatic glycogen content as a result of stimulated glycogen synthetase and suppression of glycogen phosphorylase and other gluconeogenic enzymes. The mucilage (soluble fiber) of mustard at 5, 10, and 15% dietary level had a positive effect on postprandial glucose and insulinemia (Vachon et al. 1988).

Antidiabetic Tree Plants

Syzigium cumini *(Black berry)*

Syzigium cumini (Syn. *Eugenia jambolana*), commonly known as "jamun", is widely present throughout India and is used in Indian folk medicine for the treatment of diabetes mellitus. Although there are no clinical reports, studies on animal diabetic models involving rats, mice, and rabbits have validated the antidiabetic potential of *S. cumini* seeds and lesser number of studies has also documented the antidiabetic efficacy of its fruit pulp. Oral feeding of *S. cumini* seeds (170, 240, and 510 mg/rat for 15 days) reduced blood glucose of normal, fasted rats. In addition, there was up to ninefold increase in cathepsin-B activity (proteolytic conversion of proinsulin to insulin) by *S. cumini* extract (Bansal et al. 1981). Teas prepared from leaves and seeds of *S. cumini* administered as water substitute for 95 days to normal and STZ-diabetic rats produced no detectable antihyperglycemic effect (Teixeira et al. 1997). Oral administration (2.5 and 5 g/kg) of the aqueous extract of *S. cumini* seed for 6 weeks to alloxan-diabetic rats reduced blood glucose (Prince et al. 1998). *Syzigium cumini* extract also decreased free radical formation in tissues and increased glutathione and antioxidant enzymes, showing the antioxidant property.

Daily administration of *S. cumini* seed powder (200 mg/kg) reduced plasma sugar maximally in 60 days in mild, moderate as well as severe diabetic rats (Grover et al. 2000). In addition, the treatment also partially restored altered hepatic and skeletal muscle glycogen and hepatic glucokinase, glucose-6-phosphate dehydrogenase, and phosphofructokinase. Extract of *S. cumini* seeds (200 mg/kg) administered for 50 days to STZ-diabetic mice reduced plasma glucose (Grover et al. 2002a, b). *Syzigium cumini* seeds contain 40% of water soluble gummy fiber and 15% of water insoluble neutral detergent fiber. Feeding for 21 days of the diets containing 15% seeds containing water soluble gummy fiber, 15% defatted seeds, and 6% water soluble gummy fiber from *S. cumini* seeds significantly lowered blood glucose and improved oral glucose tolerance in alloxan-diabetic rats (Pandey and Khan 2002). Feeding of the diets containing 15% degummed *S. cumini* seeds (devoid of water soluble gummy fiber but containing neutral detergent fiber) and 2.25% water insoluble neutral detergent fiber isolated from *S. cumini* seeds did not lower blood glucose levels nor improve oral glucose tolerance in both normal and diabetic rats indicating that the hypoglycemic effect of *S. cumini* seeds was due to water soluble gummy fiber.

The hypoglycemic and hypolipidemic effect of ethanolic extract of seeds of *S. cumini* (100 mg/kg) was evidenced in alloxan-diabetic rabbits (Sharma et al. 2003).

Hypoglycemic activity was evidenced by reduction in fasting blood glucose at 90 min and also improved glucose tolerance in subdiabetic and mild diabetic rabbits. When administered daily for 15 days to mild and severe diabetic rabbits, significant fall in fasting blood glucose and glycosylated hemoglobin were observed, while serum insulin, liver, and muscle glycogen were increased. *Syzigium cumini* seed extract also exhibited significant hypolipidemic effect as evident from a fall in total and LDL cholesterol. Oral administration of alcoholic extract of *S. cumini* seeds to alloxan-diabetic rats (100 mg/kg) resulted in a significant reduction in blood glucose and urine sugar and lipids in serum and tissues in diabetic rats similar to insulin (Prince et al. 2004).

Oral administration of fruit pulp of *S. cumini* to normoglycemic and STZ-diabetic rats showed hypoglycemic activity in 30 min possibly mediated by higher insulin secretion through inhibition of insulinase activity in liver and kidney (Achrekar et al. 1991). The antihyperglycemic effect of the water and ethanolic extracts of the fruit–pulp of *S. cumini* was evidenced in alloxan-induced mild diabetic as well as severely diabetic rabbits (Sharma et al. 2006). Treatment with purified water extract (25 mg/kg) in mild diabetic and severely diabetic rabbits reduced fasting blood glucose and improved glucose tolerance and increased plasma insulin in mild diabetic rabbits. Insulin release from pancreatic islets in vitro was much higher than that in untreated diabetic rabbits. The mechanism of action of this fraction appears to be both pancreatic by stimulated release of insulin and extra-pancreatic by direct action on the tissues. Tea prepared from leaves of *S. cumini* did not show any antihyperglycernic effect in volunteers submitted to a glucose blood tolerance test in a randomized, placebo-controlled trial (Teixeira et al. 2000). The treatment of leaves of *S. cumini* for 2 weeks did not produce any antihyperglycernic effect in normal and STZ-diabetic rats. An evaluation of the effect of ethanolic extracts and aqueous/butanolic fractions of the leaves of *S. cumini* on glycemia in diabetic and nondiabetic mice indicated absence of any hypoglycemic effect after acute treatment (0.2 or 2 g/kg), but the same daily treatment for 1 week reduced glycemia of nondiabetic mice (Table 14.11), which was, however, associated with a reduction of food intake and body weight (Oliveira et al. 2005).

Azadirachta indica *(Neem)*

Azadirachta indica commonly known as neem tree is found throughout India in deciduous forests and is also widely cultivated. Its leaf and seed are extremely bitter and find wide applications in folk medicines. Alcoholic extract of *A. indica* leaves was found to have antihyperglycemic and hypoglycemic property in normal, glucose-fed, and STZ-diabetic rats (Chattopadhyay et al. 1987a). Treatment with aqueous leaf extract (1 g/kg) for 6 weeks showed a fall in blood glucose in type 1 animal model (65 mg/kg, i.v. STZ-induced) and type 2 model (90 mg/kg, i.p. STZ in neonates), suggesting that neem leaf has a therapeutic potential as antihyperglycemic agent (Bajaj and Srinivasan 1999). Blood sugar lowering activity of *A. indica* leaf extract in normal and diabetic rats has been documented by other

Table 14.11. Antidiabetic potential of seeds and fruit pulp of *Eugenia jambolana*.

Experimental model	Effect demonstrated	Investigators
Normal rats	Seeds feeding produced reduction in blood glucose; increased conversion of proinsulin to insulin	Bansal et al. (1981)
	Pulp extract of fruit of *S. cumini* showed hypoglycemic activity possibly mediated by insulin secretion	Achrekar et al. (1991)
Diabetic rats	Aqueous extract of seeds feeding produced reduction in blood glucose; increased antioxidant status	Prince et al. (1998)
	Seed powder feeding reduced blood sugar; restored hepatic/muscle glycogen and activities of hepatic enzymes of carbohydrate metabolism	Grover et al. (2000)
	Aqueous extract of seeds feeding produced reduction in blood glucose	Grover et al. (2002a, b)
	Antidiabetic and antihyperlipidemic effect of alcoholic extract of seeds	Prince et al. (2004)
	Pulp extract of fruit of *S. cumini* showed hypoglycemic activity possibly mediated by insulin secretion	Achrekar et al. (1991)
Diabetic mice	Feeding soluble fiber and defatted fractions of seeds lowered blood glucose levels; improved oral glucose tolerance	Pandey and Khan (2002)
Diabetic rabbits	Hypoglycemic and hypolipidemic effect of ethanolic extract of seeds	Sharma et al. (2003)
	Antihyperglycemic effect of the water and ethanolic extracts of the fruit–pulp	Sharma et al. (2006)

investigators (Chattopadhyay 1999; Halim 2003). *Azadirachta indica* blocks the action of epinephrine on glucose metabolism, thus increasing peripheral glucose utilization (Chattopadhyay 1996). It also increased glucose uptake and glycogen deposition in isolated rat hemidiaphragm (Chattopadhyay et al. 1987b). A study in aloxan-diabetic rabbits indicated that the plant exerts its pharmacological activity independent of its time of administration (Tables 14.12 and 14.13), that is, either prior to or after alloxan injection (Khosla et al. 2000).

Pterocarpus marsupium *(Indian Malabar)*

Pterocarpus marsupium, commonly known as Indian Malabar or "vijayasar" is found in hilly regions throughout India. Hypoglycemic activity of its wood extract has been reported in different animal models. Hypoglycemic action of pterostilbene, a constituent of the bark of *P. marsupium* (10–50 mg/kg i.v.), has been observed in dogs (Haranath et al. 1958). The hypoglycemic effect of *P. marsupium* has been attributed to the presence of tannates in the extract (Joglekar et al. 1959). Oral administration of decoction of the bark (1 g/100 g) in alloxan-diabetic rats for 10 days resulted in a hypoglycemic action (Pandey and Sharma 1976). Chronic administration of the infusion of the wood powder for 5 days inhibited the rise in blood glucose in rats after glucose loading (Khandre et al. 1983). Flavonoid fraction of *P. marsupium* has been shown to cause pancreatinc β-cell regrannulation in rats, which may explain the antidiabetic activity of this plant (Chakravarthy et al. 1980). Epicatechin, a flavonoid isolated from the ethanol extract of *P. marsupium* bark, has also been shown to possess significant antidiabetic effect (Chakravarthy et al. 1982; Sheehan et al. 1983). Epicatechin has been shown to enhance insulin release and conversion of proinsulin to insulin in vitro (Sheehan et al. 1983). Phenolic constituents of *P. marsupium* such as marsupin and pterostilbene significantly lowered blood glucose level in STZ-diabetic rats analogous to metformin (Manickam et al. 1997). *Pterocarpus marsupium* has been clinically assessed for the treatment of newly diagnosed or untreated type 2 diabetes, in which the extract controlled fasting and postprandial blood glucose levels in about 70% patients by the 12th week. Oral administration of the bark extract (1 g/kg) significantly reduced the blood sugar levels 2 h after and also significantly lowered the blood glucose in alloxan-diabetic rats after daily oral administration for 21 days (Vats et al. 2002). Antidiabetic activity of various subfractions of the alcohol extract of bark was evidenced in alloxan-diabetic rats (Dhanabal et al. 2006). The effective role of *P. marsupium* in controlling diabetes and diabetes related metabolic alterations were indicated by its beneficial effect on plasma glucose and lipids.

Ficus bengalenesis *(Indian Banyan)*

Trees of *Ficus bengalenesis* are distributed throughout India. An ethanolic extract and a glycoside isolated from the bark of *F. bengalenesis* showed hypoglycemic action in normal and alloxan-diabetic rabbits (Augusti 1975a, b). Oral administration of bark extract showed significant antihyperglycemic effect in

Table 14.12. Antidiabetic potential of *Azadirachta indica* and *Pterocarpus marsupium*.

Experimental model	Effect demonstrated	Investigators
Azadirachta indica (neem)		
Normal rats	Alcoholic extract of leaves showed hypoglycemic and antihyperglycemic property	Chattopadhyay et al. (1987a)
Diabetic rats	Alcoholic extract of leaves showed hypoglycemic and antihyperglycemic property	Chattopadhyay et al. (1987a)
	Aqueous extract of leaves showed hypoglycemic property	Bajaj and Srinivasan 1999; Chattopadhyay (1999)
	Aqueous extract of leaves produced hypoglycemic effect, hypolipidemic effect and antioxidant effect	Halim (2003)
Diabetic rabbits	Hypoglycemic effect of leaves	Khosla et al. (2000)
Pterocarpus marsupium (Indian Malabar)		
Normal dogs	Hypoglycemic action of pterostilbene from bark	Haranath et al. (1958)
Normal rats	Pancreatic β-cell regeneration by flavanoid fraction	Chakravarthy et al. (1980)
	Infusion of wood powder improved glucose tolerance after a glucose load	Khandre et al. (1983)
Diabetic rats	Hypoglycemic action of decoction of the bark	Pandey and Sharma (1976)
	Epicatechin isolated from bark exhibits antidiabetic action.	Chakravarthy et al. (1982); Sheehan et al. (1983)
	Marsupin and pterostilbene lowered blood glucose	Manickam et al. (1997)
	Bark extract reduced blood sugar level	Vats et al. (2002)
	Antidiabetic activity of subfractions of the alcohol extract of the bark	Dhanabal et al. (2006)
Type 2 diabetic patients	*P. marsupium* extract controlled blood glucose level	ICMR Report (1998)

Table 14.13. Antidiabetic potential of *Ficus bengalensis* and *Aegle marmelos*.

Experimental model	Effect demonstrated	Investigators
Ficus bengalensis (Indian banyan)		
Diabetic rabbits	An ethanolic extract and a glycoside isolated from bark showed hypoglycemic potency	Augusti (1975)
Diabetic rats	Bark extract showed antihyperglycemic effect with increased serum insulin levels	Achrekar et al. (1991)
	Pelargonidin derivative from the bark showed hypo glycemic activity	Cherian et al. (1992)
	leucopelargonidin derivative from the bark showed hypoglycemic activity	Cherian and Augusti (1993)
	Hypoglycemic effect of leucopelargonidin; effectiveness of a combination of the same with insulin	Kumar and Augusti (1994)
	Hypoglycemic action of leucodelphinidin of bark	Geetha et al. (1994)
Normal rats	Hypoglycemic effect of leucopelargonidin	Kumar and Augusti (1994)
Diabetic dogs	Hypoglycemic and serum insulin raising action of leucopelargonidin isolated from the bark	Augusti et al. (1994)
Aegle marmelos (Bael)		
Diabetic rats	Aqueous extract of fruits reversed increase in blood glucose and glycosylated hemoglobin and increased hepatic glycogen	Kamalakkannan et al. (2003a)
	Aqueous extract of fruits reduced blood glucose and improved antioxidant status	Kamalakkannan and Prince (2003)
	Antilipid peroxidative activity of aqueous extract of fruits in heart and pancreas	Kamalakkannan et al. (2003)
	Hypoglycemic potential and hypolipidemic effects of aqueous extract of seeds	Kesari et al. (2006)
	Hypoglycemic efficacy of ethanolic extract of seeds; increased insulin level	Narendhirakannan et al. (2006)

STZ-diabetic rats by raising serum insulin levels or inhibiting insulinase activity in liver and kidney (Achrekar et al. 1991). A pelargonidin derivative isolated from the bark of *F. bengalenesis* decreased fasting blood glucose and improved glucose tolerance in moderately diabetic rats (Cherian et al. 1992). Treatment with the same glycoside (100 mg/kg/day) for 1 month reduced the fasting blood glucose levels to almost half of the pretreatment levels and the glucose tolerance improved. Pelargonidin was more potent than leucocyanidin in stimulating in vitro insulin secretion by β-cells (Cherian et al. 1992). Another glycoside, leucopelargonidin derivative was shown to possess significant hypoglycemic, hypolipidemic, and serum insulin raising effects in moderately diabetic rats (Cherian and Augusti 1993). Hypoglycemic effect of orally administered leucopelargonidin derivative (100 mg/kg) isolated from the bark of *F. bengalenesis* is reported in normal rats. A combination of leucopelargonidin derivative and low dose of insulin was as effective as high dose of insulin in controlling diabetes in alloxan-diabetic rats. A long-term treatment with this combination was more efficacious than insulin alone in respect of body weight, urine and blood sugar, and serum lipids (Kumar and Augusti 1994). Leucodelphinidin (250 mg/kg) also showed hypoglycemic action equal to that of glibenclamide (2 mg/kg) in normal and alloxan-diabetic rats (Geetha et al. 1994). Oral administration of leucopelargonidin (100 mg/kg) isolated from the bark showed significant hypoglycemic and serum insulin raising action in normal as also moderately diabetic dogs during a period of 2 h. The mechanism of action of the glycoside compound seems to involve stimulation of insulin secretion (Augusti et al. 1994).

Aegle marmelos *(Bael)*

The fruits of *Aegle marmelos* (*Rutaceae*) are widely used in the Indian system of medicine for the treatment of diabetes mellitus. Treatment with aqueous extract of *A. marmelos* fruit (2 × 250 mg/kg daily) for 1 month to STZ-diabetic rats reversed the increase in blood glucose and glycosylated hemoglobin and decrease in plasma insulin and liver glycogen to near normal levels (Kamalakkannan et al. 2003a). Oral administration of the water extract of *A. marmelos* fruit (125 and 250 mg/kg) twice a day for 1 month significantly reduced blood glucose, plasma thiobarbituric acid reactive substances, hydroperoxides, ceruloplasmin, and α-tocopherol and elevated plasma reduced glutathione and vitamin C in diabetic rats (Kamalakkannan and Prince, 2003). The antilipid peroxidative activity of the aqueous extract of *A. marmelos* fruits has been evidenced in STZ-diabetic rats in heart and pancreas (Kamalakkannan et al. 2003b). Oral administration of the extract for 1 month (2 × 125 and 250 mg/kg daily) decreased the elevated levels of peroxidation products, namely thiobarbituric acid reactive substances and hydroperoxides in the tissues of diabetic rats and reversed the depressed activities of antioxidant enzymes and glutathione content in the heart and pancreas of diabetic rats.

The hypoglycemic potential of aqueous extract of *A. marmelos* seeds administered orally (100–500 mg/kg) has been documented in normal and mild diabetic rats

(Kesari et al. 2006). Treatment of severely diabetic rats for 14 days (250 mg/kg) reduced the fasting blood glucose, urine sugar, serum total, and LDL cholesterol. These results indicated that *A. marmelos* seeds possess antidiabetic and hypolipidemic effects in diabetic rats. The hypoglycemic efficacy of an ethanolic extract of *A. marmelos* seeds orally administered is also reported whereas a significant decrease in the levels of blood glucose, glycosylated hemoglobin and urea, with a concomitant increase in glycogen, hemoglobin, and protein, was observed in diabetic rats (Narendhirakannan et al. 2006). The treatment also increased insulin and C-peptide levels and glucose tolerance. The decreased activities of carbohydrate-metabolizing enzymes, hexokinase, glucose-6-phosphate dehydrogenase, and glycogen synthase, the increased activities of lactate dehydrogenase, fructose-1,6-diphosphatase, glucose-6-phosphatase, and glycogen phosphorylase in STZ-diabetic rats were significantly countered by treatment with extracts of *A. marmelos*.

Cinnamomum zeylanicum *(Cinnamon)*

Beneficial influence of bark extract from *Cinnamomum zeylanicum* with respect to blood glucose and plasma insulin levels has been observed (Verspohl et al. 2005). A decrease in blood glucose was observed in a glucose tolerance test, whereas it was not obvious in rats that were not challenged by a glucose load. The elevation in plasma insulin was direct since a stimulatory in vitro effect of insulin release from insulin secreting cell line was observed. Cinnamon extract (300 mg/kg/day) is shown to improve the glucose utilization in normal rats fed a high fructose diet for 3 weeks (Qin et al. 2004). It also suggested that early cinnamon extract administration to high fructose fed rats would prevent the development of insulin resistance by enhancing insulin signaling and possibly via the nitric oxide pathway in skeletal muscle. Cinnamon and its effect on glucose and insulin sensitivity are covered in detail in Chapter 8 by Anderson.

Other Tree Plants

Isolated reports based on animal studies are also available suggesting the antidiabetic potential in respect of few other tree plants such as seed of *Terminalia chebula* (Rao and Nammi 2006), leaves of custard apple (*Annona squamosa*) (Gupta et al. 2005; Shirwaikar et al. 2004), seed extract of *Tamarindus indica* (Maiti et al. 2004), leaves of *Mangifera indica* (Muruganandan et al. 2005), and bark extract of *Ficus hispida* (Ghosh et al. 2004).

Conclusions

Plants have always been an exemplary source of drugs and, according to the available ethnobotanical information, several hundred plants may possess beneficial antidiabetic potential. A good number of such herbs, spices, and specific parts of tree plants have shown antidiabetic activity in systematic scientific studies. A wide array of active constituents derived from such plants have also demonstrated hypoglycemic potency or ability to improve glucose tolerance. Among these are alkaloids, glycosides, galactomannan gum, peptidoglycans, flavonoids, terpenoids, and amino acid derivatives. The antidiabetic attributes of these various herbs and tree plants, although largely evidenced from animal diabetic models, have the potential of therapeutic application in human type 2 diabetes mellitus. These hypoglycemic herbs, spices, and specific parts of tree plants may certainly be used in conjunction with antidiabetic drugs to achieve supplementary therapeutic effects and to minimize the requirement of oral hypoglycemic drugs or insulin. Such a strategy could serve as an effective supportive therapy in the prevention and management of long term complications of diabetes. The antidiabetic action of these herbs, spices, and tree plants seems to be mediated through one or more of the following mechanisms: (i) stimulation of the pancreas to produce and secrete insulin, (ii) interference with dietary glucose absorption, (iii) improvement of glucose utilization, and (iv) insulin-sparing action of the constituent bioactive compounds. The active constituent responsible for the proven antidiabetic potency of the plant is more or less identified in the case of a few (Figure 14.6).

Diet has been a cornerstone in the management of diabetes mellitus. The proven antidiabetic herbs and plants also include a few common spices which are regular ingredients of the diet in India and other countries. The antidiabetic efficacy of fenugreek seeds, the vegetable—bitter melon, onion, and other spices—garlic, turmeric, cumin, curry leaves, mustard, and cinnamon are well understood in numerous animal studies and limited human trials. Although the effective doses of these dietary spices and vegetables evidenced to produce the desired antidiabetic

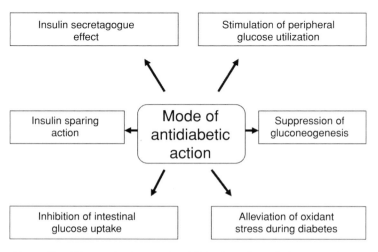

Figure 14.6. Possible mechanisms of antidiabetic action of plant constituents.

influence far exceed their normal levels encountered in our daily diet, their dietary consumption can certainly be increased without any deleterious effects to derive the health benefit.

References

Abdel-barry JA, Abdel-Hassan IA, Al-Hakiem MHH. 1997. Hypoglycaemic and anti-hyperglycaemic effects of *Trigonella foenum-graecum* leaf in normal and alloxan-induced diabetic rats. J Ethnopharmacol 58:149–155.

Abdel-Hassan IA, Abdel-Barry JA, Tariq-Mohammed S. 2000. The hypoglycaemic and anti-hyperglycaemic effect of *Citrullus colocynthis* fruit aqueous extract in normal and alloxan diabetic rabbits. J Ethnopharmacol 71:325–330.

Achrekar S, Kaklij GS, Pote MS, Kelkar SM. 1991. Hypoglycemic activity of *Eugenia jambolana* and *Ficus bengalensis*: Mechanism of action. In vivo 5:143–147.

Adeneyea AA, Amolea OO, Adeneyeb AK. 2006. Hypoglycemic and hypocholes-terolemic activities of the aqueous leaf and seed extract of *Phyllanthus amarus* in mice. Fitoterapia 77:511–514.

Agrawal P, Rai V, Singh RB. 1996. Randomized placebo-controlled, single blind trial of holy basil leaves in patients with noninsulin-dependent diabetes mellitus. Int J Clin Pharmacol Ther 34:406–409.

Ahmad M, Akhtar MS, Malik T, Gilani AH. 2000. Hypoglycemic action of flavonoid fraction of *Cuminum nigrum* seeds. Phytother Res 14:103–109.

Ahmad N, Hassan MR, Haider H, Bennoor KS. 1999. Effect of *Momordica charantia* (Karella) extracts on fasting and postprandial serum glucose levels in NIDDM patients. Bangladesh Med Res Council Bull 25:11–13.

Ahmad M, Ismail N, Ismail Z. 1995. Pharmacognistic profile of *Trigonella* seed and its hypoglycaemic activity. Nat Prod Sci 1:25–30.

Ahmed I, Adeghate E, Sharma AK, Pallot DJ, Singh J. 1998. Effects of *Momordica charantia* fruit juice on islet morphology in the pancreas of the streptozotocin diabetic rat. Diabetes Res Clin Prac 40:145–151.

Ahmed RS, Sharma SB. 1997. Biochemical studies on combined effects of garlic (*Allium sativum*) and ginger (*Zingiber officinale*) in albino rats. Indian J Exp Biol 35:841–843.

Ajabnoor MA. 1990. Effect of aloes on blood glucose levels in normal and alloxan diabetic mice. J Ethnopharmacol 28:215–220.

Ajabnoor MA, Tilmisany AK. 1988. Effect of *Trigonella foenum-graecum* on blood glucose levels in normal and alloxan diabetic mice. J Ethnopharmacol 22:45–49.

Akhtar MS, Ali MR. 1985. Study of hypoglycaemic activity of *Cuminum nigrum* seeds in normal and alloxan diabetic rabbits. Planta Med 2:81–85.

Akhtar MS, Athar MA, Yaqub M. 1981. Effect of *Momordica charantia* on blood glucose level of normal and alloxan-diabetic rabbits. Planta Med 42:205–212.

Alarcon-Aguilara FJ, Roman-Ramos R, Perez-Gutierrez S, Aguilar-Contreras A, Contreras-Weber CC, Flores-Saenz JL. 1998. Study of the anti-hyperglycemic effect of plants used as antidiabetics. J Ethnopharmacol 61:101–110.

Al-Awadi FM, Gumaa KA. 1987. Studies on the activity of individual plants of an antidiabetic plant mixture. Acta Diabetol Latina 24:37–41.

Ali L, Khan AKA, Hasssan Z. 1995. Characterization of the hypoglycaemic effect of *Trigonella foenum-graecum* seeds. Planta Med 61:358–360.

Ali L, Khan AK, Mamun MI, Mosihuzzaman M, Nahar N, Nur-e-Alam M, Rokeya B. 1993. Studies on hypoglycemic effects of fruit pulp, seed, and whole plant of *Momordica charantia* on normal and diabetic model rats. Planta Med 59:408–412.

Amin Riyad M, Abdul Ghani, Abdul Salam S, Suleiman SM. 1988. Effect of fenugreek and lupine seeds on the development of experimental diabetes in rats. Planta Med 54:286–290.

Anwar MM, Meki AR. 2003. Oxidative stress in streptozotocin-induced diabetic rats: Effects of garlic oil and melatonin. Comp Biochem Physiol 135:539–547.

Arambewelaa LSR, Arawwalaa LDAM, Ratnasooriyab WD. 2005. Antidiabetic activities of aqueous and ethanolic extracts of *Piper betle* leaves in rats. J Ethnopharmacol 102:239–245.

Arulselvan P, Senthilkumar GP, Sathish Kumar D, Subramanian S. 2006. Anti-diabetic effect of *Murraya koenigii* leaves on streptozotocin induced diabetic rats. Pharmazie 61:874–877.

Arulselvan P, Subramanian SP. 2007. Beneficial effects of *Murraya koenigii* leaves on antioxidant defense system and ultra structural changes of pancreatic beta-cells in experimental diabetes in rats. Chem Biol Interact 165:155–164.

Arun N, Nalini N. 2002. Efficacy of turmeric on blood sugar and polyol pathway in diabetic albino rats. Plant Foods Hum Nutr 57:41–45.

Augusti KT. 1973. Studies on the effects of a hypoglycaemic principle from *Allium cepa* Linn. Indian J Med Res 61:1066–1071.

Augusti KT. 1974. Effect on alloxan diabetes of allylpropyl disulfide obtained from onion. Die Naturwissenschaften 61:172–173.

Augusti KT. 1975a. Studies on the effect of allicin (diallyl disulphide oxide) on alloxan diabetes. Experientia 31:1263–1265.

Augusti KT. 1975b. Hypoglycemic action of bengalenoside, a glucoside isolated from *Ficus bengalenesis* in normal and alloxan diabetic rabbits. Indian J Physiol Pharmacol 19:218–220.

Augusti KT. 1976. Gas chromatographic analysis of onion principles and a study of their hypoglycaemic action. Indian J Exp Biol 14:110–112.

Augusti KT, Benaim ME. 1975. Effect of essential oil of onion on blood glucose, free fatty acid and insulin levels of normal subjects. Clin Chim Acta 60:121–123.

Augusti KT, Daniel RS, Cherian S, Sheela CG, Nair CR. 1994. Effect of leucopelargonin derivative from *Ficus bengalensis* Linn. on diabetic dogs. Indian J Med Res 99:82–86.

Augusti KT, Mathew PT. 1974. Lipid lowering effect of allicin (diallyl disulfide oxide) on long term feeding to rats. Experientia 30:468–469.

Augusti KT, Mathew PT. 1975. Effect of allicin on certain enzymes of liver after a short term feeding to normal rats. Experientia 31:148–149.

Augusti KT, Roy VC, Semple M. 1974. Effect of allyl propyl disulfide isolated from onion (Allium cepa) on glucose tolerance of alloxan diabetic rabbits. Experientia 30:1119–1120.

Augusti KT, Sheela CG. 1996. Antiperoxide effect of S-allyl cysteine sulfoxide, an insulin secretagogue in diabetic rats. Experientia 52:115–119.

Babu PS, Srinivasan K. 1997a. Influence of dietary capsaicin and onion on the metabolic abnormalities associated with streptozotocin induced diabetes mellitus. Mol Cell Biochem 175:49–57.

Babu PS, Srinivasan K. 1997b. Hypolipidemic action of curcumin, the active principle of turmeric (*C. longa*) in streptozotocin induced diabetic rats. Mol Cell Biochem 166:169–175.

Babu PS, Srinivasan K. 1998. Amelioration of renal lesions associated with diabetes by curcumin in streptozotocin diabetic rats. Mol Cell Biochem 181:87–96.

Babu PS, Srinivasan K. 1999. Renal lesions in streptozotocin-induced diabetic rats maintained on onion and capsaicin containing diets. J Nutr Biochem 10:477–483.

Bailey CJ, Day C, Turner SL, Leatherdale BA. 1985. Cerasee, a traditional treatment for diabetes: Studies in normal and streptozotocin diabetic mice. Diabetes Res 2:81–84.

Bajaj S, Srinivasan BP. 1999. Investigations into the anti-diabetic activity of *Azadirachta indica*. Indian J Pharmacol 31:138–141.

Bajpai MB, Asthana RK, Sharma NK, Chatterjee SK, Mukherjee SK. 1991. Hypoglycemic effect of swerchirin from the hexane fraction of *Swertia chirayita*. Planta Med 57:102–104.

Bansal R, Ahmad N, Kidwai JR. 1981. Effects of oral administration of *Eugenia jambolana* seeds and chloropropamide on blood glucose level and pancreatic cathepsin B in rat. Indian J Biochem Biophys 18:377–379.

Baskaran K, Ahamath BK, Shanmugasundaram KR, Shanmugasundaram ERB. 1990. Antidiabetic effect of a leaf extract from *Gymnema sylvestre* in non-insulin-dependent diabetes mellitus patients. J Ethnopharmacol 30:295–305.

Bolkent S, Akev N, Özsoy N, Şengezer-Inceli M, Can A, Okyar A, Yanardag R. 2004. Effect of *Aloe vera* (L.) leaf gel and pulp extracts on kidney in type-II diabetic rat models. Indian J Exp Biol 42:48–52.

Brahmachari HD, Augusti KT. 1961. Hypoglycemic agent from onions. J Pharm Pharmacol 13:128.

Brahmachari HD, Augusti KT. 1962a. Orally effective hypoglycaemic agents from plants. J Pharmacol Physiol 14:254–255.

Brahmachari HD, Augusti KT. 1962b. Effects of orally effective hypoglycaemic agents from plants on alloxan diabetes. J Pharmacol Physiol 14:617.

Brahmachari HD, Augusti KT. 1963. Hypoglycemic action of *Coccinia indica* root in rabbits. J Pharmacol 15:411–412.

Bushan S, Saxena SP, Prakash G, Nigam P, Asthavan AB. 1984. Effect of oral administration of raw onion on glucose tolerance test of diabetes: A comparison with tolbutamide. Curr Med Pract 28:712–715.

Cakici I, Hurmoglu C, Tunctan B, Abacioglu N, Kanzik I, Sener B. 1994. Hypoglycemic effect of *Momordica charantia* extracts in normoglycemic and cyproheptadine-induced hyperglycemic mice. J Ethnopharmacol 44:117–121.

Campos KE, Diniz YS, Cataneo AC, Faine LA, Alves MJ, Novelli EL. 2003. Hypoglycaemic and antioxidant effects of onion, Allium cepa:dietary onion addition, antioxidant activity and hypoglycaemic effects on diabetic rats. Int J Food Sci Nutr 54:241–246.

Chakravarthy BK, Gupta S, Gambhir SS, Gode KD. 1980. Pancreatic beta cell regeneration: A novel antidiabetic mechanism of *Pterocarpus marsupium* Roxb. Indian J Pharmacol 12:123–127.

Chakravarthy BK, Gupta S, Gode KD. 1982. Functional beta cell regeneration in the islets of pancreas in alloxan induced diabetic rats by (-) epicatechin. Life Sci 31:2693–2697.

Chakrabarti S, Biswas TK, Rokeya B, Ali L, Mosihuzzaman M, Nahar N, Azad Khan AK, Mukherjee B. 2003. Advanced studies on the hypoglycemic effect of *Caesalpinia bonducella* in type 1 and 2 diabetes in Long Evans rats. J Ethnopharmacol 84:41–46.

Chakrabarti S, Biswas TK, Seal T, Rokeya B, Ali L, Azad Khan, Nahar N, Mosihuzzaman M, Mukherjee B. 2005. Antidiabetic activity of *Caesalpinia bonducella* in chronic type 2 diabetic model in Long-Evans rats and evaluation of insulin secretagogue property of its fractions on isolated islets. J Ethnopharmacol 97:117–122.

Chandrasekhar B, Bajpai MB, Mukherjee SK. 1990. Hypoglycemic activity of *Swertia chirayita*. Indian J Exp Biol 28:616–618.

Chandrasekhar B, Mukerjee B, Mukherjee SK. 1989. Blood sugar lowering potentiality of selected *Cucurbitaceae* plants of Indian origin. Indian J Med Res 90:300–305.

Chang MLW, Johnson MA. 1980. Effect of garlic on carbohydrate metabolism and lipid synthesis in rats. J Nutr 110:931–936.

Chattopadhyay RR. 1993. Hypoglycemic effect of *Ocimum sanctum* leaf extract in normal and streptozotocin diabetic rats. Indian J Exp Biol 31:891–893.

Chattopadhyay RR. 1996. Possible mechanism of anti-hyperglycemic effect of *Azadirachta indica* leaf extract. Gen Pharmacol 27:431–434.

Chattopadhyay RR. 1999. A comparative evaluation of some blood sugar lowering agents of plant origin. J Ethnopharmacol 67:367–372.

Chattopadhyay RR, Chattopadhyay RN, Nandi AK, Poddar G, Maitra SK. 1987a. Preliminary report on anti-hyperglycemic effect of a fraction of fresh leaves of *Azadirichta indica*. Bull Calcutta Sch Trop Med 35:29–33.

Chattopadhyay RR, Chattopadhyay RN, Nandi AK, Poddar G, Maitra SK. 1987b. The effect of fresh leaves of *Azadirichta indica* on glucose uptake and glycogen content in the isolated rat hemi diaphragm. Bull Calcutta Sch Trop Med 35:8–12.

Chaturvedi P, George S, Milinganyo M, Tripathi YB. 2004. Effect of *Momordica charantia* on lipid profile and oral glucose tolerance in diabetic rats. Phytother Res 18:954–956.

Cherian S, Augusti KT. 1993. Antidiabetic effects of a glycoside of leucopelargonidin isolated from *Ficus bengalenesis* Linn. Indian J Exp Biol 31:26–29.

Cherian S, Kumar RV, August KT, Kidwai JR. 1992. Antidiabetic effect o a glycoside of pelargonidin isolated from the bark of *Ficus bengalenesis* Linn. Indian J Biochem Biophys 29:380–382.

Chitra P, Sajithlal GB, Chandrahasan G. 1998. Influence of *Aloe vera* on the healing of dermal wounds in diabetic rats. J Ethnopharmacol 59:195–201.

Dandapani S, Subramanian VR, Rajagopal S, Namasibvayam N. 2002. Hypolipidemic effect of *Cuminum cyminum* L. in alloxan-induced diabetic rats. Pharmacol Res 46:251–256.

Day C, Cartwright T, Provost J, Bailey CJ. 1984. Oral hypoglycemic activity of some medicinal plants of Srilanka. J Ethnopharmacol 11:223–231.

Day C, Cartwright T, Provost J, Bailey CJ. 1990. Hypoglycemic effect of *Momordica charantia* extracts. Planta Med 56:426–429.

Davis RH, Maro NP. 1989. *Aloe vera* and *gibberellin*: Anti-inflammatory activity in diabetes. J Am Pediatr Med Assoc 79:24–26.

Devi BA, Kamalakkannan N, Prince PS. 2003. Supplementation of fenugreek leaves to diabetic rats. Effect on carbohydrate metabolic enzymes in diabetic liver and kidney. Phytother Res 17:1231–1233.

Dhanabal SP, Kokate CK, Ramanathan M, Kumar EP, Suresh B. 2006. Hypoglycaemic activity of *Pterocarpus marsupium* Roxb. Phytother Res 20:4–8.

Dhar ML, Dhar MM, Dhawan BN, Mehrotra BN, Ray C. 1968. Screening of Indian plants for biological activity. Indian J Exp Biol 6:232–247.

Dubey DK, Biswas AR, Bapna JS, Pradhan SC. 1987. Hypoglycemic and anti-hyperglycemic effects of *M. charantia* seed extracts in albino rats. Fitoterapia 58:387–390.

El-Dakhakhny M, Mady N, Lembert N, Ammon HPT. 2002. The hypoglycemic effect of *Nigella sativa* oil is mediated by extra-pancreatic actions. Planta Med 68:465–466.

El-Demerdash FM, Yousef MI, El-Naga NI. 2005. Biochemical study on the hypo-glycemic effects of onion and garlic in alloxan-induced diabetic rats. Food Chem Toxicol 43:57–63.

Fararh KM, Atoji Y, Shimizu Y, Shiina T, Nikami H, Takewaki T. 2004. Mechanisms of the hypoglycaemic and immunopotentiating effects of *Nigella sativa* L. oil in streptozotocin-induced diabetic hamsters. Res Vet Sci 77:123–129.

Fararh KM, Atoji Y, Shimizu Y, Takewaki T. 2002. Insulinotropic properties of *Nigella sativa* oil in Streptozotocin plus Nicotinamide diabetic hamster. Res Vet Sci 73:279–282.

Farva D, Gaji IA, Joseph PK, Augusti KT. 1986. Effects of garlic oil on streptozotocin diabetic rats maintained on normal and high-fat diets. Indian J Biochem Biophys 23:24–27.

Geetha BS, Mathew BS, Augusti KT. 1994. Hypoglycemic effects of leucodelphini-din derivative isolated from Ficus bengalensis Linn. Indian J Physiol Pharmacol 38:220–222.

Genet S, Kale RK, Baquer NZ. 1999. Effects of vanadate, insulin and fenugreek (*Trigonella foenum-graecum*) on creatinine kinase levels in tissues of diabetic rat. Indian J Exp Biol 37:200–202.

Ghannam N, Kingston M, Al-Meshaal IA, Tariq M, Parman NS, Woodhouse N. 1986. The antidiabetic activity of aloes: Preliminary clinical and experimental observa-tion. Hormone Res 24:288–294.

Gholap S, Kar A. 2003. Effects of *Inula racemosa* root and *Gymnema sylvestre* leaf extracts in the regulation of corticosteroid induced diabetes mellitus: Involvement of thyroid hormones. Pharmazie 58:413–415.

Ghosh R, Sharatchandra K, Rita S, Thokchom IS. 2004. Hypoglycemic activity of Ficus hispida (bark) in normal and diabetic albino rats. Indian J Pharmacol 36:222–225.

Grover JK, Rathi SS, Vats V. 2002b. Amelioration of experimental diabetic neuropathy and gastropathy in rats following oral administration of plant (*Eugenia jambolana, Mucurna pruriens* and *Tinospora cordifolia*) extracts. Indian J Exp Biol 40:273–276.

Grover JK, Vats V, Rathi SS. 2000. Anti-hyperglycemic effect of *Eugenia jambolana* and *Tinospora cordifolia* in experimental diabetes and their effects on key metabolic enzymes involved in carbohydrate metabolism. J Ethnopharmacol 73:461–470.

Grover JK, Yadav S, Vats V. 2002a. Medicinal plants of India with anti-diabetic potential. J Ethnopharmacol 81:81–100.

Grover JK, Yadav SP. 2004. Pharmacological actions and potential uses of *Momordica charantia*: A review. J Ethnopharmacol 93:123–132.

Gupta SS. 1963. Experimental studies on pituitary diabetes: Effect of indigenous antidiabetic drugs against the acute hyperglycaemic response of anterior pituitary extract in glucose fed albino rats. Indian J Med Res 51:716–724.

Gupta A, Gupta R, Lal B. 2001. Effect of *Trigonella foenum-graecum* (fenugreek) seeds on glycaemic control and insulin resistance in type 2 diabetes mellitus: A double blind placebo controlled study. J Assoc Physicians India 49:1055–1056.

Gupta RK, Gupta S, Samuel KC. 1977. Blood sugar-lowering effect of various fractions of onion. Indian J Exp Biol 15:313–314.

Gupta RK, Kesari AN, Murthy PS, Chandra R, Tandon V, Watal G. 2005. Hypoglycemic and antidiabetic effect of ethanolic extract of leaves of *Annona squamosa* L. in experimental animals. J Ethnopharmacol 99:75–81.

Gupta D, Raju J, Baquer NZ. 1999. Modulation of some gluconeogenic enzyme activities in diabetic rat liver and kidney: Effect of antidiabetic compounds. Indian J Exp Biol 37:196–199.

Gupta S, Mediratta PK, Singh S, Sharma KK, Shukla R. 2006. Antidiabetic, antihypercholesterolaemic and antioxidant effect of *Ocimum sanctum* (Linn) seed oil. Indian J Exp Biol 44:300–304.

Gupta SK, Prakash J, Srivastava S. 2002. Validation of traditional claim of Tulsi, *Ocimum sanctum* Linn as a medicinal plant. Indian J Exp Biol 40:765–773.

Gupta SS, Seth CB. 1962. Effect of *Momordica charantia* Linn (Karela) on glucose tolerance in albino rats. J Indian Med Assoc 39:581–584.

Gupta SS, Variyar MC. 1964. Effect of *Gymnema sylvestre* and *Coccinia indica* against the hypergycaemic response of somatotropin and corticotrophin hormones. Indian J Med Res 52:200–207.

Halim ME. 2003. Lowering of blood sugar by water extract of *Azadirichta indica* and *Abroma augusta* in diabetes rats. Indian J Exp Biol 41:636–640.

Handa G, Singh J, Sharma ML, Kaul A, Zafar RN. 1990. Hypoglycemic principle of *Momordica charantia* seeds. Indian J Nat Prod 6:16–19.

Hannan JM, Marenah L, Ali L, Rokeya B, Flatt PR, Abdel-Wahab YH. 2006. *Ocimum sanctum* leaf extracts stimulate insulin secretion from perfused pancreas, isolated islets and clonal pancreatic beta-cells. J Endocrinol 189:127–136.

Hannan JM, Rokeya B, Faruque O, Nahar N, Mosihuzzaman M, Azad Khan AK, Ali L. 2003. Effect of soluble dietary fibre fraction of *Trigonella foenum-graecum* on glycemic, insulinemic, lipidemic and platelet aggregation status of Type 2 diabetic model rats. J Ethnopharmacol 88:73–77.

Haranath PS, Rao KR, Anjaneyulu CR, Ramanathan JD. 1958. Studies on the hypoglycemic and pharmacological actions of some stilbenes. Indian J Med Sci 12:85–89.

Higashino H, Suzuki A, Tanaka Y, Pootakham K. 1992. Hypoglycaemic effects of Siamese *Momordica charantia* and *Phylanthus urinaria* extracts in streptozotocin diabetic rats. Nippon Yakurigaku Zasshi 100:415–421.

Hossain MZ, Shibib BA, Rahman R. 1992. Hypoglycemic effects of *Coccinia indica*: Inhibition of key gluconeogenic enzymes, glucose-6-phosphatase. Indian J Exp Biol 30:418–420.

Indian Council of Medical Research. 1998. Annual Report, New Delhi.

Iyer UM, Mani UV. 1990. Studies on the effect of curry leaves (*Murraya koenigii*) supplementation on lipid profile, glycated proteins and amino acids in non insulin dependent diabetic patients. Plant Foods Hum Nutr 40:275–282.

Jain SC, Lohiya NK, Kapoor A. 1987. *Trigonella foenum-graecum* Linn: A hypoglycaemic agent. Indian J Pharm Sci 49:113–114.

Jain RC, Vyas CR. 1974. Hypoglycaemia action of onion on rabbits. Br Med J 2:730.

Jain RC, Vyas CR. 1975. Garlic in alloxan induced diabetic rabbits. Am J Clin Nutr 28:684–685.

Jain RC, Vyas CR, Mahatma OP. 1973. Hypoglycemic action of onion and garlic. Lancet II:1491.

Jeevathayaparan S, Tennekoon KH, Karunanayake EH, Jayasinghe SA. 1991. Oral hypo-glycaemic activity of different preparations of *Momordica charantia*. J Natl Sci Counc (Sri Lanka) 19:19–24.

Joglekar GV, Chaudhary NY, Aiaman R. 1959. Effect of Indian medicinal plants on glucose absorption in mice. Indian J Physiol Pharmacol 3:76–77.

Kaleem M, Kirmani D, Asif M, Ahmed Q, Bano B. 2006. Biochemical effects of *Nigella sativa* L seeds in diabetic rats. Indian J Exp Biol 44:745–748.

Kamalakkannan N, Prince PS. 2003. Hypoglycaemic effect of water extracts of *Aegle marmelos* fruits in streptozotocin diabetic rats. J Ethnopharmacol 87:207–210.

Kamalakkannan N, Rajadurai M, Prince PS. 2003a. Effect of *Aegle marmelos* fruits on normal and streptozotocin-diabetic Wistar rats. J Med Food 6:93–98.

Kamalakkannan N, Stanely M, Prince P. 2003b. Effect of *Aegle marmelos*. fruit extract on tissue antioxidants in streptozotocin diabetic rats. Indian J Exp Biol 41:1285–1288.

Kamble SM, Kamlakar PL, Vaidya S, Bambole VD. 1998. Influence of *Coccinia indica* on certain enzymes in glycolytic and lipolytic pathway in human diabetes. Indian J Med Sci 52:143–146.

Kameswara Rao B, Kesavulu MM, Giri R, Appa Rao C. 1999. Antidiabetic and hypolipidemic effects of *Momordica cymbalaria* fruit powder in alloxan-diabetic rats. J Ethnopharmacol 67:103–109.

Kanter M, Coskun O, Korkmaz A, Oter S. 2004. Effects of *Nigella sativa* on oxidative stress and beta-cell damage in streptozotocin-induced diabetic rats. Anat Res Discov Mol Cell Evol Biol 279:685–691.

Kanter M, Meral I, Yener Z, Ozbek H, Demir H. 2003. Partial regeneration/proliferation of the beta-cells in the islets of Langerhans by *Nigella sativa* L. in streptozotocin-induced diabetic rats. Tohoku J Exp Med 201:213–219.

Kar A, Choudhary BK, Bandyopadhyay NG. 2003. Comparative evaluation of hypoglycaemic activity of some Indian medicinal plants in alloxan diabetic rats. J Ethnopharmacol 84:105–108.

Karnick CR. 1991. A clinical trial of a composite herbal drug in the treatment of diabetes mellitus. Aryavaidyan 5:36–46.

Karunanayke EH, Jeevathayaparan S, Tennekoon KH. 1990. Effect of *Momordica charantia* fruit juice on streptozotocin induced diabetes in rats. J Ethnopharmacol 30:199–204.

Karunanayake EH, Welihinda J, Sirimanna SR, Sinnadorai G. 1984. Oral hypoglycemic activity of some medicinal plants of Srilanka. J Ethnopharmacol 11:223–231.

Kasuga S, Ushijima M, Morihara N, Itakura Y, Nakata Y. 1999. Effect of aged garlic extract on hyperglycemia induced by immobilization stress in mice. Nippon Yakurigaku Zasshi 114:191–197

Kedar P, Chakrabarti CH. 1982. Effects of bitter gourd (*Momordica charantia*) seed and glibenclamide in streptozotocin induced diabetes mellitus. Indian J Exp Biol 20:232–235.

Kelkar SM, Kaklij GS, Bapat VA. 2001. Determination of antidiabetic activity in *Allium cepa* (onion) tissue cultures. Indian J Biochem Biophys 38:277–279.

Kesari AN, Gupta RK, Singh SK, Diwakar S, Watal G. 2006. Hypoglycemic and anti-hyperglycemic activity of *Aegle marmelos* seed extract in normal and diabetic rats. J Ethnopharmacol 107:374–379.

Kesari AN, Gupta RK, Watal G. 2005. Hypoglycemic effects of *Murraya koenigii* on normal and alloxan-diabetic rabbits. J Ethnopharmacol 97:247–251.

Khan BA, Abraham A, Leelamma S. 1995. Hypoglycemic action of *Murraya koenigii* (curry leaf) and *Brassica junecea* (mustard): Mechanism of action. Indian J Biochem Biophys 32:106–108.

Khan AK, Akhtar S, Mahtab H. 1980. Treatment of diabetes mellitus with *Coccinia indica*. Br Med J 280:1044.

Khandre SS, Rajwade GG, Jangle SN. 1983. A study of the effect of bija and jumn seed extract on hyperglycemia induced by glucose load. Maharashtra Med J 30: 117.

Khanna P, Jain SC, Panagariya A, Dixit VP. 1981. Hypoglycemic activity of polypeptide-p from a plant source. J Nat Prod 44:648–655.

Khare AK, Tondon RN, Tewari JP. 1983. Hypoglycaemic activity of an indigenous drug (*Gymnema sylvestre*, "Gurmar") in normal and diabetic persons. Indian J Physiol Pharmacol 27:257–258.

Khosla P, Bhanwara S, Singh J, Seth S, Srivastava RK. 2000. A study of hypoglycemic effects of *Azadirichta indica* in normal and alloxan diabetic rabbits. Indian J Physiol Pharmacol 44:69–74.

Khosla P, Gupta DD, Nagpal RK. 1995. Effect of *Trigonella foenum-graecum* (Fenugreek) on blood glucose in normal and diabetic rats. Indian J Physiol Pharmacol 39:173–174.

Krawinkel MB, Keding GB. 2006. Bitter gourd (*Momordica charantia*): A dietary approach to hyperglycemia. Nutr Rev 64:331–337.

Kulkarni RD, Gaitande BB. 1962. Potentiation of tolbutamide action by jasad bhasma and karela (*Momordica charantia*). Indian J Med Res 50:715–719.

Kumar RV, Augusti KT. 1994. Insulin sparing action of a leucocyanidin derivative isolated from *Ficus bengalenesis* Linn. Indian J Biochem Biophys 31:73–76.

Kumar GP, Sudheesh S, Vijayalakshmi NR. 1993. Hypoglycemic effect of *Coccinia indica*: Mechanism of action. Planta Med 59:330–332.

Kumari K, Augusti KT. 2002. Antidiabetic and antioxidant effects of S-methyl cysteine sulfoxide isolated from onions (*Allium cepa* Linn) as compared to standard drugs in alloxan diabetic rats. Indian J Exp Biol 40:1005–1009.

Kumari K, Mathew BC, Augusti KT. 1995. Antidiabetic and hypolipidemic effects of S-methyl cysteine sulfoxide isolated from onion. Indian J Biochem Biophys 32:49–54.

Kuppu RK, Srivastava A, Krishnaswami CV, Vijaykumar G, Chellamariappan M, Ashabai, Babu BO. 1998. Hypoglycemic and hypotriglyceridemic effects of methica churna (Fenugreek). Antiseptic 95:78–79.

Kuppurajan K, Seshadri C, Revati R, Venkataraghavan S. 1986. Hypoglycaemic effect of C. indica in diabetes mellitus. Nagarjun 29:1–4.

Lakholia AN. 1956. The use of bitter gourd in diabetes mellitus. Antiseptic 53:608–610.

Latha M, Pari L. 2003a. Modulatory effect of *Scoparia dulcis* in oxidative stress-induced lipid peroxidation in streptozotocin diabetic rats. J Med Food 6:379–386.

Latha M, Pari L. 2003b. Antihyperglycaemic effect of *Cassia auriculata* in experimental diabetes and its effects on key metabolic enzymes involved in carbohydrate metabolism. Clin Exp Pharmacol Physiol 30:38–43.

Latha M, Pari L. 2003c. Preventive effects of *Cassia auriculata* L. flowers on brain lipid peroxidation in rats treated with streptozotocin. Mol Cell Biochem 243:23–28.

Latha M, Pari L. 2004. Effect of an aqueous extract of *Scoparia dulcis* on blood glucose, plasma insulin and some polyol pathway enzymes in experimental rat diabetes. Braz J Med Biol Res 37:577–586.

Latha M, Pari L, Sitasawad S, Bonde R. 2004. Insulin-secretagogue activity and cytoprotective role of the traditional antidiabetic plant *Scoparia dulcis*. Life Sci 75:2003–2014.

Leatherdale BA, Panesar RK, Singh G, Atkins TW, Bailey CJ, Bignell AHC. 1981. Improvement in glucose tolerance due to *Momordica charantia* (Karela). Br Med J 282:1823–1824.

Lotlikar MM, Rao MRR. 1966. Pharmacology of a hypoglycemic principle isolated from the fruits of *Momordica charantia*. Indian J Pharmacol 28:129–133.

Madar Z. 1984. Fenugreek (Trigonella foenum-graecum) as a means of reducing post-prandial glucose level in diabetic rats. Nutr Rep Int 29:1267–1272.

Madar Z, Abel R, Samish S, Arad J. 1988. Glucose lowering effect of fenugreek in non-insulin dependent diabetics. Eur J Clin Nutr 42:51–54.

Madar Z, Arad J. 1989. Effect of extracted fenugreek on post-prandial glucose levels in human diabetic subjects. Nutr Res 9:691–692.

Maiti R, Jana D, Das UK, Ghosh D. 2004. Antidiabetic effect of aqueous extract of seed of *Tamarindus indica* in streptozotocin-induced diabetic rats. J Ethnopharmacol 92:85–91.

Manickam M, Ramanathan M, Jahromi MA, Chansouria JP, Ray AB. 1997. Anti-hyperglycemic activity of phenolics from *Pterocarpus marsupium*. J Nat Prod 60:609–610.

Mathew PT, Augusti KT. 1973. Studies on the effect of allicin (diallyl disulphide oxide) on alloxan diabetes: Hypoglycaemic action and enhancement of serum insulin effect and glycogen synthesis. Indian J Biochem Biophys 10:209–212.

Mathew PT, Augusti KT. 1975. Hypoglycaemic effects of onion, *Allium cepa* Linn on diabetes mellitus—A preliminary report. Indian J Physiol Pharmacol 19: 213.

Meral I, Yener Z, Kahraman T, Mert N. 2001. Effect of *Nigella sativa* on glucose concentration, lipid peroxidation, anti-oxidant defense system and liver damage in experimentally-induced diabetic rabbits. J Vet Med Physiol Pathol Clin Med 48:593–599.

Mier P, Yaniv Z. 1985. An in vitro study on the effect of *Momordica charantia* on glucose uptake and glucose metabolism. Planta Med 1:12–16.

Mishkinsky J, Joseph B, Sulman F. 1967. Hypoglycemic effect of trigonelline. Lancet 1:1311–1312.

Miura T, Itoh C, Iwamoto N, Kato M, Kawai M, Park SR, Suzuki I. 2001. Hypoglycemic activity of the fruit of the *Momordica charantia* in Type 2 diabetic mice. J Nutr Sci Vitaminol (Tokyo) 47:340–344.

Mondal DK, Yousuf BM, Banu LA, Ferdousi R, Khalil M, Shamim KM. 2004. Effect of fenugreek seeds on the fasting blood glucose level in the streptozotocin induced diabetic rats. Mymensingh Med J 13:161–164.

Mukherjee B, Chandrasekhar B, Mukherjee SK. 1988. Sugar lowering effect of *C. indica* root and whole plant in different experimental rat models. Fitoterapia 59:207–210.

Mukherjee K, Ghosh NC, Datta T. 1972. *Coccinia indica* Linn. as potential hypo-glycemic agent. Indian J Exp Biol 10:347–349.

Mukherjee PK, Maiti K, Mukherjee K, Houghton PJ. 2006. Leads from Indian medic-inal plants with hypoglycemic potentials. J Ethnopharmacol 15:1–28.

Mukherjee PK, Pal SK, Saha K, Saha BP. 1995. Hypoglycaemic activity of *Nelumbo nucifera* gaertn. (*Nymphaeaceae*) rhizome (methanolic extract) in streptozotocin-induced diabetic rats. Phytother Res 9:522–524.

Mukherjee PK, Saha K, Pal M, Saha BP. 1997. Effect of Nelumbo nucifera rhizome extract on blood sugar level in rats. J Ethnopharmacol 58:207–213.

Murakami N, Murakami T, Kadoya M, Matsuda H, Yamahara J, Yoshikawa M. 1996. New hypoglycemic constituents in "gymnemic acid" from *Gymnema sylvestre*. Chem Pharm Bull (Tokyo) 44:469–471.

Murthy RR, Murthy PS, Prabhu K. 1990. Effects on blood glucose and serum insulin levels in alloxan-induced diabetic rabbits by fraction GII of *T. foenum-graecum*. Biomedicine 10:25–29.

Muruganandan S, Srinivasan K, Gupta S, Gupta PK, Lal J. 2005. Effect of mangiferin on hyperglycemia and atherogenicity in streptozotocin diabetic rats. J Ethnopharmacol 97:497–501.

Nahar N, Nur-e-Alam, Nasreen T, Mosihuzzaman M, Ali L, Begum R, Khan AKA. 1992. Studies of blood glucose lowering effects of *Trigonella foenum-graecum* seeds. Med Arom Plants Abstr 14:2264.

Narendhirakannan RT, Subramanian S, Kandaswamy M. 2006. Biochemical evaluation of antidiabetogenic properties of some commonly used Indian plants on streptozotocin-induced diabetes in experimental rats. Clin Exp Pharmacol Physiol 33:1150–1157.

Neeraja A, Rajyalakshmi P. 1996. Hypoglycaemic effect of processed fenugreek seeds in humans. J Food Sci Technol 33:427–430.

Ng TB, Wong CM, Li WW, Yeung HW. 1986. Isolation and characterization of a galactose binding lectin with insulinomimetic activities. From the seeds of the bitter gourd *Momordica charantia*. Int J Pept Prot Res 28:163–172.

Nmila R, Gross R, Rchid H, Roye M, Manteghetti M, Petit P, Tijane M, Ribes G, Sauvaire Y. 2000. Insulinotropic effect of *Citrullus colocynthis* fruit extracts. Planta Med 66:418–423.

Okabayashi Y, Tani S, Fujisawa T, Koide M, Hasegawa H, Nakamura T, Fujii M, Otsuki M. 1990. Effect of *Gymnema sylvestre*, R.Br. on glucose homeostasis in rats. Diabetes Res Clin Pract 9:143–148.

Okyar A, Can A, Akev N, Baktir G, Sütlüpinar N. 2001. Effect of *Aloe vera* leaves on blood glucose level in type I and type II diabetic rat models. Phytother Res 15:157–161.

Oliveira AC, Endringer DC, Amorim LA, das Graças L. Brandão MG, Coelho MM. 2005. Effect of the extracts and fractions of *Baccharis trimera* and *Syzygium cumini* on glycaemia of diabetic and non-diabetic mice. J Ethnopharmacol 102:465–469.

Pahwa ML. 1990. Effect of methi intake on blood sugar. Orient J Chem 6:124–126.

Pandey M, Khan A. 2002. Hypoglycaemic effect of defatted seeds and water soluble fibre from the seeds of *Syzygium cumini* skeels in alloxan diabetic rats. Indian J Exp Biol 40:1178–1182.

Pandey MC, Sharma PV. 1976. Hypoglycemic effect of bark of *Pterocarpus marsupium* Roxb. (Bijaka) on alloxan induced diabetes. Med Surg 16:9.

Pari L, Latha M. 2002. Effect of *Cassia auriculata* flowers on blood sugar levels, serum and tissue lipids in streptozotocin diabetic rats. Singapore Med J 43:617–621.

Pari L, Latha M. 2004a. Protective role of *Scoparia dulcis* plant extract on brain antioxidant status and lipid peroxidation in STZ-diabetic male Wistar rats. BMC Complement Altern Med 4:16.

Pari L, Latha M. 2004b. Effect of *Scoparia dulcis* (Sweet Broom weed) plant extract on plasma antioxidants in streptozotocin-induced experimental diabetes in male albino Wistar rats. Pharmazie 59:557–560.

Pari L, Latha M. 2005. Antidiabetic effect of *Scoparia dulcis*: Effect on lipid peroxidation in streptozotocin diabetes. Gen Physiol Biophys 24:13–26.

Pari L, Latha M. 2006. Antihyperlipidemic effect of *Scoparia dulcis* (sweet broom weed) in streptozotocin diabetic rats. J Med Food 9:102–107.

Pari L, Latha M, Rao CA. 2004. Effect of *Scoparia dulcis* extract on insulin receptors in streptozotocin induced diabetic rats: Studies on insulin binding to erythrocytes. J Basic Clin Physiol Pharmacol 15:223–240.

Pari L, Satheesh MA. 2004. Antidiabetic activity of *Boerhaavia diffusa* L.: Effect on hepatic key enzymes in experimental diabetes. J Ethnopharmacol 91:109–113.

Pari L, Umamaheswari J. 1999. Hypoglycaemic effect of *Musa sapientum* L. in alloxan-induced diabetic rats. J Ethnopharmacol 68:321–325.

Pari L, Umamaheswari J. 2000. Antihyperglycaemic activity of *Musa sapientum* flowers: Effect on lipid peroxidation in alloxan diabetic rats. Phytother Res 14:136–138.

Pari L, Venkateswaran S. 2003. Protective effect of *Coccinia indica* on changes in the fatty acid composition in streptozotocin induced diabetic rats. Pharmazie 58:409–412.

Patel JC, Dhirawani MK, Doshi JC. 1968. Karela in the treatment of diabetes mellitus. Indian J Med Sci 22:30–32.

Patumraj S, Tewit S, Amatyakul S, Jaryiapongskul A, Maneesri S, Kasantikul V, Shepro D. 2000. Comparative effects of garlic and aspirin on diabetic cardiovascular complications. Drug Deliv 7:91–96.

Persaud SJ, Al-Majed H, Raman A, Jones PM. 1999. *Gymnema sylvestre* stimulates insulin release in vitro by increased membrane permeability. J Endocrinol 163:207–212.

Platel K, Srinivasan K. 1995. Effect of dietary intake of freeze-dried bitter gourd (*Momordica charantia*) in streptozotocin induced diabetic rats. Nahrung 39:262–268.

Platel K, Srinivasan K. 1997. Plant foods in the management of diabetes mellitus: Vegetables as potential hypoglycaemic agents: A review. Nahrung 41:68–74.

Prakash AO, Mathur S, Mathur R. 1986. Effect of feeding *Gymnema sylvestre* leaves on blood glucose in beryllium nitrate treated rats. J Ethnopharmacol 18:143–146.

Prince PS, Kamalakkannan N, Menon VP. 2004. Antidiabetic and antihyperlipidaemic effect of alcoholic *Syzigium cumini* seeds in alloxan induced diabetic albino rats. J Ethnopharmacol 91:209–213.

Prince PS, Menon VP, Pari L. 1998. Hypoglycaemic activity of *Syzigium cumini* seeds: Effect on lipid peroxidation in alloxan diabetic rats. J Ethnopharmacol 61:1–7.

Prince P, Stanley M, Menon VP. 2003. Hypoglycaemic and hypolipidaemic action of alcohol extract of *Tinospora cordifolia* roots in induced diabetes in rat. Phytother Res 17:410–413.

Puri D, Prabhu KM, Murthy PS. 2002. Mechanism of action of a hypoglycemic principle isolated from fenugreek seeds. Indian J Physiol Pharmacol 46:457–462.

Pushpendran CK, Devasagayam TPA, Eapen J. 1982. Age related hyperglycaemic effect of diallyl disulphide in rats. Indian J Exp Biol 20:428–429.

Qin B, Nagasaki M, Ren M, Bajotto G, Oshida Y, Sato Y. 2004. Cinnamon extract prevents the insulin resistance induced by a high-fructose diet. Horm Metab Res 36:119–125.

Raghuram TC, Sharma RD, Sivakumar B, Sahay BK. 1994. Effect of fenugreek seeds on intra-venous glucose disposition in non-insulin dependent diabetic patients. Phytother Res 8:83–86.

Rai V, Iyer U, Mani UV. 1997. Effect of Tulasi (*Ocimum sanctum*) leaf powder supplementa-tion on blood sugar levels, serum lipids and tissue lipids in diabetic rats. Plant Foods Hum Nutr 50:9–16.

Rajasekaran S, Ravi K, Sivagnanam K, Subramanian S. 2006. Beneficial effects of *Aloe vera* leaf gel extract on lipid profile status in rats with streptozotocin diabetes. Clin Exp Pharmacol Physiol 33:232–237.

Raju J, Gupta D, Rao AR, Yadava PK, Baquer NZ. 2001. *T. foenum-graecum* seed powder improves glucose homeostasis in alloxan diabetic rat tissues by reversing the altered glycolytic, gluconeogenic and lipogenic enzymes. Mol Cell Biochem 224:45–51.

Rao VV, Dwivedi SK, Swarup D. 1994. Hypoglycemic effect of *Caesalpinia bonducella* in rabbits. Fitoterapia 65:245–247.

Rao NK, Nammi S. 2006. Antidiabetic and renoprotective effects of the chloroform extract of *Terminalia chebula* seeds in streptozotocin-induced diabetic rats. BMC Complement Altern Med 6:17.

Raphael KR, Sabu MC, Kuttan R. 2002. Hypoglycemic effect of methanol extract of *Phyllanthus amarus* on alloxan induced diabetes mellitus in rats and its relation with antioxidant potential. Indian J Exp Biol 40:905–909.

Rathi SS, Vats V, Grover JK. 2002. Anti-hyperglycemic effects of *Momordica charantia* and *Mucuna pruriens* in experimental diabetes and their effect on key metabolic enzymes involved in carbohydrate metabolism. Phytother Res 16:236–243.

Rchid H, Chevassus H, Nmila R, Guiral C, Petit P, Chokairi M, Sauvaire Y. 2004. *Nigella sativa* seed extracts enhance glucose-induced insulin release from rat-isolated Langerhans islets. Fundam Clin Pharmacol 18:525–529.

Ribes G, Sauvaire Y, Baccou JC, Valette G, Chenon D, Trimble ER, Loubatieres-Mariani MM. 1984. Effects of fenugreek seeds on endocrine pancreatic secretion in dogs. Ann Nutr Metab 28:37–43.

Ribes G, Sauvaire Y, Costa CD, Baccou JC, Loubatieres-Mariani MM. 1986. Antidi-abetic effects of subfractions from fenugreek seeds in diabetic dogs. Proc Soc Exp Biol Med 82:159–166.

Roman-Ramos R, Flores-Saenz JL, Alarcon-Aguilar FG. 1995. Anti-hyperglycemic effect of some edible plants. J Ethnopharmacol 48:25–32.

Sajithlal GB, Chithra P, Chandrahasan G. 1998. Effect of curcumin on the advanced gly-cation and cross-linking of collagen in diabetic rats Biochem Pharmacol 56:1607–1614.

Santhakumari P, Prakasam A, Pugalendi KV. 2006. Antihyperglycemic activity of *Piper betle* leaf on streptozotocin-induced diabetic rats. J Med Food 9:108–112.

Sarkar S, Pranava M, Marita R. 1996. Demonstration of the hypoglycemic action of *Momordica charantia* in a validated animal model of diabetes. Pharmacol Res 33:1–4.

Satheesh MA, Pari L. 2004. Antioxidant effect of *Boerhavia diffusa* L. in tissues of alloxan induced diabetic rats. Indian J Exp Biol 42:989–992.

Sauvaire Y, Petit P, Broca C, Manteghetti M, Baissac Y, Fernandez-Alvarez J, Gross R, Roye M, Leconte A, Gomis R, Ribes G. 1998. 4-hydroxyisoleucine: A novel aminoacid potentiator of insulin secretion Diabetes 47:206–210.

Saxena AM, Bajpai MB, Mukherjee SK. 1991. Swerchirin induced blood sugar lowering of streptozotocin treated hyperglycemic rats. Indian J Exp Biol 29:674–675.

Saxena AM, Bajpai MB, Murthy PS, Mukherjee SK. 1993. Mechanism of blood sugar lowering by a Swerchirin-containing hexane fraction of *Swertia chirayita*. Indian J Exp Biol 31:178–181.

Sekar BC, Mukherjee B, Chakravarti RB, Mukherjee SK. 1987. Effect of different fractions of *Swertia chirayita* on the blood sugar level of albino rats. J Ethnopharmacol 21:175–181.

Sekar DS, Sivagnanam K, Subramanian S. 2005. Antidiabetic activity of *Momordica charantia* seeds on streptozotocin induced diabetic rats. Pharmazie 60:383–387.

Shani J, Goldschmied A, Joseph B, Ahronson Z, Sulman FG. 1974. Hypoglycaemic effect of *Trigonella foenum-graecum* and *Lupinus termis* seed and their major alkaloids in alloxan diabetic and normal rats. Arch Int Pharmacodyn Ther 210:27–37.

Shanmugasundaram ER, Gopinath KL, Shanmugasundaram KR, Rajendran VM. 1990b. Possible regeneration of the islets of Langerhans in streptozotocin-diabetic rats given *Gymnema sylvestre* leaf extracts. J Ethnopharmacol 30:265–279.

Shanmugasundaram KR, Panneerselvam C, Samudram P, Shanmugasundaram ER. 1983. Enzyme changes and glucose utilisation in diabetic rabbits: The effect of *Gymnema sylvestre*. J Ethnopharmacol 7:205–234.

Shanmugasundaram ER, Rajeswari G, Baskaran K, Rajesh Kumar BR, Shanmugasundaram KR, Ahmath BK. 1990a. Use of *Gymnema sylvestre* leaf extract in the control of blood glucose in insulin-dependent diabetes mellitus. J Ethnopharmacol 30:281–294.

Sharma RD. 1986. Effect of fenugreek seeds and leaves on blood glucose and serum insulin responses in human subjects. Nutr Res 6:1353–1364.

Sharma SR, Dwivedi SK, Swarup D. 1997. Hypoglycaemic, antihyperglycaemic and hypo-lipidemic activities of *Caesalpinia bonducella* seeds in rats. J Ethnopharmacol 58:39–44.

Sharma KK, Gupta RK, Gupta S, Samuel KC. 1977. Anti-hyperglycaemic effect of onion: Effect on fasting blood sugar and induced hyperglycemia in man. Indian J Med Res 65:422–429.

Sharma SB, Nasir A, Prabhu KM, Murthy PS, Dev G. 2003. Hypoglycaemic and hypolipidemic effect of ethanolic extract of seeds of *Eugenia jambolana* in alloxan-induced diabetic rabbits. J Ethnopharmacol 85:201–206.

Sharma SB, Nasir A, Prabhu KM, Murthy PS. 2006. Antihyperglycemic effect of the fruit–pulp of *Eugenia jambolana* in experimental diabetes mellitus. J Ethnopharmacol 104:367–373.

Sharma RD, Raghuram TC. 1990. Hypoglycaemic effect of fenugreek seeds in non-insulin dependent diabetic subjects. Nutr Res 10:731–739.

Sharma RD, Raghuram TC, Sudhakar Rao N. 1990. Effect of fenugreek seeds on blood glucose and serum lipids in type I diabetes. Eur J Clin Nutr 44:301–306.

Sharma RD, Sarkar A, Hazra DK, Mishra B, Singh JB, Sharma SK, Maheshwari BB, Maheshwari PK. 1996. Use of fenugreek seed powder in the management of non-insulin dependent diabetes mellitus. Nutr Res 116:1331–1339.

Sharma VN, Sogani RK, Arora RB. 1960. Some observations on hypoglycemic activity of *Momordica charantia*. Indian J Med Res 48:471–477.

Sheehan EW, Zemaitis MA, Slatkin DJ, Schiff PL Jr. 1983. A constituent of *Pterocarpus marsupium*, (-) epicatechin, as a potential antidiabetic agent. J Nat Prod 46:232–234.

Sheela CG, Augusti KT. 1992. Antidiabetic effects of S-allyl cysteine sulfoxide isolated from garlic (*Allium sativum* Linn). Indian J Exp Biol 30:523–526.

Sheela CG, Kumari K, Augusti KT. 1995. Antidiabetic effects of onion and garlic sulfoxide amino acids in rats. Planta Med 61:356–367.

Shibib BA, Khan LA, Rahman R. 1993. Hypoglycaemic activity of *Coccinia indica* and *Momordica charantia* in diabetic rats: Depression of the hepatic gluconeogenic enzymes glucose-6-phosphatase and fructose-1,6-bisphosphatase and elevation of both liver and red-cell shunt enzyme glucose-6-phosphate dehydrogenase. Biochem J 292:267–270.

Shimizu K, Iino A, Nakajima J, Tanaka K, Nakajyo S, Urakawa N, Atsuchi M, Wada T, Yamashita C. 1997. Suppression of glucose absorption by some fractions extracted from *Gymnema selvestre* leaves. J Vet Med Sci 59:245–251.

Shirwaikar A, Rajendran K, Kumar CD, Bodla R. 2004. Antidiabetic activity of aqueous leaf extract of *Annona squamosa* in streptozotocin–nicotinamide type-2 diabetic rats. J Ethnopharmacol 91:171–175.

Sidhu GS, Mani H, Gaddipati JP. 1999. Curcumin enhances wound healing in streptozotocin induced diabetic rats and genetically diabetic mice. Wound Repair 7:362–374.

Simon OR, Singh N, Smith K, Smith J. 1987. Effect of an aqueous extract of Nichol (*Caesalpinia bonducella*) on blood glucose concentration: Evidence of an antidiabetic action presented. J Sci Res Counc 6:25–32.

Singh AK, Chaudhury R, Manohar SJ. 1991. Hypoglycaemic activity of *Curcuma longa, Phylanthus emblica* and their various extractive combinations on albino rats. Med Arom Plants Abstr 13:2260.

Singh N, Singh SB, Vrat S, Misra N, Dixit KS, Kohli RP. 1985. A study on the antidiabetic activity of *Coccinia indica* in dogs. Indian J Med Sci 39:27–29.

Singh N, Tyagi SD, Agarwal SC. 1989. Effects of long term feeding of acetone extract of *Momordica charantia* (whole fruit powder) on alloxan diabetic albino rats. Indian J Physiol Pharmacol 33:97–100.

Sitasawad SL, Shewade Y, Bhonde R. 2000. Role of bitter gourd fruit juice in STZ-induced diabetic state in vivo and in vitro. J Ethnopharmacol 73:71–79.

Srinivasan K. 2005. Plant foods in the management of diabetes mellitus: Spices as potential antidiabetic agents. Int J Food Sci Nutr 56:399–414.

Srinivasan K. 2006. Fenugreek (Trigonella foenum-graecum): A Review of health beneficial physiological effects. Food Rev Int 22:203–224.

Srinivasan M. 1972. Effect of curcumin on blood sugar as seen in a diabetic subject. Indian J Med Sci 26:269–270.

Srivastava VK, Afaq Z. 1989. Garlic extract inhibits accumulation of polyols and hydration in diabetic rat lens. Curr Sci 58:376–377.

Srivastava Y, Bhatt HV, Verma Y, Prema AS. 1987. Retardation of retinopathy by *Momordica charantia* (Bitter gourd) fruit extracts in alloxan diabetic rats. Indian J Exp Biol 25:571–572.

Srivastava Y, Bhatt HV, Verma Y, Venkaiah K, Raval BH. 1993. Antidiabetic and adaptogenic properties of *Momordica charantia* extract: An experimental and clinical evaluation. Phytother Res 7:285–289.

Srividya N, Periwal S. 1995. Diuretic, hypotensive and hypoglycemic effect of *Phyllanthus amarus*. Indian J Exp Biol 33:861–864.

Stanley M, Prince P, Menon VP. 2000. Hypoglycemic and other related actions of *Tinospora cordifolia* roots in alloxan-induced diabetic rats. J Ethnopharmacol 70:9–15.

Sugihara Y, Nojima H, Matsuda H, Murakami T, Yoshikawa M, Kimura I. 2000. Antihyperglycemic effects of gymnemic acid IV, a compound derived from *Gymnema sylvestre* leaves in streptozotocin-diabetic mice. J Asian Nat Prod Res 2:321–327.

Tank R, Sharma N, Sharma I, Dixit VP. 1990. Anti-diabetic activity of *Curcuma longa* (ethanol extract) in alloxan induced diabetic rats. Indian Drugs 27:587–589.

Teixeira CC, Pinto LP, Kessler FHP, Knijnik L, Pinto CP, Gastaldo GJ, Fuchs FD. 1997. The effect of *Syzygium cumini* (L.) seeds on post-prandial blood glucose levels in non-diabetic rats and rats with streptozotocin-induced diabetes mellitus. J Ethnopharmacol 56:209–213.

Teixeira CC, Rava CA, da Silva PM, Melchior R, Argenta R, Anselmi F, Almeida CRC, Fuchs FD. 2000. Absence of antihyperglycemic effect of jambolan in experimental and clinical models. J Ethnopharmacol 71:343–347.

Thakran S, Salimuddin, Baquer NZ. 2003. Oral administration of orthovanadate and *Trigonella foenum-graecum* seed power restore the activities of mitochondrial enzymes in tissues of alloxan-induced diabetic rats. Mol Cell Biochem 247:45–53.

Tiangda C, Mekmanee R, Praphapraditchote K, Ungsurungsie M, Paovalo C. 1987. The hypo-glycaemic activity of *Momordica charantia* Linn. in normal and alloxan induced diabetic rabbits. J Natl Res Counc (Thailand) 19:1–11.

Tjokroprawiro A, Pikir BS, Budhiarta AAG, Pranawa, Soewondo H, Donosepoetro M, Budhiarta FX, Wibowo JA, Tanuwidjaja SJ, Pangemanan M, Widodo H, Surjadhana A. 1983. Metabolic effects of onion and green beans on diabetic patients. Tohoku J Exp Med 141:671–676.

Upadhyaya GL, Pant MC. 1986. Effects of water and ether extracts of bitter gourd powder on blood sugar and serum cholesterol level in albino rats. J Diabet Assoc India 26:17–19.

Vachon C, Jones JD, Wood PJ, Savoie L. 1988. Concentration effect of soluble dietary fibre on post-prandial glucose and insulin in the rat. Can J Physiol Pharmacol 66:801–806.

Valette G, Sauvaire Y, Baccou JC, Ribes G. 1984. Hypocholesterolemic effects of fenugreek seeds in dogs. Atherosclerosis 50:105–111.

Vats V, Grover JK, Rathi SS. 2002. Evaluation of anti-hyperglycemic and hypoglycemic effect of *Trigonella foenum-graecum* Linn, *Ocimum sanctum* Linn and *Pterocarpus marsupium* Linn in normal and alloxanized diabetic rats. J Ethnopharmacol 79:95–100.

Vats V, Yadav SP, Biswas NR, Grover JK. 2004b. Anti-cataract activity of *Pterocarpus marsupium* bark and *Trigonella foenum-graecum* seeds extract in alloxan diabetic rats. J Ethnopharmacol 93:289–294.

Vats V, Yadav SP, Grover JK. 2004a. Ethanolic extract of *Ocimum sanctum* leaves partially attenuates streptozotocin-induced alterations in glycogen content and carbohydrate metabolism in rats. J Ethnopharmacol 90:155–160.

Venkateswaran S, Pari L. 2003a. Effect of *Coccinia indica* leaf extract on plasma antioxidants in streptozotocin- induced experimental diabetes in rats. Phytother Res 17:605–608.

Venkateswaran S, Pari L. 2003b. Effect of *Coccinia indica* leaves on antioxidant status in streptozotocin-induced diabetic rats. J Ethnopharmacol 84:163–168.

Venkateswaran S, Pari L, Suguna L, Chandrakasan G. 2003. Modulatory effect of *Coccinia indica* on aortic collagen in STZ-induced diabetic rats. Clin Exp Pharmacol Physiol 30:157–163.

Verspohl EJ, Bauer K, Neddermann E. 2005. Antidiabetic effect of *Cinnamomum cassia* and *Cinnamomum zeylanicum In vivo* and In vitro. Phytother Res 19:203–206.

Vinuthan MK, Girish Kumar V, Ravindra JP, Jayaprakash, Narayana K. 2004. Effect of extracts of *Murraya koenigii* leaves on the levels of blood glucose and plasma insulin in alloxan-induced diabetic rats. Indian J Physiol Pharmacol 48:348–352.

Virdi J, Sivakami S, Shahani S, Suthar AC, Banavalikar MM, Biyani MK. 2003. Antihyper-glycemic effects of three extracts from *Momordica charantia*. J Ethnopharmacol 88:107–111.

Welihinda J, Arvidson G, Gylfe E, Hellman B, Karlsson E. 1982. The insulin-releasing activity of the tropical plant *Momordica charantia*. Acta Biol Med Ger 41:1229–1240.

Welihinda J, Karunanayake EH. 1986. Extra-pancreatic effects of *Momordica charantia* in rats. J Ethnopharmacol 17:247–255.

Welihinda J, Karunanayake EH, Sheriff MH, Jayasinghe KS. 1986. Effect of *Momordica charantia* on the glucose tolerance in maturity onset diabetes. J Ethnopharmacol 17:277–282.

Willatgamuwa SA, Platel K, Saraswathi G, Srinivasan K. 1998. Antidiabetic influence of dietary cumin seeds (*Cuminum cyminum*) in streptozotocin induced diabetic rats. Nutr Res 18:131–142.

William F, Lakshminarayanan S, Hariprasad C. 1993. Effect of some Indian vegetables on glucose and insulin response in diabetic subjects. Int J Food Sci Nutr 44:191–196.

Yadav S, Vats V, Dhunnoo Y, Grover JK. 2002. Hypoglycemic and antihyperglycemic activity of *Murraya koenigii* leaves in diabetic rats. J Ethnopharmacol 82:111–116.

Yoshikawa M, Murakami T, Matsuda H. 1997. Medicinal foodstuffs. X. Structures of new triterpene glycosides, gymnemosides—c, d, e, and f, from the leaves of *Gymnema sylvestre* R.: Influence of gymnema glycosides on glucose uptake in rat small intestinal fragments. Chem Pharm Bull (Tokyo) 45:2034–2038.

Zacharias NT, Sebastian KL, Philip B, Augusti KT. 1980. Hypoglycemic and hypolipidemic effects of garlic in sucrose fed rabbits. Indian J Physiol Pharmacol 24:151–154.

Zia T, Hasnain SN, Hasan SK. 2001. Evaluation of the oral hypoglycemic effect of *T. foenum-graecum* L. (Methi) in normal mice. J Ethnopharmacol 75:191–195.

Chapter 15

Nopal (*Opuntia* spp.) and Other Traditional Mexican Plants

Rosalia Reynoso-Camacho, PhD, and Elvira González de Mejía, PhD

Overview

This chapter reviews the potential of various Mexican traditional plant species for glycemic control in diabetes. They include *Opuntia* sp. (Nopal), *Pscalium* sp. (Matarique), *Cecropia* sp. (Guarunbo), *Cucurbita ficifolia* (Chilacayote), *Equisetum myriochaetum* (Cola de caballo), *Biden pilosa* (Aceitilla), and *Phaseolus vulgaris* (Common beans). Animal studies and a limited number of clinical studies on humans support the glycemic effect of these plants. Based on available clinical studies, *Opuntia* sp. seems to be the most promising. However, other plant species that seem to have a greater hypoglycemic effect need to be further studied. The information available is not conclusive to determine the role of these plants in the management of human diabetes. Randomized control clinical studies on humans are needed. In addition, potential toxicological effects need to be evaluated. Technological approaches to maintain or improve biological activity are also needed.

Introduction

The term diabetes mellitus is a complex metabolic disorder of multiple etiologies characterized by chronic hyperglycemia with disturbances of carbohydrates, fat, and protein metabolism resulting from defects in insulin secretion, insulin action, or both (WHO 2006). The hyperglycemia can lead to macrovascular (mostly atherosclerosis) and microvascular complications (retinopathy, nephropathy, and neuropathy). These complications are the principal causes of illness and mortality in diabetic patients (Vasudevan et al. 2006).

Normalizing the plasma glucose levels is associated with the prevention and alleviation of diabetic complications (Kimura 2006). The general consensus on treatment of type 2 diabetes is lifestyle management. In addition to exercise, weight control, and medical nutrition therapy, oral glucose-lowering drugs and injections of insulin are the conventional therapies (ADA 2006). However, oral hypoglycemic drugs may lose their effectiveness in a significant percentage of patients and could produce adverse effects. Further, the main reason for their low-prescription levels is their high cost for the patients or for the social security programs (Dey et al. 2002). Therefore, the common practice in developing countries is to use traditional folk medicine as an alternative to the use of commercial drugs.

Several natural compounds such as terpenoids, alkaloids, flavonoids, and phenols have antidiabetic potential (Jung et al. 2006). In fact, many conventional drugs have been derived from prototype molecules in medicinal plants. Metformin exemplifies an efficacious oral glucose-lowering agent whose development was based on the use of *Galega officinalis* to treat diabetes. This plant is rich in guanidine, which is the hypoglycemic component. Currently, metformin is one of the most important drugs commonly used in the treatment of type 2 diabetes.

Mexico has an ancient history of traditional medicine; therefore, it will be used as an example to illustrate the utilization of plants with medicinal properties in the treatment of diabetes. Mexican people have used about 306 plant species from 235 genera and 93 families for the treatment of type 2 diabetes (Andrade-Cetto and Heinrich 2005). The most widely used plants belong to the families: Asteraceae, Cactaceae, Euphorbiaceae, Fabaceae, Laminaceae, and Solanaceae. Normally the affected individual uses traditional beverage preparations made from the plants; drinking a cup before every meal. Under this treatment scheme they obtain a hypoglycemic effect for 5–6 h. Other common scheme is to use the plant infusion as drinking water throughout the day (instead of pure water, called in this case "agua de uso") (Roman-Ramos et al. 1991). Although the use of medicinal plants is very ancient, only a small number of them (around 30%) have received scientific and medical evaluation to determine their real efficacy (Alarcón-Aguilar et al. 2003).

The following is a summary of some of the most commonly used plants for the traditional treatment of diabetes.

Nopal (*Opuntia* spp.)

Several species of the genus *Opuntia* of the Cactaceae family are called nopal in Mexico and Prickly pear cactus in Southern United States where they grow widely in desert or semidesert regions. Among the different species, *Opuntia streptacantha* is one of the hypoglycemic plants more widely used in Mexico for the treatment of diabetes, and there are several clinical studies regarding its use and antidiabetic properties (Andrade-Cetto and Heinrich 2005; Shapiro and Gong 2002). The plant is arborescent and can reach a height of 3–5 m (Figure 15.1).

Its aerial portion consists of flat green, wax-covered internodes (cladodes), usually obovate and of a size ranging between 30–60 cm long, 20–40 cm wide, and

Figure 15.1.   Mature nopal *Opuntia streptacantha.*

19–28 mm thick and has spines and glochids (Anderson 2001). In Mexico, the cactus pads are commonly known as nopal or "penca" when whole, or "nopalitos" when cut into small pieces or young cladodes from 3 to 4 weeks of age (Feugang et al. 2006). The young cladodes of *Opuntia ficus indica* are long and thin with a few spines, while other species are disc shaped. Figure 15.2 presents a picture of young and mature cladodos of *O. ficus indica*.

The young cladodes ("nopalitos") are eaten as a vegetable, and cladodes with diverse maturation stages are used by over the counter pharmaceutical industry for the treatment of diabetes and obesity. In México there are about 35 factories that produce nopal in brine or marinade, 10 factories use nopal to elaborate cosmetics, and 20 use it to elaborate medicinal products (Flores et al. 1995).

The *Opuntia* cladode is known as a source of a varied number of nutritional compounds, whose concentration depends on the cultivation site, climate, and variety. The fresh young stems are sources of proteins as well as amino acids and vitamins. In addition, cladodes are rich in pectin, mucilage, and minerals (Feugang et al. 2006). The chemical composition of cladodes of *Opuntia* spp. is presented in Table 15.1.

Other parameter that influences the nopal composition is maturation stage. The protein contents of *O. ficus indica* and acid detergent crude fiber decreased significantly as the cladodes developed, and no pattern of change in ascorbic acid has been associated with maturation stage. However, carotene content and reducing sugars increased with cladode development (Rodriguez-Felix and Cantwell 1988) (Table 15.2). This effect was also observed for *O. amyclaea* and *O. inermes*. Nopal

(a)

(b)

Figure 15.2.   Maturation stages of nopal *Opuntia ficus indica*: (a) Young cladodes and (b) mature cladodes.

powder from *O. ficus indica* also showed changes in the composition as a function of maturation stage, 60 g pads in comparison to 200 g pads. Soluble fiber decreased from 25.22 to 14.91%, while insoluble fiber increased from 29.87 to 41.65%, respectively (Rodríguez-Garcia et al. 2007). However, other *Opuntia* did not follow the same pattern (Betancourt-Dominguez et al. 2006).

The glycemic effect of *Opuntia* has been reviewed in an extensive systematic worldwide evaluation of the potential of various herbs and dietary supplements in the control of diabetes (Yeh et al. 2003). This assessment based on clinical studies ranked *Opuntia* as one of the top seven most promising supplements to treat diabetes.

Table 15.1. Main chemical constituents of cladodes from *Opuntia* spp.

Compounds	Fresh weight/100 g	Dry weight/100 g
Water	88–95 g	—
Carbohydrates	3–7	64–71 g
Ash	1–2	19–23
Fiber	1–2	18
Protein	0.5–1	4–18
Lipids	0.20	1–4
Ascorbic acid	7.22 mg	—
Niacin	0.46	—
Riboflavin	0.60	—
Thiamine	0.14	—
Calcium	—	5.64
Magnesium	—	0.19
Potassium	—	2.35
Phosphorus	—	0.15
Arginine	2.4 g	
Histidine	2.0	
Isoleucine	1.9	
Leucine	1.3	
Lysine	2.5	
Methionine	1.4	
Phenylalanine	1.7	
Valine	3.7	

Adapted from Feugang et al. (2006).

Table 15.2. Chemical composition of cladodes of *Opuntia ficus indica* at different stages of development.

Stage of development	Length (cm)	Weight (g)	Protein (% dry wt)	Crude fiber (% dry wt)	Ascorbic acid (mg/100 g fresh weight)	Carotene content (µg/100 g fresh weight)
1	7	10	16.0[a]	16.3[a]	10.9[a]	25.8[a]
2	12	31	13.0[b]	12.7[c]	11.7[a]	30.5[a]
3	15	57	12.8[b]	13.7[b]	8.8[b]	30.9[a]
4	20	87	10.3[c]	12.2[d]	8.9[b]	27.9[a]
5	31	152	10.4[c]	13.1[bc]	9.5[b]	36.7[bc]
6	31–40	293–422	9.0[d]	12.1[d]	7.1[c]	44.0[c]

Adapted from Rodriguez-Felix and Cantwell (1988).
Data within a column followed by different letters are significantly different at the 5% level.

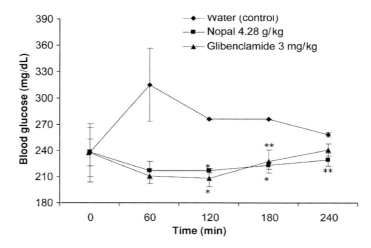

Figure 15.3. Hypoglycemic effect of nopal *Opuntia ficus indica* (4.28 g/kg body weight) on diabetic rats. The values are expressed as mean ± S.E.M. of six rats per group. *Significantly different from control $P < 0.01$. **Significantly different from control $P < 0.05$.

Opuntia streptacantha has been considered as the plant species with the most powerful hypoglycemic activity (Roman-Ramos et al. 1991). Several clinical studies have been conducted using this species in humans. Different doses (100, 300, and 500 g) of grilled stems of *O. streptacantha* were evaluated in patients with type 2 diabetes. In this study, there was a significant direct correlation between the doses and the hypoglycemic effect of the tested material (Frati-Munari et al. 1989a). Interestingly, in a different investigation, the intake of crude *O. streptacantha* (500 g of supernatant, precipitate, or complete homogenate) did not decrease glycemia in type 2 patients, but grilled *O. streptacantha* produced a significant decrease on serum glucose level. These latter results suggest that temperature may be important to activate the hypoglycemic compounds (Frati-Munari et al. 1989b, 1989c). However, Frati-Munari et al. (1989d) in a subsequent investigation found that 30 capsules of dehydrated extract (heated at 60–80°C for 24–36 h) of another nopal species (*O. ficus indica*) did not show hypoglycemic effect in patients with type 2 diabetes and could only attenuate postprandial hyperglycemia in healthy volunteers. In our laboratory, however, we studied the effect of feeding nongrilled *O. ficus indica* to diabetic rats at a level of 4.28 g/kg body weight (BW). As indicated in Figure 15.3, blood glucose levels were at all times lower than those observed in the control group. This effect was similar to glibenclamide, an oral antidiabetic drug that increases insulin secretion from pancreatic β-cells. In contrast, in other experiment, the authors found that nopal decreased blood level of insulin suggesting that this cactaceae increases insulin sensitivity (Frati-Munari et al. 1989a).

The results of our study using uncooked *Opuntia* also suggest that heat treatment may not be necessary for hypoglycemic action. Therefore, the different results between heat treated (e.g., grilled) and nonheat treated *Opuntia* may be attributed to the test model used, nopal species, the maturation stage of the cladodes, or harvest time, which was not reported in the previously mentioned human study in which grilled *Opuntia* was used. Recently, the effect of nopal added to breakfast of subjects with type 2 diabetes was studied. A reduction of up to 21 units in the glycemic index was observed, and this effect was greater in breakfast with a higher energy and protein content (Bacardi-Gascon et al. 2007).

On the other hand, an extract from *O. fuliginosa* (1 mg/kg/day) was administered orally for 15 weeks to induced diabetic rats resulting in lower levels of glucose and glycosylated hemoglobin than insulin-treated animals. Other group received an oral dose of 1 mL of *O. fuliginosa* extract plus insulin for 8 weeks, then insulin treatment was suspended, and the rats received only *O. fuliginosa* extract for 7 additional weeks. This treatment was capable of returning blood glucose to the levels of the nondiabetic rats. In this study, the oral dose of extract necessary to control diabetes contrast with the high quantities of nopal that may be required for an equivalent hypoglycemic effect and preclude a predominant role for dietary fiber (Trejo-Gonzalez et al. 1996).

The hypoglycemic activity of *O. lindheimeri* Englem extract (0, 250, or 500 mg/kg BW) was investigated in nondiabetic and streptozotocin (STZ)-induced diabetic pigs using an oral route of administration. The administration of this extract resulted in both a dose-($P < 0.001$) and time-dependent ($P < 0.001$) decrease in blood glucose only in STZ-treated pigs. These results are significant given the physiological similarities of the pig to humans regarding the long-term effects of *Opuntia* administration on the secondary problems associated with diabetes (Laurenz et al. 2003).

The effect of nopal in the presence of a glucose challenge has also been investigated. The most relevant results demonstrated that the juice of *O. streptacantha* (4 mL/kg BW) administered to rabbits reduced 21.4% the area under the glucose tolerance curve in relation to the water control (Roman-Ramos et al. 1991).

Healthy volunteers were treated with grilled *O. streptacantha* (500 g) or grilled *O. streptacantha* (100 and 500 g) plus dextrose. Only the second treatment, which included a sugar substrate, showed an effect on blood glucose and insulin concentration compared with the control group (Frati-Munari et al. 1988, 1990). This suggests that the effect of nopal only becomes relevant in hyperglycemia conditions.

These results were confirmed with different nopal species. In healthy mice, polysaccharides isolated from *O. ficus indica* (POLOF) and *O. streptacantha* (POLOS) were evaluated, and there was no significant hypoglycemic effect. However, in temporarily hyperglycemic mice, POLOF and POLOS reduced blood glucose by either oral or subcutaneous route. In alloxan diabetic mice, only the POLOS reduced the glycemia levels. These results suggest that POLOF decrease the intestinal absorption of glucose, and POLOS require the absorption of an active principle that increase insulin sensitivity and thus improve glucose utilization (Alarcón-Aguilar et al. 2003). POLOF and POLOS have been described as highly complex polymers with different composition of monosaccharides, such as arabinose, furanose, galactose, glucosa, rhamnose, abinopyranose, and galactopyranose, as main components. Nevertheless, complete chemical characterization of both polysaccharides has not yet been determined (Alarcón-Aguilar et al. 2003).

Based on the existing information it can be postulated that nopal may decrease blood glucose through two mechanisms. One is attributed to its high fiber content, which slows down the intestinal absorption of ingested glucose. The other is related to an increase in insulin sensitivity (Frati-Munari et al. 1991). Dietary fiber in cladodes has the capacity to absorb large amounts of water, forming viscous or gelatinous colloids and absorbing organic molecules. Fiber decreases blood glucose levels as a result of the increased viscosity of the stomach and small intestine gastrointestinal contents, impeding carbohydrate digestion, and absorption (Leclere et al. 1994), and it also improves insulin sensitivity in peripheral tissues. Fiber ferments in the colon resulting in increased production of short-chain fatty acids, and these have been proposed as being involved in improvement of hepatic insulin sensitivity (Thorburn et al. 1993). Increased activities of intestinal maltase, sucrase, and lactase in diabetic rats were significantly reduced in fiber-fed diabetic group (Chethankumar et al. 2002).

In relation to potential toxic effects of nopal, oral administration of a leaf extract of *O. megacantha* at 20 mg/100 g BW produced toxic effects on the kidneys with elevation in plasma creatinine and urea concentration in diabetic and nondiabetic rats (Bwititi et al. 2000). However, there are no published human toxicity studies, and adverse effects have not been reported in humans.

In conclusion, several studies support the hypoglycemic effect of *Opuntia* spp., related to either its high soluble-fiber content or to an increase on insulin sensitivity by a mechanism that still needs to be determined. Further, among *Opuntia* species, *O. streptacantha* shows the greatest potential in the treatment of diabetes. Although the clinical studies available are not conclusive to recommend the use of *Opuntia* to treat diabetes (Yeh et al. 2003), effective dose levels of nopal that have been reported in the clinical literature are 300 g and 500 g fresh or cooked nopal eaten before or with the meal.

Matarique (*Psacalium peltatum and Psacalium decompositum*)

The genus *Psacalium* comprises about 47 species (Bye et al. 1995). This herbaceous plant grows in the mountains of northern Mexico during the rainy summer months. Its green foliage resembles endive, thus making it especially easy to identify. The curative powers which are attributed to the plant are contained in an ample, intertwined root system which, after being pulled from the soft earth, readily releases its pleasant aroma and a spicy, pepsin taste (Hicks 1971).

There are several reports about the hypoglycemic activity of the root of this plant in experimental animals. However, no human studies have been conducted. A study in rabbits demonstrated that a water decoction of the root of *P. peltatum* (4 mL/kg BW) administered by gastric tube, decreased 27.9% the area under the glucose tolerance curve after a glucose challenge (Roman-Ramos et al. 1991). In addition, it produced a hypoglycemic effect similar to tolbutamide in healthy and mild diabetic rabbits (fasting glycemia 150–300 mg/dL) and mice. However, it did not cause any effect in severely diabetic rabbits (fasting glycemia higher than 400 mg/dL) or mice. These results suggest that some pancreatic function or perhaps the presence of insulin is required for the hypoglycemic activity of this plant (Contreras-Weber et al. 2002; Roman-Ramos et al. 1992). Conversely, a preparation administered by intraperitoneal injection of *P. decompositum* (200 mg dry residue/kg BW) of freeze-dried water decoction reduced blood glucose in healthy, mild-diabetic, and severe-diabetic mice. These differences, on the hypoglycemic effect, between *P. peltatum* and *P. decompositum* could be attributed to differences on either plant or route of administration used for the experiments (Alarcon Aguilar et al. 2000).

The root decoction of *P. decompositum* (i.p.) reduced the blood glucose of normal mice from 49.1 to 35.7 mg/dL and significantly lowered the hyperglycemic peak (17.1%) in rabbits with temporal hyperglycemia. These parameters were also evaluated with *P.peltatum* in both animal species, and the effects were minor (Alarcón-Aguilar et al. 1997). However, Alarcón-Aguilar et al. (2003) reported that the root extract of *P. peltatum* administered by gastric intubation to healthy rabbit caused a significant reduction of the hyperglycemic peak in impaired glucose tolerant rabbits. These results suggest that the effect of *P. peltatum* is also dependent on the route of administration.

Water and methanolic extracts of *P. peltatum* and *P.decompositum* roots, administered by i.p., exhibited the highest hypoglycemic effect in healthy mice compared with hexane and chloroform extracts (Alarcon-Aguilar et al. 2000; Contreras-Weber et al. 2002). Analysis of these extracts by thin layer chromatography has shown the presence of sesquiterpenoids such as cacalol, cacalone, and maturin (Alarcón-Aguilar et al. 1997). However, these compounds did not produce any significant hypoglycemic effect in normal mice. In contrast, two polysaccharide containing fractions, obtained from freeze-dried water extracts, significantly reduced the fasting glycemia in healthy and mild diabetic mice (Alarcon-Aguilar et al. 2000). In addition, peltalosa a new carbohydrate type ulopyranose isolated from an aqueous extracts of *P. peltatum* and administered (i.p.) to mice with mild and severe diabetes showed a significant reduction on blood sugar, 68 and 20%, respectively (Contreras et al. 2005). Peltalosa produces hypoglycemic effects similar to tolbutamide, possibly due to an enhanced secretion of insulin from the islets of Langerhans or an increased utilization of glucose by peripheral tissues (Contreras et al. 2005).

The hypoglycemic effect of this plant is promising, but clinical studies are needed to evaluate its effect in humans. In addition, there are no toxicological studies with this compound.

Guarumbo (*Cecropia obtusifolia* and *C. peltata*)

Besides guarumbo, other common names of *C. obtusifolia* are "guarumo," "Chi-flador," "Hormiguillo," "yagrumo," and "Chancarro." This tree grows as secondary vegetation in the tropical rain forest. The leaves are in spiral disposition located at the top of the branches and are simple, peltate, or deeply palmate, with a deep green color in the upper face and gray at the lower surface (Andrade-Cetto and Wiedenfeld 2001).

Traditionally, the leaves, the bark, and the root of this plant are boiled in water and the resulting infusion is drunk throughout the day for diabetes treatment (Andrade-Cetto and Wiedenfeld 2001; Argueta 1994; Pérez et al. 1984). The anti-hyperglycemic effect of leaves extract of *C. obtusifolia* has been demonstrated in healthy rabbits with temporal hyperglycemia. These animals showed a decreased area under the glucose tolerance curve (18.9%) in relation to controls (Roman-Ramos et al. 1991). Water and butanolic extracts of *C. obtusifolia* caused a hypo-glycemic effect in diabetic rats. The main active components isolated from the plant are isoorientin and chlorogenic acid (Andrade-Cetto and Wiedenfeld 2001).

Chlorogenic acid is a phenolic compound and some derivatives of chlorogenic acid are hypoglycemic agents and may also affect lipid metabolism. This effect has been demonstrated in rats treated for 3 weeks with 5 mg/kg BW/day chloro-genic acid. In these animals, this compound did not produce a sustained hypo-glycemic effect but decreased peak hyperglycemic, fasting plasma cholesterol (44%), and triacylglycerols (58%) (Rodríguez de Soltillo and Hadley 2002). It has been demonstrated that the hypoglycemic activity of *Cecropia* is dependent on the plant species. Leaf extracts, prepared in methanol (1 g/kg), from *C. peltata*, containing 19.84 mg of chlorogenic acid/g of dried leaves produced the highest decrease of plasma glucose levels (52.8%). The equivalent extract of *C. obtusifolia* had a minor chlorogenic acid content (13.2 mg/g) and minor hypoglycemic effect (34.5%). These results suggest that *C. peltata* is a better hypoglycemic plant than *C. obtusifolia* (Nicasio et al. 2005). Regarding humans, patients with type 2 dia-betes with poor response to conventional medical treatment received over 21 days a prepared infusion from *C. obtusifolia* containing 2.99 mg of chlorogenic acid/g of dried plant. The fasting blood glucose was reduced by 15.25%, while cholesterol and triglycerides decreased by 14.62 and 42.0%, respectively (Herrera-Arellano et al. 2004). On the other hand, an aqueous extract (13.5 g plant) from this species was administrated daily for 34 weeks to recently diagnosed type 2 diabetic patients, controlled only with diet and exercise. Patients showed a significant decrease in the level of glucose after 18 weeks of administration, whereas there were no significant changes in insulin and no hepatotoxicity or adverse effects (Revilla-Monsalve et al. 2007). No hypoglycemic drug was administrated throughout this study. Therefore, the hypoglycemic effect detected is due exclusively to the *C. obtusifolia* extracts.

A toxicological study in rats of the aqueous extract of *C. obtusifolia*, revealed a median lethal dose (LD_{50}) of 1,450 mg/kg BW equivalent to 11 g of plant/kg BW after i.p. administration (Pérez-Guerrero et al. 2001). This indicates that the risk of toxicity is low when consumed at traditional concentrations.

Chilacayote (*Cucurbita ficifolia*)

"Chilacayote," "cidra chayote" or white little pumpkin "calabacita blanca," is a fruit produced by *C. ficifolia*. The immature fruit is similar to a small spherical pumpkin, with 8–10 cm in diameter. The mature fruit has a thin green shell, with white spots or strips (Acosta-Patiño et al. 2001). The local healers recommend the ingestion of a crude aqueous extract of this fruit for the treatment of type 2 diabetes (Hernández-Galicia et al. 2002).

The hypoglycemic activity of traditional preparations (juice) of *Cucurbita ficifolia* obtained from mature fruits was demonstrated in healthy rabbits with temporary hyperglycemia and rabbits with moderate diabetes (fasting glycemia 150–300 mg/dL). In healthy rabbits, *C. ficifolia* (4 mL/kg BW) decreased 30.7% the area under the glucose tolerance curve, the hypoglycemic action being stronger than that obtained with 20 mg/kg of tolbutamide (Roman-Ramos et al. 1992, 1995). In type 2 diabetic patients, the juice (4 mL/kg) obtained from fresh immature fruit produced a decrease of blood glucose of 31% from the third to the fifth hour (Acosta-Patiño et al. 2001).

The freeze-dried juice from mature *C. ficifolia* administered by i.p. route (500 mg/kg BW) to healthy and diabetic mice significantly reduced glycemia. When used in healthy mice, its effect was dependent on route of administration, since i.p. produced the most important reduction of blood glucose compared with the oral route (Hernández-Galicia et al. 2002). Another study reports that when oral daily freeze-dried juice of mature *C. ficifolia* (1000 mg/kg) is administrated to diabetic rats, it produces a reduction of blood glucose levels from 297 to 53 mg/dL on the 14th day of treatment; however, at this dose some toxicity was found (Alarcon-Aguilar et al. 2002a).

Diabetic rats treated orally, with the methanolic extract of the fruit (300 and 600 mg/kg/day) for 30 days, showed blood glucose levels similar to the ones found in healthy rats, and the amount of plasma insulin was increased. The antihyperglycemic effect of this plant may be due to a potentiation of plasma insulin by increasing either the pancreatic secretion of insulin from the existing β-cells or its release from the bound form. This latter possibility is supported by the observed significant increase on insulin levels by *C. ficifolia* fruit extract in diabetic rats (Xia and Wang 2006a).

Protein-bound polysaccharides from this fruit have been evaluated in diabetic rats. These compounds produced a hypoglycemic effect, increased the levels of serum insulin, and improved glucose tolerance (Quanhong et al. 2005). The analysis of the carbohydrate content of fruit extracts from *C. ficifolia* revealed that they contained sucrose, fagopyritols, myoinositol, and Dchiroinositol (DCI) (2.9 mg/g extract). The latter compound, administered to diabetic rats, decreased plasma glucose concentration (Ortmeyer et al. 1993). The oral administration of a fruit extract from *C. ficifolia*, containing 10 or 20 mg DCI/kg BW for 30 days to diabetic rats, resulted in significantly lowered levels of blood glucose and increased plasma insulin levels. The antihyperglycemic effect of DCI could be a result of the inhibition of hepatic glucose output or enhanced glucose transport, glucose utilization, glucose disposal, or glycogen synthesis (Xia and Wang 2006b).

There are some reports about the toxicity of the juice from *C. ficifolia*. The results showed that the median lethal dose (LD_{50}), by the i.p. route to mice, is 625 mg/kg, and for oral administration it is 3,689 mg/kg. The doses tested produced the following signs: writhing, lose of locomotion activity, ataxia, increased of blood viscosity, lethargy, and hypothermia (Alarcon-Aguilar et al. 2002a; Hernández-Galicia et al. 2002). Comparing the median lethal dose (3,689 mg/kg) from freeze dried juice, a person of 70 kg would need 258 kg of fruit to reach the LD_{50}, and this is higher than the traditionally recommended dose (32 g) (Andrade-Cetto and Wiedenfeld 2001).

Cola de caballo (*Equisetum myriochaetum*)

This is an herbaceous plant that measures up to 2 m in height, with cylindrical hollow and soft stems, with rings spread on the trunk where the leaves go out.

Water decoctions of "cola de caballo" have been traditionally used to treat diabetes. The treatment beverage consists of boiling in water the aerial part of the dried plant, which is then drunk daily as "water of use." Studies in STZ induced diabetic rats showed that water (7 and 13 mg/kg) and butanol (8 and 16 mg/kg) extracts of this plant produced significant hypoglycemic-effects suggesting that *E. myriochaetum* may possess a glibenclamide like effect which stimulates insulin secretion (Andrade-Cetto et al. 2000). The administration of a single dose (0.33 g/kg) of water extract of *E. myriochaetum* aerial parts to type 2 diabetic patients produced a significant reduction of glucose levels at 90 min ($P < 0.01$), 120 min ($P < 0.001$), and 180 min ($P < 0.001$) after administration; however, there were no significant changes in insulin levels (Revilla et al. 2002).

Regarding the chemical composition of this plant, four main compounds (flavonol glycosides and one caffeoyglycoside) have been identified as the main constituents in water and butanol extracts. It can be assumed that kaempferol-3-O-sophoroside-4'-O-β-D-glucoside is one of the responsible factors for the hypoglycemic effect of *E. myriochaetum*. This compound lowers blood glucose with a similar range as the complete butanol extract (Andrade-Cetto et al. 2000).

Regarding toxicity, genotoxicity of extracts of *E. myriochaetum* has been evaluated in vivo using the wing spot test in *Drosophila melanogaster* (0.78 µg/mL and 500 µg/mL) and in the in vitro human micronucleus test (12.5 µg/mL and 500 µg/mL). No genotoxicity of the phytoextract could be determined at these concentrations, and no LD_{50} was observed (Ordaz-Tellez et al. 2007). However, more studies are needed to evaluate the toxicity of these extracts.

Aceitilla (Bidens pilosa)

This plant grows widely in tropical and subtropical areas and has been used in various disease treatments, including diabetes. For the treatment of diabetes, the healer protocol is boiling the aerial parts of *B. pilosa* in water.

The hypoglycemic effect of this plant has been demonstrated in healthy (13.8%) and mildly diabetic (32.6%) mice, treated with water ethanol extract of *B. pilosa* (Alarcón-Aguilar et al. 2002b).

Analysis of an aqueous ethanolic extracts of *B. pilosa* yielded a mixture of polyacetylene glucosides identified as 2-β-D-glucopyranosyloxy-1-hydroxy-5(E)-tridecene-7,9,11-triyne (1) and 3-β-D-glucopyranosyloxy-1-hydroxy-6(E)-tetradecene-8,10, 12 triyne (2). This mixture produced a significant decrease in blood glucose concentration and in food intake in obese-diabetic mice. The results were attributed to an anorexic effect of the mixture of compounds. A mechanism of antidiabetic action was demonstrated with a butanol fraction from *B. pilosa*, which inhibited differentiation of Th0 cells into Th1 cells, but enhanced their transition into type Th2 cells in nonobese diabetic (NOD) mice. Th1 cells are characterized by the secretion of Th1 cytokines, and these cells are implicated in autoimmune diabetes. Therefore, changing the ratio of Th2 to Th1 cells may suppress the disease. The compounds (1) and (2) above were also detected in this extract and evaluated in NOD mice. The results were similar to the ones produced by the butanol extract, showing the latter compound has a stronger activity (Chang et al. 2004, 2005). Cytopiloyne, other polyacetylenic glucoside, also prevented the development of diabetes in NOD mice, but it was only able to inhibit the differentiation of Th0 cells into Th1. This modulatory effect on T cell differentiation may provide a possible mechanism for the ethnopharmacologic effect of *B. pilosa* on preventing diabetes (Chiang et al. 2007).

Common Beans (*Phaseolus vulgaris*)

Phaseolus vulgaris is an annual leguminose plant, erect and bushy, 20–60 cm tall, or twining with stems 2–3 m long; with a taproot and nitrogenous nodules. It is widely cultivated in temperate and in semitropical regions. In temperate regions, the green immature pods are cooked and eaten as a vegetable. Immature pods are marketed fresh, frozen or canned, whole, cut, or french cut. Mature ripe beans are widely consumed by the Mexican population (Castellanos et al. 1997).

Various parts of the plant have been extensively used for the treatment of diabetes, and there are different human and animal studies that demonstrate its hypoglycemic effect (Panlasigui et al. 1995; Roman-Ramos et al. 1995; Venkateswaran et al. 2002; Venkateswaran and Pari 2002).

The antihyperglycemic effect of decoction pods from *P. vulgaris* has been demonstrated in healthy rabbits submitted to subcutaneous glucose tolerance tests. The decrease in glucose caused by this preparation was significant in every test point with a 22.6% reduction in the hyperglycemic peak in relation to the water control (Roman-Ramos et al. 1995).

Experiments on alloxan-diabetic rabbits showed that the *P. vulgaris* plant complex (PC) reduced the level of glycemia by 27–32% after glucose administration. Since in the subchronic treatment, the PC returned the blood level of glucose to normal on the 11th day, the authors suggested that the PC exerted its sugar reducing action by extrapancreatic means (Khaleeva et al. 1987). However, when tested in diabetic rats, an aqueous extract of *P. vulgaris* had a hypoglycemic effect related to an increase in the level of plasma insulin. *Phaseolus vulgaris* pods contain high amounts of flavonoids, which have been reported to regenerate damaged pancreatic β-cells in diabetic animals and stimulate insulin secretion. In addition, these compounds also decrease glucose-6-phosphatase and fructose 1, 6-bisphosphatase activities that are regulatory enzymes of gluconeogenesis, which are increased in diabetes. Negative modulation of these enzymes could result in lower concentration of glucose in blood (Pari and Venkateswaran 2003).

On the other hand, the antidiabetic effects of beans could also be explained by its high content of dietary fiber. In this regard, it is known that fiber slows carbohydrate absorption. The white bean also contains a high level of amylase inhibitor, and this compound reduced blood glucose levels by delaying digestion and absorption of carbohydrates (Tormo et al. 2006).

The oral administration of an aqueous extract of *P. vulgaris* pods (200 mg/kg BW) for 45 days to diabetic rats decreased the level of cholesterol, triglycerides, and saturated fatty acids (palmitic and stearic) and increased unsaturated fatty acids (oleic, linolenic, and arachidonic) concentrations in liver, kidney, and brain as compared with the control (Pari and Venkateswaran 2004). This extract also significantly reduced serum triglycerides, free fatty acids, total cholesterol, very low-density lipoprotein cholesterol, and low-density lipoprotein cholesterol. It has also been reported that *P. vulgaris* pods increase insulin levels, which inhibits hormone-sensitive lipase and therefore decreases mobilization of free fatty acids from peripheral depots (Venkateswaran et al. 2002).

Induced hyperglycemia increases glucose autoxidation and enhances production of reactive oxygen species (ROS). The levels of ROS are regulated by a variety of cellular defense mechanisms consisting of enzymes (catalase, superoxide dismutase, and glutathione peroxidase) and nonenzymes systems (glutathione, vitamins E, and C). The levels of these compounds are altered in diabetes, and oxidative stress may be the cause of tissue damage and therefore a primary cause of diabetes (Limaye et al. 2003).

Phaseolus vulgaris pod extract (PPEt) caused a significant decrease in plasma thiobarbituric acid-reactive substances (TBARS), hydroperoxides, and vitamin E.

Table 15.3. Summary of *Opuntia* and other traditional plants from Mexico.

Plant name	Active part of the plant	Possible mechanism	Route of administration	References
Opuntia spp.	Stem	Fiber slows down the digestion and absorption of glucose	Oral	Frati-Munari et al. (1988)
		Increase insulin sensitivity		Trejo-Gonzalez et al. (1996)
Psacalium peltatum and *P. decompositum*	Roots	Increase secretion and sensitivity of insulin	Oral	Roman-Ramos et al. (1991)
			Intraperitoneal	Alarcon-Aguilar et al. (2000)
Cecropia obtusifolia and *C. peltata*	eaves	Increase secretion and sensitivity of insulin	Oral	Roman-Ramos et al. (1991)
				Andrade-Cetto and Wiedenfeld (2001)
				Rodriguez de Sotillo et al. (2006)
Cucurbita ficifolia	Fruit	Increase secretion and sensitivity of insulin.	Oral	Roman-Ramos et al. (1995)
		Decrease gluconeogenesis	Intraperitoneal	Hernández-Galicia et al. (2002)
				Xia and Wang (2006b)
Equisetum myriochaetum	Aerial part	Increase secretion of insulin in rats	Oral	Andrade-Cetto et al. (2000)
		Not increase secretion of insulin in humans.		Revilla et al. (2002)

(continued)

Table 15.3.   (*continued*)

Plant name	Active part of the plant	Possible mechanism	Route of administration	References
Bidens pilosa	Whole plant	T cell modulator	Intraperitoneal	Alarcón-Aguilar et al. (2002b)
				Chang et al. (2004) and (2006)
Phaseolus vulgaris	Pods	Increase secretion of insulin	Oral	Roman-Ramos et al. (1995)
		Decrease gluconeogenesis		Pari and Venkateswaran (2003)

The decreased serum levels of high-density lipoprotein cholesterol, antiatherogenic index, plasma insulin, vitamin C, and glutathione in the diabetic rats were also reversed toward normalization. It also produced a significant increase in reduced glutathione, superoxide dismutase, catalase, glutathione peroxidase, and glutathione-S-transferase in liver and kidneys of rats with STZ induced diabetes. These results clearly show the antioxidant property of PPEt in addition to its antidiabetic action (Venkateswaran et al. 2002; Venkateswaran and Pari 2002).

These results suggest that the consumption of *P. vulgaris* pods decreases blood glucose levels as well as oxidative stress biomarkers related to diabetes complications. Therefore, it would be beneficial to include this legume in the diet.

Conclusions

All the traditional plants mentioned in this chapter have shown to have hypoglycemic activity and thus potential for the treatment of diabetes. However, there are differences in activity among different genera and species of plants. This is particularly true for *Opuntia* species among which *O. streptacantha* is the most active. Further, the mode of administration, different parts of the plants (bark extracts, fruit extracts, stem extracts), and various growth stages of the plant influence the hypoglycemic activity. Table 15.3 summarizes the information on hypoglycemic activity of the plants discussed.

Clinical studies in humans with these plants are limited or non existing. Nopal has been the most studied; however, its therapeutic effect is low compared to other more promising plant species. Besides the need of additional clinical studies on efficacy and safety of these plants, it would be interesting to investigate their synergistic effect when used in combination with commercial drugs and/or diet and exercise.

References

Acosta-Patiño JL, Jiménez-Balderas E, Juárez-Oropeza MA, Diaz-Zagoya JC. 2001. Hypoglycemic action of *Cucurbita ficifolia* on Type 2 diabetic patients with moderately high blood glucose levels. J Ethnopharmacol 77:99–101.

Alarcon-Aguilar FJ, Hernandez-Galicia E, Campos-Sepulveda AE, Xolalpa-Molina S, Rivas-Vilchis JF, Vazquez-Carrillo, Roman-Ramos, R. 2002a. Evaluation of the hypoglycemic effect of *Cucurbita ficifolia* Bouche (Cucurbitaceae) in different experimental models. J Ethnopharmacol 82:185–199.

Alarcon-Aguilar FJ, Jimenez-Estrada M, Reyes-Chilpa R, González-Paredes B, Contreras CC, Roman-Ramos R. 2000. Hypoglycemic activity of root water decoction, sesquiterpenoids, and one polysaccharide fraction from *Pcalium decompositum* in mice. J Ethnopharmacol 63:207–215.

Alarcón-Aguilar FJ, Roman-Ramos R, Flores-Saenz JL, Aguirre-García F. 2002b. Investigation on the hypoglycaemic effects of extracts of four Mexican medicinal plants in normal and alloxan-diabetic mice. Phytotherapy 16(4):383–386.

Alarcón-Aguilar FJ, Roman-Ramos R, Jiménez-Estrada M, Reyes-Chilpa R, Gónzalez-Paredes B, Flores-Saenz JL. 1997. Effects of three Mexican medicinal plants (Asteraceae) on blood glucose levels in healthy mice and rabbits. J Ethnopharmacol 55:171–177.

Alarcón-Aguilar F, Valdés-Arzate A, Xolalpa-Molina S, Banderas-Dorantes T, Jiménez-Estrada M, Hernández-Galicia E, Román-Ramos R. 2003. Hypoglycemic activity of two polysaccharides isolated from *Opuntia ficus-indica* and *O. streptacantha*. Proc West Pharmacol Soc 46:139–142.

American Diabetes Association (ADA). 2006. Standards of medical care in diabetes—2006. Diabetes Care 29:S4–S42.

Anderson EF. 2001. The Cactus Family, 776 pp. Timber Press, Portland, OR, USA.

Andrade-Cetto A, Heinrich M. 2005. Mexican plants with hypoglycaemic effect used in the treatment of diabetes. J Ethnopharmacol 99:325–348.

Andrade-Cetto A, Wiedenfeld H. 2001. Hypoglycemic effect of *Cecropia obtusifolia* on streptozotocin diabetic rats. J Ethopharmacol 78:145–149

Andrade-Cetto A, Wiedenfeld H, Revilla MC, Segio IA. 2000. Hypoglycemic effect of *Equisetum myriochaetum* aerial parts on streptozotocin diabetic rats. J Ethnopharmacol 72:129–133.

Argueta VA. 1994. Atlas of the Traditional Mexican Medicinal Plants, I, pp. 486–489. México: National Indigenous Institute.

Bacardi-Gascon M, Dueñas-Mena D, Jimenez-Cruz A. 2007. Lowering effect on postprandial glycemic response of nopal added to Mexican breakfast. Diabetes Care 30:1264–1265.

Betancourt-Dominguez MA, Hernandez-Perez T, Garcia-Saucedo P, Cruz-Hernandez A, Paredes-Lopez O. 2006. Physico-chemical changes in cladodes (nopalitos) from cultivated and wild cacti (*Opuntia* spp.). Plant Foods Hum Nutr 61(3):115–119.

Bwititi P, Musabayane CT, Nhachi CF. 2000. Effects of *Opuntia megacantha* on blood glucose and kidney function in streptozotocin diabetic rats. J Ethnopharmacol 69:247–252.

Bye R, Linares E, Estrada E. 1995. Biological diversity of medicinal plants in Mexico. In Phytochemistry of Medicinal Plants, edited by Arnason JT, Mata R, Romeo JT, p. 65. New York: Plenum Press.

Castellanos JZ, Guzman Maldonado H, Jimenez A, Mejia C, Munoz Ramos JJ, Acosta Gallegos JA, Hoyos G, Lopez Salinas E, Gonzalez Eguiarte D, Salinas Perez R, Gonzalez Acuna J, Munoz Villalobos JA, Fernandez Hernandez P, Caceres B. 1997. Preferential habits of consumers of common bean (*Phaseolus vulgaris* L.) in Mexico. Arch Latinoam Nutr 47:163–167.

Chang SL, Chang CL, Chiang YM, Hsieh RH, Tzeng CR, Wu TK, Sytwu HK, Shyur LF, Yang WC. 2004. Polyacetylenic compounds and butanol fraction from *Bidens pilosa* can modulate the differentiation of helper T cells and prevent autoimmune diabetes in non-obese diabetic mice. Planta Med 70:1045–1051.

Chang CL, Kuo HK, Chang SL, Chiang YM, Lee TH, Wu WM, Shyur LF, Yang WC. 2005. The distinct effects of a butanol fraction of *Bidens pilosa* plant extract on the development of Th1-mediated diabetes and Th2-mediated airway inflammation in mice. J Biomed Sci 12:79–89.

Chethankumar M, Salimath PV, Sambaiah K. 2002. Butyric acid modulates activities of intestinal and renal disaccharidases in experimentally induced diabetic rats. Nahrung 46:345–348.

Chiang YM, Chang CL, Chang SL, Yang WC, Shyur LF. 2007. Cytopiloyne, a novel polyacetylenic glucoside from *Bidens pilosa*, functions as a T helper cell modulator. J Ethnopharmacol 110:532–538.

Contreras C, Roman R, Pérez C, Alarcón F, Zavala M, Pérez S. 2005. Hypoglycemic activity of a new carbohydrate isolated from the roots of *Psacalium peltatum*. Chem Pharmacol Bull 53:1408–1410.

Contreras-Weber C, Perez-Gutierrez S, Alarcon-Aguilar F, Roman-Ramos R. 2002. Anti-hyperglycemic effect of *Psacalium peltatum*. Proc West Pharmacol Soc 45:134–136.

Dey L, Attele, AS, Yuan CS. 2002. Alternative therapies for type 2 diabetes. Altern Med Rev 7:45–58.

Feugang JM, Konarski P, Zou D, Stintzing FC, Zou C. 2006. Nutritional and medicinal use of Cactus pear (*Opuntia* spp.) cladodes and fruits. Front Biosci 1:2574–2589.

Flores FC, Luna EJ, Ramírez MP. 1995. Mercado mundial del nopalito. Manual de apoyos y servicios a la comercialización agropecuaria and Universidad Autónoma de Chapingo, p. 115. Chapingo, Mexico: ASERCA-UACh-CIESTAAM.

Frati-Munari AC, Altamirano BE, Rodríguez-Barcenas N, Araiza-Andraca R, López-Ledesma R. 1989b. Hypoglycemic effect of *Opuntia streptacantha* lemaire: Research with crude extracts. Arch Invest Med (Mex) 20:321–325.

Frati-Munari AC, De León C, Araiza-Andraca R, Bañales-Ham MB, López-Ledesma R, Lozoya X. 1989d. Influence of a dehydrated extract of the nopal (*Opuntia ficus indica* mill) on glycemia. Arch Invest Med (Mex) 20:211–216.

Frati-Munari AC, Del Valle-Martínez LM, Araiza-Andraca CR, Islas-Andrade S, Chavez-Negrete A. 1989a. Hypoglycemic action of different doses of nopal (*Opuntia streptacantha* Lemaire) in patients with type II diabetes mellitus. Arch Invest Med (Mex) 20(2):197–201.

Frati-Munari AC, Del Valle-Martínez LM, Araiza-Andraca CR, Islas-Andrade S, Chavez-Negrete A. 1989c. Duration of hypoglycemic action of *Opuntia streptacantha* Lem. Arch Invest Med (Mex) 20:297–300.

Frati-Munari AC, Gordillo BE, Altamirano P, Araiza CR, Cortes-Franco R, Chavez-Negrete A, Islas-Andrade S. 1991. Influence of nopal intake upon fasting glycemia in type II diabetics and healthy subjects. Arch Invest Med (Mex) 22:51–56.

Frati-Munari AC, Licona-Quesada R, Araiza-Andraca CR, López-Ledesma R, Chavez-Negrete A. 1990. The action of *Opuntia streptacantha* on healthy subjects with induced hyperglycemia. Arch Invest Med (Mex) 21:99–102.

Frati-Munari AC, Quiroz-Lazaro JL, Altamirano-Bustamante P, Bañales-Ham M, Islas-Andrade S, Araiza-Andraca CR. 1988. The effect of different doses of prickley pear cactus (*Opuntia streptacantha* lemaire) on the glucose tolerance test in healthy individuals. Arch Invest Med (Mex) 19:143–147.

Hernández-Galicia E, Campos-Sepulveda AE, Alarcón-Aguilar FJ, Vazquez-Carrillo LI, Flores-Saenz JL, Roman-Ramos R. 2002. Acute toxicocological study of *Curcubita ficifolia* juice in mice. Proc West Pharmacol 45:42–43.

Herrera-Arellano A, Aguilar-Santamaria L, García-Hernández B, Nicasio-Torres P, Tortoriello J. 2004. Clinical trial of *Cecropia obtusifolia* and *Marrubium vulgare* leaf extracts on blood glucose and serum in type 2 diabetes. Phytomedicine 11:561–566.

Hicks S. 1971. Desert Plant and People, 74 p. San Antonio, TX: The Naylor Company.

Jung M, Park M, Lee HC, Kang YH, Kang ES, Kim SK. 2006. Antidiabetic agents from medicinal plants. Curr Med Chem 13:1203–1218.

Khaleeva LD, Maloshtan LN, Sytnik AG. 1987. Comparative evaluation of the hypoglycemic activity of the vegetal complex of *Phaseolus vulgaris* and chlorpropamide in experimental diabetes. Probl Endokrinol (Mosk) 33:69–71.

Kimura I. 2006. Medical benefits of using natural compounds and their derivatives having multiple pharmacological actions. Yakugaku Zasshi 126:133–432.

Laurenz JC, Collier CC, Kuti JO. 2003. Hypoglycaemic effect of *Opuntia lindheimeri* Englem in a diabetic pig model. Phytother Res 17:26–29.

Leclere CL, Champ M, Boillot J, Guille G, Lecannu G, Molis C, Bornet F, Krempf M, Delort-Laval J, Galmiche JP. 1994. Role of viscous guar gums in lowering the glycemic response after a solid meal. Am J Clin Nutr 59:914–921.

Limaye PV, Raghuram N, Sivakami S. 2003. Oxidative stress and gene expression of antioxidant enzymes in the renal cortex of streptozotocin-induced diabetic rats. Mol Cell Biochem 243:147–152.

Nicasio P, Aguilar-Santamaría L, Aranda E, Ortiz S, González M. 2005. Hypoglycemic effect and chlorogenic acid content in two *Cecropia* species. Phytother Res 19:661–664.

Ordaz-Tellez MG, Rodríguez HB, Olivares GQ, Sortibran AN, Cetto AA, Rodríguez-Arniz R. 2007. A phytotherapeutic extract of *Equisetum myriochaetum* is not genotoxic either in the in vivo wing somatic test of *Drosophila* or in the vitro human micronucleus test. J Ethnopharmacol 111:182–189.

Ortmeyer HK, Huang LC, Zhang L, Hansen BC, Larner J. 1993. Chiroinositol deficiency and insulin resistance. II. Acute effects of D-chiroinositol administration in streptozotocin-diabetic rats, normal rats given a glucose load, and spontaneously insulin-resistant rhesus monkeys. Endocrinology 132:646–651.

Panlasigui LN, Panlilio LM, Madrid JC. 1995. Glycaemic response in normal subjects to five different legumes commonly used in the Philippines. Int J Food Sci Nutr 46:155–160.

Pari L, Venkateswaran S. 2003. Effect of an aqueous extract of *Phaseolus vulgaris* on plasma insulin and hepatic key enzymes of glucose metabolism in experimental diabetes. Pharmazie 58:916–919.

Pari L, Venkateswaran S. 2004. Protective role of *Phaseolus vulgaris* on changes in the fatty acid composition in experimental diabetes. J Med Food 7:204–209.

Pérez G, Ocegueda A, Muñoz JL, Avila JG, Morrow WW. 1984. A study of the hypoglycemic effect of some Mexican plants. J Ethnopharmacol 12:253–262.

Pérez-Guerrero C, Herrera MD, Ortiz R, Alvarez de Sotomayor M, Angeles-Fernández M. 2001. A pharmacological study of *Cecropia obtusifolia* Bertol aqueous extract. J Ethnopharmacol 76:279–284.

Quanhong L, Caili F, Yukui R, Guanghui H, Tongyi C. 2005. Effects of protein-bound polysaccharide isolated from pumpkin on insulin in diabetic rats. Plant Foods Hum Nutr 60:13–16.

Revilla MC, Andrade-Cetto A, Islas S, Wiedenfeld H. 2002. Hypoglycemic effect of *Equisetum myriochaetum* aerial parts on type 2 diabetic patients. J Ethnopharmacol 81:117–120.

Revilla-Monsalve MC, Andrade-Cetto A, Palomino-Garibay MA, Wiedenfeld H, Islas-Andrade S. 2007. Hypoglycemic effect of *Cecropia obtusifolia* Bertol aqueous extracts on type 2 diabetic patients. J Ethnopharmacol 111:636–640.

Rodríguez de Soltillo DV, Hadley M. 2002. Cholorogenic acid modifies plasma and liver concentrations of cholesterol, triacylglycerol, and minerals in (fa/fa) Zucker rats. J Nutr Biochem 13:717–726.

Rodriguez de Sotillo DV, Hadley M, Sotillo JE. 2006. Insulin receptor exon 11± is expressed in Zucker (fa/fa) rats, and chlorogenic acid modifies their plasma insulin and liver protein and DNA. J Nutr Biochem 17:63–71.

Rodriguez-Felix A, Cantwell M. 1988. Developmental changes in composition and quality of prickly pear cactus cladodes (nopalitos). Plant Foods Hum Nutr 38:83–93.

Rodríguez-Garcia ME, de Lira C, Hernández-Becerra E, Cornejo-Villegas MA, Palacios-Fonseca AJ, Rojas-Molina I, Reynoso R, Quintero LC, Del-Real A, Zepeda TA, Muñoz-Torres C. 2007. Physicochemical characterization of Nopal Pads

(*Opuntia ficus indica*) and dry vacuum Nopal powders as a function of the maturation. Plant Foods Hum Nutr 62(3):107–112.

Roman-Ramos R, Flores-Saenz JL, Alarcón-Aguilar F. 1995. Anti-hyperglycemic effect of some edible plants. J Ethnopharmacol 48:25–32.

Roman-Ramos R, Flores-Saenz JL, Partida-Hernández G, Lara-Lemus A, Alarcón-Aguilar F. 1991. Experimental study of the hypoglycemic effect of some antidiabetic plants. Arch Invest Med (Mex) 22:87–93.

Roman-Ramos R, Lara-Lemus A, Alarcón-Aguilar F, Flores-Saenz JL. 1992. Hypoglycemic activity of some antidiabetic plants. Arch Med Res 23:105–109.

Shapiro K, Gong WC. 2002. Natural products used for diabetes. J Am Pharm Assoc (Wash) 42:217–226.

Thorburn A, Muir J, Proietto J. 1993. Carbohydrate fermentation decreases hepatic glucose output in healthy subjects. Metabolism 42:780–785.

Tormo MA, Gil-Exojo I, Romero de Tejada A, Campillo JE. 2006. White bean amylase inhibitor administered orally reduces glycaemia in type 2 diabetic rats. Br J Nutr 96:539–544.

Trejo-Gonzalez A, Gabriel OG, Puebla PA, Huízar CM, Munguía MM, Mejía AS, Calva E. 1996. A purified extract from prickly pear cactus (*Opuntia fuliginosa*) controls experimentally induced diabetes in rats. J Ethnopharmacol 55:27–33.

Vasudevan AR, Burns A, Fonseca VA. 2006. The effectiveness of intensive glycemic control for the prevention of vascular complications in diabetes mellitus. Treat Endocrinol 5:273–286.

Venkateswaran S, Pari L. 2002. Antioxidant effect of *Phaseolus vulgaris* in streptozotocin-induced diabetic rats. Asia Pac J Clin Nutr 11:206–209.

Venkateswaran S, Pari L, Saravanan G. 2002. Effect of *Phaseolus vulgaris* on circulatory antioxidants and lipids in rats with streptozotocin-induced diabetes. J Med Food 5:97–103.

World Health Organization. 2006. Definition and diagnosis of diabetes mellitus and intermediate hyperglycaemia: Report of a WHO/IDF consultation. WHO Document Production Services, Geneva, Switzerland.

Xia T, Wang Q. 2006a. Antihyperglycemic effect of *Cucurbita ficifolia* fruit extract in streptozotocin-induced diabetic rats. Fitoterapia 77:530–533.

Xia T, Wang Q. 2006b. D-chiro-inositol found in *Cucurbita ficifolia* (Cucurbitaceae) fruit extracts plays the hypoglycaemic role in streptozotocin-diabetic rats. J Pharm Pharmacol 58:1527–1532.

Yeh G, Eisenberg DM, Kaptchuk TJ, Phillips RS. 2003. Systemic review of herbs and dietary supplements for glycemic control in diabetes. Diabetes Care 26(4):1277–1294.

Chapter 16

Natural Resistant Starch in Glycemic Management: From Physiological Mechanisms to Consumer Communications

Rhonda S Witwer, BS, MBA

Overview

Natural RS2 resistant starch (RS) from high-amylose corn has been utilized as an ingredient in better-for-you foods targeting glycemic management around the world. This chapter summarizes 32 studies (31 published and one unpublished) on the glycemic benefits of this ingredient by segregating them according to study design. This chapter also briefly considers options in communicating glycemic messages to American consumers. In summary, one type of RS (type 2 RS) is a promising functional food ingredient with numerous applications in foods for managing blood sugar levels.

Introduction

The concept and benefits of glycemic management (moderating or managing blood glucose levels) are relatively new for most Americans, with the notable exception of diabetics. Over the past several years, however, consumer interest has grown significantly. This awareness was sparked by the Atkins diet fad, but is now being driven by the continued publicity surrounding the two major categories of benefits for glycemic management: Potential disease risk reduction and energy management.

Over the past decade, it has been suggested that consuming foods with lower impact on blood glucose may reduce the risk of developing diabetes, obesity, and heart disease. In the research and public health communities, there is widespread research underway. The glycemic index (GI) has been proposed as an analytical method to compare different types of carbohydrates for their impact on blood sugar, and numerous studies and reviews have been published correlating the consumption of reduced GI foods to reduced risk of diseases (Ludwig 2002). However, other studies found a correlation between dietary fiber consumption and insulin sensitivity, insulin secretion, and adiposity but no correlation between GI and these conditions (Liese et al. 2005). Additional research is needed to more clearly identify the glycemic impact of different types of carbohydrates, different methods of measuring this impact of foods, and the resulting health consequences.

This chapter will examine the evidence and published studies on the glycemic impact of one ingredient utilized in the formulation of reduced glycemic foods: Natural RS2 resistant starch (RS) from high-amylose corn. Between 1987 and 2007, a total of 31 studies have been published examining the glycemic and insulin impact of foods containing this ingredient. In addition, data from one unpublished study supplied by communication from National Starch Food Innovation have been included. In summary, 24 studies have shown a reduced glycemic response in at least one type of glycemic measurement, 18 studies have shown a reduced insulin response, and 3 studies have shown increased insulin sensitivity from dietary consumption of RS2 from high-amylose corn.

This chapter also examines the challenges in communicating these benefits to American consumers and presents numerous opportunities for marketers in the development of better-for-you foods containing natural RS2 from high-amylose corn.

Resistant Starch

Historically, it was assumed that all starch was digested in the small intestine to produce glucose. Underlying that assumption was the oversimplified view that starch, after all, is nothing more than chains of glucose with two types of chain linkages and different sizes of storage granules depending upon which type of plant produced it. Over the past 30 years, however, researchers have discovered that some types of starch escape small intestinal digestion and can be found in the large intestine. In 1992, RS was officially defined: "Resistant starch is the sum of starch and products of starch degradation not absorbed in the small intestine of healthy individual" (Asp 1992).

In contrast to most ingredient definitions, which are defined by their chemistry (i.e., omega-3 fatty acids or ascorbic acid), RS has a physiological definition. Many different types of starches can be RS under different circumstances, health conditions, or physiological variances. Historically, grains, legumes, and starch-rich fruits and vegetables delivered a significant quantity of RS because processing had not yet made the starch readily available for digestion. Since the industrial revolution, however, modern processing conditions have made foods smaller in particle size, seedless, lighter, and consequently more digestible. In the case of starch, the physiological consequences of convenience and texture preference are just beginning to be discovered. Research is showing that the loss of RS in our diets is contributing to our modern health problems (including obesity and diabetes).

Classes of Resistant Starch

Three classes of RS were initially identified to characterize types of RS that can be found in foods (Englyst et al. 1992). A fourth class was added in 1995 to describe resistant starches that are not naturally found in foods (Brown et al. 1995) (Table 16.1).

Many types of foods deliver class 1RS (RS1). The starch escapes digestion in the small intestine because it is protected by a food matrix—either a hull or shell in the case of unprocessed whole grains, or a seed casing in the case of legumes and sweet corn. Because it takes time for the body to break down the protective barrier, the starch, or a portion of the starch, escapes digestion. Thus, one of the bioactive components of unprocessed whole grains is RS. Processed whole grains have lost their protective barrier and will likely lose their RS component (if the starch is not also resistant due to its amylose content). Thus, whole grain wheat delivers RS1, but processed whole-grain wheat flour does not.

The RS2 class of RS is not encased in a protective physical barrier, but it still retains the natural starch granule. Green bananas deliver a significant quantity of RS2. As bananas ripen, however, their natural starch component turns to sugar, reducing the naturally occurring RS. In addition, because banana RS2 gelatinizes at a relatively low temperature, the RS in bananas is lost by cooking or processing. Uncooked potatoes also contain a significant quantity of RS2, but like bananas, the RS is lost upon cooking or processing. Specialty grains and corn that are naturally high in amylose (the linear type of starch polymer) also have RS2, but they have a higher gelatinization temperature due to the difference in starch structures. This natural property enables them to retain their natural RS2 through common food processing and makes them ideal for use as food ingredients.

The RS3 class of RS is retrograded starch polymers that may result from starch gelatinization. The most common food sources of RS3 are cooked and cooled potato, rice, and pasta. In this case, the starch granule has been cooked or broken apart, and glucose polymers have formed crystalline structures in a process known as "retrogradation." Commercially available RS3 resistant starches have also been created by cooking or enzymatically breaking apart the starch granule and retrograding the starch under controlled conditions. The characteristics, size, and properties of commercially produced RS3s from a variety of starch sources vary.

The RS4 class of RS is not naturally present in foods. RS4s are made by chemically treating a starch to introduce bonds that cannot be broken down by digestive enzymes in the body. The properties of RS4 resistant starches depend upon the type and level of chemical treatment. They can be soluble or insoluble, and may or may not test as dietary fiber under AOAC methods 985.29 and 991.43.

Table 16.1. Classes of resistant starch (RS).

RS type	Characteristic	Properties	Food sources	Commercial ingredient sources	Other
RS1	Trapped in food matrix	Not completely dispersed, readily swollen or gelatinized; not easily penetrated by digestive enzymes	Whole or partially milled grain; sweet corn, parboiled rice, legumes	*Hi-maize*® whole grain corn flour	RS content varies in food: Reduced if food matrix is degraded by chewing, milling, or pureeing
RS2	Granular, with highly crystalline regions	Starch retains its natural granular structure	Raw potato, green banana, high-amylose corn	*Hi-maize*® whole grain corn flour and *Hi-maize*® RS from high-amylose corn	Insoluble fiber; reduced if starch granule is ruptured or gelatinized
RS3	Retrograded amylose and amylopectin	Nongranular. Can vary in size of retrograded fractions	Cooked and cooled starches such as potatoes, rice or tortillas; occurs readily in the crumb of products like bread	*NOVELOSE*® RS from high-amylose corn, RS from tapioca	May or may not test as dietary fiber according to AOAC 985.29 or 991.43.
RS4	Chemically modified starches	Properties will vary depending upon chemical treatment. May be soluble or insoluble	Does not occur in nature. Does not exist in foods naturally produced	Modified RS from wheat, modified RS from tapioca	Studies needed to determine physiological impact

Table 16.2.   Glycemic effects of RS2 resistant starch from high-amylose corn.

General effect	Specific effect	Proposed mechanism and site responsible for effect
Glycemic response	Low glucose response	Digestion in small intestine
Insulin response	Low insulin response	Digestion in small intestine
Energy content	Lower available energy than fully digestible carbohydrates	Digestion in small intestine and fermentation in large intestine.
Glycemic response of subsequent meal	Lower glucose response	Fermentation in large intestine
Insulin sensitivity	Increased insulin sensitivity of fat tissue and muscles	Fermentation in large intestine
Lipid oxidation	Increased oxidation of stored fat and decreased oxidation of dietary carbohydrates in liver	Fermentation in large intestine
Satiety	Increased satiety hormones (GLP-1 and PYY) in the bloodstream	Fermentation in large intestine

Abbreviations: GLP-1 = glucagon-like peptide-1, PYY = peptide YY.

Categories of Resistant Starch Benefits

RS impacts the body in three ways: (1) glycemic impact, (2) colonic impact, and (3) metabolism. Glycemic and colonic impact certainly contribute to health of the entire system (body), but they are categorized separately because there is a localized impact that can be measured and analyzed. The third category, metabolic impact, is likely the result of a cascade of physiological pathways that appear to be driven by dietary consumption of RS. The metabolic impacts of RS are thus more difficult to isolate, track, and understand. It is clear, however, that it is difficult to separate out the health consequences of a reduced glycemic diet containing RS without also considering the health consequences of the fermentation of RS.

This chapter will focus on the glycemic impact and systemic impact that pertain to glycemic management, and will not review the colonic benefits. These glycemic effects are summarized in Table 16.2. Numerous publications are available that reviewed the colonic health benefits of natural RS (Topping and Clifton 2001; Young and Le Leu 2004).

This review examines the different effects and mechanisms separately and is based upon 31 published clinical studies, 10 published animal studies, and one unpublished GI report on RS2 from high-amylose corn. Tables 16.3 and 16.4 summarize the findings and important details about the design of these studies.

In total, more than 160 studies have been published with RS2 from high-amylose corn on a wide range of health benefits. It has been commercially available around the world as a standardized ingredient (branded *Hi-maize*® resistant starch) since 1993. Because it is easily incorporated into great-tasting foods such as bread, muffins, and biscuits, researchers have incorporated it into many studies over the past 15 years. The combination of benefits demonstrated in these research studies makes natural RS2 from high-amylose corn an ideal ingredient for use in the formulation of better-for-you foods.

Research has shown that different types and sources of RS behave differently in the body. As a result, resistant starches are not interchangeable. They must be considered individually. This results from a definition that is physiological instead of chemical in which many starches can be found in the large intestine for a variety of reasons. However, many of the benefits are the result of a biochemical cascade of events. This cascade is just beginning to be identified and is not well understood. Biomarkers for fermentation may be utilized, but even the linkages between the biomarkers and the physiological benefits are not fully established. In other words, it is not possible to specify that one measurable biomarker (i.e., glycemic response or butyrate) represents the biological pathway for the full range of benefits without significant additional research. Benefits demonstrated with RS2 from high-amylose corn cannot be assumed for other types and sources of RS without substantiating evidence.

Glycemic Response

Definitions

Different methods have been utilized to measure the glycemic impact of foods. In order to understand the inherent differences in the glycemic methodology, a review of the different definitions is required (Table 16.5)

The GI was first defined in 1981 by Dr. David Jenkins and others, as cited in Table 16.5. While the GI has been widely cited, the data presented as "glycemic index" have not always been based upon 50 g of available carbohydrates. Instead, it has been based on another quantity of available carbohydrates, or on the total carbohydrate content instead. For instance, research studies have calculated a GI for a food, but an analysis of the method reveals that the authors calculated a glycemic response based upon a different standard instead. In addition, a very low GI for certain dietary fibers and/or nonglycemic carbohydrates has been promoted, when it would far exceed the dose tolerance of the ingredient to consume the quantity

Table 16.3. Animal studies on the effect of high-amylose corn RS2 on glycemic and insulin responses

Reference	Starch	Subject and study characteristics	Glucose response (GR)	Insulin response (IR)
Brown et al. (2003)	High-amylose corn RS2 from National Starch. Waxy corn starch control	Wistar rats. Meal test. Starch was uncooked and cooked	NSD peak or AUC, uncooked or cooked	Uncooked: ↓ AUC ↓ peak Cooked: NSD peak ↓ AUC
Byrnes et al. (1995)	*Hi-maize* RS2. Waxy corn starch control	Wistar rats: IVGTT. Acute up to 12 weeks. Sprague-Dawley rats. IVGTT; 9 weeks	Acute: ↓ GR ↓ peak Chronic: ↓ GR Differences in GR emerged at 12 weeks	Acute: ↓ IR. ↓ peak Chronic: ↓ AUC ↓ IR ↓ peak. Differences in AUC emerged at 8 weeks. Differences in IR emerged at 12 weeks
Higgins et al. (1996)	*Hi-maize* RS2. Waxy corn starch control	Wistar rats. IVGTT. Up to 52 weeks		↓AUC up to 26 weeks, NSD 52 weeks Insulin resistance (AUC) delayed to 26 weeks versus 12 weeks for the control. Fasting hyperinsulinemia developed at 26 weeks versus 16 weeks for the control
Morand et al. (1992)	High-amylose corn RS. Wheat starch control	Wistar rats; 3 weeks	↓ GR	↓ IR
Morand et al. (1994)	High-amylose corn RS. Wheat starch control	Wistar rats; 21 days		↓IR in the fed and post-absorptive phase

(continued)

Table 16.3. (continued)

Reference	Starch	Subject and study characteristics	Glucose response (GR)	Insulin response (IR)
Pawlak et al. (2001)	High-amylose corn RS. Waxy corn starch control	Wistar rats. EU clamp and IVGTT; 7 weeks	Basal: NSD IVGTT: NSD peak NSD AUC	Basal: NSD IVGTT: ↓ peak ↓ IR ↓ AUC Euglycemic clamp: NSD insulin sensitivity
Pawlak et al. (2004)	High-amylose corn RS. Waxy corn starch control	(1) Sprague-Dawley rats, partial pancreatectomy. Parallel design. 18 weeks. OGTT. (2) As above. Cross-over design. 10 weeks. OGTT	(1) ↓ AUC by week 5 (2) ↓ AUC	(1) ↓ AUC by week 5 (2) ↓ AUC
Scribner et al. (2007)	*Hi-maize* RS2. 100% amylopectin corn starch control	Mice. Parallel design. 25 week	Week 15: NSD GR Week 21: NSD GR Week 25: NSD GR	Week 15: ↓ IR Week 21: ↓ IR Week 25: ↓ IR
Wiseman et al. (1996)	*Hi-maize* RS2. Waxy corn starch control	Wistar rats. Cross-over (8 weeks) or parallel (8 weeks, 16 weeks). IVGTT	8 weeks: NSD GR 16 weeks: NSD GR Cross-over: NSD GR	8 weeks: ↓ IR ↓ peak NSD AUC 16 weeks: ↓ AUC ↓ peak ↓ IR Cross-over: NSD IR (feeding RS after the control starch did not reverse insulin resistance)
Zhou and Kaplan (1997)	Acid-treated (modified) high-amylose corn RS4. Corn starch control	Sprague-Dawley rats; 4 weeks	↑GR	NSD IR NSD insulin/glucagon ratio

Abbreviations: AUC = area under the curve, GR = glycemic response, IVGTT = intravenous glucose tolerance test, IR = insulin response, NSD = no significant difference, OGTT = oral glucose tolerance test."

408

Table 16.4. Human studies on the effect of high-amylose corn RS2 on glucose and insulin responses.

	Reference	Starch	Subject and study characteristics	Glucose response	Insulin response
1	Anderson et al. (2002)	*Hi-maize* RS2. Polycose, sucrose and amylopectin controls	RS in beverage containing 75 g carbohydrate. Healthy men. One meal	↓AUC	
2	Behall et al. (1988)	High-amylose corn RS2. Corn starch control	RS in a cracker. Healthy. One meal	↓peak NSD AUC, ↑ GR (sustained release)	↓IR ↓ AUC
3	Behall et al. (1989)	High-amylose corn RS2. Corn starch control	RS in a meal. Healthy. Glucose test at 4 weeks. Starch test at 5 weeks	Glucose test: NSD GR NSD AUC Starch test: NSD AUC ↓ GR for first hour ↑ GR at 3 and 4 h (sustained release)	Glucose test: NSD AUC ↓ IR Starch Test: ↓ IR ↓ AUC
4	Behall and Howe (1995)	High-amylose corn RS2. Corn starch control	RS in bread, muffins, cereal, cookies, and cheese puffs. Healthy and hyperinsulinemic; 14 weeks	Bread test: NSD GR NSD AUC	Fasting: NSD IR Bread test: ↓ IR ↓ AUC
5	Behall and Hallfrisch (2002)	High-amylose corn RS2 with 30–70% amylose Glucose control	RS in bread. Healthy. One meal	30–70%: ↓ AUC ↓ GR 50–70%: ↓ peak 70%: lowest AUC	60–70%: ↓ IR ↓ AUC Only >70% amylose corn affected both glucose and insulin

(continued)

409

Table 16.4. *(continued)*

	Reference	Starch	Subject and study characteristics	Glucose response	Insulin response
6	Behall and Scholfield (2005)	High-amylose RS2 cornstarch or cornmeal. Corn starch control	RS in corn chips and corn muffins. Standardized to 1 g carb/kg body weight. Healthy and hyperinsulinemic (HI). One meal	RS2 starch versus cornstarch: ↓ GR ↓ AUC Muffins versus chips: ↓ GR NSD AUC Cornmeal versus cornstarch: ↓ GR Normal versus HI: ↓ GR, ↓ AUC	RS2 muffins and chips: ↓ IR Muffins versus chips with cornstarch or cornmeal: ↓ IR RS2 muffins versus control muffins: ↓ AUC Normal versus HI: ↓ IR
7	Behall et al. (2006a)	High-amylose corn RS2 and oat beta-glucan. Glucose control	RS and beta-glucan in muffins. Standardized to 1 g carb/kg body weight. Normal and overweight women. One meal	Normal weight versus overweight: NSD High or moderate RS and high beta-glucan: ↓ GR ↓ AUC	Normal weight versus overweight: ↓ IR High or moderate RS and high beta-glucan: ↓ IR ↓ AUC
8	Behall et al. (2006b)	*Hi-maize* RS2 and barley beta-glucan. Glucose control	RS and beta-glucan in muffins. Standardized to 75 g glycemic carb. Normal and overweight/obese men. One meal	Normal weight versus overweight: ↓ GR High RS versus low RS: NSD GR, NSD AUC High beta-glucan versus low beta-glucan: ↓ GR ↓ AUC	Normal weight versus overweight: ↓ IR Overweight with high beta-glucan: ↓ IR compared to other treatments. High RS versus low RS: ↓ IR ↓ AUC High beta-glucan versus low beta-glucan: ↓ IR ↓AUC

#	Study/Control	Method	Result 1	Result 2
9	Brighenti et al. (2006)	*Hi-maize* RS2 Amylopectin = starch control. Cellulose = fiber control. Lactulose = fermentation control	RS in sponge cake for breakfast. Standardized to 75 g available carb including RS. Healthy. MTT at lunch with identical meal. Two meals	Breakfast: Low glycemic with RS2 vs. high glycemic: ↓ GR Lunch: Fermentable lactulose (high glycemic) and RS (low glycemic) versus nonfermentable cellulose: ↓ GR	Breakfast: Low glycemic with RS2 vs. high glycemic: ↓ IR Lunch: NSD IR
10	Brown et al. (1995)	*Hi-maize* RS2. White bread control	RS in bread. Healthy. One meal	↓AUC	
11	Giacco et al. (1998)	*Hi-maize* RS2. Waxy corn starch control	RS in biscuit consumed as part of a meal. Type 2 diabetics. One meal	↓GR ↓ AUC No second meal effect	NSD IR No second meal effect
12	Goodpaster et al. (1996)	High-amylose corn RS2. Waxy corn starch control	RS in a beverage. Healthy cyclists. One meal	NSD GR	NSD IR
13	Granfeldt et al. (1995)	High-amylose corn RS2. Corn starch control	RS in arepas. Healthy. Matched/unmatched for available carb. One meal	Matched: ↓AUC ↓ GR Unmatched: ↓ AUC ↓ GR	Matched: ↓ AUC ↓ IR Unmatched: ↓ AUC ↓ IR
14	Higgins et al. (2004a)	High-amylose corn RS2 from National Starch. Low RS control foods.	RS in spaghetti, drink and fruit snack. Healthy; 6 h postprandial measurements. One meal	NSD AUC	NSD AUC
15	Hoebler et al. (1999)	High-amylose corn RS2. White bread control	RS in bread. Healthy. One meal	↓AUC ↓ GR	↓AUC ↓ IR

(continued)

Table 16.4. (*continued*)

	Reference	Starch	Subject and study characteristics	Glucose response	Insulin response
16	Hospers et al. (1994)	High RS2 from National Starch. Corn starch and wheat flour controls	RS in pasta. Healthy men. One meal	↓response at 0.5 h and 1 h ↓AUC for 1st h 3 h: NSD GR NSD AUC	↓AUC ↓ IR
17	Howe JN (1996)	High-amylose corn RS2. Corn starch control	RS in bread, muffins, cookies, cereal, and cheese puffs. Healthy and hyperinsulinemic. MTT; 14 weeks	↓AUC	↓AUC ↓ IR
18	Jenkins et al. (1998)	High-amylose corn RS2 and RS3. Low RS starch control	RS in muffins and breakfast cereal. Healthy; 2 weeks	NSD AUC	
19	Krezowski et al. (1987)	High-amylose corn RS2. Waxy corn starch control	RS in muffins. Type 2 diabetics. One meal	↓peak ↓ AUC	NSD peak, NSD AUC
20	Noakes et al. (1996)	*Hi-maize* RS2. Low amylose corn starch control	RS in bread, cereal, pasta, and muffins. Hypertri-glyceridemic. MTT; 4 weeks	Fasting: NSD GR Postprandial: ↓ GR, NSD AUC	Fasting: NSD IR Postprandial: ↓ AUC ↓ IR
21	NSFI (2006)	*Hi-maize* RS2. White bread control	RS in white bread. Healthy; one meal	↓GI. GI for Hi-maize bread = 63.6. GI of control bread = 71	

22	Olesen EJCN (1994)	Hi-maize RS2. High-amylose corn RS3. Glucose control	RS cooked and cooled in water. Healthy. One meal	↓peak (no statistics shown)	↓IR compared to both plain muffin and bread
23	Quílez et al. (2007)	Hi-maize RS2. Flour in bread and plain muffin as controls	RS in low calorie muffin. (Fat and sugar were also replaced with carboxymethyl-cellulose, guar gum and maltitol in test food.) Healthy. One meal	NSD GR compared to plain muffin. ↓Glycemic response compared to bread	NSD AUC
24	Ranganathan et al. (1994)	High-amylose corn. Cellulose control	RS in a "lintner." Healthy. One meal	NSD AUC	NSD AUC
25	Reiser et al. (1989)	High-amylose corn RS2. Fructose control	RS in a muffin. Fructose in a fondant. Healthy and hyperinsulinemic. MTT; 5 weeks	MTT: ↑ GR ↑AUC	MTT: ↑ IR ↑ AUC ↓insulin:glucose ratio. ↑ RBC insulin binding
26	Robertson et al. (2003)	Hi-maize RS2. Waxy corn starch control	RS in jelly. Healthy. RS intake on day 1. MTT; 2 days	MTT: ↓ GR	MTT:↓ IR ↑insulin sensitivity. ↑ C-peptide/insulin ratio. NSD C-peptide
27	Robertson et al. (2005)	Hi-maize RS2. Waxy corn starch control	RS incorporated into everyday foods. Healthy. MTT and hyperinsulinemic-euglycemic clamp; 4 weeks	MTT: NSD GR NSD AUC	MTT: NSD fasting insulin sensitivity as measured by HOMA. MTT: ↓ AUC ↑insulin sensitivity during clamp study, ↑ C-peptide:insulin ratio, ↑ total glucose uptake by adipose tissue.

(continued)

413

Table 16.4. (*continued*)

Reference	Starch	Subject and study characteristics	Glucose response	Insulin response
28 van Amelsvoort and Weststrate (1992)	*Hi-maize* RS2 + high-amylose rice. Waxy corn starch and low amylose rice control	RS cooked in a hot dish meal. Healthy. One meal	↓/↑GR, ↑ AUC	↓/↑IR, ↓ AUC
29 Vonk et al. (2000)	*Hi-maize* RS2 and *NOVELOSE* RS3.	Uncooked RS in gelatin drink. Healthy. One meal	↓ AUC	
30 Weickert et al. (2005)	Corn starch control *Hi-maize* RS2. white bread control	RS in bread. Healthy. One meal. Next day: control bread following 10-h fast	NSD GI. Next day: NSD glucose AUC (trend toward reduction)	NSD IR. Next day: NSD IR
31 Weststrate and van Amelsvoort (1993)	High-amylose corn. Low-amylose corn starch control	RS in baguette and apple almond compote for breakfast, RS in pizza for lunch. Healthy. Two meals	Breakfast: NSD GR. NSD AUC Lunch: ↓ GR. ↓ AUC Second meal: NSD AUC	Breakfast: NSD IR ↓ AUC Lunch: NSD IR ↓ AUC Second meal: NSD AUC
32 Zhang et al. (2007)	*Hi-maize* RS2. Wheat flour control	RS in steamed bread and noodles. Type 2 diabetics. Cross-over design; 4 weeks	↓Fasting Blood Glucose ↓ GR	↑insulin sensitivity index

Abbreviations: AUC = area under the curve, GI = Glycemic index, GR = glycemic response, HI = hyperinsulinemic, HOMA = homeostatic model assessment, IR = insulin response, MTT = meal tolerance test, NSD = no significant difference, RBC = red blood cell.

Table 16.5. Glycemic definitions.

Method	Definition	Source (year)
Glycemic index	The area under the curve (AUC) for the increase in blood glucose after the ingestion of 50 g of "available" or "glycemic" carbohydrate in a food during the 2 h postprandial period, relative to the same amount of carbohydrate from a reference food (white bread or glucose) tested in the same individual under the same conditions and using the initial blood glucose concentration as a baseline.	Jenkins (1981)
Glycemic response	The change in blood glucose concentration induced by ingested food	AACC (2006)
Available carbohydrate	Carbohydrate that is released from a food in digestion and which is absorbed as monosaccharides and metabolized by the body	AACC (2006)
Glycemic carbohydrate	Carbohydrate in a food which elicits a measurable glycemic response after ingestion	AACC (2006)
Glycemic impact	The weight of glucose that would induce a glycemic response equivalent to that induced by a given amount of food	AACC (2006)

necessary to conduct the GI test. Thus, significant confusion over this concept has emerged over the past 25 years, and it is difficult to determine the health benefits of a low GI diet when the data are not clearly presented or consistently evaluated.

In 2006, AACC International adopted four definitions, cited in Table 16.5, related to glycemic carbohydrates (AACC 2006). This effort was designed to help food manufacturers communicate how the carbohydrate content of a product will affect blood glucose levels. Representatives from academia, industry as well as government contributed to the creation of these definitions, which will be utilized in this review.

While the GI is narrowly defined, glycemic response has a much broader definition. It can be standardized upon a quantity of available carbohydrates, a quantity of total carbohydrates, or on a quantity of food regardless of the macronutrient content. Because it is not standardized to a specific type of comparison, careful attention must be given to the design of the studies describing its effects.

One of the primary differences between GI and glycemic response can be illustrated with the example of bread containing 5 g of dietary fiber per 50 g serving size. The GI would compare the glycemic response of approximately 125 g of fiber-fortified bread to 100 g of control bread, while the glycemic response could compare 100 g of fiber-fortified bread to 100 g of control bread. In this instance, the GI may provide meaningful data for research purposes, but it does not provide meaningful data for consumers who evaluate food based upon portion sizes. In other words, consumers would not increase the quantity of bread they include in a sandwich because it contained more dietary fiber—they would consume two slices of bread in either case.

A different type of glycemic testing extends beyond the immediate postprandial hours of the meal and tests the glycemic response of a subsequent standardized meal. For instance, a breakfast containing RS is consumed, and the glycemic response to a standardized lunch is examined.

For the purposes of this review, there are five types of glycemic testing: (1) GI, (2) glycemic response based upon a portion size of available carbohydrates that is not 50 g, (3) glycemic response based upon total carbohydrate quantity, (4) glycemic response based upon portion size, and (5) the glycemic response of a subsequent meal.

Measuring Glycemic Response of RS2 Resistant Starch

Between 1987 and 2007, 31 clinical studies were published on the glycemic response of RS2 from high-amylose corn. In addition, data from a 2006 unpublished study (Internal National Starch Food Innovation Communication) are included. Table 16.3 outlines the specifics of each study. The results will be presented according to the different designs: postprandial glycemic response based upon available carbohydrate, postprandial glycemic response based upon total carbohydrates, and postprandial glycemic response of a standardized meal preceded by RS2 consumption.

Glycemic Response of RS2 Compared to an Equivalent Quantity of Available Carbohydrates

Eight studies have been published on the glycemic response of RS2 from high-amylose corn standardized on quantity of available carbohydrates, with mixed results. In addition, data from one unpublished study (Internal National Starch Food Innovation Communication) have been included in which the GI of a white bread containing 5 g of fiber from natural *Hi-maize* RS2 was clinically tested. Five studies found significant reductions in glycemic response (Behall et al. 2005; Brighenti et al. 2006; Granfeldt et al. 1995; NSFI 2006; Quílez et al. 2007), but four studies did not (Behall et al. 2006a, b; Higgins et al. 2004; Jenkins et al. 1998; Weickert et al. 2005). It is difficult to identify the particular factors that cause these differences—it could be differences in portion size, food types, macronutrients, or relative amounts of RS. Brighenti et al. (2006) may have found significant reductions because he standardized portion sizes as well as available carbohydrate content.

The changing portion size of the food required by this method is difficult for consumers to understand, especially when they would likely compare a fiber-fortified portion of food with its nonfiber-fortified equivalent. For instance, they may compare the glycemic impact of a sandwich containing two slices of bread to a sandwich containing two slices of fiber-fortified bread. They would not generally be able to calculate and compare the differences based upon available carbohydrates. For this reason, the meaningfulness of these data to consumers is unclear.

For the food formulator with a wide range of ingredient options, the GI methodology appears to discourage the use of dietary fiber and RS, as the method requires the consumption of greater quantities of fiber-fortified and RS-fortified food, and less certain reductions in glycemic response, as indicated by the contradictory results from the nine studies examined in this section (Table 16.6).

In summary, when comparing standardized quantities of available carbohydrates, more food containing RS is consumed in the test than the control food. For instance, Granfeldt's (1995) study fed people 71% more food and still measured a significant reduction in glycemic and insulin response. In the NSFI (2006) study, 21% more food was consumed, but still delivered a significant reduction in GI. In studies demonstrating no significant reduction in glycemic response, the interpretation of the results is confounded due to the inability to exclude the impact of differences in portion size, macronutrient content, food types, or the relative quantity of RS.

Glycemic Response of RS2 Compared to an Equivalent Quantity of Carbohydrates

This standardization enables the comparison of the glycemic response of fiber-fortified foods with their nonfiber containing counterparts. As food manufacturers consider adding dietary fiber to food, this type of comparison more clearly distinguishes the difference in glycemic response and comes closer to comparing foods on a portion size basis that will be understandable to consumers.

Twenty-one studies, seen in Table 16.7, have shown that when RS2 from high-amylose corn substitutes for an equivalent quantity of glycemic carbohydrates such as cornstarch or wheat flour, the glycemic response is reduced. In general, more substitution produces a higher reduction in glycemic response. Only two studies showed no effect under these conditions (Goodpaster et al. 1996; Weststrate and

Table 16.6. Glycemic response of RS2 from high-amylose corn standardized to available carbohydrates.

Reference	Quantity of food consumed	Glycemic carbohydrates	Result
Granfeldt et al. (1995)	180 g of arepas from high-amylose corn compared to 105 g of corn arepas	45 g	glycemic response: ↓ GR, ↓ AUC Insulin Response: ↓ IR, ↓ AUC
Jenkins et al. (1998)	Cereal: 59 to 83 g portions. Muffins: 204 to 264 g portions. White bread: 50 g glycemic carbohydrate control	50 g	NSD GI
Higgins et al. (2004)a	Spaghetti ranged from 197 to 218 g. Baked goods ranged from 36 to 44 g. Fiber content was equivalent	84 g	NSD GR (AUC)
Behall and Scholfield (2005)	70 g carbohydrate for RS corn chips and muffins compared to 60 g carbohydrate for control corn chips and muffins	60 g	RS2 chips versus control chips: ↓ GR RS2 muffins versus control muffins: NSD GR RS2 chips and muffins: ↓ AUC
Brighenti et al. (2006)	Portion size equivalent. Starch and fiber content were equivalent	75 g	Low glycemic with RS versus high glycemic: ↓ GR
Behall et al. (2006b)	9 test foods with beta-glucan and RS: 81.8 g for high beta-glucan/low RS treatment to 105.9 g for low beta-glucan/high RS treatment	75 g	High RS versus low RS: NSD GR NSD AUC
NSFI (2006)	128 g of *Hi-maize* bread versus 104 g of control white bread	50 g	↓GI
Quílez et al. (2007)	101.8 g low calorie muffin versus 106.6 g plain muffin versus 90.9 g bread	50 g	↓GR and ↓ AUC versus bread NSD GR versus plain muffin NSD AUC versus plain muffin
Weickert et al. (2005)	123 g of *Hi-maize* bread versus 103 g of control white bread	50 g	NSD GI

Abbreviations: NSD = no significant difference, GR = glycemic response, GI = glycemic index, AUC = area under the curve, IR = insulin response.

418

Table 16.7. Glycemic response of RS2 from high-amylose corn standardized to equivalent quantity of rapidly digested carbohydrates

Study	Subjects	Food	Effect
Anderson et al. (2002)	Healthy	Uncooked starch	↓GR
Behall et al. (1988)	Healthy	Crackers	↓GR
Behall et al. (1989)	Healthy	Pudding, muffins, corn chips, shortbread cookies, and brownies	↓GR
Behall and Hallfrisch (2002)	Healthy	Bread	↓GR
Behall and Scholfield (2005)	Healthy and hyperinsulinemic	Corn chips and corn muffins	↓GR
Behall et al. (2006a)	Healthy—normal weight and overweight	Muffins	↓GR
Brighenti et al. (2006)	Healthy	Sponge cake	↓GR
Brown et al. (1995)	Healthy	Bread	↓GR
Giacco et al. (1998)	Type 2 diabetics	Cheesecake made with RS biscuits	↓GR
Goodpaster (1996)	Healthy elite athletes	Beverage	NSD GR
Granfeldt et al. (1995)	Healthy	Arepas	↓GR
Hoebler et al. (1999)	Healthy	Bread	↓GR
Hospers et al. (1996)	Healthy	Pasta	↓GR
Howe et al. (1996)	Healthy and hyperinsulinemic	Bread, cereal, cheese puffs, muffins, cookies	↓GR
Krezowski et al. (1987)	Type 2 diabetics	Muffins	↓GR
Noakes et al. (1996)	Dyslipidemic	Bread, cereal, pasta, muffins	↓GR
Olesen et al. (1994)	Healthy	Cooked starch, porridge	↓GR
van Amelsvoort and Weststrate (1992)	Healthy	Hot dish meal	↓GR
Vonk et al. (2000)	Healthy	Uncooked starch in gelatin drink	↓GR
Weststrate and van Amelsvoort (1993)	Healthy	Breakfast: bread; lunch: pizza	Breakfast: NSD GR Lunch: ↓ GR
Zhang et al. (2007)	Type 2 diabetics	Steamed bread and noodles	↓GR

van Amelsvoort 1993). This may have been due to the excellent physical conditioning of the test population (Goodpaster et al. 1996), and to additional variables in the breakfast meal (Weststrate and van Amelsvoort 1993).

Many types of foods containing RS2 have been tested and have demonstrated a reduced glycemic response, as illustrated in Table 16.7. The effect is not dependent upon the type of food delivered as long as the RS2 replaces high glycemic carbohydrate.

The reduced glycemic response has also been demonstrated in subjects with a variety of health conditions, also seen in Table 16.7. These data suggest that the effect is relevant to individuals with a broad spectrum of health conditions.

Three studies have demonstrated reductions in glycemic response in the first hour combined with increased glycemic response at 2–5 h (Behall et al. 1988, 1989; van Amelsvoort and Westrate 1992). This pattern can be described as sustained or delayed energy release. In these cases, the area under the curve (AUC) may not be significantly reduced. However, this is a beneficial effect as delayed energy release has a beneficial impact on insulin response.

Glycemic Response of RS2 Compared to Other Low Glycemic Carbohydrates

Two studies compared the glycemic response of RS2 from high-amylose corn to other low-glycemic carbohydrates; both studies found no significant difference in glycemic response. One study compared RS2 from high-amylose corn to cellulose (Ranganathan et al. 1994). The second study compared RS2 from high-amylose corn to fructose (Reiser et al. 1989). This suggests that there is a generic "substitution effect" in the glycemic impact of foods. Because fiber is not digested in the small intestine, it does not cause an increase in blood glucose. When dietary fiber or other nonglycemic carbohydrates replace glycemic ingredients in a portion of food, the glycemic response of that food will be reduced if the same portion of food is consumed.

Glycemic Response of RS2 via Meal Tolerance Testing

Five studies, seen in Table 16.8, tested postprandial glycemic response of a standardized food or quantity of glucose as a subsequent meal instead of feeding the test foods. They fed individuals RS2 from high-amylose corn for varying periods of time and tested the glycemic response of a standardized quantity of glucose or a standardized test meal. In some cases, the glycemic response of the standardized meal was reduced. In other cases, it was not.

The authors who found reduced glycemic response of subsequent meals have attributed the effect to the fermentation of the RS in the large intestine. The physiological cascade or mechanism causing this response is not well understood. Additional research is needed to determine if the different effects may be related to the duration of intervention.

Table 16.8.   Glycemic response of RS2 from high-amylose corn via meal tolerance testing.

Study	Time frame for RS2 consumption	Test challenge	Results
Behall et al. (1989)	4 weeks prior	1 g glucose/kg body weight	NSD GR NSD AUC
Brighenti et al. (2006)	one meal prior	715 kcal meal with 93 g glycemic carbs	↓GR
Robertson et al. (2003)	1 day prior	2.1 kJ chocolate drink delivering 59 g carb, 21 g fat, and 19 g protein	↓GR
Robertson et al. (2005)	4 weeks prior	500 kcal beverage delivering 60 g carb and 21 g fat	NSD GR NSD AUC
Weickert et al. (2005)	1 day prior	240 kcal bread delivering 50 g available carb, 7.2 g protein, 0.8 g fat	NSD GR – trend toward lower glucose values

Abbreviations: NSD = no significant difference, GR = glycemic response, AUC = area under the curve.

Insulin Response

Twenty-six studies have been published analyzing the insulin response of RS2 from high-amylose corn.

Measuring Insulin Response of RS2 Resistant Starch

A variety of methods were utilized which are identical to the variety of methods utilized in glycemic testing: postprandial insulin response based upon available carbohydrate, postprandial insulin response based upon total carbohydrates, and postsprandial insulin response of a standardized meal preceded by RS2 consumption. In addition, change in insulin sensitivity has been analyzed.

Insulin Response of RS2 Standardized to Quantity of Glycemic Carbohydrates

Seven studies, seen in Table 16.9, have been published on the insulin response of RS2 from high-amylose corn standardized on quantity of available carbohydrates. Five out of the seven studies found significant reductions in insulin response (Behall et al. 2006b; Behall and Scholfield 2005; Brighenti et al.

Table 16.9. Insulin response of RS2 from high-amylose corn standardized to glycemic carbohydrates.

Reference	Quantity of food consumed	Glycemic carbohydrates	Insulin response
Granfeldt et al. (1995)	180 g of arepas from high-amylose corn compared to 105 g of corn arepas	45 g	↓IR ↓ AUC
Higgins et al. (2004a)	Spaghetti ranged from 197 to 218 g. Baked goods ranged from 36 to 44 g. Fiber content was equivalent	84 g	NSD AUC
Behall and Scholfield (2005)	70 g carbohydrate for RS corn chips and muffins compared to 60 g carbohydrate for control corn chips and muffins	60 g	RS2 muffins and chips versus control muffins and chips: ↓ IR ↓ AUC
Brighenti et al. (2006)	Portion sizes of sponge cake were equivalent. Starch and fiber content were equivalent	75 g	Low glycemic with RS versus high glycemic: ↓ IR
Behall et al. (2006b)	9 test foods with varying amount of beta-glucan and RS: Ranging from 81.8 g for high beta-glucan/low RS treatment to 105.9 g for low beta-glucan/high RS treatment	75 g	↓IR ↓ AUC
Quílez et al. (2007)	101.8 g low calorie muffin versus 106.6 g plain muffin versus 90.9 g bread	50 g	↓AUC versus both plain muffin and bread NSD IR/GR ratio
Weickert et al. (2005)	123 g of *Hi-maize* bread compared to 103 g of control white bread. Energy content was equivalent (240–241 kcal).	50 g	NSD IR. Time of maximal insulin concentration was earlier for RS

Abbreviations: NSD = no significant difference, IR = insulin response, AUC = area under the curve.

2006; Granfeldt et al. 1995; Quílez et al. 2007). The Higgins et al. (2004) and the Weickert et al. (2005) studies did not find a significant reduction. In contrast to the glycemic response under the same conditions, the majority of these studies found a beneficial effect. These data may be of particular relevance to consumers because the decreased insulin response was achieved despite the consumption of larger portion sizes.

Insulin Response of RS2 Compared to an Equivalent Quantity of Carbohydrates

Seventeen studies, seen in Table 16.10, have been published comparing the post-prandial insulin response based upon an equivalent quantity of carbohydrates. Fourteen studies demonstrated reduced insulin response. Three studies showed no effect: two were in diabetic populations (Giacco et al. 1998; Krezowski et al. 1987) and one was in trained cyclists in excellent physical condition (Goodpaster et al. 1996). The finding in cyclists is not surprising as exercise is known to increase insulin sensitivity. The Giacco 1998 study noted a trend toward reduced insulin response, but the effect was not statistically significant.

Many types of foods containing RS2 have been tested and have demonstrated a reduced insulin response, as illustrated in Table 16.10. The effect is not dependent upon the type of food delivered as long as the RS2 replaces high glycemic carbohydrate.

The reduced insulin response has also been demonstrated in healthy, hyperinsulinemic, and dyslipidemic subjects, also seen in Table 16.10. Two studies found no difference in insulin response in type 2 diabetics (Giacco et al. 1998; Krezowski et al. 1987). A third study in type 2 diabetics found increased insulin sensitivity index, but the insulin response data were not specified (Zhang et al. 2007).

Insulin Response of RS2 Compared to Other Low Glycemic Carbohydrates

Two studies have compared the insulin response of RS2 from high-amylose corn to other low glycemic carbohydrates. One study compared RS2 from high-amylose corn to cellulose and found no significant difference (Ranganathan et al. 1994). The second study compared RS2 from high-amylose corn to fructose and found increased insulin response (Reiser et al. 1989). It is difficult to draw conclusions from these two studies, as there are significant differences in the digestion profile, and the likely mechanisms of the carbohydrates utilized in these studies.

Insulin Response of RS2 via Meal Tolerance Testing

Five studies (Table 16.11) examined postprandial insulin response of a standardized meal or quantity of glucose instead of feeding the test foods. They fed individuals RS2 from high-amylose corn for varying periods of time and measured the insulin

Table 16.10. Insulin response of RS2 from high-amylose corn standardized to carbohydrate quantity.

Study	Subjects	Food	Effect
Behall et al. (1988)	Healthy	Crackers	↓IR
Behall et al. (1989)	Healthy	Pudding, muffins, corn chips, shortbread cookies, and brownies in meals	↓IR
Behall and Howe (1995)	Healthy and hyperinsulinemic	Bread, muffins, cereal, cookies, and cheese puffs	↓IR
Behall and Hallfrisch (2002)	Healthy	Bread	↓IR
Behall and Scholfield (2005)	Healthy and hyperinsulinemic	Corn chips and corn muffins	↓IR
Behall et al. (2006a)	Healthy—normal weight and overweight	Muffins	↓IR
Brighenti et al. (2006)	Healthy	Sponge cake	↓IR
Giacco et al. (1998)	Type 2 diabetics	Cheesecake made with RS biscuits	NSD
Goodpaster et al. (1996)	Healthy elite athletes	Beverage	NSD
Granfeldt et al. (1995)	Healthy	Arepas	↓IR
Hoebler et al. (1999)	Healthy	Bread	↓IR
Hospers et al. (1994)	Healthy	Pasta	↓IR
Howe et al. (1996)	Healthy and hyperinsulinemic	Bread, cereal, cheese puffs, muffins, cookies	↓IR
Krezowski et al. (1987)	Type 2 diabetics	Muffins	NSD
Noakes et al. (1996)	Dyslipidemic	Bread, cereal, pasta, muffins	↓IR
van Amelsvoort and Weststrate (1992)	Healthy	Hot dish meal	↓IR
Weststrate and van Amelsvoort (1993)	Healthy	Breakfast: Bread; lunch: Pizza	Breakfast: ↓ AUC IR lunch: ↓ AUC IR

Abbreviations: NSD = no significant difference, IR = insulin response, AUC = area under the curve.

Table 16.11. Insulin response of RS2 from high-amylose corn via meal tolerance testing.

Study	Time frame for RS2 consumption	Test challenge	Results
Behall et al. (1989)	4 weeks prior	1 g glucose/kg body weight	NSD AUC↓ IR
Brighenti et al. (2006)	one meal prior	715 kcal meal with 93 g glycemic carbs	NSD IR
Robertson et al. (2003)	1 day prior	2.1 kJ chocolate drink delivering 59 g carb, 21 g fat, and 19 g protein	↓IR↑ insulin sensitivity
Robertson et al. (2005)	4 weeks prior	500 kcalorie beverage delivering 60 g carb and 21 g fat	↓AUC ↑ insulin sensitivity during hyperinsulinemic-euglycemic camp study, ↑ total glucose uptake by adipose tissue. NSD fasting insulin sensitivity as measured by HOMA
Weickert et al. (2005)	1 day prior	240 kcal bread delivering 50 g available carb, 7.2 g protein, 0.8 g fat	NSD IR

Abbreviations: NSD = no significant difference, IR = insulin response, AUC = area under the curve, HOMA = homeostatic model assessment.

response of a standardized quantity of glucose or a standardized test meal. A reduced insulin response was demonstrated in three out of the five studies. Two studies feeding RS2 over longer periods of time demonstrated reduced insulin response, suggesting that the effect may be sustained with chronic consumption. Additional research would be useful to confirm that the effects extend beyond 4 weeks.

Glycemic Benefits Emanating from the Large Intestine

The third major category of RS benefits to be examined in this review is metabolic benefits particularly related to glycemic management. It should be noted that all the studies previously examined in this review have been clinical trials. With regard to glycemic benefits emanating from the large intestine, animal studies will also be considered. Table 16.4 summarizes the relevant published animal studies.

Numerous studies have demonstrated that RS2 from high-amylose corn is fermented in the large intestine and increases the production of short-chain fatty acids (Topping and Clifton 2001; Young and Le Leu 2004). Traditionally, it has been assumed that the benefits from this fermentation applied primarily to colonic tissue. While the dietary consumption of RS and reduced risk of colon cancer is under active investigation (Burn et al. 1998; Mathers et al. 2003), recent studies with RS2 from high-amylose corn demonstrate that fermentation has important metabolic benefits as well. The potential for fermentable carbohydrates to impact metabolism and help in the management of food intake, body weight, and metabolic syndrome has been recently summarized (Delzenne and Cani 2005).

The short-chain fatty acids (SCFAs) produced by the fermentation are biologically active throughout the body. Butyrate is the primary energy source utilized by the colonic mucosa. In addition, butyrate, acetate, and propionate are absorbed and distributed via the bloodstream. Additional products of fermentation are also produced: hydrogen, carbon dioxide, methane, lactate, succinate, formate, ethanol, and others. Acetate and propionate are utilized as energy sources by muscles throughout the body, but it is clear that the biological cascade provides additional benefits beyond the delivery of energy, as described below. Additional research is needed to identify the particular bioactive, or combination of bioactives, responsible for the particular metabolic benefits. It may be individual SCFAs, a ratio of SCFAs, or a combination of as yet known or unknown compounds.

Insulin Sensitivity

Three clinical studies found that dietary consumption of *Hi-maize* RS increased insulin sensitivity (Robertson et al. 2003, 2005; Zhang et al. 2007). The 2003 study was conducted over one day, with individuals consuming 100 g of *Hi-maize* RS, delivering 60 g of dietary fiber throughout one day. The control diet included 40 g of glycemic starch to match the amount of glycemic starch delivered from the RS2. In this way, the postprandial glycemic response was controlled and the large intestinal fermentation effects were evaluated. The following day, insulin sensitivity was tested via a standardized meal tolerance test. They found that prior RS2 consumption led to lower postprandial plasma glucose and insulin response as well as higher insulin sensitivity. The 2005 study found similar results: daily consumption of 50 g of *Hi-maize* RS delivering 30 g of dietary fiber consumed over 4 weeks also resulted in significant increases in insulin sensitivity in healthy individuals. The Zhang study fed individuals with type 2 diabetics 30 g of *Hi-maize* RS per day for 4 weeks. They found significant reductions in fasting blood sugar, postprandial glycemic response, and a significant increase in insulin sensitivity, as measured by an Insulin Sensitivity Index.

The authors have suggested that this effect is caused by fermentation in the large intestine. This finding is nutritionally important as insulin sensitivity is one of the underlying biomarkers for metabolic syndrome, and increases the risk of developing diabetes and heart disease (Grundy et al. 2005). Thus, dietary consumption of RS2 from high-amylose corn may assist in reducing the risk of diabetes, metabolic syndrome, and other physiological consequences of insulin resistance. Additional randomized, double blind, clinical studies should be performed to investigate these long-term benefits.

Animal studies are supportive (Table 16.4). Three animal studies have shown that RS2 from high-amylose corn reduced the glycemic response when replacing highly digestible starches (Byrnes et al. 1995; Morand et al. 1992; Pawlak et al. 2004). In addition, eight animal studies have shown that RS2 from high-amylose corn reduced the insulin response when replacing highly digestible starches (Brown et al. 2003; Byrnes et al. 1995; Higgins et al. 1996; Morand et al. 1992, 1994; Pawlak et al. 2001, 2004; Wiseman et al. 1996). One study found that plasma insulin levels were significantly lower following consumption of RS2 from high-amylose corn compared to high glycemic starch (Scribner et al. 2007).

Three studies have demonstrated that RS2 from high-amylose corn delayed the development of insulin resistance in rats (Byrnes et al. 1995; Higgins et al. 1996; Wiseman et al. 1996). The Wiseman study also showed that once insulin resistance develops in animals, dietary consumption of RS2 from high-amylose corn did not reverse it.

These studies provide supporting evidence that consumption of RS2 from high-amylose corn helps to maintain healthy blood sugar levels over the longer term and may assist in reducing the onset of insulin resistance when used to replace highly digestible carbohydrates in the diet.

Lipid Oxidation

One clinical trial has demonstrated that dietary consumption of RS2 from high-amylose corn significantly increased lipid oxidation throughout the day (Higgins et al. 2004). The author has suggested that SCFAs interfere with the production of carbohydrate-derived acetyl CoA as a fuel source in the liver (Higgins 2004). With this metabolic change, the liver switches to fat-derived acetyl CoA as its fuel source. This shift in lipid oxidation (fat burning) probably contributes to and may be responsible for the changes in abdominal fat described below.

The study in healthy individuals provides clinical evidence that RS2 from high-amylose corn has the potential to impact body composition in humans. In other words, the health consequences of RS2 from high-amylose corn in weight management extend significantly beyond its reduced caloric content (Behall and Howe, 1996) and its reduced glycemic response.

Satiety

One clinical trial has demonstrated that low-calorie muffins containing RS2 from high-amylose corn had significantly higher satiety response than plain muffins (Quílez et al. 2007). These data are difficult to extrapolate primarily because multiple ingredients were altered in the test food: sugar was replaced with maltitol, fat was replaced with carboxycellulose and guar gum, and flour was replaced with RS2 from high-amylose corn.

Animal trials are supportive and point to fermentation as a mechanism. They have shown increases in the genetic expression of glucagon-like peptide-1 (GLP-1) and Peptide YY (PYY) following consumption of RS2 from high-amylose corn (Keenan et al. 2006a, b; Zhou et al. 2006). GLP-1 is an important biomarker because it has multiple benefits: stimulation of insulin, restoration of the rapid response of insulin, suppression of glucagon, enhanced satiety, appetite control, and weight loss and long-term improvements in insulin sensitivity. This natural peptide has a half-life of minutes due to rapid degradation by the dipeptidyl peptidase IV (DPP-4) enzyme and is the target of several new pharmaceuticals recently introduced for diabetics (Exenatide LAR® and Januvia® from Merck, and Vildagliptin® from Novartis). Additional research is needed to determine the potential for synergistic effects and/or range of effectiveness between natural RS and the pharmaceutical agents.

Not all fiber produces the same effect on GLP-1 and PYY. Experiments compared the metabolic effects of energy dilution of cellulose (a nonfermentable fiber) with the metabolic effects of RS2 from high-amylose corn (Keenan et al. 2006b). When measuring GLP-1 and PYY, they found that the fermentation effects were more significant than the energy dilution effects or the bulking effects. Much research has focused on the energy dilution effects and the bulking effects in weight management. Energy dilution effects are exemplified by public health recommendations that individuals consume lower calorie foods to lose weight. Bulking effects are exemplified by fibers that provide bulk in the large intestine (i.e., wheat fiber and cellulose). Their physiological impact is characterized by their trait of significant water absorption and lack of fermentation. This research is a strong indication that the fermentable RS may have satiety advantages over nonfermentable fiber.

Abdominal Fat

Two types of studies suggest that consumption of RS2 from high-amylose corn affects abdominal fat: animal studies showing changes in body composition and human studies with biomarkers. To date, clinical studies confirming changes in body composition have not been published.

Animals fed RS2 from high-amylose corn have reduced fat content and reduced abdominal fat compared to animals fed highly digestible carbohydrates (Keenan et al. 2006a; Pawlak et al. 2004; Scribner et al. 2007; Williamson et al. 1999). Many mechanisms probably contribute to this effect—many of which are reviewed in this chapter.

In 2004, Dorota Pawlak, Jake Kushner, and David Ludwig (Pawlak et al. 2004) utilized *Hi-maize* RS2 as a low-GI starch compared to amylopectin, a high-GI starch. Rats fed the high-GI diet for 17 weeks had 71% more body fat and 8% less lean body mass than the low-GI group containing RS2. Because rats in the high-GI group were gaining weight at a faster rate, their food was restricted starting at week 8 to maintain similar weight as the low-GI group. An autopsy, at week 18, the high-GI animals had more combined epididymal and retroperitoneal fat than the low-GI animals. Blood glucose AUC had increased significantly more in the high-GI than in the low-GI group at week 5 and the difference persisted for the duration of the experiment. Plasma insulin AUC also increased significantly more in the high-GI group by week 5 and the difference persisted at week 14. Insulin sensitivity, assessed by insulin tolerance test at week 16, did not differ between the groups.

In 2007, Kelly Scribner, Dorota Pawlak, and David Ludwig (Scribner et al. 2007) extended these findings. Mice fed high-glycemic carbohydrates had twice the normal amount of fat in their bodies, blood, and livers, compared to mice fed *Hi-maize* RS2 in a 25 week study.

Published Studies with Other Sources of Resistant Starch

Over the past several years, chemically modified resistant starches (RS4s) have been introduced into the food industry for the purpose of fiber fortification. Published studies confirm that different types and sources of RS may be different than natural RS2 from high-amylose corn in glycemic response, insulin response, digestion profile, fermentation profile (rate, location, and by-products), and satiation profile.

In addition, the analytical methods utilized to measure the content of dietary fiber do not reliably predict physiological impact. A RS may be resistant physiologically but may or may not measure as dietary fiber according to official AOAC methods. Research is needed to identify how chemically modified RS4s are processed in the body and the potential benefits that may result from their addition to foods.

Beyond RS2 from high-amylose corn, the physiological consequences from various types of RS are not known. Studies are just beginning to be published comparing different sources of resistant starches. Regarding glycemic response, RS2 from high-amylose corn reduces the glycemic response across a wide range of healthy individuals, hyperinsulinemic individuals, type 2 diabetics, and dyslipidemic individuals (Table 16.3), but other resistant starches have shown contradictory effects. One animal study with RS4 from high-amylose corn demonstrated an increased glycemic response (Zhou and Kaplan 1997), while a separate human study with enzymatically treated RS3 from tapioca found a reduced glycemic response in only a subgroup with higher baseline levels of blood glucose and did not find a significant effect in a larger population of 20 individuals (Yamada et al. 2005). Based upon this evidence, it cannot be assumed that all classes of resistant starches will deliver reduced glycemic response in various populations.

Regarding insulin response, while RS2 from high-amylose corn reduces the insulin response of healthy individuals, there are little data from other types of RS. One animal study with chemically modified RS4 from high-amylose corn demonstrated no effect on insulin response (Zhou and Kaplan 1997). A separate human study with enzymatically treated RS3 from tapioca found a reduced insulin response in 12 individuals with higher/borderline levels of glucose response, but not in eight individuals with normal glucose response (Yamada et al. 2005). Based upon this evidence, the insulin response of different resistant starches may have different degrees of physiological effects.

Three direct in vitro comparisons and one animal model have demonstrated significant differences between different sources and types of resistant starch (Fassler et al. 2006a, b, 2007; Rideout et al. 2007). Caroline Fassler and her colleagues (Fassler et al. 2006b) demonstrated different kinetics of digestion, digestion residues, and degrees of variability in starch digestion when comparing RS2 from high-amylose corn to RS3 from tapioca. A second study (Fassler et al. 2006a) demonstrated that the quantities of short-chain fatty acids, ammonia, total gas, and hydrogen varied by source and type of RS. A third 2007 study (Fassler 2007) showed differences in the colonic impact of fermentation by-products in several models. Finally, a recent study in pigs found that the nutrient utilization and intestinal fermentation are differentially affected by different resistant starch varieties and different resistant starch types (Rideout et al. 2007). These researchers compared RS2 from high-amylose corn with RS2 from potatoes, RS3 from high-amylose corn, guar gum, and cellulose.

Chemically modified RS4 from high-amylose corn changes the digestion and fermentation profile of RS as compared to natural RS2 from high-amylose corn. The type of modification significantly changed the fermentation by-products. Morita et al. (2005) demonstrated in animals that chemically modified RS4 from high-amylose corn that has been modified with acetate produces more acetate in the large intestine than unmodified RS2. Similarly, Bajka et al. (2006) demonstrated in animals that chemically modified RS4 from high-amylose corn that has been treated with butyrate produces more butyrate in the large intestine than unmodified RS2. The physiological consequences of these different fermentation products are not known. Thus, the type of RS can significantly change the physiological impact.

Finally, the impact on satiety of different types of RS has been shown to vary. Natural RS2 from high-amylose corn induced more satiety for 6 h after eating compared to corn with low-amylose content (van Amelsvoort and Westrate 1992). Another study found that natural RS2 from potato decreased satiety scores (Raben et al. 1994).

The scientific evidence supporting glycemic management benefits from natural RS2 from high-amylose corn is strong and consistent, especially when the differences in methods and studies are considered. These studies provide the substantiation to support the communication of glycemic benefits of foods containing natural RS2 from high-amylose corn to consumers.

The physiological impact of different types and sources of RS can vary significantly. It has been shown that other types and sources of RS do not deliver the same physiological impact as natural RS2 from high-amylose corn. Research is needed to confirm the physiological effects of dietary consumption of additional types and sources of resistant starches. It cannot be assumed that benefits delivered with natural RS2 from high-amylose corn apply to other categories of RS. Just as different types and sources of dietary fiber are not interchangeable, different types and sources of RS are not interchangeable. The evidence supporting each type of ingredient must be considered individually.

Consumer Communications

Most American consumers are unfamiliar with the benefits of reduced glycemic foods or the importance of maintaining insulin sensitivity. These technical and medical terms may be familiar to diabetics, but not to the general population. This presents challenges as well as opportunities for food companies seeking to develop better-for-you foods delivering glycemic management benefits.

The Atkins diet significantly contributed to increasing consumer understanding of the weight reduction benefits of extreme glycemic management. In the past several years, however, consumers are beginning to understand that more moderate reductions in the glycemic response of foods may be beneficial as well, not only for potentially reducing the risk of diseases but also for managing their energy levels. While the potential for reducing the risk of obesity, diabetes, and cardiovascular disease may be the most important factor for public health authorities, the energy management benefits may drive consumer demand and early adoption of reduced glycemic foods.

One prominent proponent for reducing glycemic impact for the purpose of reducing the risk of disease is Dr. Walter Willett from the Harvard School of Public Health. He has proposed a "Healthy Eating Pyramid" which portrays processed flour, white bread, potatoes, white rice, and pasta as harmful and lists them at the top of the pyramid with admonitions to "use sparingly." Whole grain and high-fiber carbohydrates are portrayed as beneficial, with whole-grain bread and brown rice listed at the bottom of the pyramid with "consume at every meal" (Willett et al. 2001, 2006). Willett links high intake of starch from refined grains and potatoes to a high risk of type 2 diabetes and coronary heart disease, and fiber consumption to a lower risk of these illnesses (Willett et al. 2003). As additional studies are published and the critical differences in methods measuring glycemic impact of foods are examined, this perspective may be gaining momentum and will likely be widely debated for many years. Communication to consumers regarding these potential benefits may be indirect or delayed until a scientific consensus emerges and health claims are approved to communicate these claims.

In the meantime, consumer interest in glycemic management appears to be driven by appetite control and energy management benefits. Consumers can feel the difference between consuming high glycemic carbohydrates and low glycemic carbohydrates in a few hours. The effect seems to be stronger as individuals get older, and may be accentuated in individuals with insulin resistance. These appetite and energy benefits are highly relevant to consumers and are experienced every day. Numerous articles in consumer magazines and other news sources highlight these benefits of glycemic management and point out that the low glycemic foods can help smooth out or moderate the energy swings experienced by individuals throughout the day.

For instance, an article from *O, the Oprah Magazine* described the effect of blood sugar swings:

> All you did was pour a little too much syrup on your pancakes a couple of hours ago, and here you are with the sirens blaring, red lights flashing, and then come the jitters, racing heart-beat, cold sweats, that I-could-kill-someone feeling . . . Preventing a sugar crash in the first place is clearly a much saner approach. Fortunately, if you take greater care with your carb intake and change a few other habits, it's not hard to even the seesaw
> (Davis 2002).

Similar messages have been repeated in countless consumer magazines over the past 5 years. For marketers, the challenge of communicating the benefits remains. There are three main approaches: description of the dietary ingredients contained in the food (i.e., RS, or dietary fiber content), description of the physiological impact or biomarker (i.e., GI or glycemic response), and description of the benefits delivered by the food (i.e., appetite control, energy management, or reduced risk of disease). Each of these approaches is valid and may appeal to different segments of consumers. Depending upon the consumer awareness of these relationships and benefits, some approaches may be more successful than others.

The first approach has been utilized by food marketers for many years. Label claims announcing the inclusion of dietary fiber are widely utilized today, as consumers may have some understanding of the benefits of dietary fiber. Consumers have little knowledge of natural RS as a particular type of dietary fiber with specific health benefits to date. Most consumers link dietary fiber with regularity, although many also identify additional benefits to dietary fiber, including cholesterol reduction, cancer prevention, and weight management.

The second approach (description of the physiological impact or biomarker) has educational challenges as most American consumers have a low awareness of the glycemic impact of foods. Given the scientific debate around the different methodology and potential benefits, it is not surprising that the technical terminology has not reached American consumers to a significant degree. Some companies are already communicating information on the GI of their products compared to competitive products. Over time, consumers may eventually link the technical "glycemic" term to physiological benefits, but an indication of glycemic response of a food may not be very meaningful for American consumers at this time.

The third approach focuses more specifically on benefits resulting from glycemic management. These benefits may include energy management, appetite control, weight management, and/or potentially reduced risk of developing diabetes, obesity, or cardiovascular disease. These types of claims require substantiation to ensure that the food delivers on the explicit or implicit benefit. For instance, a reduced risk of disease claim is a "health claim" in the U.S. market and would require prior approval by the U.S. Food & Drug Administration. On the other hand, an energy management claim or a hunger management would be easier to substantiate. Structure/function claims, such as "helps maintain healthy blood sugar levels," are possible in the U.S. market, if they are substantiated and do not imply that the food prevents, mitigates, treats, or diagnoses a disease.

Around the world, natural RS2 from high-amylose corn is being utilized in foods specifically designed for glycemic management. Some of these foods are positioned for their dietary fiber content. Some are positioned for their reduced or low GI and/or glycemic response. Others are positioned as foods for diabetics, foods for energy management, foods for blood sugar management, or simply better-for-you foods.

Summary

RS2 from high-amylose corn is an ideal food ingredient for the development of foods targeting glycemic management. It naturally impacts many mechanisms in the body and can be easily communicated to consumers. It delivers fewer energy and calories than digestible carbohydrates. It reduces the glycemic and insulin response of foods when replacing flour or other digestible carbohydrates. Its fermentation in the large intestine has also been shown to improve insulin sensitivity, lipid oxidation (fat burning), and satiety.

Natural RS2 from high-amylose corn is easily utilized in the formulation of commonly consumed foods without impacting the taste or texture of the food—a tremendous advantage versus some other sources of fiber and a critical component of success in the development of functional foods. With the significant body of evidence demonstrating glycemic benefits, and multiple marketing opportunities, it is an ideal ingredient for use in the formulation of better-for-you foods.

Note

Hi-maize® and *NOVELOSE®* are registered trademarks of National Starch, LLC, New Castle, DE. Exenatide LAR® and Januvia® are registered trademarks of

Merck, Rahway, New Jersey. Vildagliptin® is a registered trademark of Novartis, Basel, Switzerland.

Acknowledgments

The author wishes to acknowledge numerous contributions from the team of research professionals at National Starch Food Innovation. In particular Dr. Anne Birkett significantly contributed over the past 5 years to the discovery and organization of information utilized in the preparation of this manuscript.

References

AACC International. 2006. AACC International Approves Definitions Related to Glycemic Carbohydrates, Sept. 13, AACC Press Release. http://www.aaccnet.org/news/06glycemicdefinitions.asp

Anderson GH, Catherine NLA, Woodend DM, Wolever TMS. Inverse association between the effect of carbohydrates on blood glucose and subsequent short-term food intake in young men. 2002. Am J Clin Nutr 76:1023–1030.

Asp NG. 1992. Resistant starch. Proceedings from the second plenary meeting of EURESTA: European FLAIR Concerted Action No. 11 on physiological implications of the consumption of resistant starch in man. Eur J Clin Nutr. 46(Suppl 2):S1.

Bajka BH, Topping DL, Cobiac L, Clarke JM. 2006. Butyrylated starch is less susceptible to enzymic hydrolysis and increases large-bowel butyrate more than high-amylose maize starch in the rat. Br J Nutr. 96:276–282.

Behall KM, and Hallfrisch J. 2002. Plasma glucose and insulin reduction after consumption of breads varying in amylose content. Eur J Clin Nutr 56(9):913–920.

Behall KM, Hallfrisch JG, Scholfield DJ, Liljeberg-Elmstahl HGM. 2006a. Consumption of both resistant starch and beta-glucan improves postprandial plasma glucose and insulin in women. Diabetes Care. 29(5):976–981.

Behall KM, Howe JC. 1995. Effect of long-term consumption of amylose vs amylopectin starch on metabolic variables in human subjects. Am J Clin Nutr 61(3):334–340.

Behall KM, Howe JC. 1996. Resistant starch as energy. J Am Coll Nutr 15(3):248–254.

Behall KM, Howe JC, Anderson RA. 2002. Apparent mineral retention is similar in control and hyperinsulinemic men after consumption of high-amylose cornstarch. J Nutr 132:1886–1891.

Behall KM, Scholfield DJ. 2005. Food amylose content affects postprandial glucose and insulin responses. Cereal Chem 82(6):654–659.

Behall KM, Scholfield DJ, Canary J. 1988. Effect of starch structure on glucose and insulin responses in adults. Am J Clin Nutr 47(3):428–432.

Behall KM, Scholfield DJ, Hallfrisch JG. 2006b. Barley beta-glucan reduces plasma glucose and insulin responses compared with resistant starch in men. Nutr Res. 26(12):644–650.

Behall KM, Scholfield DJ, Yuhaniak I, Canary J. 1989. Diets containing high amylose vs amylopectin starch: Effects on metabolic variables in human subjects. Am J Clin Nutr 49(2):337–344.

Brighenti F, Benini L, Del Rio D, Casiraghi D, Pellegrini N, Spazzina F, Jenkins DJA, Vantini I. 2006. Colonic fermentation of indigestible carbohydrates contributes to the second-meal effect Am J Clin Nutr. 83:817–822.

Brown IL, McNaught KJ, Moloney E. 1995. *Hi-maize*™: New directions in starch technology and nutrition. Food Aust. 47(6):272–275.

Brown MA, Storlien LH, Brown IL, Higgins JA. 2003. Cooking attenuates the ability of high-amylose meals to reduce plasma insulin concentrations in rats. Br J Nutr. 90:823–827.

Burn J, Chapman PD, Bishop DT, Mathers J. 1998. Diet and cancer prevention: The concerted action polyp prevention (CAPP) studies. Proc Nutri Soc. May; 57(2):183–186.

Byrnes SE, Miller JC, Denyer GS. 1995. Amylopectin starch promotes the development of insulin resistance in rats. J Nutr. 125(6):1430–1437.

Davis L. 2002. The Good-Mood Diet. O The Oprah Magazine 3(7):101–104.

Delzenne NM, Cani PD. 2005. A place for dietary fibre in the management of the metabolic syndrome. Curr. Opin. Clin Nutr Metab Care. 8:636–640.

Englyst HN, Kingman SM, Cummings JH. 1992. Classification and measurement of nutritionally important starch fractions. Eur J Clin Nutr. 46(Suppl. 2):533–550.

Fassler C, Arrigoni E, Venema K, Brouns F, Amado R. 2006a. In vitro fermentability of differently digested resistant starch preparations. Mol. Nutr Food Res. 50(12):1220–1228.

Fassler C, Arrigoni E, Venema K, Hafner V, Brouns F, Amado R. 2006b. Digestibility of resistant starch containing preparations using two in vitro models. Eur J Nutr. 45(8):445–453.

Fassler C, Gill CIR, Arrigoni E, Rowland I, Amado R. 2007. Fermentation of resistant starches: Influence of in vitro models on colon carcinogenesis. Nutr Cancer. 58(1):85–92.

Giacco R, Clemente G, Brighenti F, Mancini M, D'Avanzo A, Coppola S, Ruffa G, La sorella G, Rivieccio AM, Rivellese AA, Riccardi G. 1998. Metabolic effects of resistant starch in patients with type 2 diabetes. Diabetes Nutr Metab. 11:330–335.

Goodpaster BH, Costill DL, Fink WJ, Trappe TA, Jozsi AC, Starling RD, Trappe SW. 1996. The effects of preexercise starch ingestion on endurance performance. Int J Sports Med. 17:366–372.

Granfeldt Y, Drews A, Bjorck I. 1995. Arepas made from high-amylose corn flour produce favorably low glucose and insulin responses in healthy humans. J Nutr. 125(3):459–465.

Grundy SM, Cleeman JI, Daniels SR, Donato KA, Eckel RH, Franklin BA, Gordon DJ, Krauss RM, Savage PJ, Smith SC, Spertus JA, Costa F. 2005. Diagnosis and Management of the Metabolic Syndrome – An American Heart Association/National Heart, Lung, and Blood Institute Scientific Statement. Circulation 112:e1–e6.

Higgins JA. 2004. Resistant Starch: Metabolic effects and potential health benefits. J AOAC Int 87(3):761–768.

Higgins JA, Brand Miller J, Denyer GS. 1996. Development of insulin resistance in the rat is dependent on the rate of glucose adsorption from the diet. J Nutr. 126:596–602.

Higgins JA, Brown MA, Storlien LH. 2006. Consumption of resistant starch decreases postprandial lipogenesis in white adipose tissue of the rat. Nutr J. 5:25.

Higgins JA, Higbee DR, Donahoo WT, Brown IL, Bell ML, Bessesen DH. 2004. Resistant starch consumption promotes lipid oxidation. Nutr Metab 1:8.

Hoebler C, Karinthi A, Chiron H, Champ M, Barry J-L. 1999. Bioavailability of starch in bread rich in amylose: Metabolic responses in healthy subjects and starch structure. Eur J Clin Nutr 53(5):360–366.

Hospers JJ, van Amelsvoort JMM, Weststrate JA. 1994. Amylose-to-amylopectin ratio in pastas affects postprandial glucosee and insulin responses and satiety in males. J Food Sci 59(5):1144–1149.

Howe JC, Rumpler WV, Behall KM. 1996. Dietary starch composition and level of energy intake alter nutrient oxidation in "carbohydrate-sensitive" men. J Nutr. 126(9):2120–2129.

Jenkins DJ, Vuksan V, Kendall CW, Wursch P, Jeffcoat R, Waring S, Mehling CC, Vidgen E, Augustin LS, Wong E. 1998. Physiological effects of resistant starches on fecal bulk, short chain fatty acids, blood lipids and glycemic index. J Am Coll Nutr. 17(6):609–616.

Jenkins DJ, Wolever TM, Taylor RH, Barker H, Fielden H, Baldwin JM, Bowling AC, Newman HC, Jenkins AL, Goff DV. 1981. Glycemic index of foods: A physiological basis for carbohydrate exchange. Am J Clin Nutr 34(3):362–366.

Keenan MJ, Zhou J, McCutcheon KL, Raggio AM, Bateman HG, Todd E, Jones CK, Tulley RT, Melton S, Martin RJ, Hegsted M. 2006a. Effects of resistant starch, a non-digestible fermentable fiber, on reducing body fat. Obesity 14(9):1523–1534.

Keenan MJ, Zhou J, Raggio AM, McCutcheon KL, Newman SS, Tulley RT, Martin RJ, Brown I, Birkett A, Hegsted M. 2006b. Preliminary microarray analysis of genes from cecal cells of resistant starch fed rats. Abstract presented at 2006 Annual meeting of NAASO, The Obesity Society, Boston, MA, October 23, 2006.

Krezowski PA, Nuttall FQ, Gannon MC, Billington CJ, Parker S. 1987. Insulin and glucose responses to various starch-containing foods in Type II diabetic subjects. Diabetes Care 10(2):205–212.

Liese AD, D'Agostino Jr, RB, Schulz M, Sparks KC, Fang F, Mayer-Davis EJ, Wolever TMS. 2005. Dietary glycemic index and glycemic load, carbohydrate and fiber intake, and measures of insulin sensitivity, secretion, and adiposity in the insulin reistance atherosclerosis study. Diabetes Care. 28:2832–2838.

Ludwig, DS. 2002. The Glycemic Index— Physiological mechanisms relating to obesity, diabetes and cardiovascular disease. J Am Med Assoc 287:2414–2423.

Mathers JC, Mickleburgh I, Chapman PC, Bishop DT, Burn J; Concerted Action Polyp Prevention (CAPP) 1 Study. 2003. Can resistant starch and/or aspirin prevent the development of colonic neoplasia? The Concerted Action Polyp Prevention (CAPP) 1 Study. Proc Nutr Soc 62(1):51–57.

Morand C, Levrat MA, Besson C, Demigné C, Rémésy C. 1994. Effects of a diet rich in resistant starch on hepatic lipid metabolism in the rat. J Nutri Biochem 5:138–144.

Morand C, Remesy C, Levrat MA, Demigne C. 1992. Replacement of digestible wheat starch by resistant cornstarch alters splanchnic metabolism in rats. J Nutr 122(2):345–354.

Morita T. Kasaoka S, Kiriyiama S, Brown IL, Topping DL. 2005. Comparative effects of acetylated and unmodified high-amylose maize starch in rats. Starch 57:246–253.

Noakes M, Clifton PM, Nestel PJ, Le Leu R, McIntosh G. 1996. Effect of high-amylose starch and oat bran on metabolic variables and bowel function in subjects with hypertriglyceridemia. Am J Clin Nutr 64(6):944–951.

NSFI 2006. Internal National Starch Food Innovation Communication, Determination of glycemic index of white bread.

Olesen M, Rumessen JJ, Gudmand-Hoyer E. 1994. Intestinal transport and fermentation of resistant starch evaluated by the hydrogen breath test. Eur J Clin Nutr 48(10):692–701.

Pawlak DB, Bryson JM, Denyer GS, Brand-Miller JC. 2001. High glycemic index starch promotes hypersecretion of insulin and higher body fat in rats without affecting insulin sensitivity. J Nutr. 131:99–104.

Pawlak DB, Kusher JA, Ludwig DS. 2004. Effects of dietary glycaemic index on adiposity, glucose homeostasis, and plasma lipids in animals. Lancet 364:778–785.

Quílez J, Bulló M, Salas-Salvadó J. 2007. Improved postprandial response and feeling of satiety after consumption of low-calorie muffins with maltitol and high-amylose corn starch. J Food Sci 72(6):S407–S411.

Raben A, Tagliabue A, Christensen NJ, Madsen J, Holst JJ, Astrup A. 1994. Resistant starch: The effect on postprandial glycemia, hormonal response, and satiety. Am J Clin Nutr 60:544–551.

Ranganathan S, Champ M, Pechard C, Blanchard P, Nguyen M, Colonna P, Krempf M. 1994. Comparative study of the acute effects of resistant starch and dietary fibers on metabolic indexes in men. Am J Clin Nutr 59(4):879–883.

Reiser S, Powell AS, Scholfield DJ, Panda P, Fields M, Canary JJ. 1989. Day-long glucose, insulin, and fructose responses of hyperinsulinemic and nonhyperinsulinemic men adapted to diets containing either fructose or high-amylose cornstarch. Am J Clin Nutr 50:1008–1014.

Rideout TC, Liu Q, Wood P, Fan MZ. 2007. Nutrient utilization and intestinal fermentation are differentially affected by the consumption of resistant starch varieties and conventional fibres in pigs. Br J Nutr. Published online Nov 16; e1–9.

Robertson MD, Bickerton AS, Dennis AL, Vidal H, Frayn KN. 2005. Insulin-sensitizing effects of dietary resistant starch and effects on skeletal muscle and adipose tissue metabolism. Am J Clin Nutr 82:559–567.

Robertson MD, Currie JM, Morgan LM, Jewell DP, Frayn KN. 2003. Prior short-term consumption of resistant starch enhances postprandial insulin sensitivity in healthy subjects. Diabetologia 46(5):659–665.

Scribner KB, Pawlak DB, Ludwig DS. 2007. Hepatic steatosis and increased adiposity in mice consuming rapidly vs. slowly absorbed carbohydrate. Obesity 15(9):2190–2199.

Topping DL, and Clifton PM. 2001. Short-chain fatty acids and human colonic function: Roles of resistant starch and nonstarch polysaccharides. Physiol Rev 81:1031–1064.

van Amelsvoort JMM, and Westrate JA. 1992. Amylose-amylopectin ratio in a meal affects postprandial variables in male volunteers. Am J Clin Nutr. 55:712–718.

Vonk RJ, Hagedoorn RE, de Graaff R, Elzinga H, Tabak S, Yang Y-X, Stellaard F. 2000. Digestion of so-called resistant starch sources in the human small intestine. Am J Clin Nutr 72:432–438.

Weickert MO, Mohlig M, Koebnick C, Holst JJ, Namsolleck P, Ristow M, Osterhoff M, Rochlitz H, Rudovich N, Spranger J, Pfeiffer AFH. 2005. Impact of cereal fibre on glucose-regulating factors. Diabetologia. (48):2343–2353.

Weststrate JA, van Amelsvoort JMM. 1993. Effects of the amylose content of breakfast and lunch on postprandial variables in male volunteers. Am J Clin Nutr 58:180–186.

Willett WC. Skerrett PJ. 2001. Eat, Drink, and be Healthy. New York: Simon & Schuser.

Willett WC., Stampfer MJ. 2003. "Rebuilding the Food Pyramid" In Scientific American, pp. 64–71. New York: Scientific American Inc.

Willett WC, Stampfer MJ. 2006. Rebuilding the Food Pyramid. Eating to Live. In Scientific American Reports, Special Edition on Diet and Health, pp. 12–21. New York: Scientific American Inc.

Williamson SLH, Kartheuser A, Coaker J, Kooshkghazi MD, Fodde R, Burn J, Mathers JC. 1999. Intestinal tumorigenesis in the Apc1638N mouse treated with aspirin and resistant starch for up to 5 months. Carcinogenesis 20(5):805–810.

Wiseman CE, Higgins JA, Denyer GS, Miller JC. 1996. Amylopectin starch induces nonreversible insulin resistance in rats. J Nutr 126(2):410–415.

Yamada Y, Hosoya S, Nishimura S, Tanaka T, Kajimoto Y, Nishimura A, Kajimoto O. 2005. Effect of bread containing resistant starch on postprandial blood glucose levels in humans. Biosci Biotechnol Biochem 69(3):559–566.

Young GP, Le Leu RK. 2004. Resistant starch and colorectal neoplasia. J AOAC Int 87(3):775–786.

Zhang W, Wang H, Zhang Y, and Yang Y. 2007. Effects of resistant starch on insulin resistance of type 2 diabetes mellitus patients. Chin J Prev Med 41(2):101–104.

Zhou J, Hegsted M, McCutcheon KL, Keenan MJ, Xi X, Raggio AM, Martin RJ. 2006. Peptide YY and Proglucagon mRNA expression patterns and regulation in the gut. Obesity 14(4):683–689.

Zhou X, Kaplan ML. 1997. Soluble amylose cornstarch is more digestible than soluble amylopectin potato starch in rats. J Nutr 127:1349–1356.

Chapter 17

Proteins, Protein Hydrolysates, and Bioactive Peptides in the Management of Type 2 Diabetes

Joris Kloek, PhD, Vijai K Pasupuleti, PhD,
and Luc JC van Loon, PhD

Overview

A complete overview of amino acids, proteins, protein hydrolysates, or bioactive peptides and their role in glucose management is presented in this chapter. The research in this field is emerging rapidly. Based on the human clinical trials, it is clearly demonstrated that proteins, amino acids, protein hydrolysates, or bioactive peptides are playing an active role independently and/or synergistically with each other in managing glucose levels directly or indirectly. Some of the possible mechanisms are: increasing insulin production, stimulating incretins, increased skeletal muscle mass, and regulating satiety.

From safety point of view amino acids, proteins, protein hydrolysates, or bioactive peptides exist in the food chain, and there are no known toxicological effects if taken in moderation. Further research and randomized control clinical studies in humans are needed to establish documented evidence of their beneficial effects in managing glucose levels especially for type 2 diabetic patients.

Introduction

Type 2 diabetes mellitus is a widespread disease that seriously impacts quality of life and life expectancy. Due to the ever-increasing obesity epidemic and increased life expectancy, the number of diabetes patients is increasing at an alarming rate. In a recent publication from the Framingham Heart Study (Fox et al. 2006), a doubling of the incidence of type 2 diabetes over the past 30 years was reported, and a similar rise is expected to take place in the next 25 years (Wild et al. 2004). Type 2 diabetes is a multi factorial disease that is associated with both microvascular and macrovascular complications (Anonymous 1998).

In nondiabetic individuals, blood glucose levels are tightly regulated (5–7 mM) despite variations in nutrient availability and glucose disposal rates throughout the day. When blood glucose levels rise, β-cells in the pancreas secrete insulin, promoting glucose uptake in peripheral tissues, and normalizing blood glucose levels. In diabetes patients, however, this regulation is impaired, leading to excessively high blood glucose levels, especially in the postprandial state. Both genetic and environmental factors (most notably obesity and physical inactivity) contribute to the etiology of type 2 diabetes. Pathophysiologically, two important defects seem to be associated with the dysregulation of blood glucose levels, namely, inadequate insulin production as a result of β-cell dysfunction (Poitout and Robertson 1996), and/or peripheral insulin resistance (DeFronzo 2004).

Both in the early and later stages of type 2 diabetes, lifestyle changes are considered the first line of action to halt or delay further progression of the disease. Dietary advice represents a key element in improving lifestyle and generally aims to limit (saturated) fat intake, restrict energy intake, and stimulate energy expenditure with the intention to reduce fat mass and lower the prevalence of obesity which has been shown to be associated with insulin resistance and the development into type 2 diabetes (American Diabetes Association Task Force for Writing Nutrition Principles and Recommendations for the Management of Diabetes and Related Complications 2002; International Diabetes Federation 2007). In addition to that approach, evidence is accumulating that some food elements may exert a specific beneficial effect beyond the traditional nutritional adequacy (Riccardi et al. 2005). There are several food ingredients that are known to be beneficial for diabetes patients. Some of these are covered in other chapters. Here we specifically address amino acids, proteins, protein hydrolysates or bioactive peptides, and the combination thereof in diabetes management.

Protein, Protein Hydrolysates or Bioactive Peptides, and Amino Acids: Their Impact on Insulin Secretion and Blood Glucose Management

Until the 1960s, the only known dietary components to stimulate insulin secretion were carbohydrates. Only with the advent of new analytical techniques, it was made possible to study the regulation of insulin secretion in more detail. As a result, a variety of pharmacological agents and natural food ingredients were described to affect endogenous insulin secretion. The possibility that substances other than glucose could stimulate insulin secretion was first reported by Cochrane et al. (1956). They observed that casein ingestion could induce acute hypoglycemia in children with familial idiopathic hypoglycemia. Cochrane et al. described a child who was presented at the hospital with convulsions after having been switched from human breast milk to cow milk. Two more cases include the father of the above and an unrelated 5-month-old baby also presented with convulsions that had already started when it was breast fed. The convulsions were attributed to episodes of hypoglycemia. The observation that some of the patients (first two cases) had more fits when on a high-protein diet, prompted the authors to investigate their blood sugar after feeding a test dose of casein. The result established a clear causative relationship between protein feeding and the incidence of hypoglycemia.

Protein As an Insulin Secretagogue in Healthy Subjects

It has long been known that protein ingestion does not raise plasma glucose levels in normoglycemic subjects or type 2 diabetes patients (Nuttall and Gannon 1990). This is an interesting finding in view of the fact that 50–70% of the ingested protein is believed to be de aminated and converted to glucose via gluconeogenesis (Conn and Newburgh 1936b; Nuttall and Gannon 1990). It is assumed that protein ingestion does not increase glucose concentrations because it also stimulates endogenous insulin release. One of the first studies to address the effects of protein on insulin and glucose concentration was published by Rabinowitz et al. in 1966 (Rabinowitz et al. 1966). In four female, normoglycemic volunteers, the investigators compared changes in plasma insulin and glucose concentration after ingesting either 100 g glucose or a protein load (450 g beef steak), or a combination of both. As expected, ingestion of glucose led to a rise in plasma insulin concentration. A small but significant rise in plasma insulin levels was observed after meat ingestion, and the combined ingestion of meat and glucose led to a substantially greater postprandial rise in plasma insulin concentrations, with a corresponding attenuation of the postprandial glucose response.

Rather than adding protein to a glucose load, Nuttall and co wokers varied the proportion of macronutrients in an equicaloric diet and followed insulin and glucose responses after the meal. Using this approach, the 12 h integrated insulin response was 74% greater following the high-protein diet (41 En% protein) when compared to a standard diet (11 En% protein). In accordance, the glucose response was significantly lower when using the high-protein diet (Nuttall et al. 1985).

Insulin Secretion and Glycemic Effects of Protein in Diabetes Patients

Obviously, studies investigating the effects of proteins, bioactive peptides, or amino acids co ingestion on postprandial insulin secretion have not been limited to normoglycemic subjects but have also been extended to diabetes populations. It has been suggested that the response to protein may be somewhat different in diabetes patients compared to healthy subjects. Two studies report that in diabetes patients, protein actually induces larger insulin responses when compared to normoglycemic subjects (Pallotta and Kennedy 1968) even though insulin responses to glucose were similar (Berger and Vongaraya 1966). However, Fajans has pointed out that this increased effect following protein ingestion may only be true for obese diabetes patients and that the insulinotropic response to protein in nonobese diabetes patients may actually be lower (Conn and Newburgh 1936a). Of course, it would be evident to assume that there are large differences in the absolute and relative insulinotropic response to protein co ingestion between diabetes patients based on the duration of the diabetes state and the presence or absence of compensatory hyperinsulinemia which is generally present during the early stages of type 2 diabetes. Nuttall et al. (1984) tested the insulinotropic and glycemic responses for 5 h following ingestion of 50 g protein, 50 g glucose, or the combination of both in type 2 diabetes patients. The insulin response following glucose ingestion was only modestly greater than after protein ingestion. The combined ingestion of both glucose and protein produced a synergetic stimulation of endogenous insulin release, with an insulinotropic response that was greater than the sum of the insulin response reported after ingestion of either glucose or protein only. The plasma glucose area under the curve was 34% lower when protein was co ingested. This study is also interesting because a dose–effect relationship was established using increasing amounts (10–50 g) of protein. Insulin and glucose responses increased in a linear fashion with the amount of protein ingested.

Insulinotropic Potential of Specific Amino Acids

In his study in children with familial idiopathic hypoglycemia, Cochrane (1956) also aimed to investigate whether the casein-induced hypoglycemia was attributed to a specific amino acid composition or specific amino acids. To that end, one subject was fed a test dose of gelatin, a protein with a very different amino acid composition from that of casein. Because a much smaller fall in the blood glucose concentration was reported following gelatin ingestion as opposed to casein, the authors concluded that some particular amino acid or combination of amino acids was likely responsible for the hypoglycemia. Tyrosine (almost absent in gelatin but abundant in casein) was examined first but proved to have no blood glucose lowering effect. Leucine was tried next and had a dramatic effect. In their paper (Cochrane 1956), the hypoglycemic response following protein co ingestion could be entirely accounted for by leucine. The relative importance of leucine has, however, been a matter of some debate as a study in healthy subjects reported that the protein-mediated stimulation of insulin release could not be attributed to the leucine content of a protein (Berger 1964). In line with these findings, Floyd et al. (1966b) noted that the magnitude of the increase in plasma insulin concentration observed after the ingestion of either ground beef or chicken liver in healthy subjects could not be entirely attributed to the leucine content of the meal. Based on these and other findings (Charlton and Nair 1998; Floyd 1964; Floyd et al. 1966b), Floyd and co workers concluded that amino acids other than leucine or amino acids in combination with leucine effectively stimulate insulin release. Thereafter, the authors aimed to assess the insulin response to the intravenous administration of single amino acids and combinations of amino acids with and without leucine in healthy adults (Floyd et al. 1966a). In this study, the intravenous administration of a 30 g mixture of 10 essential amino acids induced a prompt and large increase in plasma insulin level. Peak increases averaged 120 μU/mL and the mean sum of increments during the first hour averaged 312 μU/mL. When the same amount of leucine was administered in the absence of the other amino acids in the mixture, the mean sum of increments was only 32 μU/mL. Consequently, the authors concluded that the stimulation of insulin secretion following amino acid administration was not fully attributed to the amount of leucine contained in the amino acid mixture. Indeed, when tested separately, each of the amino acids tested in the study (except histidine) stimulated endogenous insulin release. Although the plasma amino nitrogen levels resulting from infusion of the individual amino acids were very similar, their insulin responses varied considerably, in order of decreasing insulin response arginine > lysine > leucine > phenylalanine > valine > methionine (threonine, isoleucine, and histidine were not tested individually).

Based on additional studies by the same group, a synergetic effect was suggested between pairs of amino acids. Their studies indicated that out of a number of combinations tested, the combined administration of arginine–leucine and arginine–phenylalanine resulted in the strongest insulinotropic effects (Floyd et al. 1970a).

van Loon et al. (2000a) built on the data from Floyd and investigated the insulinotropic response following the in vivo co ingestion of a variety of amino acid mixtures and/or protein hydrolysates with carbohydrate . To define an amino acid and protein (hydrolysate) mixture with an optimal insulinotropic effect, insulin responses were assessed for a 2-h period after ingestion of different beverage compositions in fasted healthy, lean males. Subjects ingested 0.8/kg/h carbohydrate and 0.4/kg/h of an amino acid and/or protein (hydrolysate) mixture. Whereas Floyd observed intravenous administration of arginine to be a particularly good strategy to enhance insulin plasma levels, van Loon et al. observed that the oral route of administration disqualified the use of arginine due to gastrointestinal side effects as an effective dietary strategy to augment endogenous insulin release. Addressing combinations of different amino acids, the authors reported a mixture of free leucine, phenylalanine, and arginine to produce a large insulinotropic effect (100% greater than with the intake of carbohydrate only). The addition of leucine and phenylalanine to a protein hydrolysate produced a similar insulinotropic effect, without gastrointestinal discomfort.

The Importance of the Presence of Glucose for Eliciting Amino Acid Induced Insulin Responses

In a series of studies (Fajans et al. 1963; Floyd et al. 1966a), Floyd et al. demonstrated that the elevations in blood glucose observed during the infusion of some amino acids could not be the major cause of the insulin release that was observed. There was not a strong correlation between increase in blood glucose and plasma insulin concentrations. However, some of the infusions of individual amino acids and mixtures of amino acids that were accompanied by the larger and more consistent increase in blood glucose level were also the more potent in stimulating insulin release. The latter suggested that the insulin-releasing effect of certain amino acids might be potentiated by the concomitant elevation of blood glucose concentration. Therefore, the authors determined if certain amino acids might act synergistically with glucose to stimulate endogenous insulin release. Three essential amino acids (arginine, leucine, and histidine) and glucose were each administered intravenously. The effects of the separate amino acid and glucose infusions were compared with the effects following the infusion of each of the amino acids combined with glucose. The results demonstrated synergistic effects of glucose and each of the amino acids. The synergy was strongest in the case of arginine + glucose, and weakest in the case of histidine + glucose (Floyd et al. 1970b). The data on orally administered glucose and protein from Rabinowitz et al. (Rabinowitz et al. 1966) support such a synergistic effect of protein and glucose on insulin secretion. Synergism was also reported in mild untreated type 2 diabetes patients given protein and glucose in a meal (Berger and Vongaraya 1966; Nuttall et al. 1984). To address this issue further, Krezowski et al. (1986) studied the effects of the ingestion of either 50 g glucose, 50 g protein (provided as a cooked lean hamburger), or the combination of both on the plasma insulin and glucose response in normoglycemic subjects. The resulting area under the insulin curve measured for 4 h using the fasting value as baseline was considerably smaller following the ingestion of protein than following ingestion of glucose, which is in line with other data (Berger and Vongaraya 1966; Rabinowitz et al. 1966). No synergistic effect of protein and glucose ingestion could be confirmed, as the sum of the areas following glucose and protein ingestion together were 100% of that after the combined intake. However, since lean hamburger contains carbohydrate, these data do not necessarily contradict the results studies that do report a synergistic effect of protein and carbohydrate on insulin secretion. Indeed, in an experiment in which the amounts of individual macronutrients were strictly controlled, Pallotta and Kennedy did report synergy. Following an overnight fast, 10 male, nondiabetic subjects received 100 g glucose, or 60 g protein, or 700 g starch, or combinations of these. Blood glucose and insulin levels were measured for up to 6 h. The insulin response to the combined intake of protein and glucose was greater than the sum of the responses to the ingestion of protein or glucose alone. These responses were also reflected in the blood glucose curves, with the peak glucose concentration being substantially lower (and returning to baseline faster) following ingestion of both glucose and protein when compared to the ingestion of glucose only (Pallotta and Kennedy 1968). Therefore, most data suggest that although protein administration can induce some insulin release, the secretory effects of protein on insulin are more potent in the presence of dietary carbohydrate.

Protein Hydrolysates or Bioactive Peptides for Blood Glucose Management

By definition, protein hydrolysates or bioactive peptides are the resulting products of the hydrolysis of proteins and this can be achieved by enzymes, acid or alkali, and fermentation process. The degree to what extent a protein is hydrolyzed is measured by several analytical techniques and is commonly reported as a percent of amino nitrogen over total nitrogen (AN/TN) or degree of hydrolysis (DH). Acid hydrolysis of proteins results in pretty much individual amino acids and minimum amount of smaller peptides and is widely used in the manufacturing of hydrolyzed vegetable proteins/hydrolyzed acid proteins as flavor enhancers. Alkaline hydrolysis is used seldom in the industry; one of the specific functions of alkaline hydrolysis is that it provides functional properties such as foaming. Enzyme hydrolysis is relatively mild and offers several advantages including controlled processing and ease of obtaining a desired DH. For this reason majority of the protein hydrolysates are manufactured via enzyme hydrolysis (endo as well as exoproteases and peptidases) to obtain optimal breakdown of the entire protein chain. The choice of a proteolytic enzyme or enzymes used in the hydrolysis of proteins is crucial in determining the characteristics of the protein hydrolysates because each enzyme has its own unique action site to break down large chains of proteins into smaller peptides and amino acids.

In a very broad sense the protein hydrolysates manufactured by the food industry may be classified based on their DH.

Protein hydrolysates with a low DH typically lower than 10% are used to improve the functional properties like solubility, emulsifying capacity, foaming, etc. The applications of these products are found in bakery products, ice cream, mayonnaise, and meat products.

Protein hydrolysates with a variable DH typically greater than 20–40% are used for enhancing the taste profile. The applications of these products are found in soups, meat stocks, sauces, meats, and precooked meals.

Protein hydrolysates with a high DH typically greater than 40% are used in medical diets and nutritional supplements. The applications of these products are found in the formulas to treat phenylketonuria, tyrosinemia, and colic babies. Interestingly, smaller peptides ranging from 2 to 10 are also finding their way into nutraceuticals. For example, ACE inhibitors and sometimes these are referred as bioactive peptides.

As the name suggests, bioactive peptides exert specific physiological function and may ultimately influence the health positively. Typically, bioactive peptides contain two–five amino acids but can range in tens of amino acids. After controlled liberation of bioactive peptides from proteins by enzymatic hydrolysis, they are separated by using centrifugation, filtration, and chromatography. Depending on the amino acid sequence, these bioactive peptides may exert a number of different physiological reactions in vivo, affecting the cardiovascular, endocrine, immune, and nervous systems in addition to nutrient utilization. Sometimes bioactive peptides are interchangeably used with protein hydrolysates and vice versa. Technically bioactive peptides are pure peptides, whereas protein hydrolysates may contain a mixture of peptides and amino acids.

Protein hydrolysates offer a very suitable delivery form of peptides and amino acids. They can be considered predigested proteins, and it has been reported that the gastrointestinal transit time of a protein hydrolysate is shorter than that of its native protein (Mihatsch et al. 2001). Also, faster restoration of plasma amino acid levels after intake of protein hydrolysates as compared to intact proteins has been reported during recovery of surgery in patients (Ziegler et al. 1990). Although protein hydrolysates are often associated with poor taste, new technologies can prevent that drawback and the resulting hydrolysates taste better than a mixture of amino acids they are composed of. It is not surprising, therefore, that protein hydrolysates have been tested for their effects on insulin secretion and blood glucose management. As reported earlier, van Loon et al. (2000a, b) demonstrated previously that the in vivo insulinotropic potential of the combined ingestion of carbohydrate with protein hydrolysate can be enhanced by the addition of free leucine and phenylalanine in normoglycemic subjects. Following up on these results, they compared the insulinotropic response to the ingestion of a protein hydrolysate/amino acid mixture in both long-standing type 2 diabetes patients and matched normoglycemic controls. Subjects received beverages providing 0.7 g/kg/h carbohydrate (50% glucose and 50% maltodextrin) with or without an additional 0.35 g/kg/h of an amino acid/protein hydrolysate mixture consisting of a wheat protein hydrolysate, free leucine, and free phenylalanine (2:1:1 by weight). Plasma insulin and glucose responses were assessed over a 2-h period. The long-standing type 2 diabetes patients in this study were not yet on exogenous insulin treatment but no longer showed any compensatory hyperinsulinemia. In accordance, the plasma insulin response following carbohydrate ingestion was less than half in the diabetes patients when compared to their normoglycemic controls. However, the insulin response in these patients could be increased approximately threefold simply by co ingesting the amino acid/protein mixture. The additional increase in insulin response following co ingestion of the amino acid/protein hydrolysate mixture was similar between groups, which seems to suggest that even in long-standing type 2 diabetic patients, insulin secretion in response to stimuli other than glucose is not impaired (van Loon et al. 2003). In a follow-up study, the authors applied a casein hydrolysate with leucine and phenylalanine to increase insulin release in type 2 diabetes patients and normoglycemic controls and used stable isotope tracer methodology to show that the greater insulin release accelerated blood glucose disposal and, as such, attenuated the postprandial rise in blood glucose concentrations (Manders et al. 2005). In both studies, a large amount of protein hydrolysate/amino acid mixture and carbohydrate was ingested repeatedly for methodological considerations. A more relevant approach in terms of a potential therapeutic point of view would be to provide such an insulinotropic protein hydrolysate and/or amino acid mixture as a single meal like bolus. Therefore, a follow-up study compared the effects of the co ingestion of a casein hydrolysate with and without additional leucine (no phenylalanine was added in any condition) on the postprandial blood glucose and insulin response in type 2 diabetes patients and healthy controls. Co ingestion of the casein hydrolysate with carbohydrate strongly increased insulin secretion, and a trend was observed for an even greater insulin response following additional leucine co ingestion. In accordance, postprandial glucose responses were lower following co ingestion of the casein hydrolysate with or without added leucine when compared to the ingestion of carbohydrate only (Manders et al. 2006a).

In the latest study in this series, the authors assessed the impact of protein hydrolysate/amino acid co ingestion with each main meal on daily glycemic control by using continuous glucose monitoring in type 2 diabetes patients (Manders et al. 2006b). Total 24 h glycemic responses were recorded under strict dietary standardization but otherwise free-living conditions while subjects were kept on their usual medication. In one condition, subjects were provided with a standardized diet and in the other subjects received the same standardized diet with the addition of a casein hydrolysate with additional free leucine being ingested after each main meal. In the latter trial, the prevalence of postprandial hyperglycemia was shown to be reduced by 25% and mean glucose levels were ~10% lower observed over the entire 24 h period. These data indicate that protein hydrolysate/amino acid co ingestion forms an effective and feasible nutritional strategy to improve daily blood glucose homeostasis in type 2 diabetes patients.

Mechanisms of Action

Several mechanisms of action have been demonstrated or suggested that could explain the impact of food-derived proteins and/or peptides on blood glucose management. Amino acids from proteins and peptides are known to directly stimulate β-cell insulin secretion. In addition, the combined increase of plasma insulin and amino acids concentrations in blood have been associated with net muscle protein anabolism. Specific amino acids, like leucine, have also been reported to activate the mTOR pathway through insulin-independent mechanisms. Furthermore, some data support the notion that incretins mediate the insulin secretory response following protein (hydrolysate) ingestion. Each of these mechanisms will be addressed in more detail below.

Amino Acid-Induced β-Cell Insulin Secretion

The mechanisms that could explain amino acid-induced insulin secretion in the β-cell have recently been reviewed by Newsholme et al. (2006) and van Loon (2007). Various mechanisms that could explain the impact of amino acids on β-cell insulin secretion are illustrated in Figure 17.1. The way in which arginine elicits insulin secretion is probably the most straightforward of all insulin

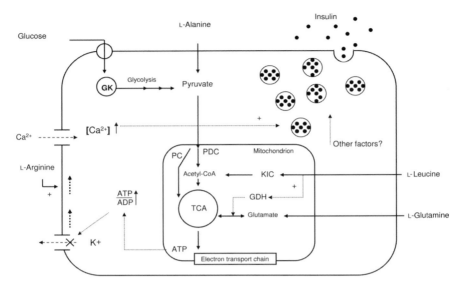

Figure 17.1.   A simplified overview on some of the proposed mechanisms by which amino acids stimulate insulin secretion in the pancreatic β-cell. Glucose entering the cell is phoshorylated by glucokinase (GK). Pyruvate is formed by glycolysis, after which it becomes a substrate for pyruvate dehydrogenase complex (PDC) or pyruvate carboxylase (PC) before entering the TCA cycle. Increased TCA cycle activity and oxidative phosphorylation will result in an increased ATP/ADP ratio, which will lead to the closing of ATP-sensitive K^+ channels. The latter will lead to the depolarization of the plasma membrane, thereby opening up voltage activated Ca^{2+} channels, resulting in Ca^{2+} activated insulin exocytosis. Arginine has been reported to be able to directly depolarize the plasma membrane. Metabolizable amino acids are catabolysed to generate ATP and, as such, to increase intracellular ATP/ADP ratios and activate insulin exocytosis. Numerous interactions and co-factors are evident in the complex regulation of amino acid induced insulin secretion. For example, leucine-induced insulin secretion is mediated by its oxidative decarboxylation as well as by allosterically activating glutamate dehydrogenase (GDH). Figure based on Newsholme et al. (2006) and derived from van Loon et al. (2007).

stimulating amino acids: due to its positive charge at a physiologic pH, arginine induces depolarization of the β-cell (Blachier et al. 1989a, b). This in turn opens voltage-dependent Ca^{2+} channels, leading to the influx of Ca^{2+}, which prompts the β-cells to release insulin. In addition, metabolizable and nonmetabolizable amino acids co-transported with Na^+ act synergistically with membrane depolarization to exert or enhance their actions on $[Ca^{2+}]_i$ and subsequent stimulation of insulin secretion (McClenaghan et al. 1998). Finally, metabolizable amino acids can be taken up in the β-cell's TCA cycle, which results in an increased ATP/ADP ratio in the cell. The increased ratio closes ATP-dependent potassium (K_{ATP}) channels, promoting membrane depolarization (Brennan et al. 2002; Dunne et al. 1990; New-shome et al. 2006) and subsequent events that ultimately lead to insulin secretion as outlined above. Interestingly, this mechanism of amino acid induced insulin secretion is similar to the pharmaceutical class of insulin secretagogues, sulfonyl urea derivatives, which also close K_{ATP} channels (Rosak 2002). For leucine, an additional mechanism to the one mentioned above is through its stimulatory effect on glutamate dehydrogenase (Newshome et al. 2006; Sener and Malaisse 1980; Xu et al. 2001). The latter effect (Figure 17.1) has also been suggested for pheny-lalanine (Kofod et al. 1986).

Amino Acid-Induced Muscle Protein Synthesis

The loss of skeletal muscle mass with aging is proportionally related to the loss of blood glucose disposal capacity. The latter is attributed to the fact that muscle tissue is responsible for the majority of glucose taken up from the blood. Therefore, aging and the concomitant loss of muscle mass are a risk factor for the development of insulin resistance and/or type 2 diabetes. Moreover, the loss of muscle mass and strength seems to be even further accelerated in the type 2 diabetes state (Park et al. 2007). Therefore, dietary strategies to improve glucose homeostasis in type 2 diabetes should also focus on the impact on muscle protein anabolism.

It has been firmly established that amino acid and/or protein ingestion stim-ulate muscle protein synthesis and inhibit proteolysis (Biolo et al. 1995, 1997, 1999; Rasmussen et al. 2000; Volpi et al. 1998). The presence of insulin seems to be essential in this process. In accordance, hyperinsulinemia has been shown to promote net muscle protein accretion under conditions of hyperaminoacidemia (Fryburg et al. 1995; Gelfand and Barrett 1987; Hillier et al. 1998). However, recent studies indicate that insulin merely plays a permissive role in stimulating postprandial muscle protein synthesis in healthy subjects (Rennie et al. 2006). Nonetheless, it might be speculated that a greater postprandial insulin response might improve the net muscle protein anabolic response to food intake in insulin resistant and/or type 2 diabetes patients. In support, it has been observed that the anabolic response to increased insulin and energy availability is maintained in the type 2 diabetes state. Consequently co ingestion of insulinotropic amino acids and/or protein (hydrolysates) might improve the protein anabolic response to food intake in the insulin resistant or type 2 diabetes state. More studies are warranted to address the potential impairments in muscle protein metabolism in type 2 diabetes and to define nutritional strategies that can effectively compensate for the pro-posed impairments. The latter includes more research investigating the properties of specific amino acids to act as potent signaling molecules in muscle tissue.

Amino Acids as Signaling Molecules

Amino acids do not merely function as precursors for the de novo protein synthesis, but can also play key roles as nutritional signaling molecules regulating multiple cellular processes. Changes in amino acid availability can activate signal trans-duction pathways, resulting in the up- or down-regulation of mRNA translation initiation. Mammalian target of rapamycin (mTOR) is a key component of such a pathway. It plays an important role in regulating protein synthesis in response to nutritional conditions, and leucine is one of its potent activators. It has been suggested that the actions of amino acids, especially leucine, on β-cell insulin secretion are mediated by the same secondary signals that also activate mTOR sig-naling (McDaniel et al. 2002; Xu et al. 2001). Interestingly, the activation of mTOR by leucine may also affect β-cell function through improved maintenance of β-cell mass (Xu et al. 2001). Stimulation of the mTOR pathway by leucine has been sug-gested to be responsible for the greater muscle protein synthetic response to food intake as has recently been observed following either protein (Rieu et al. 2006) or amino acid ingestion (Katsanos et al. 2006) in the elderly. Such nutritional inter-ventions might prove to be clinically relevant to maintain or even increase muscle mass in the elderly and/or type 2 diabetes patients, especially considering the fact that uncontrolled diabetes is often accompanied by a negative skeletal muscle pro-tein balance (Charlton and Nair 1998). Since muscle mass is vital for maintenance of blood glucose disposal capacity, the gradual decline in skeletal muscle mass with aging has been suggested to be an important factor contributing to the preva-lence of insulin resistance (Dela and Kjaer 2006). More research is warranted to develop effective nutritional interventions that can contribute to attenuate or even counteract the loss of muscle mass in type 2 diabetes patients. It seems likely that specific proteins, protein hydrolysates or bioactive peptides, and/or amino acids will be defined and/or developed for these purposes.

Incretins

It has been demonstrated that proteins and protein hydrolysates or bioactive peptides promote the release of incretins, and there are indications that protein hydrolysates are more potent in doing so than their corresponding intact proteins or individual amino acids. Incretins (e.g., glucagon-like peptide-1 (GLP-1) and gastric inhibitory peptide (GIP)) are signaling molecules from the gut that prime the pancreas to produce insulin in response to elevated plasma glucose (Meier and Nauck 2006). Incretin signaling has increasingly received attention as a therapeutic target (Gautier et al. 2005), especially because it could provide a way to stimulate insulin release without the chance of hypoglycemia that is associated with existing insulin secretory therapies (Burcelin 2005; Jennings et al. 1989). The implications of incretins in protein-mediated reduction in blood glucose levels have recently been demonstrated in a real-life setting. Diabetic volunteers received breakfast and lunch, both of which were supplemented with whey on one day and lean ham on the other treatment day. Blood glucose responses were significantly reduced after lunch when whey was added compared to ham, and this effect was accom-panied by greater postprandial GIP (but not GLP-1) levels (Frid et al. 2005). In another study by the same group, the effect of common dietary sources of animal or vegetable proteins on concentrations of postprandial levels of incretin hormones in healthy subjects was evaluated. A correlation was found between postprandial insulin responses and early increments in plasma amino acids; the strongest correlations were seen for leucine, valine, lysine, and isoleucine. A correlation was also obtained between responses of insulin and GIP concentrations. The authors concluded that food proteins differ in their capacity to stimulate insulin release, possibly in part by differently affecting the early release of incretin hormones (Nilsson et al. 2004). Interestingly, in another study, the effects of intact whey protein on postprandial insulin, glucose, and incretin release were compared to the effects of a mixture of amino acids. Both interventions produced similar glycemic effects, but these were paralleled by an increase in plasma GIP levels in the hydrolysate intervention only. It has been suggested that protein hydrolysates may be more prominent activators of GIP release than their corresponding intact proteins. In most studies on protein or peptide mediated incretin release, the effect on GIP release appears to be more prominent than the effect of GLP-1 release. The clinical relevance of these findings remains to be established, since diabetes patients are suggested to lose their sensitivity towards GIP but not GLP-1 (Burcelin 2005). Besides stimulating the incretin response as such, inhibition of the breakdown of incretins is another way to improve glycemic control. Inhibitors of DPP-IV, the primary enzyme responsible for breakdown of incretins, form an emerging class of antidiabetic medication (Langley et al. 2007). Several studies indicate that hydrolysates (Rouanet et al. 1990) as well as single or combined oligopeptides from food compounds (Augustyns et al. 2005; Berger et al. 1987; Harada et al. 1982; Piazza et al. 1989) display DPP-IV inhibitory effects as well.

Safety Aspects of Protein, Bioactive Peptides, or Protein Hydrolysates Administration

Obviously, functional foods for any condition, including diabetes, should be absolutely safe. A couple of potential safety issues therefore deserve some attention.

Hypoglycemia

Adverse effects of several types of blood glucose-lowering medication include hypoglycemia in case insulin action outweighs glycemic load (Jennings et al. 1989). Since some of the mechanisms that could explain the insulinotropic action of protein (hydrolysate), peptides, or amino acid co ingestion are comparable to many oral blood glucose lowering medications, it could be speculated that there could be some risk of hypoglycemia when co ingesting these food compounds for glucose management. However, hypoglycemia following administration of protein, peptides, or amino acids has never been observed to our knowledge, except in the familial idiopathic hypoglycemia cases (Cochrane et al. 1956) discussed earlier. Of course, insulinotropic protein and/or amino acid mixtures should be ingested in combination with carbohydrate to allow the augmented postprandial insulin response, therefore co ingestion of these food compounds in type 2 diabetes patients can only attenuate the rise in postprandial blood glucose concentration following food intake. Furthermore, the absence of any hypoglycemia following amino acid induced insulin secreting might also be associated with the concomitant stimulation of incretin release, since incretins exert their hypoglycemic action only in the presence of hyperglycemia (Burcelin 2005).

Insulin Resistance

Although the co ingestion of protein, protein hydrolysates, leucine, and/or other amino acids has been shown to stimulate endogenous insulin secretion, there are also reports showing leucine or amino acid administration to impair insulin stimulated glucose uptake (Krebs 2005). Continuous intravenous amino acid infusion has been shown to reduce insulin stimulated glucose uptake in vivo in human skeletal muscle tissue (Abumrad et al. 1982; Flakoll et al. 1992; Krebs et al. 2002; Krebs 2005; Pisters et al. 1991; Tessari et al. 1985). These amino acid induced effects on glucose metabolism are likely attributed to the downstream effects of amino acid induced activation of the mTOR signaling pathway (Krebs et al. 2002; Krebs 2005; Tremblay et al. 2005; Tremblay and Marette 2001). A simplified overview of the proposed mTOR/P70S6 kinase mediated negative feedback loop within the insulin signaling cascade (Tremblay et al. 2005) is provided in Figure 17.2. In the context of the application of insulinoptropic and/or anabolic amino acids as pharmaco-nutrients in the treatment of type 2 diabetes (van Loon 2007), it is important to consider these reports on amino acid induced insulin resistance. The global increase in diabetes prevalence is strongly associated with our Western lifestyle, in which overfeeding and obesity are instrumental to the development of insulin resistance, leading to type 2 diabetes. With the continuous intravenous infusion of fatty acids (Dresner et al. 1999; Itani et al. 2002; Roden et al. 1996) or amino acids (Abumrad et al. 1982; Krebs et al. 2002; Pisters et al. 1991; Tessari et al. 1985) as a model for overfeeding, it has become clear that there are various mechanisms by which prolonged nutrient oversupply will induce insulin resistance in skeletal muscle, liver, and/or adipose tissue. Excess provision of any of the macronutrients (leading to hyperlipidemia, hyperglycemia, and/or hyperaminoacidemia) will induce a (temporary) state of insulin resistance. Therefore, it seems fair to assume that, within the context of an energy-balanced diet, merely the supplementation with protein and/or specific amino acids is unlikely to induce insulin resistance (Figure 17.2).

Protein Load

Nephropathy is a relatively common complication in diabetes. Given that relatively high protein or peptide loads have been used in most studies demonstrating

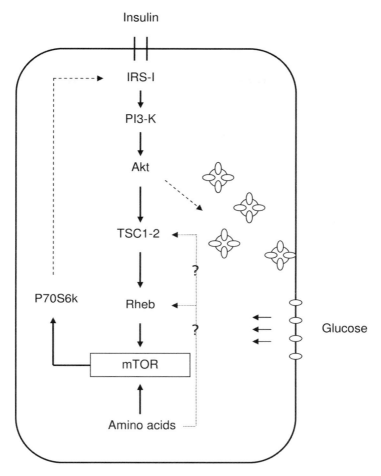

Figure 17.2. An overview on the proposed mechanism by which amino acids might impair insulin signaling and reduce insulin stimulated glucose uptake through activating the mTOR signal transduction pathway. Insulin binds to its receptor, leading to tyrosine phosphorylation of the insulin receptor substrate 1 (IRS-I). This activates phospho-inositide 3-kinase (PI3-kinase). The latter will lead to the activation of Akt (protein kinase B), thereby promoting translocation of the glucose transporter 4 (GLUT-4) to the sarcolemma. A negative feedback inhibition has been proposed through the subsequent phosphorylation of the tuberous sclerosis 1 and 2 complex (TSC1-2), which through the Ras homolog enriched in brain (Rheb) can activate mTOR. As amino acids like leucine can activate mTOR (possibly acting through TSC1-2 or Rheb) this could likely represent the mechanism by which amino acids could interact with the insulin signaling cascade. It has been proposed that (continuous) overactivation of mTOR/P70S6kinase can lead to an increased phosphorylation of IRS-I on its serine residues, thereby inhibiting PI3-kinase and inducing insulin resistance. Figure based on Tremblay et al. (2005) and derived from van Loon et al. (2007).

improved postprandial blood glucose homeostasis following protein and/or amino acid ingestion, it is important to assess the risk associated with high protein intake in the context of kidney function in type 2 diabetes patients. The American Diabetes Association (ADA) consensus prescribes a protein intake equivalent to the Recommended Dietary Allowance (RDA) of 0.8 g/kg/d ($\sim$10% of daily calories) in the adult patient with overt nephropathy (Franz et al. 2004). Although it has been suggested that once the glomerular filtration rate (GFR) begins to fall, protein restriction to 0.6 g/kg/d may prove useful in slowing the decline in GFR in selected patients, the current ADA recommendations do not describe a protein intake lower than 0.8 g/kg/d, even in later stages of kidney disease (Molitch et al. 2004). The Harvard Medical School-affiliated Joslin Diabetes Center has advocated the use of 20–30 En% protein on a daily basis (Rosenzweig et al. 2006). For now, it seems reasonable to assume that protein (hydrolysate) or peptide interventions should fit within the daily recommended amount of dietary protein. It may, therefore, be useful to assess the impact of co ingesting lower amounts of insulinotropic protein and/or amino acids mixtures than those amounts applied in previous nutritional intervention studies.

Conclusions

An increasing body of evidence shows that protein, protein hydrolysates or bioactive peptides, and/or amino acids can be used in effective nutritional strategies to improve postprandial blood glucose homeostasis in type 2 diabetes patients. Protein, protein hydrolysate, and/or amino acid co ingestion with carbohydrate can strongly enhance endogenous insulin secretion, accelerate blood glucose disposal, and, as such, attenuate the postprandial rise in blood glucose concentration in type 2 diabetes patients. Furthermore, protein hydrolysates or bioactive peptides that combine an ample supply of amino acids with strong insulinotropic properties can be used to stimulate the protein anabolic response to food intake in type 2 diabetes patients. The latter could likely contribute to dietary interventions aiming to attenuate the gradual decline in skeletal muscle mass with aging and, as such, prevent a further decline in blood glucose disposal capacity.

The biggest challenge facing by the peptide manufacturers is to identify and characterize the metabolic role of bioactive peptides and to further prove their efficacy in clinical trials. At the same time the challenge to the food industry is to incorporate these bioactive peptides without hurting the taste/sensory profile, convenience, and safety. This scenario will change gradually as the manufacturers of bioactive peptides are going to great lengths in getting a user-friendly label, performing research/clinical studies, applications, and concept ideas to use in functional foods.

The use of bioactive peptides for health benefits of diabetic patients will increase only when there is a collaborative effort by the peptide manufacturers, food industry, health insurance companies, and government to educate and promote the health benefits of bioactive peptides in functional foods. This will in the long run improve the health and reduce the burden on failing health of individuals, insurance companies, and government.

References

Abumrad NN, Robinson RP, Gooch BR, Lacy WW. 1982. The effect of leucine infusion on substrate flux across the human forearm. J Surg Res 32(5):453–463.

American Diabetes Association Task Force for Writing Nutrition Principles and Recommendations for the Management of Diabetes and Related Complications. 2002. American diabetes association position statement: Evidence-based nutrition principles and recommendations for the treatment and prevention of diabetes and related complications. J Am Diet Assoc 102(1):109–118.

Augustyns K, Van Der Veken P, Senten K, Haemers A. 2005. The therapeutic potential of inhibitors of dipeptidyl peptidase IV (DPP IV) and related proline-specific dipeptidyl aminopeptidases. Curr Med Chem 12(8):971–998.

Berger S. 1964. Effect of protein ingestion on plasma insulin concentration. J Lab Clin Med 64:842.

Berger E, Fischer G, Neubert K, Barth A. 1987. Stepwise degradation of the hexapeptide met-ala-ser-pro-phe-ala by dipeptidyl peptidase IV. Biomed Biochim Acta 46(10):671–676.

Berger S, Vongaraya N. 1966. Insulin response to ingested protein in diabetes. Diabetes 15(5):303–306.

Biolo G, Declan Fleming RY, Wolfe RR. 1995. Physiologic hyperinsulinemia stimulates protein synthesis and enhances transport of selected amino acids in human skeletal muscle. J Clin Invest 95(2):811–819.

Biolo G, Tipton KD, Klein S, Wolfe RR. 1997. An abundant supply of amino acids enhances the metabolic effect of exercise on muscle protein. Am J Physiol 273(1, Pt 1):E122–E129.

Biolo G, Williams BD, Fleming RY, Wolfe RR. 1999. Insulin action on muscle protein kinetics and amino acid transport during recovery after resistance exercise. Diabetes 48(5):949–957.

Blachier F, Leclercq-Meyer V, Marchand J, Woussen-Colle MC, Mathias PC, Sener A, Malaisse WJ. 1989b. Stimulus-secretion coupling of arginine-induced insulin release. Functional response of islets to L-arginine and L-ornithine. Biochim Biophys Acta 1013(2):144–151.

Blachier F, Mourtada A, Sener A, Malaisse WJ. 1989a. Stimulus-secretion coupling of arginine-induced insulin release. Uptake of metabolized and nonmetabolized cationic amino acids by pancreatic islets. Endocrinology 124(1):134–141.

Brennan L, Shine A, Hewage C, Malthouse JP, Brindle KM, McClenaghan N, Flatt PR, Newsholme P. 2002. A nuclear magnetic resonance-based demonstration of substantial oxidative L-alanine metabolism and L-alanine-enhanced glucose metabolism in a clonal pancreatic beta-cell line: Metabolism of L-alanine is important to the regulation of insulin secretion. Diabetes 51(6):1714–1721.

Burcelin R. 2005. The incretins: A link between nutrients and well-being. Br J Nutr 93(1):S147–S156.

Charlton M and Nair KS. 1998. Protein metabolism in insulin-dependent diabetes mellitus. J Nutr 128(2):323S–327S.

Cochrane WA, Payne WW, Simpkiss MJ, Woolf LI. 1956. Familial hypoglycemia precipitated by amino acids. J Clin Invest 35(4):411–422.

Conn JW and Newburgh LH. 1936a. The glycemic response to isoglucogenic quantities of protein and carbohydrate. J Clin Invest 15(6):665–671.

Conn JW and Newburgh LH. 1936b. The glycemic response to isoglucogenic quantities of protein and carbohydrate. J Clin Invest 15(6):665–671.

DeFronzo RA. 2004. Pathogenesis of type 2 diabetes mellitus. Med Clin North Am 88(4):787,835, ix.

Dela F and Kjaer M. 2006. Resistance training, insulin sensitivity and muscle function in the elderly. Essays Biochem 42:75–88.

Dresner A, Laurent D, Marcucci M, Griffin ME, Dufour S, Cline GW, Slezak LA, Andersen DK, Hundal RS, Rothman DL, et al. 1999. Effects of free fatty acids on glucose transport and IRS-1-associated phosphatidylinositol 3-kinase activity. J Clin Invest 103(2):253–259.

Dunne MJ, Yule DI, Gallacher DV, Petersen OH. 1990. Effects of alanine on insulin-secreting cells: Patch-clamp and single cell intracellular Ca2+ measurements. Biochim Biophys Acta 1055(2):157–164.

Fajans SS, Knopf RF, Floyd JC, Power L, Conn JW. 1963. The experimental induction in man of sensitivity to leucine hypoglycemia. J Clin Invest 42(2):216–229.

Flakoll PJ, Wentzel LS, Rice DE, Hill JO, Abumrad NN. 1992. Short-term regulation of insulin-mediated glucose utilization in four-day fasted human volunteers: Role of amino acid availability. Diabetologia 35(4):357–366.

Floyd JC. 1964. Postprandial hyperacidemia and insulin secretion, a physiological relationship. J Lab Clin Med 64:858.

Floyd JC, Jr, Fajans SS, Conn JW, Knopf RF, Rull J. 1966a. Stimulation of insulin secretion by amino acids. J Clin Invest 45(9):1487–1502.

Floyd JC, Jr, Fajans SS, Conn JW, Knopf RF, Rull J. 1966b. Insulin secretion in response to protein ingestion. J Clin Invest 45(9):1479–1486.

Floyd JC, Jr, Fajans SS, Pek S, Thiffault CA, Knopf RF, Conn JW. 1970a. Synergistic effect of certain amino acid pairs upon insulin secretion in man. Diabetes 19(2):102–108.

Floyd JC, Jr, Fajans SS, Pek S, Thiffault CA, Knopf RF, Conn JW. 1970b. Synergistic effect of essential amino acids and glucose upon insulin secretion in man. Diabetes 19(2):109–115.

Fox CS, Pencina MJ, Meigs JB, Vasan RS, Levitzky YS, D'Agostino RB S. 2006. Trends in the incidence of type 2 diabetes mellitus from the 1970s to the 1990s: The framingham heart study. Circulation 113(25):2914–2918.

Franz MJ, Bantle JP, Beebe CA, Brunzell JD, Chiasson JL, Garg A, Holzmeister LA, Hoogwerf B, Mayer-Davis E, Mooradian AD, et al. 2004. Nutrition principles and recommendations in diabetes. Diabetes Care 27(1):S36–S46.

Frid AH, Nilsson M, Holst JJ, Bjorck IM. 2005. Effect of whey on blood glucose and insulin responses to composite breakfast and lunch meals in type 2 diabetic subjects. Am J Clin Nutr 82(1):69–75.

Fryburg DA, Jahn LA, Hill SA, Oliveras DM, Barrett EJ. 1995. Insulin and insulin-like growth factor-I enhance human skeletal muscle protein anabolism during hyper-aminoacidemia by different mechanisms. J Clin Invest 96(4):1722–1729.

Gautier JF, Fetita S, Sobngwi E, Salaun-Martin C. 2005. Biological actions of the incretins GIP and GLP-1 and therapeutic perspectives in patients with type 2 diabetes. Diabetes Metab 31(3 Pt 1):233–242.

Gelfand RA and Barrett EJ. 1987. Effect of physiologic hyperinsulinemia on skeletal muscle protein synthesis and breakdown in man. J Clin Invest 80(1):1–6.

Harada M, Fukasawa KM, Fukasawa K, Nagatsu T. 1982. Inhibitory action of proline-containing peptides on xaa-pro-dipeptidylaminopeptidase. Biochim Biophys Acta 705(2):288–290.

Hillier TA, Fryburg DA, Jahn LA, Barrett EJ. 1998. Extreme hyperinsulinemia unmasks insulin's effect to stimulate protein synthesis in the human forearm. Am J Physiol 274(6 Pt 1):E1067–E1074.

International Diabetes Federation. 2007. http://www.idf.org/webdata/docs/GGT2D 05 Lifestyle management.pdf . Accessed April 2007

Itani SI, Ruderman NB, Schmieder F, Boden G. 2002. Lipid-induced insulin resistance in human muscle is associated with changes in diacylglycerol, protein kinase C, and IkappaB-alpha. Diabetes 51(7):2005–2011.

Jennings AM, Wilson RM, Ward JD. 1989. Symptomatic hypoglycemia in NIDDM patients treated with oral hypoglycemic agents. Diabetes Care 12(3):203–208.

Katsanos CS, Kobayashi H, Sheffield-Moore M, Aarsland A, Wolfe RR. 2006. A high proportion of leucine is required for optimal stimulation of the rate of muscle protein synthesis by essential amino acids in the elderly. Am J Physiol Endocrinol Metab 291(2):E381–E387.

Kofod H, Lernmark A, Hedeskov CJ. 1986. Potentiation of insulin release in response to amino acid methyl esters correlates to activation of islet glutamate dehydrogenase activity. Acta Physiol Scand 128(3):335–340.

Krebs M. 2005. Amino acid-dependent modulation of glucose metabolism in humans. Eur J Clin Invest 35(6):351–354.

Krebs M, Krssak M, Bernroider E, Anderwald C, Brehm A, Meyerspeer M, Nowotny P, Roth E, Waldhausl W, Roden M. 2002. Mechanism of amino acid-induced skeletal muscle insulin resistance in humans. Diabetes 51(3):599–605.

Krezowski PA, Nuttall FQ, Gannon MC, Bartosh NH. 1986. The effect of protein ingestion on the metabolic response to oral glucose in normal individuals. Am J Clin Nutr 44(6):847–856.

Langley AK, Suffoletta TJ, Jennings HR. 2007. Dipeptidyl peptidase IV inhibitors and the incretin system in type 2 diabetes mellitus. Pharmacotherapy 27(8):1163–1180.

Manders RJ, Koopman R, Sluijsmans WE, Van Den Berg R, Verbeek K, Saris WH, Wagenmakers AJ, van Loon LJ. 2006a. Co ingestion of a protein hydrolysate with or without additional leucine effectively reduces postprandial blood glucose excursions in type 2 diabetic men. J Nutr 136(5):1294–1299.

Manders RJ, Praet SF, Meex RC, Koopman R, de Roos AL, Wagenmakers AJ, Saris WH, van Loon LJ. 2006b. Protein hydrolysate/leucine co ingestion reduces the prevalence of hyperglycemia in type 2 diabetic patients. Diabetes Care 29(12):2721–2722.

Manders RJ, Wagenmakers AJ, Koopman R, Zorenc AH, Menheere PP, Schaper NC, Saris WH, van Loon LJ. 2005. Co ingestion of a protein hydrolysate and amino acid mixture with carbohydrate improves plasma glucose disposal in patients with type 2 diabetes. Am J Clin Nutr 82(1):76–83.

McClenaghan NH, Barnett CR, Flatt PR. 1998. Na+ cotransport by metabolizable and nonmetabolizable amino acids stimulates a glucose-regulated insulin-secretory response. Biochem Biophys Res Commun 249(2):299–303.

McDaniel ML, Marshall CA, Pappan KL, Kwon G. 2002. Metabolic and autocrine regulation of the mammalian target of rapamycin by pancreatic beta-cells. Diabetes 51(10):2877–2885.

Meier JJ and Nauck MA. 2006. Incretins and the development of type 2 diabetes. Curr Diab Rep 6(3):194–201.

Mihatsch WA, Hogel J, Pohlandt F. 2001. Hydrolysed protein accelerates the gastro-intestinal transport of formula in preterm infants. Acta Paediatr 90(2):196–198.

Molitch ME, DeFronzo RA, Franz MJ, Keane WF, Mogensen CE, Parving HH, Steffes MW, American Diabetes Association. 2004. Nephropathy in diabetes. Diabetes Care 27(1):S79–S83.

Newshome P, Brennan L, Bender K. 2006. Amino acid metabolism, β-cell function, and diabetes. Diabetes 55(2):S39–S46.

Nilsson M, Stenberg M, Frid AH, Holst JJ, Bjorck IM. 2004. Glycemia and insulinemia in healthy subjects after lactose-equivalent meals of milk and other food proteins: The role of plasma amino acids and incretins. Am J Clin Nutr 80(5):1246–1253.

Nuttall FQ and Gannon MC. 1990. Metabolic response to egg white and cottage cheese protein in normal subjects. Metabolism 39(7):749–755.

Nuttall FQ, Gannon MC, Wald JL, Ahmed M. 1985. Plasma glucose and insulin profiles in normal subjects ingesting diets of varying carbohydrate, fat, and protein content. J Am Coll Nutr 4(4):437–450.

Nuttall FQ, Mooradian AD, Gannon MC, Billington C, Krezowski P. 1984. Effect of protein ingestion on the glucose and insulin response to a standardized oral glucose load. Diabetes Care 7(5):465–470.

Pallotta JA and Kennedy PJ. 1968. Response of plasma insulin and growth hormone to carbohydrate and protein feeding. Metabolism 17(10):901–908.

Park SW, Goodpaster BH, Strotmeyer ES, Kuller LH, Broudeau R, Kammerer C, de Rekeneire N, Harris TB, Schwartz AV, Tylavsky FA, et al. 2007. Accelerated loss of skeletal muscle strength in older adults with type 2 diabetes: The health, aging, and body composition study. Diabetes Care 30(6):1507–1512.

Piazza GA, Callanan HM, Mowery J, Hixson DC. 1989. Evidence for a role of dipep-tidyl peptidase IV in fibronectin-mediated interactions of hepatocytes with extra-cellular matrix. Biochem J 262(1):327–334.

Pisters PW, Restifo NP, Cersosimo E, Brennan MF. 1991. The effects of euglycemic hyperinsulinemia and amino acid infusion on regional and whole body glucose disposal in man. Metabolism 40(1):59–65.

Poitout V and Robertson RP. 1996. An integrated view of beta-cell dysfunction in type-II diabetes. Annu Rev Med 47:69–83.

Rabinowitz D, Merimee TJ, Maffezzoli R, Burgess JA. 1966. Patterns of hormonal release after glucose, protein, and glucose plus protein. Lancet 2(7461):454–456.

Rasmussen BB, Tipton KD, Miller SL, Wolf SE, Wolfe RR. 2000. An oral essential amino acid-carbohydrate supplement enhances muscle protein anabolism after resistance exercise. J Appl Physiol 88(2):386–392.

Rennie MJ, Bohe J, Smith K, Wackerhage H, Greenhaff P. 2006. Branched-chain amino acids as fuels and anabolic signals in human muscle. J Nutr 136(1):264S–268S.

Riccardi G, Capaldo B, Vaccaro O. 2005. Functional foods in the management of obesity and type 2 diabetes. Curr Opin Clin Nutr Metab Care 8(6):630–635.

Rieu I, Balage M, Sornet C, Giraudet C, Pujos E, Grizard J, Mosoni L, Dardevet D. 2006. Leucine supplementation improves muscle protein synthesis in elderly men independently of hyperaminoacidaemia. J Physiol 575(Pt 1):305–315.

Roden M, Price TB, Perseghin G, Petersen KF, Rothman DL, Cline GW, Shulman GI. 1996. Mechanism of free fatty acid-induced insulin resistance in humans. J Clin Invest 97(12):2859–2865.

Rosak C. 2002. The pathophysiologic basis of efficacy and clinical experience with the new oral antidiabetic agents. J Diabetes Complications 16(1):123–132.

Rouanet JM, Zambonino Infante JL, Caporiccio B, Pejoan C. 1990. Nutritional value and intestinal effects of dipeptides and tripeptides, comparison with their issuing bovine plasma protein in rats. Ann Nutr Metab 34(3):175–182.

Sener A and Malaisse WJ. 1980. L-Leucine and a nonmetabolized analogue activate pancreatic islet glutamate dehydrogenase. Nature 288(5787):187–189.

Tessari P, Inchiostro S, Biolo G, Duner E, Nosadini R, Tiengo A, Crepaldi G. 1985. Hyperaminoacidaemia reduces insulin-mediated glucose disposal in healthy man. Diabetologia 28(11):870–872.

Tremblay F and Marette A. 2001. Amino acid and insulin signaling via the mTOR/p70 S6 kinase pathway. A negative feedback mechanism leading to insulin resistance in skeletal muscle cells. J Biol Chem 276(41):38052–38060.

Tremblay F, Jacques H, Marette A. 2005. Modulation of insulin action by dietary proteins and amino acids: Role of the mammalian target of rapamycin nutrient sensing pathway. Curr Opin Clin Nutr Metab Care 8(4):457–462.

van Loon LJ. 2007. Amino acids as pharmaco-nutrients for the treatment of type 2 diabetes. Immun Endoc Metab Agents Med Chem 7(1):39–48.

van Loon LJ, Kruijshoop M, Verhagen H, Saris WH, Wagenmakers AJ. 2000b. Ingestion of protein hydrolysate and amino acid-carbohydrate mixtures increases postexercise plasma insulin responses in men. J Nutr 130(10):2508–2513.

van Loon LJ, Kruijshoop M, Menheere PP, Wagenmakers AJ, Saris WH, Keizer HA. 2003. Amino acid ingestion strongly enhances insulin secretion in patients with long-term type 2 diabetes. Diabetes Care 26(3):625–630.

van Loon LJ, Saris WH, Verhagen H, Wagenmakers AJ. 2000a. Plasma insulin responses after ingestion of different amino acid or protein mixtures with carbohydrate. Am J Clin Nutr 72(1):96–105.

Volpi E, Ferrando AA, Yeckel CW, Tipton KD, Wolfe RR. 1998. Exogenous amino acids stimulate net muscle protein synthesis in the elderly. J Clin Invest 101(9):2000–2007.

Wild S, Roglic G, Green A, Sicree R, King H. 2004. Global prevalence of diabetes: Estimates for the year 2000 and projections for 2030. Diabetes Care 27(5):1047–1053.

Xu G, Kwon G, Cruz WS, Marshall CA, McDaniel ML. 2001. Metabolic regulation by leucine of translation initiation through the mTOR-signaling pathway by pancreatic beta-cells. Diabetes 50(2):353–360.

Ziegler F, Ollivier JM, Cynober L, Masini JP, Coudray-Lucas C, Levy E, Giboudeau J. 1990. Efficiency of enteral nitrogen support in surgical patients: Small peptides v non-degraded proteins. Gut 31(11):1277–1283.

Chapter 18

Future Trends and Directions

James W Anderson, MD, and Vijai K Pasupuleti, PhD

Introduction

As well-summarized by the expert authors of different chapters, diabetes is a problem of epidemic proportions worldwide. In the United States alone, the estimated number with diabetes, prediabetes, and the metabolic syndrome totals approximately 150 million individuals. The increase in average body weight is the major contributor to this enormous problem. Creative approaches to prevention are urgently needed to stem this tsunami. These approaches, to be affordable and utilized worldwide, must be practical and readily available. Nutrition and nutraceutical approaches currently appear to most likely meet these needs. Probably these approaches will need to be tested in developed countries where individuals have the resources and then disseminated to less affluent areas after cost-effective strategies are established.

Informed consumers in developed countries want to have an active role in their health and disease prevention and attenuation. They spend a substantial percentage of their health management resources on nutrition and nutraceutical approaches to health preservation. Many countries—such as China, India, and Mexico—have traditional medicines that are widely used and accepted. The challenge is to identify safe and effective nutraceutical approaches, provide high-quality products, and educate the population about the impact of these products on protection and reversal of diabetic tendencies.

Lifestyle Changes and Prevention

Currently, lifestyle changes—energy-restricted diets and increased physical activity—are the only proven-effective approaches to prevention of diabetes (Greenway 2008). Until additional nutritional and nutraceutical approaches that enhance these lifestyle approaches are established, high-risk individuals should be educated in these measures and societal institutions—government, education, religious groups—should join forces to disseminate this information. Persons at high risk for developing diabetes, those with a history of gestational diabetes, pre-diabetes, the metabolic syndrome, or a strong family history of diabetes, should work toward these goals: achieve and maintain a desirable body weight (BMI <25 kg/m^2); walk 30 min daily or perform a similar level of physical activity daily; and eat a diet rich in whole grains, soy foods, dry beans, and limit the intake of red meat.

Nutritional Approaches to Diabetes Prevention

Individuals at high risk for developing diabetes should give the highest nutrition priority to achieving and maintaining an ideal body weight since obese individuals have the risk of developing diabetes that is 90-fold higher than the risk for persons with body mass indices (BMIs) of 22–23 kg/m^2 (Anderson et al. 2003). These individuals should focus on plant-based diets rich in whole grains, soy foods, dry beans, vegetables, fruit, and minimize the intake of red meat. Intake of high fructose sweetened beverages may contribute to weight gain and risk for diabetes. Moderate use of coffee and alcohol (up to 7 drinks/week for women and 10 drinks/week for men) may protect from development of diabetes (Anderson 2008). While many diets have been and currently are advocated for weight loss and protection from diabetes, none of the high-protein or low-carbohydrate diets has epidemiological or clinical trial support related to diabetes prevention and should not be recommended until a strong scientific evidence supports their use. Currently, the consensus recommendations for health promotion and protection from chronic diseases encourage a generous intake of plant products and limitations of the intake of fat from beef, pork, and dairy sources. Health-promoting diets to suit the ethnic and regional preferences can be developed provided attention is focused on energy requirements and appropriate choices of carbohydrates and fats (Fung et al. 2001; Fung and Hu 2003; Hu and Willett 2002; Kerver et al. 2003; McCullough et al. 2000).

Supplementation

Certain supplements have a strong association with lower risk for diabetes and can be recommended for persons at high risk for developing diabetes. Using levels of intake that have well-documented safety, the following minerals, listed at the elemental dose of the mineral, should be considered for daily intake: magnesium, 300 mg; chromium 400 µg (as the picolinate salt); vanadium, 100 µg; and zinc, 15 mg (Chapter 10). The role of antioxidants for diabetes prevention is still under investigation but the following vitamins can be recommended for daily intake: folate, 400 µg and vitamin C, 1,000 mg (Chapter 11). Although conclusive data are not available, emerging studies would support the enjoyment of cinnamon seasoning at about 1 g/day (Chapter 8) and cocoa (dark chocolate) about 5 g/day

(Brand-Miller et al. 2003). Using the supplements in divided doses with meals is likely to facilitate absorption and effectiveness. However, placebo-controlled, randomized, clinical trials are required to document the efficacy of all of these supplements since epidemiological and observational studies are not adequate to provide persuasive evidence regarding efficacy.

Herbal Use

While more than 800 plants or herbal extracts appear to affect blood glucose regulation (Grover et al. 2002), the clinical evidence to support their use is limited. No herbal extracts can be endorsed as being safe and effective at the current time. Table A1 in the Appendix provides a summary of the best characterized herbal products with documented glycemic effects. The following herbal extracts, listed alphabetically, have potential and require further study: Aloe vera (Shane-McWhorter 2005, Yeh et al. 2003); bitter melon fruit or juice (Yeh et al. 2003); fenugreek products (Chapter 14); ginseng, American (Sievenpiper et al. 2008, Chapter 12; Yeh et al. 2003; ginseng, Korean red (Sievenpiper et al. 2008, Chapter 12; gymnema (Yeh et al. 2003); and nopal or prickly pear (Yeh et al. 2003) (Chapter 15).

As outlined by Sievenpiper and colleagues (2008), the specific active compound(s) of any herbal product must be identified and supplied in a reliable manner. After developing a reliable and reproducible method of extracting the active components, the safety and efficacy must be established. Specific herbal compounds should be tested in placebo-controlled, randomized, clinical trials as outlined below.

Clinical Trial Considerations

Safety

The safety of any herbal product needs to be established before human trials can be initiated. Usually this is done by short-, intermediate-, and long-term studies in animals, which examine weight gain, organ weights, gross examination, and histological studies of selected organs, and blood hematological and biochemical measurements. Usually short- and intermediate-term open trials in human volunteers are done with careful assessment of adverse effects and hematological and biochemical measurements of serum and urine. Once these preliminary safety measurements are completed, clinical trials can be done.

Clinical Trials

Two placebo-controlled, randomized clinical trials providing information on 15–20 completed subjects in the placebo and control groups are usually necessary to establish an assertion that a new compound has efficacy. Often dose-ranging studies are first done to determine the minimally and maximally effective doses. Most nutritional products need to be given at least twice daily for full efficacy. Important considerations for trial design—as detailed elsewhere (Anderson et al. 2008)—are these: subject selection (inclusion of high-risk subjects without prior use of the product); randomization (following a rigorous randomization process); blinding (ensuring double-blinding including subjects and all study personnel); description of therapeutic intervention (full details of product, diet, procedures, and efforts made to maintain stable body weight during study); obtaining baseline measurements after screening visit; description of subjects not completing the study; acceptable subject completion rates; appropriate measurements (weight, blood pressure, safety laboratory measurements, and outcome measurements such as blood glucose, insulin, lipoproteins, and hemoglobin A1c); data analysis (preferably blinded to the identity of the test groups); information on adverse effects; and information on funding or support.

In clinical trials of nutraceuticals for prevention of diabetes the following enrollment criteria are commonly used: selection of women or men aged 30–65 years; BMI ranges of 25–35 kg/m^2 (to avoid moderate and severely obese subjects who desire weight loss during the trial); fasting plasma glucose values of 100–125 mg/dL (defined as impaired fasting glucose); and persons with no past history of a diagnosis of diabetes or treatment with an antidiabetes agent. Usually, detailed nutrition instruction is not provided but volunteers are encouraged to avoid dieting to lose weight and not use any nutrition supplements that might affect blood glucose values. After screening to establish diagnosis of impaired fasting glucose and rule out significant biochemical abnormalities such as hypothyroidism, liver or kidney disease, and severe dyslipidemia, volunteers are invited back for randomization. A 12-week trial period often is used with baseline, 4-, 8-, and 12-week visits for weight, blood pressure, and outcome measures (fasting plasma glucose and lipoproteins). Often plasma insulin and blood hemoglobin A1c measurements are done at baseline, 6, and 12 weeks. Significant reductions in serum glucose, insulin, and hemoglobin A1c values for the test product compared to placebo represent a favorable effect of the test product.

Future Needs

Reductions in the development rates for diabetes will require major consumer education and changes in food availability in schools, work places, restaurants, and other places. To be effective these changes will probably require availability of new foods, functional foods, and nutraceuticals. Educational efforts related to foods probably should follow current efforts related to "five-a-day" efforts for fruit and vegetables and "three-a-day" for whole grains. Probably food labels with a seal, like the Whole Grain seal, that indicate that a food has "prevent diabetes" properties or application of some user-friendly label would be useful.

Functional foods designed to be "prevent diabetes" should include the well-established minerals—magnesium, chromium, vanadium, and zinc—and consider including other ingredients such antioxidant vitamins (folate and vitamin C), cinnamon, cocoa, and soy peptides.

Nutraceuticals should have excellent safety credentials and documented efficacy with well-defined and assayable compounds—such as those from American ginseng as outlined by Sievenpiper et al. (2008).

Conclusions

It is relatively easy to prevent or delay the onset of diabetes if it is diagnosed early. The future studies should focus on early biomarkers that may be used to more accurately predict future incidence of diabetes among apparently healthy people. Recent study by Liu et al. (2007) in the first large-scale, multiethnic study has identified interleukin-6 (IL-6) and high-sensitivity C-reactive protein (hs-CRP) as early biological markers for predicting diabetes in still-healthy people, years ahead of the traditional risk factors of obesity or insulin resistance. The proinflammatory state is often linked to obesity, which can lead to insulin resistance. Therefore, identifying these markers by a simple blood test well before a disease begins can not only help improve mechanistic understanding of diabetes, but also motivate the individuals to change their lifestyle to prevent diabetes.

Lifestyle changes—modest weight loss and moderate daily physical activity—can reduce the emergence of clinical diabetes by more than 50% over a 3-year period in high-risk individuals (Greenway 2008). More effective weight loss programs are likely to reduce this risk to an even greater extent (Anderson et al. 2003). Working through different mechanisms, supplemental mineral, and vitamin intakes should substantially increase the protective effects. The efficacy of certain herbal extracts, such as American ginseng, are documented in persons with type 2 diabetes but are not well-tested in persons at risk for diabetes. Again, since these herbal extracts operate through a variety of different mechanism, their use is likely to further reduce risk for emergence of clinical diabetes in high-risk individuals.

References

Anderson JW. 2008. Dietary fiber and associated phytochemicals in prevention and reversal of diabetes. In Nutraceuticals, Glycemic Health and Diabetes, edited by Pasupuleti VK, Anderson JW. Ames, IA: Wiley Blackwell Publishing Professional.

Anderson JW, Kendall CWC, Jenkins DJA. 2003. Importance of weight management in type 2 diabetes: review with meta-analysis of clinical studies. J Am Coll Nutr 22:331–339.

Anderson JW, Liu C, Kryscio RJ. 2008. Blood pressure response to transcendental meditation: a meta-analysis. Am J Hypertens. In press.

Brand-Miller J, Holt SHA, de Jong V, Petocz P. 2003. Cocoa powder increases post-prandial insulinemia in lean young adults. J Nutr 133:3149–3152.

Fung TT, Hu FB. 2003. Plant-based diets: what should be on the plate? Am J Clin Nutr 78:357–358.

Fung TT, Rimm EB, Spiegelman D, Rifai N, Tofler GH, Willett WC, Hu FB. 2001. Association between dietary patterns and plasma biomarkers of obesity and cardiovascular disease risk. Am J Clin Nutr 73:61–67.

Greenway F. 2008. Preventing type 2 diabetes mellitus. In Nutraceuticals, Glycemic Health and Diabetes, edited by Pasupuleti VK, Anderson JW. Ames, IA: Wiley Blackwell Publishing Professional.

Grover JK, Yadav S, Vats V. 2002. Medicinal plants of India with anti-diabetic potential. J Ethnopharmacol 81:81–100.

Hu FB, Willett WC. 2002. Optimal diets for prevention of coronary heart disease. J Am Med Assoc 288:2569–2578.

Kerver JM, Yang EJ, Bianchi L, Song WO. 2003. Dietary patterns associated with risk factors for cardiovascular disease in healthy US adults. Am J Clin Nutr 78:1103–1110.

Liu S, Tinker L, Song Y, Rifai N, Bonds DE, Cook NR, Heiss G, Howard BV, Hotamisligil GS, Hu FB, Kuller LH, Manson JE. 2007 Arch Intern Med 167:1676–1685.

McCullough ML, Feskanich D, Stampfer MJ, Rosner B, Hu FB, Hunter DJ, Variyam JN, Colditz GA, Willett WC. 2000. Adherence to the dietary guidelines for Americans and risk of major chronic diseases in women. Am J Clin Nutr 72:1214–1222.

Shane-McWhorter L. 2005. Update on complementary therapies in diabetes. Diabet Microvasc Complications Today 40–45.

Sievenpiper JL, Jenkins AL, Descalu A, Stavro M, Vuksan V. 2008. Ginseng in type 2 diabetes mellitus: a review of the evidence in humans. In Nutraceuticals, Glycemic Health and Diabetes, edited by Pasupuleti VK, Anderson JW. Ames, IA: Wiley Blackwell Publishing Professional.

Yeh GY, Eisenberg DM, Kaptchuk TJ, Phillips RS. 2003. Systematic review of herbs and dietary supplements for glycemic control in diabetes. Diabetes Care 26:1277–1294.

Appendix

James W Anderson, MD, and Manan Jhaveri, MBBS

Table A1. Herbals for diabetes references.

Herbal product		Major country of identity	Animal studies			Human studies		Good clinical evidence[a]
Common name	Latin name		Proposed mechanism	Reference	Clinical trials reference	Reference		
African potato	*Hypoxis hemorocallidea*	Africa	A5	Mahomed and Ojewole (2003)				
Aloe vera	*Aloe vera*	India	D1	Srinivasan (2008)	Shane-McWhorter (2005)	Yeh et al. (2003)		Yes
Aristolochia	*Aristolochia longa*		D1	Srinivasan (2008)		Haddad et al. (2001)		
Artemisia	*Artemisia herba alba*		A5	Srinivasan (2008)		Haddad et al. (2001)		
Artichoke	*Cynara scolymus*		A1, A2			Haddad et al. (2001)		
Avaram	*Cassia auriculata*	India	A1	Srinivasan (2008)				
Ayurveda	Many	India	A5		Hsia et al. (2004)	Grover et al. (2002) and Saxena and Vikram (2004)		
Bael	*Aegle marmelos*	India	A5, F1	Srinivasan (2008)				

(continued)

Table A1. (*continued*)

Herbal product		Major country of identity	Animal studies		Clinical trials reference	Human studies	
Common name	Latin name		Proposed mechanism	Reference		Reference	Good clinical evidence[a]
Banaba, crepe myrtle	*Lagerstroemia speciosa*	Phillippines	B1		Shane-McWhorter (2005)		
Banana	*Musa sapientum*	India	B1, J1	Srinivasan (2008)			
Berberine	*Rhizoma coptidis*	China	A1, A2	Lee (2006)	Lankarani-Fard and Li (2008)		
Bilberry, European	*Vaccinium myrtillus*		A5			Haddad et al. (2001)	
Bitter apple	*Citrullus colocynthis*	India	E1	Srinivasan (2008)			
Bitter melon, Indian	*Momordica charantia*	India, China	A1, D1, F1	Srinivasan (2008)	Agrawal et al. (2002)	Saxena and Vikram (2004) and Yeh et al. (2003)	Yes
Blackberry	*Eugenia jambolana*	India	A5, B1	Grover et al. (2000)	Agrawal et al. (2002)		
Black cumin	*Nigella sativa*	India	A5, E1	Le et al. (2004)		Haddad et al. (2001)	
Blueberry	*Vaccinium angustifolia*		A5	Martineau et al. (2006)		Haddad et al. (2001)	
Caiapo, sweet potato	*Ipomomoea batabus*	Japan	A5, C1		Ludvic et al. (2002)	Shane-McWhorter (2005)	
Chenopodiaceae	*Salicornia herbacea*	Korea	A5	Park et al. (2006)			

Common name	Scientific name	Country	Code	Reference 1	Reference 2	Reference 3	
Chinese Herbal Medicine			A5		Collins and McFarlane (2006)	Bian et al. (2006) and Liu et al. (2004)	
Chinese magnolia vine fruit	*Schizandra chinensis*	China	B1	Lankarani-Fard and Li (2008)			
Chocolate, dark	Cocoa powder		A1	Ruzaidi et al. (2005)	Grassi et al. (2005)	Brand-Miller et al. (2003)	
Cinnamon	*Cinnamomum cassia*		A4, B1	Anderson et al. (2004)	Khan et al. (2003) and Mang et al. (2006)	Bradley et al. (2007)	Yes
Cogent db	Indian herbal preparation		A5		Shekhar et al. (2002)		
Cordyceps sinensis	*Cordyceps sinensis*		A5	Lo et al. (2006)			
Cortex Moutan polysaccharide	*Paeonia suffruticosa*	China	A5, C1	Hong et al. (2003)			
Cumin	*Cuminum cyminum*	India	A1, C1	Srinivasan et al. (2008)			
Curry leaf	*Murraya koenigii*	India	A2, A3	Grover et al. (2003)	Srinivasan (2008)	Grover et al. (2002)	
Dandelion	*Taraxacum officinale*		A5			Haddad et al. (2001)	
Dianex	Polyherbal	India	C1				
Dogwood fruit	Fructus corni	China	A1, C1	Lankarani-Fard and Li (2008)	Sudha et al. (2005)		
Dwarf lilyturf tuber	Radix Ophiopogonis	China	A1	Lankarani-Fard and Li (2008)			

(*continued*)

Table A1. (*continued*)

Herbal product		Major country of identity	Animal studies		Clinical trials reference	Human studies	Good clinical evidence[a]
Common name	Latin name		Proposed mechanism	Reference		Reference	
Elder	*Sambucus nigra*		D1	Gray et al. (2000)			
Enicostemma littorale	*Enicostemma littorale*		A5		Upadhyay and Goyal (2004)		
Fenugreek	*Trigonella foenum graecum*	India	A1, C1	Srinivasan (2008)	Agrawal et al. (2002)	Haddad et al. (2001) and Saxena and Vikram et al. (2004)	Yes
Foxglove root, Chinese	Radix Rehmanniae	China	A1	Lankarani-Fard and Li (2008)			
Garlic	*Allium sativum*	Many	A1	Srinivasan (2008)	Srinivasan (2008)	Milner and Rivlin (2001)	
Glossy Privet Fruit	*Ligustrum lucidum*	China	A1, B1	Lankarani-Fard and Li (2008)			
Ginseng, American	*Panax quinquefolium*	Canada, US	A5	Sievenpiper et al. (2008)	Sievenpiper et al. (2008)	Sievenpiper et al. (2008)	Yes
Ginseng, Chinese or Korean	*Panax ginseng*	China, Korea	A5	Dey et al. (2003)		Sievenpiper et al. (2008)	Yes
Ginseng, Japanese	*Panax japonicus*	Japan	A5	Lankarani-Fard and Li (2008)		Sievenpiper et al. (2008)	
Ginseng, Siberian	*Eleutherococcus senticosus*	Siberia	A5	Lankarani-Fard and Li (2008)		Sievenpiper et al. (2008)	
Ginseng, Southern	*Gynostemma pentaphyllum*	China	A1,C1	Lankarani-Fard and Li (2008)		Sievenpiper et al. (2008)	

Globularia	*Globularia alypum*		A5			Haddad et al. (2001)	
Goshajinkigan		China	A5, C1		Uno (2005)		
Gymnema	*Gymnema sylvester*	India	A1	Srinivasan (2008) and Xie et al. (2003)	Agrawal et al. (2002)	Bradley et al. (2007) and Grover et al. (2002)	Yes
Herbal tea			A5		Ryan et al. (2000)		
Holy basil	*Ocimum sanctum*	India	D1	Vats et al. (2004)	Srinivasan (2008)	Grover et al. (2002)	Yes
Horsewood	*Clausena anisata*	Africa	A5	Ojewole (2002)			
Indian banyan	*Ficus bengalenesis*	India	B1, D1	Srinivasan (2008)		Grover et al. (2002)	
Indian gentian	*Swertia chirata*	India	A5, B1	Srinivasan (2008)		Grover et al. (2002)	
Indian Malabar	*Pterocarpus marsupiam*	India	A5	Srinivasan (2008)	Saxena and Vikram (2004)	Grover et al. (2002)	Yes
Inolter		India	A5		Agrawal et al. (2002)	Liu et al. (2004)	
Ivy gourd	*Coccinia indica*	India	A1, A5, B1, C1	Srinivasan (2008)	Srinivasan (2008)	Yeh et al. (2003)	Yes
Jambul	*Syzgium cumini*		A5			Haddad et al. (2001)	
Kampo (Japanese Herbal Medicines)			A5			Borchers et al. (2000)	
Kothala Himbutu tea	*Salacia reticulate*	India	A5		Jayawardena et al. (2005)		

(continued)

Table A1. *(continued)*

| Herbal product | | Major country of identity | Animal studies | | Clinical trials reference | Human studies | |
Common name	Latin name		Proposed mechanism	Reference		Reference	Good clinical evidence[a]
Licorice	*Glycyrrhiza glabra*		A5			Haddad et al. (2001)	
Lotus	*Nelumbo nucifera*	India	A5	Srinivasan (2008)			
Lupin, white	*Lupinus albus*		A5			Haddad et al. (2001)	
Milk thistle	*Silybum marianum*	India	A5		Shane-McWhorter (2005)	Haddad et al. (2001)	
Mustard (Rai)	*Brassia juncea*	India	A3, A5	Yadav et al. (2004)			
Neem	*Azadirachta indica*	India	A5	Srinivasan (2008)			
Nitobegiku	*Tithonia diversifolia*		A5, C1	Miura et al. (2005)			
Nigella	*Nigella sativa*		A5			Haddad et al. (2001)	
Nopal or prickly pear	*Opuntia streptacantha*	Mexico	A5, C1	Wolfram et al. (2002)	Shane-McWhorter (2005)	Yeh et al. (2003)	Yes
Onion	*Allium cepa*		A5, B1	Srinivasan (2008)	Srinivasan (2008)	Grover et al. (2002)	
Okchun-San		Korea	A5, C1	Chang et al. (2006)			
Oregano	*Origanum compactum*		A5, E1			Haddad et al. (2001)	

(*continued*)

Phyllanthus amarus	*Phyllanthus amarus*	India	A5, E1	Srinivasan (2008)
Poria	Poria	China	A4	Lankarani-Fard and Li (2008)
Piper betle	*Piper betle*	India	A5	Srinivasan (2008)
Punarnava	*Boerhaavia diffusa*	India	A5, D1	Srinivasan (2008)
Punica granatum flower extract	*Punica granatum*		C1	Huang et al. (2005) and Li et al. (2005)
Radix Astragali	Radix Astragali	China	A1	Lankarani-Fard and Li (2008)
Radix Puerariae	Radix Puerariae	China	A1, A4, B1	Lankarani-Fard and Li (2008)
Radix Trichosanthis	Radix Trichosanthis	China	A1	Lankarani-Fard and Li (2008)
Red vine	*Vitis vinifera*		A5	Haddad et al. (2001)
Rhizoma Anemarrhenae	Rhizoma Anemarrhenae	China	A4	Lankarani-Fard and Li (2008)
Rhizoma Atractyloidis	Rhizoma Atractyloidis	China	A1	Lankarani-Fard and Li (2008)
Rhizoma Polygonati	Rhizoma Polygonati	China	A1, A4	Lankarani-Fard and Li (2008)
Rosemary	*Rosmarinus officinalis*		A5	Haddad et al. (2001)
Shikonin			B1, C1	Kamei (2002)

473

Table A1. (*continued*)

Herbal product		Major country of identity	Animal studies		Human studies		
Common name	Latin name		Proposed mechanism	Reference	Clinical trials reference	Reference	Good clinical evidence[a]
Soy (gold bean) components	*Phaseolus aureus*		A5	Anderson and Pasupuleti (2008)	Kim et al. (2005)	Anderson and Pasupuleti (2008)	
Stevia	*Stevia rebaudiana*		A5	Chen et al. (2005)			
Sutherlandia frutescens	*Tetrapleura tetraptera*		A5	Ojewole (2004)		Ojewole (2004)	
Sweet broom weed	*Scoparia dulcis*	India	A1, B1	Srinivasan (2008)			
Tinospora cordifolia	*Tinospora cordifolia*	India	A5	Grover et al. (2000)		Grover et al. (2002)	
Turmeric	*Curcuma longa*	India	A1, B1, C1	Srinivasan (2008)			
Velvet bean seed	*Caesalpinia bonducella*	India	A1, C1	Srinivasan (2008)			
White lupin	*Lupinus alba*		A5			Haddad et al. (2001)	
White mulberry	*Morus alba*	China	A1, A5, B1, C1	Lankarani-Fard and Li (2008)	Lankarani-Fard and Li (2008)		
Wolfberry	*Lycium barbarum*	China	A1, B1, C1	Lankarani-Fard and Li (2008)			

Key: A1, increases glycolysis; A2, increases glycogen synthesis; A3, decreases glycogenolysis; A4, decreases gluconeogenesis; A5, hypoglycemic effect; B1, increases glucose uptake (transport); C1, increases insulin sensitivity; D1, increases insulin secretion; E1, antioxidant properties; F1, anti-inflammatory properties.
[a]The clinical trial evidence appears to support use of the herbals indicated with "yes," as safe and efficacious for use as supplements for persons with diabetes or risk for diabetes, but additional randomized clinical trials are required to support these claims.

References

Agrawal RP, Sharma A, Dua AS, Chandershekhar A, Kochar DK, Kothari, RP. 2002. A randomized placebo controlled trial of Inolter (herbal product) in the treatment of type 2 diabetes. J Assoc Physicians India 50:391–393.

Anderson JW, Pasupuleti VK. 2008. Soybean and soy component effects on obesity or diabetes. In Nutraceuticals, Glycemic Health, and Diabetes, pp. 161–188. Ames, IA: Blackwell Publishing Professional.

Anderson RA, Broadhurst CL, Polansky MM, Schmidt WF, Khan A, Flanagan VP, Schoene NW, Graves DJ. 2004. Isolation and characterization of polyphenol type-A polymers from cinnamon with insulin-like biological activity. J Agric Food Chem 52:65–70.

Bian ZX, Moher D, Dagenais S, Li YP, Liu L, Wu TX, Miao JX. 2006. Improving the quality of randomized controlled trials in Chinese herbal medicine. Part II: Control group design. Zhong Xi Yi Jie He Xue Bao 4:130–136.

Borchers AT, Sakai S, Henderson GL, Harkey MR, Keen CL, Stern JS, Terasawa K, Gershwin ME. 2000. Shosaiko-to and other Kampo (Japanese herbal) medicines: A review of their immunomodulatory activities. J Ethnopharmacol 73:1–13.

Bradley R, Oberg EB, Calabrese C, Standish LJ. 2007. Algorithms for complementary and alternative medicine practice and research in type 2 diabetes. J Altern Complement Med 13:159–175.

Brand-Miller J, Holt SHA, de Jong V, Petocz P. 2003. Cocoa powder increases postprandial insulinemia in lean young adults. J Nutr 133:3149–3152.

Chang MS, Oh MS, Kim DR, Jung KJ, Park S, Choi SB, Ko BS, Park SK. 2006. Effects of Okchun-San, a herbal formulation, on blood glucose levels and body weight in a model of Type 2 diabetes. J Ethnopharmacol 103:491–495.

Chen TH, Chen SC, Chan P, Chu YL, Yang HY, Cheng JT. 2005. Mechanism of the hypoglycemic effect of stevioside, a glycoside of *Stevia rebaudiana*. Planta Med 71:108–113.

Collins M, McFarlane JR. 2006. An exploratory study into the effectiveness of a combination of traditional Chinese herbs in the management of type 2 diabetes. Diabetes Care 29:945–946.

Dey L, Xie JT, Wang A, Wu J, Maleckar SA, Yuan CS. 2003. Anti-hyperglycemic effects of ginseng: Comparison between root and berry. Phytomedicine 10:600–605.

Grassi D, Lippi C, Necozione S, Croce G, Valeri L, Pasqualetti P, Desideri G, Blumberg JB, Ferri C. 2005. Short-term administration of dark chocolate is followed by a significant increase in insulin sensitivity and a decrease in blood pressure in healthy persons. Am J Clin Nutr 81:611–614.

Gray AM, Abdel-Wahab HA, Flatt PR. 2000. The traditional plant treatment, *Sambucus nigra* (elder), exhibits insulin-like and insulin-releasing actions in vitro. J Nutr 130:15–20.

Grover JK, Vats V, Rathi SS. 2000. Anti-hyperglycemic effect of *Eugenia jambolana* and *Tinospora cordifolia* in experimental diabetes and their effects on key metabolic enzymes involved in carbohydrate metabolism. J Ethnopharmacol 73:461–470.

Grover JK, Yadav SP, Vats V. 2002. Medicinal plants of India with anti-diabetic potential. J Ethnopharmacol 81:81–100.

Grover JK, Yadav SP, Vats V. 2003. Effect of feeding *Murraya koeingii* and *Brassica juncea* diet on [correction] kidney functions and glucose levels in streptozotocin diabetic mice. J Ethnopharmacol 85:1–5.

Haddad PS, Depot M, Settaf A, Cherrah Y. 2001. Use of antidiabetic plants in Morocco and Quebec. Diabetes Care 24:608–609.

Hong H, Wang QM, Zhao ZP, Liu GQ, Shen YS, Chen GL. 2003. Studies on antidiabetic effects of cortex Moutan polysaccharide-2b in type 2 diabetes mellitus rats. Yao Xue Xue Bao 38:255–259.

Hsia SH, Bazargan M, Davidson MB. 2004. Effect of Pancreas Tonic (an ayurvedic herbal supplement) in type 2 diabetes mellitus. Metabolis 53:1166–1173.

Huang TH, Peng G, Kota BP, Li GQ, Yamahara J, Roufogalis BD, Li, Y. 2005. Anti-diabetic action of *Punica granatum* flower extract: Activation of PPAR-gamma and identification of an active component. Toxicol Appl Pharmacol 207:160–169.

Jayawardena MH, de Alwis NM, Hettigoda V, Fernando DJ. 2005. A double blind randomised placebo controlled cross over study of a herbal preparation containing *Salacia reticulata* in the treatment of type 2 diabetes. J Ethnopharmacol 97:215–218.

Kamei R, Kitagawa Y, Kadokura M, Hattori F, Hazeki O, Ebina Y, Nishihara T, Oikawa S. 2002. Shikonin stimulates glucose uptake in 3T3-L1 adipocytes via an insulin-independent tyrosine kinase pathway. Biochem Biophys Res Commun 292:642–651.

Khan A, Safdar M, li Khan MM, Khattak KN, Anderson RA. 2003. Cinnamon improves glucose and lipids of people with type 2 diabetes. Diabetes Care 26:3215–3218.

Kim JI, Kim JC, Kang MJ, Lee MS, Kim JJ, Cha IJ. 2005. Effects of pinitol isolated from soybeans on glycaemic control and cardiovascular risk factors in Korean patients with type II diabetes mellitus: A randomized controlled study. Eur J Clin Nutr 59:456–458.

Lankarani-Fard A, Li Z. 2008. Traditional chinese medicine in the management and treatment of the symptoms of diabetes. In Nutraceuticals, Glycemic Health, and Diabetes, edited by Pasupuleti VK, Anderson JW, pp. xxx–xxx. Ames, IA: Blackwell Publishing Professional.

Le PM, haddou-Andaloussi A, Elimadi A, Settaf A, Cherrah Y, Haddad PS. 2004. The petroleum ether extract of *Nigella sativa* exerts lipid-lowering and insulin-sensitizing actions in the rat. J Ethnopharmacol 94:251–259.

Lee YS, Kim WS, Kim KH, Yoon MJ, Cho HJ, Shen Y, Ye JM, Lee CH, Oh WK, Kim CT, Hohnen-Behrens C, Gosby A, Kraegen EW, James DE, Kim JB. 2006. Berberine. Diabetes 55:2256–2264.

Li Y, Wen S, Kota BP, Peng G, Li GQ, Yamahara J, Roufogalis BD. 2005. *Punica granatum* flower extract, a potent alpha-glucosidase inhibitor, improves postprandial hyperglycemia in Zucker diabetic fatty rats. J Ethnopharmacol 99:239–244.

Liu JP, Zhang M, Wang WY, Grimsgaard S. 2004. Chinese herbal medicines for type 2 diabetes mellitus. Cochrane Database Syst Rev, CD003642.

Lo HC, Hsu TH, Tu ST, Lin KC. 2006. Anti-hyperglycemic activity of natural and fermented *Cordyceps sinensis* in rats with diabetes induced by nicotinamide and streptozotocin. Am J Chin Med 34:819–832.

Ludvic BH, Mahdjoobian K, Waldhaeusl W, Hofer A, Prager R, Kautzky-Willer A, Pacini G. 2002. The effect of *Ipomoea batatas* (Caiapo) on glucose metabolism and serum cholesterol in patients with Type 2 diabetes. Diabetes Care 25:239–240.

Mahomed IM, Ojewole JA. 2003. Hypoglycemic effect of *Hypoxis hemerocallidea* corm (African potato) aqueous extract in rats. Methods Find Exp Clin Pharmacol 25:617–623.

Mang B, Wolters M, Schmitt B, Kelb K, Lichtinghagen R, Stichtenoth DO, Hahn A. 2006. Effects of a cinnamon extract on plasma glucose, HbA, and serum lipids in diabetes mellitus type 2. Eur J Clin Invest 36:340–344.

Martineau LC, Couture A, Spoor D, haddou-Andaloussi A, Harris C, Meddah B, Leduc C, Burt A, Vuong T, Mai LP, Prentki M, Bennett SA, Arnason JT, Haddad PS. 2006. Anti-diabetic properties of the Canadian lowbush blueberry *Vaccinium angusti-folium* Ait. Phytomedicine 19:612–623.

Milner JA, Rivlin RS. 2001. Recent advances on the nutritional effects associated with the use of garlic as a supplement. J Nutr 131:951S –1123S.

Miura T, Nosaka K, Ishii H, Ishida T. 2005. Antidiabetic effect of Nitobegiku, the herb *Tithonia diversifolia*, in KK-Ay diabetic mice. Biol Pharm Bull 28:2152–2154.

Ojewole JA. 2002. Hypoglycaemic effect of *Clausena anisata* (Willd) Hook methanolic root extract in rats. J Ethnopharmacol 81:231–237.

Ojewole JA. 2004. Analgesic, antiinflammatory and hypoglycemic effects of *Sutherlandia frutescens* R. BR. (variety incana E. MEY.) [Fabaceae] shoot aqueous extract. Methods Find Exp Clin Pharmacol 26:409–416.

Park SH, Ko SK, Choi JG, Chung SH. 2006. *Salicornia herbacea* prevents high fat diet-induced hyperglycemia and hyperlipidemia in ICR mice. Arch Pharm Res 29:256–264.

Ruzaidi A, Amin I, Nawalyah AG, Hamid M, Faizul HA. 2005. The effect of Malaysian cocoa extract on glucose levels and lipid profiles in diabetic rats. J Ethnopharmacol 98: 55–60.

Ryan EA, Imes S, Wallace C, Jones S. 2000. Herbal tea in the treatment of diabetes mellitus. Clin Invest Med 23:311–317.

Saxena A, Vikram NK. 2004. Role of selected Indian plants in management of type 2 diabetes: A review. J Altern Complement Med 10:369–378.

Shane-McWhorter L. 2005. Update on complementary therapies in diabetes. Diabet Microvasc Complications Today 40–45.

Shekhar KC, Achike FI, Kaur G, Kumar P, Hashim R. 2002. A preliminary evaluation of the efficacy and safety of Cogent db (an ayurvedic drug) in the glycemic control of patients with type 2-diabetes. J Altern Complement Med 8:445–457.

Sievenpiper JL, Jenkins AL, Descalu A, Stavro M, Vuksan V. 2008. Ginseng in type 2 diabetes mellitus: A review of the evidence in humans. In Nutraceuticals, Glycemic Health, and Diabetes, edited by Pasupuleti VK, Anderson JW, pp. xxx–xxx. Ames, IA: Blackwell Publishing Professional.

Srinivasan K. 2008. Fenugreek and traditional antidiabetic herbs from Indian origin. In Nutraceuticals, Glycemic Health, and Diabetes, edited by Pasupuleti VK, Anderson JW, pp. xxx–xxx. Ames, IA: Blackwell Publishing Professional.

Sudha V, Bairy KL, Shashikiran U, Sachidananda A, Jayaprakash B, Shalini S. 2005. Efficacy and tolerability of Dianex in Type 2 diabetes mellitus: A non-randomized, open label non-comparative study. Med J Malaysia 60:204–211.

Uno T, Ohsawa I, Tokudome M, Sato Y. 2005. Effects of Goshajinkigan on insulin resistance in patients with type 2 diabetes. Diabetes Res Clin Pract 69:129–135.

Upadhyay UM, Goyal RK. 2004. Efficacy of *Enicostemma littorale* in Type 2 diabetic patients. Phytother Res 18:233–235.

Vats V, Yadav SP, Grover JK. 2004. Ethanolic extract of *Ocimum sanctum* leaves partially attenuates streptozotocin-induced alterations in glycogen content and carbohydrate metabolism in rats. J Ethnopharmacol 90:155–160.

Wolfram RM, Kritz H, Efthimiou Y, Stomatopoulos J, Sinzinger H. 2002. Effect of prickly pear (*Opuntia robusta*) on glucose- and lipid-metabolism in non-diabetics with hyperlipidemia—a pilot study. Wien Klin Wochenschr 114:840–846.

Xie JT, Wang A, Mehendale S, Wu J, Aung HH, Dey L, Qiu S, Yuan CS. 2003. Antidiabetic effects of *Gymnema yunnanense* extract. Pharmacol Res 47:323–329.

Yadav SP, Vats V, Ammini AC, Grover JK. 2004. *Brassica juncea* (Rai) significantly prevented the development of insulin resistance in rats fed fructose-enriched diet. J Ethnopharmacol 93:113–116.

Yeh GY, Eisenberg DM, Kaptchuk TJ, Phillips RS. 2003. Systematic review of herbs and dietary supplements for glycemic control in diabetes. Diabetes Care 26:1277–1294.

Index